TECHNICAL ADVANCEMENTS IN SPILLWAY DESIGN

PROGRESS AND INNOVATIONS FROM 1985 TO 2020

The purpose of this bulletin is to provide the dam engineering community with information on the technical characteristics of the various types of spillways that have been developed and implemented over the past four decades, as well as the hydraulic characteristics associated with these structures.

Although spillways as such can be built with different types of structures and arrangements, the approach of this bulletin has been to retain some specific types of structures – step spillways, labyrinth spillways, PKW and tunnel spillways – and to analyze more conventional types of spillways operating under special conditions – very high flows, very high heads, very cold climates.

In each chapter, the topics covered focus on the major issues related to the specific problems and structures considered, with the goal of providing information to the designer and owner to assist in the choices to be made during project development.

Recent technological developments in the design, construction, and operation of spillways have focused primarily on economic and safety issues. Spillways are a key element of a dam project, providing the necessary protection of the project from the destructive effects of floods. This has led to the development of different approaches to selecting and defining the best type of spillway for each project. The purpose of this bulletin is to present and discuss the latest results of this evolution.

TECHNICAL ADVANCEMENTS IN SPILLWAY DESIGN

PROGRESS AND INNOVATIONS FROM 1985 TO 2020

B 172

INTERNATIONAL COMMISSION ON LARGE DAMS
COMMISSION INTERNATIONALE DES GRANDS BARRAGES
6 Quai Watier – 78400 Chatou (France)
http://www.icold-cigb.org.

Cover illustration: Cross section and photo of Ertan Dam – China (personal photo)

CRC Press/Balkema is an imprint of the Taylor & Francis Group, an informa business

© 2025 ICOLD/CIGB, Paris, France

Typeset by CodeMantra

Published by CRC Press/Balkema
4 Park Square, Milton Park, Abingdon, Oxon, OX14 4RN
and by CRC Press/Balkema
2385 NW Executive Center Drive, Suite 320, Boca Raton FL 33431

NOTICE – DISCLAIMER:

For Product Safety Concerns and Information please contact our EU representative GPSR@taylorandfrancis.com. Taylor & Francis Verlag GmbH, Kaufingerstraße 24, 80331 München, Germany.

Original text in English
Translation by the French and Swiss Committees
Layout by Nathalie Schauner

ISBN: 978-1-041-08378-8 (Pbk)
ISBN: 978-1-003-64875-8 (eBook)

COMMITTEE MEMBERS
(2009-2016)

Chairman

    B. P. MACHADO   Brazil

Vice-Chairperson

    J. GUO   China

Former Committee Chairmen

    A. LEJEUNE   Belgium

    B. PETRY   Netherlands

Members

R. WARK		Australia
G. RODRIGUEZ ROCA		Bolivia
Z. MICOVIC	From Feb. 2015	Canada
T. NZAKIMUENA	Until Jan. 2015	
J. HODAK	From Jan. 2010	Czech
B. TAQUET	Until Dec. 2016	France
F. LAUGIER	From Apr. 2017	
R. HASELSTEINER	From Feb. 2015	Germany
H. B. HORLACHER	Until Jan. 2015	
SIBA PRASAD SEN		India
C. FOULADI		Iran
A. PIETRANGELI	From Oct. 2010	Italy
T. TAKASUKA	From Aug. 2015	Japan
T. HINO	Until Jul. 2015	
I. KO	From Mar. 2016	Korea
S. LEE	Until Feb. 2016	
H. MARENGO	From May 2012	Mexico
R. DE JONG		Netherland
H-M. KJELLESVIG		Norway
A. BASHIR		Pakistan
C. M. RAMOS	Until May 2010	Portugal
R. SHAKIROV		Russia
H-J. WRIGHT	From Feb. 2015	South Africa
D. I. VAN WYK	From Mar. 2010 until Jan. 2015	
E. F. SNELL	Until Feb. 2010	
K. LAKSIRI		Sri Lanka
A. GRANADOS	From June, 2012	Spain
J. YANG	From May 2012	Sweden
A. WÖRMAN	From Mar. 2010 until Apr. 2012	
A. SCHLEISS		Switzerland

P. MASON		United Kingdon
J. E. LINDELL	From May 2012	United States
S. HUI	Until Apr. 2012	
A. MARCANO		Venezuela

Co-opted Members

J. E. LINDELL	Until May 2012	United States
A. PINHIERO		Portugal
R. BOES	From October, 2013	Switzerland

Other contributors to the Bulletin

B. CROOKSTON		United States
B. TULLIS		United States
S. BARFUSS		United States
S. ERPICUM		Belgium
L. LIA		Norway
F. LEMPÉRIÈRE		France
HO TA KHANH		Vietnam
P. MANSO		Switzerland

CONTENTS

TABLE OF CONTENTS

TABLES & FIGURES

TABLES

FIGURES

17

ACKNOWLEGEMENTS

This Bulletin was prepared by the members of the ICOLD Technical Committee on Hydraulics for Dams during the period 2009-2017. The authors and writers of the 10 chapters are listed in continuation and in certain cases included specialists not formally members of the Committee who agreed to furnish up to date information to enhance the contents of the Bulletin.

Chapter 1 – INTRODUCTION was drafted by B. P. Machado (Brazil)

Chapter 2 – LARGE SPILLWAYS was drafted by B.P. Machado (Brazil) and complemented with inputs by A. Schleiss (Switzerland), S. P. Sen (India), R. Wark (Australia), Z. Micovic (Canada), B. Taquet (France), H. Marengo (Mexico) and A. Marcano (Venezuela).

Chapter 3 – HIGH HEAD SPILLWAYS & ENERGY DISSIPATION was drafted by A. Schleiss (Switzerland) with contributions by P. Manso (Switzerland) and inputs by P. Mason (United Kingdom).

Chapter 4 – STEPPED SPILLWAYS was drafted by A. Granados (Spain) with contributions by P. Mason (United Kingdom), A. Schleiss (Switzerland), R. Boes (Switzerland) and inputs by H-J Wright (South Africa).

Chapter 5 – LABYRINTH SPILLWAYS was drafted by D. van Wyk (South Africa) together with B. Crookston (United States) and B. Tullis (United States).

Chapter 6 – PKW SPILLWAYS was drafted by B. Taquet (France) together with S. Erpicum (Belgium) and F. Laugier (France).

Chapter 7 – TUNNEL, VORTEX AND SHAFT SPILLWAYS was drafted by J. Guo (China) with extensive contributions by R. Shakirov (Russia) and inputs by S. P. Sen (India).

Chapter 8 – SPILLWAYS IN VERY COLD CLIMATE was drafted by H-M. Kjellesvig (Norway) with inputs by T. Nzakimuena (Canada), R. Shakirov (Russia) and L. Lia (Norway).

Chapter 9 – COMPOSITE MODELLING OF HYDRAULIC STRUCTURES was drafted by J. Lindell (United States) together with S. Barfuss (United States), with inputs of R. Wark (Australia) R. Shakirov (Russia) and Z. Micovic (Canada).

Chapter 10 – ECONOMICS, RISK AND SAFETY IN SPILLWAY DESIGN was drafted by C. Fouladi (Iran) with contributions by B. P. Machado (Brazil) and B. Taquet (France), and inputs by H-J Wright (South Africa), S. P. Sen (India), F. Lempérière (France) and Ho Ta Khanh (Vietnan).

Initially, the Bulletin included an additional chapter dedicated to the discussions of Debris and High Sediment Flow in Spillway Operation. This chapter was moved to another bulletin in a combined preparation by the Committee on Hydraulics for Dams and the Committee on Flood Evaluation and Dam Safety. The original chapter was written by R. Wark (Australia) and received inputs by S. P. Sen (India). This other bulletin was co-ordinated by R. Wark and is denominated *Blockage of Spillways and Outlet Works.*

The whole English text of the Bulletin was revised for typographical corrections and coherence by H-J Wright and B. P. Machado.

1. INTRODUCTION

1.1. PURPOSE AND SCOPE OF THE BULLETIN

The purpose of this bulletin is to provide to the dam engineering community information about the technical features of different types of spillways developed and implemented during the last four decades and some hydraulic features associated with these structures. The information presented in this bulletin does not supersede the contents of previous ICOLD bulletin on this issue, but rather complements that information with recent technical developments.

Although spillways as such, can be built with many different types of structures and arrangements, the approach used for this bulletin selected some specific types of structures – stepped spillways, labyrinth, PKW and tunnel spillways – and analysed conventional types of structures operating in special conditions – very large flows, very high head, very cold climate. The information was complemented with a chapter on application of design techniques using numerical and physical models, and a chapter on economics and cost issues. The very important issue of debris and sedimentation control affecting spillway design and operation was moved to another ICOLD bulletin that will include dam safety issues resulting from floods.

In each chapter, the subjects treated focused on the main questions that affect the specific issues or structures considered with the objective of providing information for the designer and for the project owner for the choice and development of the project. The subjects exposed were generally based on recent physical and experimental hydraulic studies and on real examples of works built in different parts of the world.

The recent evolution of the technologies related to the design, construction and operation of spillways has focused mainly on safety and economics. Spillways are the key elements of a dam project to ensure the necessary protection of the project against the destructive action of floods and flows in excess to the discharge capacity of other dam features. However, each dam project has individual characteristics that require the selection of specific types of spillways and this has led to the development of different approaches to select and define the best type of spillway for each project. This bulletin aims to present and discuss the most recent results of this evolution.

1.2. RELATION TO EARLIER BULLETINS

ICOLD has produced various bulletins related to the design and operation of spillways. As mentioned, the present bulletin does not intend to supersede these publications, although it is recognised that certain subjects like energy dissipation deserved a more ample treatment. In any case, the aggregated of the following bulletins comprise a significant volume of valuable technical information on the hydraulic design of spillways:

N°	BULLETIN TITLE	YEAR
49A	OPERATION OF HYDRAULIC STRUCTURES OF DAMS	1986
58	SPILLWAY FOR DAMS	1987
81	SPILLWAYS – SHOCK WAVES AND AIR ENTRAINEMENT	1992
82	SELECTION OF THE DESIGN FLOOD – CURRENT METHODS	1992
108	COST OF FLOOD CONTROLS IN DAMS	1997
125	DAMS AND FLOODS – GUIDELINES AND CASE HISTORIES	2003
130	RISK ASSESSMENT IN DAM SAFETY MANAGEMENT	2005
142	SAFE PASSAGE OF EXTREME FLOODS	2012
156	INTEGRATED FLOOD RISK MANAGEMENT	2014

1.3. CONTENT OF THE BULLETIN

A summary description of the content of the bulletin follows:

Chapter 2 – LARGE CAPACITY SPILLWAYS

This chapter addresses the design and specific arrangements of spillways with capacities varying from 20 000 m³/s to the present largest existing spillway designed for 110 000 m³/s. It includes references and questions related to gated and ungated spillways highlighting the issue of the general use of very large radial gates. It also addresses the questions of structures for energy dissipation and some references of operation of large spillway and the intrinsic risk associated with these structures.

Chapter 3 – HIGH HEAD SPILLWAYS – THE CHALLENGE OF ENERGY DISSIPATION AND SCOUR CONTROL DOWNSTREAM

This chapter deals with the problem of energy dissipation by jets issuing from high head spillways and the corresponding scour process developed in the downstream rock floor. A physical description of the scour process is presented as well as information on semi-empirical formulae and physically based evaluation methods used for estimating the size of the scour hole. Methods for controlling the scour are also presented including the discussion of the pre-excavation of plunge pools, its sizing and a discussion on the conditions and criteria related to its use.

Chapter 4 – STEPPED SPILLWAYS

This chapter describes the main applications of stepped spillways, their hydraulics and expected performance as well as the aspects that require specific attention during the design phase. The chapter presents a comprehensive review of the hydraulics of stepped spillways with information of recent investigations and from these, criteria for design purposes. Case examples are also presented to illustrate the performance of real structures.

Chapter 5 – LABYRINTH SPILLWAYS

Design information for labyrinth spillways is presented in this chapter. An ample review of different structure geometries and corresponding hydraulic parameters is presented. Hydraulic design criteria and suggested design procedures are included. Literature references for additional details and significant case examples are also included.

Chapter 6 – PKW SPILLWAYS

This chapter contains information on Piano-key Weirs (PKW), a variation of traditional labyrinth spillways developed in the last decade. Information comprises description of the structure, hydraulic performance as observed in laboratories and in actual projects, hydraulic design criteria and design parameters, some information on structural and construction issues and actual cases and variations in the arrangements. Extensive literature references are included.

Chapter 7 – TUNNEL, SHAFT AND VORTEX SPILLWAYS

This chapter considers different kinds of tunnel spillways, such as high-level, low-level (or low outlet), vortex, shaft and orifice tunnels. The general arrangement strategies, control structure, conveyance structure, terminal structure, operation issues, risk analysis and considerations are discussed. Some new developments, innovations and applications on tunnel spillway design for high dams which have been achieved in the past 20 to 30 years, are specially presented. Some cases are presented as examples.

Chapter 8 – SPECIAL PROBLEMS OF SPILLWAYS IN VERY COLD CLIMATE

This chapter deals mainly with the serious problems of snow and ice blocking, either partly or fully, spillways passages and thus reducing their water conveyance capacity in cold regions. It also covers the issue of negative temperatures during construction and operation of spillway structures that impose specific requirements to the design of structures and the use of construction materials. In addition, other challenges concerning regulation of gates and special loads and impacts from ice on structures, are discussed as well.

Chapter 9 - COMPOSITE MODELLING OF HYDRAULIC STRUCTURES

Composite modelling is the effective utilization of both physical and numerical models, used either in series or in parallel, to solve difficult hydraulic problems. This chapter discusses the use, advantages and limitations of physical and numerical modelling in hydraulic engineering practice and describes the processes of using composite modelling in the design of spillways and outlet dam structures.

Chapter 10 – ECONOMICS, RISK AND SAFETY ISSUES IN SPILLWAY DESIGN

Spillways are the main safety-assurance feature of any type of dam. For this reason, as a rule, direct cost reduction attempts in these structures will normally mean increases in the risk of lowering the safety of the overall dam project. However, in the formulation of the overall dam project layout the analysis of different types of spillway alternatives, for the same safety criteria, may have a significant impact on the overall cost of the project. In addition, different spillway configuration, such as single gated structure versus combination of gated and ungated spillways, for example, may lead as well to important cost savings. This chapter discusses these issues.

2. LARGE CAPACITY SPILLWAYS

2.1. INTRODUCTION

The purpose of this chapter is to present and discuss the engineering issues associated with the design, construction and operational features of spillways designed to discharge very large flood flows. Although there are no major differences between the criteria used for the hydraulic design of different sizes of spillways, construction and cost considerations play a significant role in the optimisation of design solutions for large spillways. In addition, it is recognised that these cases deserve special considerations due to the inherent operational risks associated with the management of large flows from the reservoir water level to the lower downstream reaches of the river.

For the purpose of this Bulletin, very large spillways are arbitrarily defined as structures designed for discharging flows larger than 20 000 m³/s and/or with specific discharge in the downstream energy dissipation facility, larger than 130 m³/s/m.

Most of present and future dam projects incorporating very large spillways are being built in Asia, Latin America and Africa, in rivers draining very large basins. Tables shown at the end of this chapter depict data for projects incorporating very large spillways built in selected countries. Projects of similar magnitude exist in other countries as well. These projects allowed gathering important experience which forms the basis of the discussions presented hereinafter.

2.2. GENERAL ARRANGEMENT STRATEGIES FOR LARGE CAPACITY SPILLWAYS

2.2.1. The spillway design flood

It is not the purpose of the present document to discuss methodologies for determining the spillway design flood. However, in discussing large capacity spillways, it is pertinent to observe the background on which the design flood magnitude is generally defined.

The characterisation of a spillway design flood generally requires:

- the return period (T_R) of different peak values;

- the volume of water associated with the peak flood values; and

- the PMF or Probable Maximum Flood, a deterministic value assumed to be the absolute maximum flood possible to be produced at the project site.

Many modern large spillways were designed to discharge the 1:10 000-year flood, which theoretically corresponds to the natural river flow whose peak has the probability of occurrence of 1% in an expected project life of 100 years. As a rule, these spillways are capable of discharging this flood peak with a reservoir level at its maximum allowable level, which often is higher than the normal maximum operating level of the project but does not encroach into the dam freeboard. If the reservoir area is large enough, the reservoir volume created by this level difference may attenuate the peak flood hydrograph and represents a certain margin of safety which is specifically considered in the strategy for managing the flood to be disposed of.

Instead of using the statistical 10 000-yr flood, the PMF is also widely used as the basic element to define the spillway design flow. However, as a rule, for large projects, both flood values are normally computed and with these, discharge conditions defined. It is common to define the spillway design flood as the 10 000-yr flood associated with the maximum allowable reservoir level and verify the spillway capacity to discharge the PMF from a reservoir level encroaching into the dam freeboard.

Although these criteria are widely used for the design of large spillways there is an inherent risk of underestimation of the flood for which discharge capacity should be provided, due to the possible non-representativeness of the data used for computation of probabilistic values and for the PMF determination. In fact, these criteria do not take into consideration the statistical simplifications of using a limited range of data to compute a return period many times larger, as well as the physical future changes of the site, such as climate changes and watershed modifications affecting runoff. These questions are extensively discussed in the ICOLD Bulletin No.142 *Safe Passage of Extreme Floods* (ICOLD, 2012) and deserve careful attention in the design of large spillways. The determination of confidence intervals for the computed flood value and the provision of additional means of discharge such as emergency spillways may be a prudent consideration.

2.2.2. Location, type and size

The technical characteristics and the cost of a spillway are not only related to the structures and equipment that control the flood releases themselves but are also heavily affected by its location and by the arrangement of other project features, particularly the dam type, height and the overall site characteristics. Many variations of arrangements are possible, and cost optimisations are worked out for each project and site, considering the flow magnitude, the operational requirements and the topographical and geological characteristics, as the most relevant parameters.

Environmental regulations in many countries require that a continuous steady flow, compatible with the environmental conditions of the downstream stretch of the river, be released from the dam barrier. This may affect the arrangement and operation of the spillway, especially in cases where the structures are built in large rivers. The present trend, however, seems to favour locating such release facilities in a separate entity. As an example, the Belo Monte Project presently being built in Brazil, which bypasses a long river curve, will have a spillway designed for 62 000 m³/s and must release in the bypassed stretch during the dry period, a minimum continuous flow of 700 m³/s and, during the wet season, flow values that may reach 8 000 m³/s. Instead of doing this through the spillway these flows will be released through a separate structure which will include an additional powerhouse to use the "environmental" flow. Many projects in various countries are following these criteria.

Spillways discharging large flows are necessarily large structures irrespective of the relative size of the dam or other appurtenant structures. If concrete dams are contemplated, the natural location of the spillway is over or across the dam. Generally, the spillway will be placed near or close to the river area, to facilitate the flow return to the river, a location that may have to be optimised when there is a powerhouse foreseen for the same location. In the Itaipu Project, Brazil-Paraguay border, (Figure 2.1), for example, the spillway was located in one abutment since the whole river valley was fully taken by the 18-generating unit powerhouse.

Fig. 2.1
Itaipu Project (Brazil-Paraguay border) and its 62 200 m³/s spillway

Depending on the height and type of dam – embankment or concrete gravity, arch-gravity or arch – many different arrangements and configurations are possible, and the creativity of engineers worldwide has provided innumerable examples.

For dams in wide valleys the spillway will normally be built over or through a concrete portion of the dam, whether or not the rest of the structure is an embankment. An example of this case is the Tucurui Project (Figure 2.2), in Brazil. In all these cases the width of the valley was compatible with the width of the spillway.

Fig. 2.2
View of Tucurui Project (Brazil), 110 000 m³/s spillway

However, if the dam site is not wide enough, it may be difficult to fit the whole spillway over the dam. Different solutions have been employed, from locating the spillways completely outside the dam structure, to combining surface spillways with intermediate and bottom outlets, as in Ertan Dam (Figure 2.3), in China. This project includes seven passages in the dam crest discharging 6 260 m³/s, six mid-level outlets with capacity of 6 930 m³/s, four bottom outlets for 2 084 m³/s and two tunnels discharging 7 400 m³/s. The total design capacity is 22 674 m³/s.

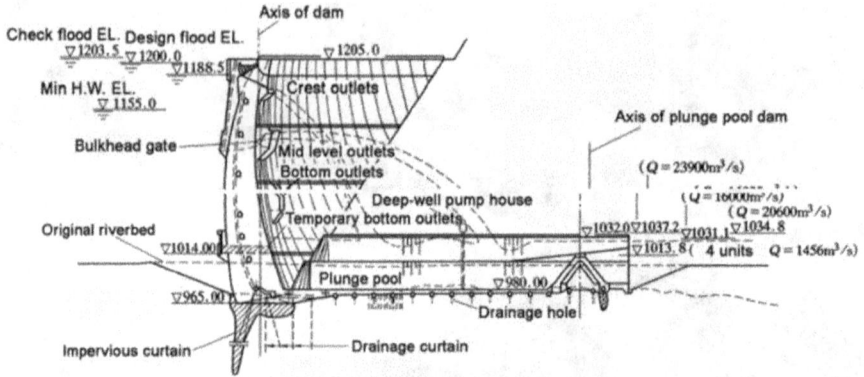

Fig. 2.3
Cross section and photo of Ertan Dam – China

2.2.3. Flood peak attenuation by reservoir

The use of part of the reservoir volume to store part of the flood hydrograph and allow the reduction of the flood peak and reduce the spillway design flood is a common practice for large spillway projects. Table 2.1 depicts some cases where this practice have been used (compare the columns "Peak inflow" and "Spillway design flood"). In such a case, particular attention has to be taken by the engineer to the elevation reached inside the reservoir to discharge the peak flow of the natural hydrograph.

This practice, however, increases the reservoir area and besides the cost impact – which has to be compared with the spillway and dam cost – may increase the environmental impact of the project. For large reservoirs in rather flat areas, and depending on the environmental legislation of each country, this may pose unsurpassable difficulties.

2.2.4. Single structure vs. different functional components

Spillways designed as a single structure to discharge very large flows will necessarily be long structures and in certain cases it will be economically difficult to fit the whole spillway in the axis of the dam. In these cases, it may be convenient to divide the spillway structure in two or more parts, one, to be operated more frequently, discharging the smaller more common floods (referred to as service spillways), and the others (auxiliary spillways) to complement the full capacity of the spillway system. Even when there is no problem of space, the division of the spillway into separate structures can be used to allow savings by building the auxiliary spillways with less stringent technical features since its use is not as frequent as the main structure. Of course, this requires that in any case the overall safety of the project is not affected, but it may require continuous inspections on its performance. Some examples that illustrate this possibility are discussed below.

In the Itá Project in Brazil, with two spillways, the total capacity is 52 800 m³/s. The 275-m long chute of the auxiliary spillway was lined only over the initial 120 m (capacity 20 000 m³/s, specific discharge 234 m³/s/m). The key element supporting the decision to use this cost reducing feature was the accessibility to the area and the possibility of remedial work if and when necessary. In fact, initial tests carried out on this structure showed that erosion risks were higher than anticipated and the concrete slab was extended 55 m further downstream to cover the affected area as shown in Figure 2.4 (Andrzejewski, 2002).

(a) Initial lining extension   (b) Extended lining after erosion tests

Fig. 2.4
Auxiliary spillway of the Itá Project (Brazil)

Another example is the Xingó Project, also in Brazil. The project has two parallel spillways for a combined capacity of 33 000 m³/s. The auxiliary spillway 252-m long chute, designed for a maximum flow of 15 500 m³/s has a lined concrete slab only along the first 90 m. This spillway has operated in test conditions up to 4 000 m³/s. Some localised erosion was observed and backfilled with concrete. After the tests the chute was approved for operation in view of the fact that if higher flows cause additional erosion, full access to eventual repair will be available (Eigenheer *et al.* 2002).

These two examples show that the use of an auxiliary spillway for infrequent operation with some flexibility in their technical standards can provide initial cost savings but must secure safe operation and this is dependent on continuous inspection and available unrestricted access to repair eventual damages.

The use of an auxiliary spillway for discharge part of the total design flood should not be confused with the provision of an emergency spillway. This structure is recommended when the confidence on the basic hydrological data and studies, as indicated before, is limited in extension or quality. In that case emergency spillway provisions or other type of facility to retain or divert the excessive flood may be included into the project. This question is also discussed in ICOLD Bulletin 142 (ICOLD, 2012) where there are considerations and recommendations concerning alternative means of handling floods larger than the spillway design flood.

2.3. CONTROL STRUCTURES

2.3.1. *Surface spillways*

Spillways designed to pass large flows will generally be surface spillways mainly because of the magnitude of the flood to be released. However, low level bottom outlets have been used as well, especially when the structure was also used for river diversion and/or if discharge of accumulated sediment in the reservoir was required.

Most of the large spillways built for hydroelectric or water supply projects are gate controlled because otherwise the length of the spillway and the loss of useful head would impair the benefits of the project. However, the use of gated spillways has sometimes generated concerns about the possibility of gate failure or mal operation in critical circumstances and creating conditions for dam overtopping and collapse. Clearly ungated spillways eliminate this risk.

Whatever the purpose of the project, it is not easy to fit physically and economically a total ungated spillway to discharge very large flows in a project site. Nevertheless, this has been done in some projects where it was possible to use most of the site width for the spillway, locating the water intake or other project facilities in a place not interfering with the spillway. Where the physical possibility exists, the balance between economy and risk is the key decision factor. Figure 2.5 shows a view of the Burdekin Dam, in Australia, designed for a flow of 64 600 m^3/s with a maximum head over the spillway crest of 17 m. If this were a gated spillway it would be possible to have the operation level of the reservoir 17 m higher than in the actual ungated solution. This indicates the economic limitation of large ungated spillways for hydroelectric projects. Additional data on other Australian large ungated spillways are depicted in Table 2.3.

Fig. 2.5
The 64 600 m^3/s ungated spillway of the Burdekin Dam, Australia

Probably most of the modern gated spillways designed to discharge large flows are equipped with large radial or tainter gates. Large vertical lift gates can and are, of course being used but they require a large overhead structure to allow the raising operation and also require lateral slots that cause some disturbance of the issuing flow. These gates are normally operated with cable hoists because if hydraulic cylinders would be used, the height of the overhead structure would have to double. On the other hand, vertical-lift gates suspended by cables are prone to generate serious vibration problems when the gate is lowered in high velocity flow.

The configuration of the control structure for surface spillways for both types of gates is essentially the same for small and large structures. The orientation of the spillway axis in relation to the approaching flow for large spillways will quite often require hydraulic model studies to avoid flow disturbances and unequal discharge in different passes.

A combination of gated spillways and outlets with ungated or emergency spillways to avoid or minimize the gate failure risk is sometimes used. This is also not easy to do for projects involving very large flood flows. A recent study for the Inga Project in the Congo River in Africa, where the design flood is 60 000 m³/s, proposes a combination of a conventional surface gated spillway with a PK labyrinth spillway (see chapter 6) to deal with such a large discharge (Lempérière et al. 2012). The combination of PK spillways and gated bottom outlets has already been used for projects with smaller flood flows.

2.3.2. Radial Gates

The use of radial gates for controlling large capacity spillways is an established trend. Very large radial gates up to more than 440 square metres in area were installed and are operating successfully in some Brazilian surface spillways. Gates twenty metres wide by more than twenty metres high were used in Itaipu, Tucurui, Estreito, Santo Antonio and Jirau projects, in Brazil as indicated in Table 2.1. This will cause, of course, higher values for the specific discharge in the downstream spillway channel but designs had cope successfully with this.

In general, the main reason to use such large gates is to achieve a maximum reduction of the dam length. In many very low head hydroelectric projects built in very large rivers, there is a limitation on the power output of the generating units, which means increasing the number of units to meet the total available hydraulic power, and therefore the length of the powerhouse becomes significant, leaving a reduced space for the spillway. The spillway design floods in these projects are also very large and the result is the need of saving space by using large gates. As an example, the Jirau Project on the Madeira River, in Brazil (3 750 MW), has 50 generating units operating under a maximum gross head of 15.7 m, and a spillway design flow of 82 600 m³/s with 18 radial gates, 20.0 m wide by 22.8 m high (Figure 2.6). The spillway has a length of 445.0 m and forced the powerhouse to be divided into two parts located one on each bank of the river.

In the Yaciretá-Apipé Project, between Argentina and Paraguay, the power-plant houses 20 units and is 816 m long, while the spillway design flood of 95 000 m³/s had to be divided into two independent structures (Figure 2.7) one with 18 radial gates 15.0 m wide by 19.5 m height and the other with 16 gates 15.0 m wide by 15.5 m high.

Fig. 2.6
82 600 m³/s Jirau Project Spillway under construction – Madeira River, Brazil

Fig. 2.7
Yaciretá-Apipé Project – 816 m long powerhouse on the Paraná River next to the 55 000 m³/s main spillway - Argentina/Paraguay Border

Although radial gates are generally considered to be very reliable and safe equipment, there have been some events in which radial gates failed causing concerns in the wide use of this type of equipment. A general discussion of issues related to gates in general was the subject of Question 79 of ICOLD Twentieth Congress. The General Report for this question (Cassidy, 2000) pointed out failure of gates caused by design deficiencies related to earthquakes, flow induced vibrations, fabrication and erection errors. Considering the size of gates and the importance of flows in large spillways, special attention to these issues by designers and owners is very important.

It is not, of course, the object of this report to treat extensively the issue of gate operation and safety, but to call the attention of designers and owners of large spillways to the need of considering in the design and operation of these structures, details that many times are not included in the conception of large projects.

As an illustration of the kind of problems that may occur with radial gates, the following five cases indicate the diversity of situations that can happen:

- Folson Dam Spillway, in the United States - One of its 8 spillway radial gates failed in 1995. The gate was 12.8 m wide by 15.2 m high and failed when being raised with the reservoir practically full (Todd, 1997). Investigations examined the possibility of vibrations affecting the gate structure but concluded that the gate failed because of friction of the trunnion causing the failure of a lower strut transmitting the water load to the trunnion. It was discovered that the friction force in the trunnion was not considered in the structural design of the gate and besides that the reduced frequency of lubrication and lack of weather protection resulted in corrosion which increased the friction over time. The failure of the gate with reservoir full released 1 132 m³/s.

- Itaipu Dam Spillway, in Brazil (Lima da Silva *et al.* 2000) – The rod of the one hydraulic hoist cylinder used to move one of the gates broke in July 1994 after 12 years of safe operation. There are 14 gates at Itaipu, each 20.0 m wide by 21.4 m high, each operated by two cylinders. The rod breakage occurred while lowering the gate during a programmed maintenance operation with no water flowing in the passage. Investigations of the failed equipment concluded that the failure originated from a transverse crack associated with a corrosion process and triggered by the vibration originated in the friction of the side seals of the gate to the lateral walls without water. After the rod failure, the broken cylinder was discarded and replaced by a new one and all the 27 remaining hydraulic cylinders were dismantled, machined and polished to remove cracks and corrosion. A detailed program of maintenance was then formulated and implemented.

- Salto Osório Dam Spillway, in Brazil – One of the 9 radial gates of this project, each 15.3 m wide by 20.77 m high, failed in September 2011, after 37 years of safe operation. The gate movement was carried out by cables moved by winches and automatically controlled by switch relays. A command for closure was issued but a failure of the limit switch relay did not stop the hoist and continue to order the turning of the winch drum, winding the cable in reverse direction in the hoist pulley, and fully opening the gate. The winding occurred in more than one layer and cause rubbing the cable against the concrete, severing the cable by friction. The gate closed violently by gravity and was then carried by the flow with the bearings being ripped off the concrete piers. Two thousand cubic meters per second were released, fortunately without any specific consequence. The reservoir level was lowered, a stop-log placed in the passage and a new gate installed.

- Shiroro Dam Spillway, in Nigeria (Epko and Adegunwa, 2011) – This project, commissioned in 1984 has a spillway with 4 radial gates each 15.0 m wide by 16.85 m high, designed for 7 500 m³/s. The gates are moved with hydraulic hoists. In spite of the existence of slots for stop-logs, no stop-logs were provided. In 2005 maintenance actions were carried out including replacement of the seals. However, gate No.4 could not be opened more than ¼ of its course. Investigation at this time detected that the top right-hand pier had moved and closed the width at the top by 57 mm, at half height by 46 mm and at the base by 9 mm. This prevented the full opening of the gate and of course limited the capacity of discharging the spillway design flood. Reports up to 2011 indicate that the reason for this problem had not yet been found, although it is probably related to alkali-aggregate reaction. No solution has known to be proposed or implemented so far.

- Tarbela Spillway, in Pakistan (Khalio Khan and A-Siddiqui, 1994) – One of the 7 gates of the service spillway of this project, each 15.2 m wide by 18.6 m high, failed after 17 years of problem-free operation. The accident was described as the gate getting stuck during a lowering operation and then falling down breaking two hoist ropes and damaging the hoist deck. The event happened when the gates were open with the spillway discharging 2 475 m³/s and the operator started the closing operation of the extreme right gate. As reported, the motor of this gate hoist device tripped and the gate, after falling from an undetermined height, got stuck leaving an opening of 112 mm from the bottom sill resulting in high velocity discharge. After detailed investigation of the accident, it was concluded that the gate got stuck because insufficient clearances between the side sealing plates on the pier and the clamping bar of rubber seal on the gate. The original clearance was 17 mm and over the years was reduced to as little as 2 mm. The investigation of the cause of this reduction concluded that it was the combination of the gate thermal expansion and slippage of the seal holding device due to the loosening of the bolts along the years of operation. The gate fell down by its weight when freed by hydraulic vibrations after cooling of the skin plate by the evening lower temperatures.

In spite of such problems, the percentage of radial gates that suffered failure is relatively small. The US Bureau of Reclamation, which is the owner of Folson Dam informed that they have 314 radial gates in their works with 18 000 gate-years of operation and only one gate (Folson) failed (USBR, 2011). The US Army Corps of Engineers have 90 radial gates in various projects and although some technical incidents and design inadequacies have been found, there were no major failures in any of them (Ebner and Craig, 2012). In Brazil, there are 330 radial gates installed in spillway projects whose capacity is larger than 20 000 m³/s and besides the Itaipu and Salto Osório gate problems mentioned above, there was only one other problem in one gate of the Furnas project, which was related to crystalline corrosion of the steel rods anchoring the trunnion bean. This problem had no influence on the gate operation.

Very important feature of a spillway controlled by radial gates is the provision for the use of stop-logs in all gate bays and of course their availability. The cases listed above illustrate the important role of stop-logs to overcome an accident with the gate, but they are also very important to allow maintenance operation of the gate in the dry. As mentioned in the general report of ICOLD Question 79 (Cassidy, 2000) the lack of stop-logs in Folson Dam, created a very difficult problem to correct the gate failure accident. Other example of problems due to unavailability of stop-logs is the recovering of a gate in La Villita Dam in Mexico, damaged by the sudden failure of the gate steel reinforcement.

Present practice for large gates indicates that gate movers based on servomotors (hydraulic cylinders) rather than cables or chains have an increased reliability. Many modern large radial gates are designed in this way, with two cylinders per gate. As per Itaipu Project considerations that only one of the cylinders should be able to close or open the gate is a sound criterion.

The available registers of incidents and accidents involving radial gates indicate that in a significant number of cases, faulty operation of the gate, and not a problem with the gate proper, was the primary cause. These include failure in the electricity supply to move the gates and difficulties in access to the spillway area under emergency conditions. This kind of problems (for radial and vertical-lift gates) occurring during heavy rains and flooding conditions, were the cause of very well-known dam collapsing cases by overtopping (among which, Euclides da Cunha Dam in Brazil, Tous Dam in Spain and Belci Dam in Bulgaria). In large spillways, designed to control very large flows, it is of utmost importance to build redundant facilities in electric supply and alternative means of unrestricted access to gates even under heavy rain or flood conditions.

It has also been suggested, and sometimes applied, the so-called N-1 criterion in which the number of gated passages of the spillway is determined so that the total spillway design flood can be discharged by all gates except one. This rule does not have a universal consensus mainly because of its economic impact. However, one rational approach to the issue would be related to the size of the incoming flood and the speed of increase of the reservoir level. For the very large reservoirs generally associated with large spillway projects the raising of the reservoir level caused by the jamming of one gate is normally slow and takes one or more days before it reaches a critical level. During this time, it should be possible to reach the jammed gate and manually force its opening. Of course, this implies that access to the damaged gate is available at all times. In such conditions there would be no need to provide an additional gate bay to the spillway. However, for reservoirs with small areas in which large floods would cause a rapid increase in the reservoir level elevation, the application of the N-1 criterion seems to be a justifiable approach if no other economically alternative emergency discharge feature can be provided.

2.3.3. Bottom and intermediate level spillways

The exclusive use of intermediate level orifice spillways and/or bottom outlets to discharge very large flows is relatively limited mainly because of the limitation in capacity of the maximum individual passages, where the flow is proportional to the square root of the head, as compared to surface weirs where the flow varies with the 1.5 power of the head. This of course requires a larger head in orifice spillways to offset the flow capacity difference. However, in many projects, a large head orifice is necessary anyway to allow a wider control of the reservoir level which may be convenient for project operation and also for the control of flood routing in the reservoir.

Large bottom outlets are normally used when the flushing of sediment is required or when the river diversion during construction requires a low-level passage which can be more easily controlled during river closure and reservoir impounding. The exclusive use of bottom outlets for discharging large floods is limited to dams of small or medium height, and generally where the combination of spillway function with sediment disposal or river diversion control during construction, is required or convenient.

However, in many projects with concrete dams a combination of bottom and orifice outlets or with surface spillways is rather common. The Ertan Project, in China, mentioned above and the Three Gorges Project in China are significant examples of the use in large rivers of permanent and provisional (used for river diversion) orifices and bottom outlets.

Figure 2.8 depicts a cross section of the spillway block of the Three Gorges Dam, which incorporates a very large flood discharge system able to pass a maximum flood of 102 500 m³/s with the reservoir at the check flood level at El. 180.4 m. As shown in the figure it contains two levels of permanent discharging facilities (surface [7] and a deep orifice outlet [8]) in addition to a bottom diversion outlet [9] used during construction and lately plugged.

The surface spillway is formed by 22 surface vertical lift gates, 8 m wide, with a sill level at El. 158. The deep orifice outlet ([8], in the figure) comprises 23 passages controlled by radial gates, with intake at El.90, each 7 m wide by 9 m high. The designed single bottom diversion outlet capacity is 2 117 m³/s, operating under 85 m of head, and produces an outgoing flow with a specific discharge of 302 m³/s/m. The total designed discharge capacity of the deep orifice outlet is 48 691 m³/s, with reservoir level at El.175. This is one of the largest intermediate orifice outlet facilities worldwide and is clearly the result of fitting the project layout to the various needs of the undertaking, both during and after construction.

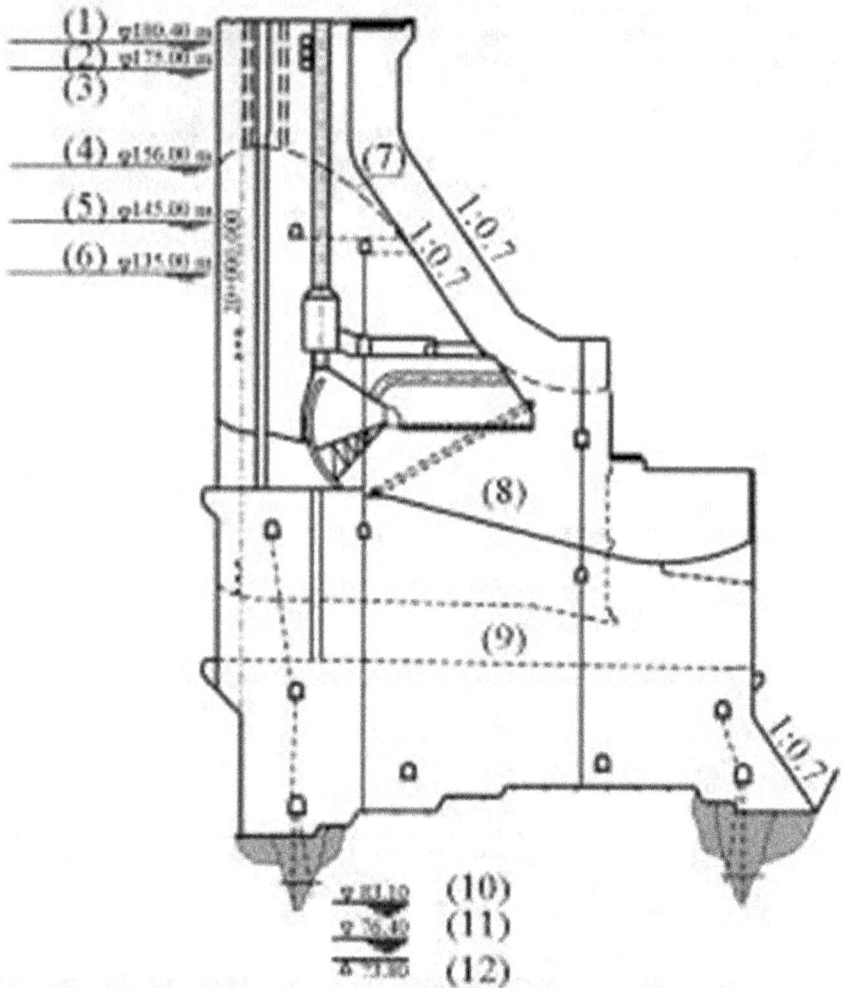

(1)—Check flood level; (2)—Normal storage level;
(3)—Design flood level; (4)—Initial normal storage level;
(5)—Flood control limit level; (6)—Initial flood control
limit level; (7)—Surface outlet; (8)—Deep outlet;
(9)—Bottom diversion outlet;
(10)—Discharge of check flood, $Q=102\ 500\ m^3/s$;
(11)—Discharge of design flood, $Q=69\ 800\ m^3/s$;
(12)—$P=1\ \%$, flood discharge, $Q=56\ 700\ m^3/s$

Fig. 2.8
Three Gorges Project – Cross section of the spillway block (Wang et al, 2011)

Very large spillways have also being built in India. Representative of these structures is the Chamera I Project spillway designed for 26 500 m³/s (Figure2.9), which contains 8 orifice outlets 10.0 m wide by 12.8 m high, controlled by radial gates and 4 bottom outlets 4.0 m wide by 5.4 m high controlled by sluice gates.

Fig. 2.9
Orifice and bottom outlets of the spillway of Chamera I Project, in India

Another large bottom outlet spillway is the on Jupiá Project, in the Paraná River, in Brazil (Figure 2.10). As indicated in Table 2.1, this project contains four surface spillway passages and 37 bottom spillway passages, each 10.00 m in width by 7.61 m height, with a nominal capacity of 44 000 m³/s. The reason for this kind of solution was to facilitate the control of the river during construction and the impounding of the reservoir.

Fig. 2.10
Jupiá Project – 44 000 m³/s bottom outlet spillway during construction (1967)

The bottom outlet passages in this project are controlled by radial gates. The project was commissioned in 1968 and during 47 years of operation its performance has been good, with localised problems of erosion particularly in the downstream part of the floor slab. The maximum daily flow recorded since 1968 was 28 943 m³/s but about 20% of this flow was discharged by the four surface spillway passages.

The hydraulic design of bottom outlets is extensively treated in technical literature, both in their general terms and for specific cases. It is, however, appropriate to stress the importance of paying thorough attention to the risk of cavitation and the need of providing adequate air supply to the region immediately downstream of the controlling gate.

Bottom and intermediate dam outlets can also be provided through tunnels. Specific references to these works are given in chapter 7 of this bulletin. In many cases these outlets are built in tunnels that were used for river diversion during construction. In these cases, special attention should be paid to the schedule of the overall works, since transformation from a provisional construction feature into a permanent outfit of the project after its use as a river diversion feature, may incur time delays and adverse economic impacts.

2.4. CONVEYANCE STRUCTURES

The return of the flow to the river depends on the type and height of the dam, the form of the valley and its geological configuration. In general, for medium-height dams, very large spillways discharging large flows will be surface spillways, returning the flow to the river through a short steep chute or long chute channels. In both cases the chutes normally end either in a flip bucket, if the height of the dam allows it, or through a hydraulic jump or roller-bucket stilling basins. Very low dams will normally have a hydraulic jump stilling basin immediately following the crest structure. As a rule, underground tunnel chutes, as a single outflow facility, will not fit the flow magnitude of large spillways.

As discussed above, large spillways will quite often be gated structures resulting in a rather high specific discharge for the issuing flow. This affects the design criteria used for the dissipation of the energy of the outgoing flow.

2.4.1. Chute Channels

In projects with high embankment dams built in valleys with narrow to moderate widths, long chute spillways built in the dam abutments, ending in a ski jump and discharging over a plunge pool, is the rather standard solution both for small or large spillways. There are some examples of ungated chute spillways built over embankment dams, but only for small design floods, among which Crotty Dam, in Australia, Ahning Dam, in Malaysia (Cooke, 1985) and Tongbai Dam, in China (Zhao and Yuyan, 2006).

When the valley width is wide enough, a concrete stretch of the dam to accommodate the spillway, with a steep short chute and flip bucket, can be designed, as in the Tucuruí Project (Figure 2.2). In these cases, a hydraulic jump stilling basin could also have been provided as in Sardar-Sarovar Project (Figure 2.11) overflow concrete dam. This 163 m high dam has a very large spillway system including a service spillway with 23 surface radial gates, 16.78 m wide by 18.30 m high, an auxiliary spillway with 7 surface radial gates 18.30 m wide by 18.30 m high and 4 bottom outlets 2.4 m wide by 3.6 m high. The dam also had 10 bottom sluices 2.15 m wide by 2.75 m high used for construction and ultimately closed.

Fig. 2.11
Sardar-Sarovar 87 000 m³/s spillway in India (piers and radial gates not yet installed)

The design of a long chute spillway in an embankment dam abutment depends on the magnitude of the flood and on the topography and geology of the site. For large spillways this scheme often requires very large excavations which have to be wasted or may be used as a source of material for the embankment. However, sometimes the balance of quantities or the planning schedule of the project or even the geotechnical stability of excavation pits, create difficulties to fit a single very large spillway in one abutment. This may lead to the division of the spillway in two or more separate structures, a service and auxiliary spillways, a solution that may have already been envisaged by other reasons, as indicated before.

The location and the profile of the chute will, for these reasons, be designed to minimise excavations. In plan, the best hydraulic solution is to have a straight chute with constant width although converging chutes are also common. It is important to locate the flip bucket of the ski jump above the maximum tail-water level and in such way that the river swirling flow does not affect the downstream toe of the dam. It is often convenient and most of the time necessary to carry out hydraulic model investigations to define the final design of this feature, and especially the geometry and performance of the plunge pool.

2.4.2. Hydraulic jump stilling basins

Hydraulic jump stilling basins are necessarily used for very low dams but can also be used in higher dams, as indicated above. For medium-high dams there is no universally accepted criteria for selecting a stilling basin solution against flip bucket and plunge pool. Economic considerations, local geological setting and the usual practice of different countries, usually define the type of solution.

For large spillways with large specific discharges, the basin design must consider the possibility of downstream erosion caused by the occurrence of the jump outside the basin and the asymmetrical operation of the gates causing swirling currents depositing rocks and debris into the basin causing abrasion of the concrete. This is a rather common problem with stilling basins. One illustrative example of this kind of damage occurred in the stilling basin of the 21 400 m^3/s Marimbondo Project spillway, (Carvalho, 2002), in Brazil (Fig 2.12(a) and 2.12(b)).

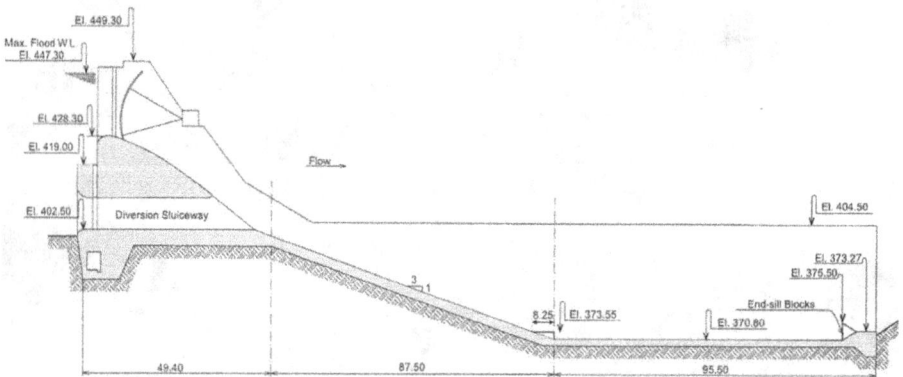

Fig. 2.12
(a) Marimbondo Spillway stilling basin

42

Fig. 2.12
(b) –Marimbondo Spillway stilling basin: damages from asymmetric operation

- Spillway profile

- Solid material deposited in the basin due to asymmetrical gate openings

- Damaged steelwork in basin

The use of chute or baffles blocks, as it is illustrated in the Marimbondo Project, is not recommended specially for basins with high specific discharge flows, since they are prone to produce serious cavitation damages. An illustrative example of this fact occurred in the Porto Colombia Project, also in Brazil. The 16 000 m³/s spillway was originally built with a chute and end-sill baffle blocks, as illustrated in Figure 2.13 (a). After 10 years in operation, very serious cavitation damage occurred adjacent to the chute blocks and to a lesser degree on the end-sill blocks. To correct these problems the blocks were removed and the end-sill made continuous, as also shown on Figure 2.13(b). After these measures, no further damages were observed (Carvalho, 2002a).

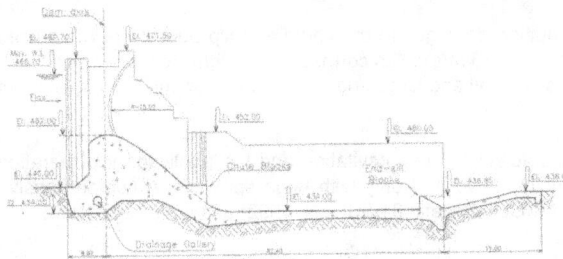

Fig. 2.13
(a) - Porto Colombia spillway: cavitation damage

- Original spillway profile

- Cavitation damage downstream of chute blocks

43

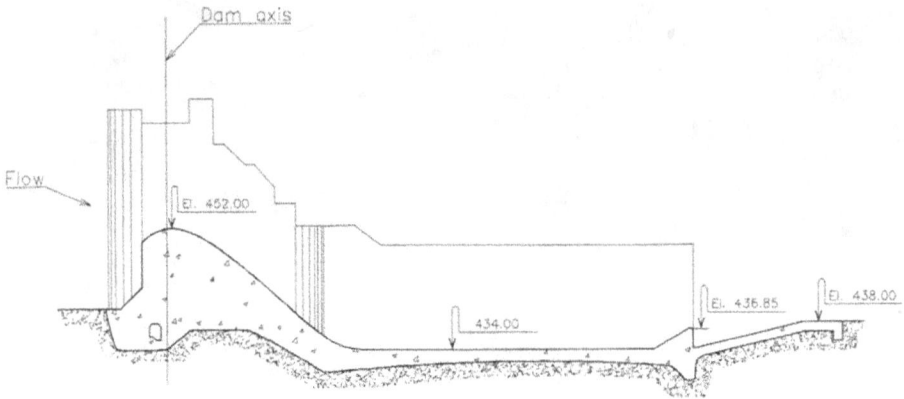

Fig. 2.13
(b) - Porto Colombia spillway.

- Refurbished spillway profile without chute and end-sill blocks

Besides these types of problems, the integrity of the basin may also be affected by fluctuating uplift forces whose peaks overcome the weight of the concrete slab and the resistance of the steel anchors that are provided to incorporate the weight of the underlying rock. Examples of serious accidents caused by these fluctuating forces are Malpaso, Mexico (Sanches-Bribiesca and Viscaino, 1973) and Karnafully, Bangladesh (Bowers and Toso, 1988). The effect of these fluctuating uplift forces results from the transfer of dynamic pressures generated by the turbulent hydraulic jump to the underside of the basin slabs, through joints and drainage system pipes. A comprehensive summary of present knowledge of the problem and recommendations for design procedures can be found in Bollaert, (2009).

In many cases seeking economy, designs consider the occurrence of the jump within the basin only for frequent floods, allowing it to happen downstream for very large and infrequent floods. Of course, this should only be allowed if the downstream rock is considered sound and resistant to erosion caused by such non frequent floods.

2.4.3. Aerator facilities

Aerator facilities to prevent cavitation damages to the chute and flip bucket are often used when the average flow velocity reaches about 30 m/s which corresponds roughly to a cavitation index of 0.25. This of course applies to spillways small and large and is dependent on the velocity and not on flow.

It is not the purpose of this chapter to discuss cavitation and the applicability of aeration facilities, since this question is not strictly associated only with large spillways. A comprehensive treatment of the subject can be found in ICOLD Bulletin 81 (ICOLD, 1992) and in Falvey, (1990).

2.4.4. Energy dissipation and downstream erosion

The dissipation of the energy from the flow outgoing from a spillway and the control of the downstream erosion that may be caused by the remaining energy of the flow, are major issues in the design and operation of large spillways. This is especially true due to the normally high values of the specific discharge of the outgoing flows. It can be seen in the data shown in the Tables presented at the end of this chapter, and from examples mentioned above, that values of the specific discharge above 150 m³/s/m and up to 300 m³/s/m, are rather frequent in large spillways outflows. In general, high specific discharge in large spillways results from the need of reducing the gated surface spillway length, as discussed before, or in dams built in narrow valleys with surface and bottom outlets.

The main common types of energy dissipaters are the hydraulic jump and/or roller bucket stilling basin, the ski jump and plunge pool arrangement, and the free fall jet into a plunge pool.

In many cases where the evaluation of the quality of the rock allows it, instead of a stilling basin as such, there is simply a concrete slab leading to a bare rock channel. This solution has been used in Brazil in auxiliary spillways in some projects, as mentioned before for the Itá (Figure 2.4) and Xingó Projects, while in Canada there are a good number of projects in which the single main spillway follows this trend. It is of course an economic solution where the rock is sound, but in any case, some erosion damage can occur, and it is necessary to have access to the damaged area and a flow regime with time intervals to allow eventually needed repairs. An interesting example of this case is the LG-2 spillway, part of the James Bay Project, in Canada, which has a spillway for 17 500 m³/s discharging in a bare rock channel excavated in steps (Figure 2.14). Although in operation since September 1979, its maximum discharge along 30 years has been only 3 500 m³/s (Nzakimuena and Zulfiquar, 1999). Nevertheless, there has been significant scour immediately downstream of the protecting upstream slab.

Fig. 2.14
LG-2 spillway discharging in a bare rock channel and erosion damages downstream
from the control structure.

The issuing of a high specific discharge flow from a stilling basin may create erosion by dislodging and displacing pieces of fractured rock, some of very large size, and creating scour holes. It is very difficult to evaluate the magnitude and extension of this erosion process, which is different from the erosion caused by a plunging jet, which is also difficult to foresee but nevertheless has been better studied. Hydraulic model studies to infer the possibility and extension of scour holes are often necessary together with a careful evaluation of the geological condition of the site. It is of course very important to prevent the evolution of the scour hole into the foundation of the stilling basin. In the Macagua Project 30 000 m³/s spillway, in Venezuela, the originally designed concrete slab downstream from the control structure, had to be extended 40 m after deep erosion occurred next to the stilling basin (Marcano, 2009). In many cases, especially in very low dams, after the start of operation of the project this additional construction work may be difficult and expensive, because it requires coffer-damming the area where the slab will be extended. It is therefore prudent, in these cases, to be conservative in the original dimensioning of the extension of the stilling basin slab.

For higher embankment and gravity concrete dams, a ski jump at the end of the spillway channel is the normal rule. The flow is discharged from the ski jump creating a jet that impacts on the riverbed thereby producing a scour hole, which assists in the process of energy dissipation. The prediction of the scour evolution and the control of the scour hole dimensions is one of the most difficult tasks of the designer of large spillways with high specific discharge, because of the costs and risks associated with it. Chapter 3 of the present bulletin deals specifically with this subject and covers the various theoretical and practical aspects of this question.

Among the issues generally present in most cases of large spillways is the need of a pre-excavated plunge pool if the depth of the downstream natural level is considered to be insufficient to dissipate the outgoing flow energy and the project does not allow the construction of a tail pond dam. For large gated spillways with high specific discharges, the amount of energy to be dissipated is substantial.

The determination of the need of a pre-excavated plunge pool and especially its adequate stable dimensions is based on the estimated depth and lateral expansion of the scour hole due to jet action. As also discussed in Chapter 3, this issue has been the object of many studies and researches, with both empirical and physical approaches. The development of the jet erosion scour must avoid the risk of affecting the foundation of the permanent structures and the stability of the valley slopes, but the difficulties of anticipating the length and width of the scour pit and establishing the horizontal dimensioning of the plunge pool, especially in the case of large spillways with high specific discharge values, lead to a defensive and conservative approach. However, as pointed out in the following chapter of this bulletin, to use the maximum spillway flow for this seems to be too conservative, and it is recommended using a serviceability design discharge with a probability of occurrence of about 50 % during the useful lifetime of a dam. Nevertheless, the stability of the dam itself including its abutments and foundation should not be endangered up to the safety check flood of the dam.

As of today, in current practice, probably most cases of estimating scour depths are determined by empirical formulae. The computation with these empirical formulae may give an approximation of the real depth of a scour pit associated with a given flow for the project and provide information for the location of outlet structures and the design of pre-excavated plunge pools.

An interesting example illustrating the approximations and anticipating the depth and extension of jet erosion in a large spillway, is shown in Figure 2.15 (Sucharov and Fiorini, 2002) produced as part of the analysis of the Itaipu Project spillway erosion. The data reflects the field information surveyed in 1988, six years after the exceptional floods of 1982. During this whole period the spillway discharged continuously all the incoming flow since the power generating units were still being assembled. The maximum discharge was 40 000 m³/s roughly equal to the 500-yr flood. In the graph of Figure 2.16 the model and prototype data are shown in comparison with projections based on Veronese empirical formula – which was used for the prediction of erosion depth – with different values of the formula K coefficient, and data from other projects.

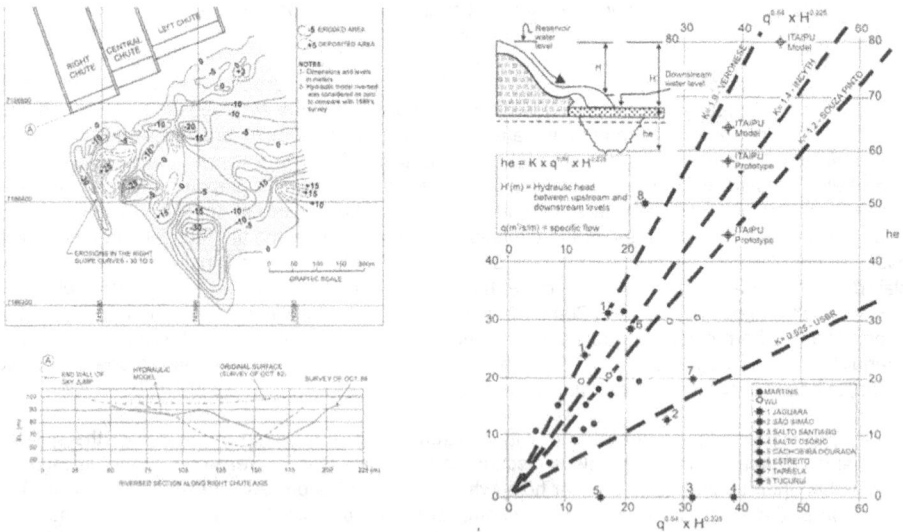

Fig. 2.15
Itaipu Spillway erosion data. (Sucharov & Fiorini, 2002)

One important and controversial issue which substantially affects the cost and schedule of a project is related to the need of lining the plunge pool. If the rock is sound and free of fractures the general practice has been to leave the excavated plunge pool unlined. Plunge pools at the Tucuruí and Itaipu Projects, mentioned before, were unlined with good results. However, in order to prevent the evolution of the scour to the permanent structures, concrete lining and or rock bolting and anchoring have been considered necessary in some projects. As discussed in Chapter 3 of this bulletin, the requirements to obtain a problem-free lining are very stringent, difficult and expensive. However, in some projects plunge pools have been totally lined with concrete slabs anchored to the rock, as shown in Figure 2.16, for the 15 000 m³/s Karun III spillway, in Iran, to prevent any detrimental consequences on the high arch dam. In this case the plunge pool lining required 250 000 m³ of reinforced concrete and 135 000 m of rock bolting, anchoring and doweling (IWPC, 2004).

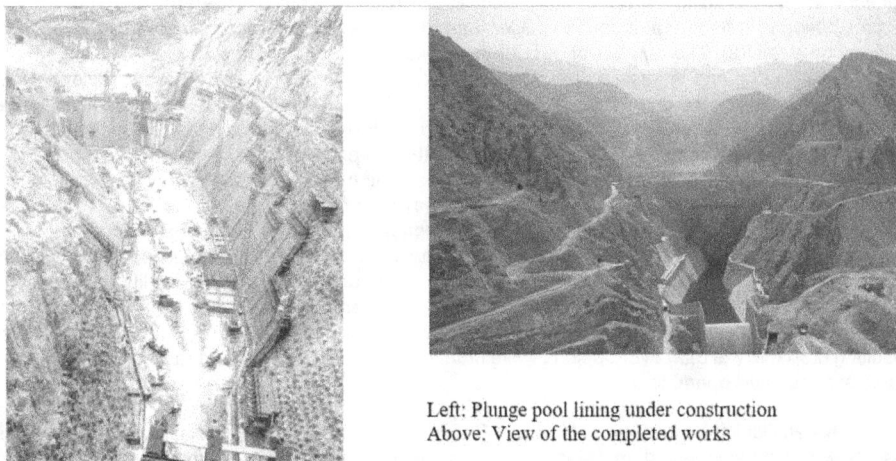

Left: Plunge pool lining under construction
Above: View of the completed works

Fig. 2.16
Karun III spillway (15 000 m³/s) concrete lined plunge-pool

2.5.   OPERATIONAL ISSUES AND RISK MANAGEMENT

Very large spillways are associated with large rivers and with large floods. As discussed above these spillways will normally be gated structures and, as such, require strict operation rules to avoid incidents and accidents related to the passage of extreme floods over the dam barrier. This of course is true for any size of gated spillway, but the magnitude of the damage caused by operational failure of a large spillway can have much more serious consequences.

The technical literature and specially ICOLD documents – bulletins, reports and papers presented in congresses and workshops – have dealt extensively with this subject. Bulletins 49 (1986) and 142 (2012) and the general reports of ICOLD Questions 71 (Pinto, 1994) and 79 (Cassidy, 2000) present information and comprehensive summaries of operational issues of spillways.

In most cases, and specially, in projects incorporating very large spillways, the project owner produces an Operation & Maintenance Manual that includes, for a specific project, rules and directives to be followed by operators in normal and in emergency situations. It is of course very important that these rules be enforced and that the operating staffs be trained and qualified to carry out the corresponding activities.

The proper operation of gates in large spillways is the key issue to prevent the risk of dam being overtopped and eventually breached or destroyed. Some of the possible problems with gates have been mentioned before in this document but in general the causes of malfunctioning can be classified as follows:

- failure of the operator to issue the commanding order to open the gates on time;

- failure of the gates to move as commanded or the supporting systems to provide the power or the commanding signals to correctly move the gate; and

- clogging of the passage due to large size debris brought by the flood.

However, the evaluation of the true risk of dam overtopping in gated spillway projects is more complex. For instance, the commonly used Dam Safety concept of Inflow Design Flood (IDF) does not address the possibility of "operational flood" in which a dam could be overtopped and fail due to a combination of a flood that is much smaller than the IDF and one or more operational faults (e.g. spillway gate failure). The number of possible combinations of unfavourable events causing such a failure is very large and increases with the complexity of the dam or system of dams. Micovic *et al.* (2015) presented results of stochastic flood simulation for a system of three dams and reservoirs in Western Canada, where the focus of flood hazard analysis was shifted from reservoir inflows to corresponding reservoir levels. The authors derived the full probability domain of the peak reservoir level as the combination of probabilities of all factors leading to it, including reservoir inflows, initial reservoir level, reservoir(s) operating rules and spillway gate failures. The results showed that dams were much more likely to be overtopped (and consequently fail) due to a combination of otherwise somewhat common individual events than due to a single extreme flood event such as the IDF. Micovic *et al.* (2015) also concluded that the importance of spillway gate reliability increases as the size of available surcharge storage decreases. For example, in the case of a small reservoir surcharge storage, simulating possibility of spillway gate failures during the flood increased the resulting probability of dam overtopping by five orders of magnitude over the case in which all spillway gates are assumed operable.

A statistical survey cited by Hinks and Charles (2004) based on Foster *et al* (2000), indicates that 46% of embankment dam failures were due to overtopping, and among these 13% were associated with spillway gate operation. Although there seems to be no figure to indicate which percentage is related to human errors in the operation of gates, this factor is generally recognised as being very important. The risk associated with human errors is generally considered to be possible to mitigate by training of operators and by simple, clear and unambiguous operating rules. There are anyway recommendations that a team of two people be assigned the task of operating the gate system for occupational safety reasons (Barker *et al.* 2006). Irrespectively of the type of error, in a flood occurring situation it is obvious that adverse physical and psychological conditions may affect the correct action of operators.

Besides the need of comprehensive and extensive training of operators, there are practical recommendations to minimise the risk of operation errors and deficiencies. Among these are the need of avoid overstressing of operators due to the occurrence of failure in operating equipment, and to ensure unobstructed access to gates and reliable communications when an emergency condition happens. To prevent that, the provision of redundancy in equipment, access and communication is mandatory. A systematic inspection of these operational procedures, in which the operators had been included, is also very important (Bister, 2000). ICOLD Bulletin 154 (2013) discusses in more general terms practical measures to deal with operation failures and mitigation measures.

The risk of failure to move the gates or to follow command instructions due to mechanical and electrical problems, were illustrated by the examples mentioned before in this chapter. Redundancy, as mentioned above and specific design solutions may help to minimise this issue. However, the provision of unobstructed access to the gates and the assurance of electric supply to operate them even under extreme flood condition, is very important specially in large gated spillways.

There are examples of complete dam failures which could have possibly been avoided if these conditions existed when the accident occurred.

In the Euclides da Cunha Dam accident, in the Pardo River in Brazil, in January 1977, a spillway with 2 surface gates designed for 2 040 m³/s and one bottom outlet for 300 m³/s, were not operated properly during a sudden flood peaking 3 670 m³/s which generated an incoming flood wave in the reservoir of approximately 2.000 m³/s. This happened because the operator lost communication with the central office of the owner from whom he used to receive orders to operate the gates and was afraid to produce a downstream flood affecting some riverside population and, when he decided to open the gates, the gate command was inoperative because of power supply failure, and he could not reach the spillway because of the flooded access. Overtopping destroyed the Euclides da Cunha Dam and the next downstream Armando de Salles Oliveira Dam, in the Pardo River. After these events the owner conducted an overall revision of all its practices for dam operation (Siqueira, 1978).

Besides the electro-mechanical issues, civil structures must be operated and maintained as well to remain in good conditions. The questions that generally deserve special attention are cavitation action in chutes and stilling basins, hydraulic problems associated with asymmetrical operation of gates, the transport and deposition of sediment in stilling basins and the expansion of downstream erosion into the foundation of the structures. These questions are extensively treated in the reference papers mentioned in this chapter.

2.6. REFERENCES

Andrzejewski, R. H. (2002) – "The two spillways of Itá Hydroelectric Powerplant", *Large Brazilian Spillways*, CBDB, Rio de Janeiro, 53-64.

Barker, M., B. Vivian and D. Bowels (2006) – "Gate reliability assessment for a spillway design in Queensland, Australia", *Proceedings of the 26th USSD Conference*, USSD, San Antonio, TX, USA, 1-22.

Bister, D. (2000) – "Practical guidelines for improvement of dam safety during floods", Q.79, R. 29, *Transactions of the 20th Congress on Large Dams*, ICOLD, Beijing, China, Vol. IV, 497-513.

Bollaert, E. F. R. (2009) – "Dynamic uplift of concrete linings: theory and case studies", *Proceedings of the 29th Annual USSD Conference*, USSD, Nashville, Tenn. USA, 149-164.

Bowers, E. and J. Toso (1988) – "Karnafuli Project, model studies of spillway damage", *Journal of Hydraulic Engineering*, ASCE, Vol. 14, No. 5, 469-483.

Carvalho, E. (2002) – "Marimbondo Spillway – Performance and repair of the stilling basin", *Large Brazilian Spillways*, CBDB, Rio de Janeiro, 99-108

Carvalho, E. (2002a) – "Porto Colombia Spillway – Performance and remedial works in the stilling basin", *Large Brazilian Spillways*, CBDB, Rio de Janeiro, 123-132.

Cassidy, J. (2000) – "Gated spillways and other controlled release facilities, and dam safety", General Report Question 79, *Transactions of the 20th Congress on Large Dams*, ICOLD, Beijing, Vol. IV, 735-781

Cooke, J. B. (1985) – *Spillways over embankment dams*, Memo No. 80, Personal Communication.

Ebner, L. and M. Craig (2012) – "Comprehensive spillway tainter gate assessment and identification of interim risk reduction measures", *Proceedings of the 32nd Annual USSD Conference*, USSD, New Orleans, Lo, USA, 1257-1271.

Eigenheer, L. P., A. Vasconcelos, A. Conte and J. A. Souza (2002) – "Design, construction and performance of Xingó Spillway", *Large Brazilian Spillways*, CBDB, Rio de Janeiro, 177-184.

Epko, I. and A. Adegunwa (2011) – "Shiroro Hydroelectric Dam: operation, performance and safety monitoring of the rockfill dam (2004-2010)". *Transactions of the II International Symposium on Rockfill Dams*. CBDB-CHINCOLD, Rio de Janeiro.

Falvey, H. T. (1990) – *Cavitation in chutes and spillways*, US Bureau of Reclamation, Engineering Monograph No. 42, Denver. CO. USA.

Foster M. and M. Spannagle (2000) – "The statistics of embankment dam failures and accidents" *Canadian Geotechnical Journal*, Vol. 37, No.5, 1000-1024.

Hinks, J. L. and J. A. Charles (2004) – "Reservoir management, risk and safety considerations", *Long term benefits and performance of dams*, Tomas Telford, London.

ICOLD (1992) – *Spillways. Shock waves and air entrainment*. ICOLD Bulletin No. 81, Paris,

ICOLD (2012) – *Safe Passage of Extreme Floods,* ICOLD Bulletin No. 142, Paris.

ICOLD (2013) – Dam safety management: operational phase of dam life cycle. ICOLD Bulletin No. 154, Paris

IWPC (2004) – *Karun III Development*, Data sheet on construction works presented at the ICOLD 73rd Annual Meeting, Teheran.

Khaliq-Khan, A. and N. A. Siddiqui (1993) – "Malfunction of a spillway gate at Tarbela after 17 years of normal operation", Q.71, R.27, *Transactions of the 18th Congress on Large Dams*, ICOLD, Durban, South Africa, Vol.IV, 411-428.

Lempérière, F., J-P Vigni and L. Deroo (2012) – "New methods and criteria for designing spillways could reduce risks and costs significantly", *The International Journal on Hydropower & Dams*, Volume 19, Issue 3, 120-128.

Lima da Silva, C.A., R. Garcete, E. da Rosa, E. Fancello and P. Bernardini (2000) – "Failure and repair of hoist rod of a very large radial spillway gate – Itaipu Hydroelectric Powerplant". Q.79, R.43, *Transactions of the 20th Congress on Large Dams*, ICOLD, Beijing, Vol. IV, 709-732

Marcano, A. (2009) – *Lower Caroni developments - Some features of large spillways*. Contribution to the ICOLD Committee on Hydraulic for Dams

Micovic, Z., D. Hartford, M. Schaefer and B. Barker (2015) – "A non-traditional approach to the analysis of flood hazard for dams". *Stochastic Environmental Research and Risk Assessment*, Springer Verlag, Berlin, Germany

Nzakimuena, T. and A. Zulfiquar (1999) - *Rock erosion downstream of the Spillways at Hydro-Quebec Installations, Quebec, Canada – Some case histories*, Contribution presented to the ICOLD Committee on Hydraulic for Dams.

Pinto, N.L.S. (1994) – "Deterioration of spillways and outlet works". General Report Question 71, *Transactions of the 18th Congress on Large Dams*, ICOLD, Durban, South Africa, 1101-1208

Sanchez-Bribiesca, J. L. and A. C. Viscaino (1973) – Turbulence effects on the lining of stilling basins", Q. 41, R. 83, *Transactions of the 11th Congress on Large Dams*, ICOLD, Madrid, Vol.2, 1575-1592.

Siqueira, G. (1978) – "As lições do Pardo" ("The lessons of the Pardo" – in Portuguese), *Atas do XII Seminário Nacional de Grandes Barragens*, CBGB, São Paulo, Brazil, 141-170.

Sucharov, M. and A. Fiorini (2002) – "Itaipu spillway", *Large Brazilian Spillways*, CBDB, 65-78

Todd, R. V. (1997) – "Failure of spillway radial gate at Folson Dam, California", Q.75, R.9, *Transactions of the 19th Congress on Large Dams*, ICOLD, Florence, Vol. IV, 113-126.

USBR (2011) – *Managing water in the West – Gate failures – Best practices*. Powerpoint presentation, US Bureau of Reclamation. Denver, Colo. USA.Wang, X., Xu L. and Liao R. (2011) – "The dam design of the Three Gorges Project", *Engineering Sciences*, Vol. 9, No.3, 57-65.

Zhao, X. and S. Yuyab (2006) – "Exploration into safety and economy of choosing the over dam spillway on lower reservoir's concrete-face rock-fill dam of the Tongbai Pumped Storage power plant", Q.84, R.55, *Transactions of the 22nd Congress on Large Dams*, ICOLD, Barcelona, Spain, Vol. I, 959-967.

Table 2.1
Data on Brazilian Large Spillways

PROJECT NAME	RIVER	YEAR	MAX RESERVOIR FLOOD ELEV (m)	MAX TAIL WATER FLOOD ELEV (m)	DESIGN FLOOD CRITERION	PEAK INFLOW (m³/s)	SPLW DESIGN FLOW (m³/s)	WIDTH OF SPLW EXIT (m)	SPILLWAY TYPE	GATES Type	No.	Width (m)	Height (m)	TYPE OF ENERGY DISSIPATION DEVICE	SPECIFIC DISCH. FLOW (m³/s/m)
TUCURUÍ	Tocantins	1984	75.30	6.80	PMF	114 300	110 000	482.50	Chute	Radial	23	20.00	20.75	Flip bucket & plunging pool	228.00
ST ANTONIO - MAIN	Madeira	2012	72.00	65.25	1:10 000 yr	84 000	70 000	370.00	Crest overflow	Radial	15	20.00	22.00	Hydr.jump stilling basin	189.19
ST ANTONIO - AUX		2012					14 000	70.00	Crest overflow	Radial	3	20.00	22.00	Hydr.jump stilling basin	200.00
JIRAU	Madeira	2012	90.00	74.00	1:10 000 yr	82 600	82 600	445.00	Crest overflow	Radial	18	20.00	22.77	Hydr.jump stilling basin	185.62
ITAIPU	Paraná	1982	223.10	142.15	PMF	70 020	62 300	335.00	Chute	Radial	14	20.00	21.34	Flip bucket & plunging pool	185.67
ESTREITO – TOC.	Tocantins	2011	158.00	152.00	1:10 000 yr	62 719	62 719	332.60	Crest overflow	Radial	14	19.10	22.50	Hydr.jump stilling basin	188.69
FOZ DO CHAPECÓ	Uruguai	2010	266.60	240.00	PMF	62 190	62 190	343.50	Crest overflow	Radial	15	18.70	20.80	Flip bucket & plunging pool	181.05
PORTO PRIMAVERA	Paraná	1966	259.70	244.60	PMF	62 040	52 800	315.00	Crest overflow	Radial	16	14.96	22.86	Hydr.jump stilling basin	167.62
JUPIÁ - MAIN	Paraná	1968	280.58	270.00	PMF	60 790	44 000	505.00	Bottom outlet	Radial	37	7.61	10.00	Hydr.jump stilling basin	87.13
JUPIÁ - AUX		1968					5 000	70.00	Crest overflow	Radial	4	12.80	15.00	Hydr.jump stilling basin	83.30
ILHA SOLTEIRA	Paraná	1973	329.00	286.00	PMF	55 230	40 000	355.00	Crest overflow	Radial	19	18.50	21.50	Hydr.jump stilling basin	112.68
ITÁ - MAIN	Uruguai	2000	375.00	292.40	PMF	52 800	29 964	131.00	Chute	Radial	6	18.00	21.86	Flip bucket	228.73
ITÁ - AUX		2000					19 976	85.50	Part. Lined Chute	Radial	4	18.00	21.86	Flow over unlined rock	293.64
SALTO CAXIAS	Iguaçu	1999	326.00	268.50	1:10 000 yr	52 400	49 600	231.00	Crest overflow	Radial	14	16.50	70.00	Flip bucket	214.70
LAJEADO	Tocantins	2001	212.30	201.50	1:10 000 yr	49 870	49 870	321.00	Crest overflow	Radial	14	17.00	21.50	Hydr.jump stilling basin	154.39
PEIXE ANGICAL	Tocantins	2006	265.21	249.25	1:10 000 yr	42 500	37 044	200.00	Crest overflow	Radial	9	17.00	22.82	Hydr.jump stilling basin	185.20
MACHADINHO	Uruguai	2002	485.36	398.00	PMF	39 750	37 874	129.50	Chute	Radial	8	18.00	20.00	Flow over unlined rock	213.61
PAULO AFONSO IV	S. Francisco	1979	253.00	151.00	1:10 000 yr	35 000	10 000	105.00	Chute	Radial	8	11.50	19.60	Flow over concrete slab	95.24
MOXOTÓ	S. Francisco	1974	253.00	230.30	1:10 000 yr	35 000	28 000		Bottom outlet	Radial	70	10.00	8.00		
XINGÓ MAIN	S. Francisco	1994	139.00	29.70	1:10 000 yr	33 000	16 500	109.00	Chute	Radial	6	14.83	20.76	Flip bucket & plunging pool	151.38
XINGÓ AUX		1994					16 500	109.00	Part. lined Chute	Radial	6	14.83	20.76	Flow over unlined rock	151.38
ITAPARICA	S. Francisco	1986	305.40	263.00	1:1 500-yr peak + 1:1 000-yr volume	35 500	26 415	375.00	Crest overflow	Radial	9	15.00	19.70	Roller bucket	150.94
SLT. OSÓRIO - MAIN	Iguaçu	1975	398.00	328.00	1:10 000 yr	28 000	15 000	94.50	Chute	Radial	5	15.30	20.77	Flip bucket & plunging pool	158.73
SLT. OSÓRIO - AUX		1975					12 000	73.20	Chute	Radial	4	15.30	20.77	Flip bucket & plunging pool	161.93
SALTO SANTIAGO	Iguaçu	1980	509.00	419.00	1:10 000 yr	26 000	24 530	149.50	Chute	Radial	8	15.30	21.57	Flip bucket & plunging pool	164.08
CAPIVARA	Paranapan	1977	336.00	295.00	1:10 000 yr	24 500	17 100	144.40	Chute	Radial	8	15.00	15.56	Flip bucket & deflectors	118.42
BARRA GRANDE	Pelotas	2005	649.17	480.00	PMF	23 840	21 810	111.00	Chute	Radial	6	15.00	20.8	Flip bucket & plunging pool	196.48
ITUMBIARA	Paranaíba	1980	521.20	449.50	PMF	22 100	16 270	118.00	Chute	Radial	6	15.00	18.55	Flip bucket & stilling basin	137.90
MARIMBONDO	Grande	1975	447.36	403.00	1:10 000 yr	21 400	21 400	163.00	Crest overflow	Radial	9	15.00	18.85	Hydr.jump stilling basin	131.30
ÁGUA VERMELHA	Grande	1978	386.00	333.80	1:10 000 yr	20 000	20 000	156.00	Crest overflow	Radial	8	15.00	19.00	Flip bucket & plunging pool	128.21

Table. 2.2
(A) Data on Chinese Large Spillways

NAME	RIVER	YEAR OF COMPLETION	DESIGN FLOOD CRITERION	PEAK INFLOW (m³/s)	SPILLWAY DESIGN FLOW (m³/s)	SPILLWAY TYPE	GATES TYPE	NUMBER	WIDTH (m)	HEIGTH (m)	TYPE OF ENERGY DISSIPATION DEVICE	SPECIFIC DISCHARGE OF EXIT FLOW (m³/s/m)
Three Gorges - surface	Yangtze	2009		124 300	102 500	Crest orifice	Vertical lift	22	8.0	17.0		134.0
Three Gorges - mid-outlet						Orifice	Vertical lift	2	10.0	12.0	Ski-jump flip bucket	254.0
Three Gorges - deep outlet						Orifice	Radial	23		9.0		302.0
Longtan - surface	Hongshui	2009		35 500	27 194	Crest	Radial	7	15.0	20.0	Ski-jump flip bucket	223.0
Longtan - deep outlet						Orifice	Vertical lift	2	5.0	8.0		
Ertan - surface	Yalong	1990		23 900	6 260	Crest	Radial	7	11.0	11.5		125.0
Ertan - mid-outlet					6 930	Orifice	Vertical lift	6	6.0	5.0	Ski-jump flip bucket and plunge pool	186.0
Ertan - deep outlet					2 084	Orifice	Vertical lift	4	3.0	5.0		285.0
Ertan - tunnel spillway					7 400	Two tunnels	Vertical lift	2	13.0	13.0		
Daoxiaoshan - surface	Lancang	2003		23 800	23 800	Crest	Radial	5	14.0	17.8	Flaring pier + stepped spillway	193.6
Daoxiaoshan - deep outlet						Orifice	Radial	3	7.5	10.0	Ski-jump and flip bucket	267.0
Tianshengqiao I	Nanpan	2000		21 750	21 750	Chute	Radial	5	13.0	20.0	Ski jump and flip bucket	335.0
Wujiangxi - surface	Yuan	1994	0.1% flood frequency	55 967	40 132	Crest	Radial	9	19.0	23.0	Stilling basin with flaring piers	288.00
Wujiangxi - middle outlet					3 244	Orifice	Vertical lift	1	9.0	13.0	Stilling basin	287.00
Wujiangxi - deep outlet					2 915	Orifice	Vertical lift	5	3.5	7.0	Stilling basin	172.00
Panjiakou - surface	Luan	1992		56 200	42 600	Crest	Radial	18	15.0	15.0	Flaring piers and flip bucket	142.00
Panjiakou - deep outlet					3 000	Orifice	Vertical lift	4	4.0	6.0		
Ankang - surface	Han	1995	0.2% design 0.02% check	45 000	37 000	Crest	Radial	5	15.0	17.0	Stilling basin	
Ankang - middle outlet						Orifice	Vertical lift	5	11.0	12.0	Stilling basin	209.3
Ankang - deep outlet						Orifice	Vertical lift	4	5.0	8.0	Stilling basin	
Yantan - surface	Hongshui	1994		33 400	33 400	Crest	Radial	7	15.0	21.0	Flaring piers and stilling basin	306
Yantan - deep outlet						Orifice	Vertical lift	1	5.0	8.0	Flaring piers and stilling basin	210
Geheyan - surface	Qingjiang	1995		27 800	23 458	Crest	Radial	7	12.0	18.2	Flaring piers and stilling basin	231
Geheyan - deep outlet						Orifice	Vertical lift	4	4.5	6.5		275
Geheyan - bottom outlet						Orifice	Vertical lift	2				

Table 2.2
(B) Data on Chinese Large Spillways

NAME	RIVER	YEAR OF COMPLE-TION	DESIGN FLOOD CRITERION	PEAK INFLOW (m³/s)	SPILLWAY DESIGN FLOW (m³/s)	SPILLWAY TYPE	GATES				TYPE OF ENERGY DISSIPATION DEVICE	SPECIFIC DISCHARGE OF EXIT FLOW (m³/s/m)
							TYPE	NUMBER	WIDTH (m)	HEIGHT (m)		
Manwan - surface					12 025	Crest	Radial	5	13.0	20.0	Ski jump & plunge pool	262
Manwan - low level outlet	Lancang	1995		20 910	2 436	Orifice	Vertical lift	2	5.0	8.0		225
Manwan - tunnel					2 344	Tunnel	Vertical lift	1	12.0	12.0		
Wujiangdu -surface			Design 1:500-yr			Crest		6	13.0	19.0	Ski jump & plunge pool	144
Wujiangdu - middle outlet	Wujiang	1983	Check	21 350	21 350	Orifice		2	4.0	4.0		201
Wujiangdu - tunnel			1:5000-yr			Tunnel		2	9.0	10.0		240
Xiluodu - surface					33 278	Crest orifice		7	12.5	13.5	Ski jump & plunge pool	207
Xiluodu - low level outlets	Jinshia			50 311		Orifice		8	6.0	6.7	Ski jump & plunge pool	267
Xiluodu - tunnel					17 600	Tunnel		4	14.0	12.0	Ski jump & plunge pool	283
Xiangjiaba - surface	Jinshia	2015		48 680				12	8.0	26.0	Flaring gate & stilling basin	300
Xiangjiaba - middle outlet								10	6.0	9.5		331
Nuozhadu - chute					19 814	Chute	Radial	8	15.0	20.0	Ski-jump and plunge pool	162
Nuozhadu - left tunnel	Lancang	2012	PMF	31 318	3 211	Tunnel	Vertical lift	2	5.0	8.0	Ski-jump and plunge pool	308
Nuozhadu - right tunnel					3 154	Tunnel	Vertical lift	2	5.0	8.0	Ski-jump and plunge pool	393
Guopitan - surface								6	12.	13.0		129
Guopitan - middle outlet				26 950				7	6.0	8.0		228
Guopitan - tunnel								1	11.0	12.0		254
Xiaowan - surface						Crest orifice	Radial	5	11.0	15.0	Ski-jump and plunge pool	146
Xiaowan - middle outlet	Lancang	2009		23 600	20 709	Orifice	Vertical lift	6	6.0	6.5	Ski-jump and plunge pool	223
Xiaowan - bottom outlet						Orifice	Vertical lift	2			Ski-jump and plunge pool	
Xiaowan - tunnel						Tunnel	Vertical lift	1	15.0	16.5	Ski-jump and plunge pool	238

Table 2.3
(A) Data on Australian Large Spillways

NAME	RIVER	YEAR	MAX RESER-VOIR FLOOD ELEV (m)	MAX TAIL WATER FLOOD ELEV (m)	DESIGN FLOOD CRITERION	PEAK INFLOW (m³/s)	SPLW DESIGN FLOW (m³/s)	WIDTH OF SPLW EXIT (m)	SPILLWAY TYPE	GATES				TYPE OF ENERGY DISSIPATION DEVICE	SPECIFIC DISCH FLOW (m³/s/m)
										TYPE	NUMBER	W (m)	H (m)		
BURDEKIN FALLS	Burdekin	1987	171.9	154.5	1:9,000 (AEP for PMPOF)	97 900	64 600	504.0	Ungated	—	—	—	—	Apron slab with splitter piers	128
WARRAGAMBA	Warragamba	1960	131.2	74	PMF	52 100	42 200	94.5 (Main) 182.0 (Aux)	Gated & fuseplug	Radial Drum	4 1	12.19 27.43	13.3 7.62	Hydraulic Jump (main) Flip Bucket (Aus)	225 (Main) 1000(Aus)
KUNUNURRA DIVERSION	Ord	1963	43.6	42.8	1x10⁶	35 400	35 400		Gated plus auxiliary						118
PARADISE	Burnett River	2005	87.69	79	1:30,000 (AEP for PMPOF)	94 861	33 000	315.0 (Main) 485.8 (Aux)	Ungated	—	—	—	—	Stilling Basin	41
TALLOWA	Shoalhaven	1976	67.0	48	PMPOF	33 000	32 000	352.0	Ungated	—	—	—	—	Roller Bucket	91
BURRINJUCK	Murrumbidgee	1928	380.7		PMF	44 400	29 000		Gated & ungated	Radial	3	2x15.2 1x24.4	4.6	Natural Rocks	124
ORD RIVER	Ord	1972	111.5	48	PMF	200 000	27 750		Main chute plus auxiliary						
FITZROY RIVER BARRAGE	Fitzroy	1970					23 800		Gated						
PINDARI	Severn	1969	527.56		PMF	22 980	21 900	200.0	Ungated	—	—	—	—	Natural Rocks	
HARDING	Harding	1985					21 500		Ungated	—	—	—	—		
FAIRBAIRN	Nogoa	1972	218.59	195.54	1:61,275 (AEP for PMPOF)	27 777	15 580	158.5	Ungated	—	—	—	—	Sill block apron with baffles	98
COPETON	Gwydir	1976	575.4		PMF	37 300	14 800	156.0	Gated & fuseplug	Radial	9	14.6	13.0	Natural Rocks	95
WYANGALA	Lachlan	1971	382.51		PMF	35 320	14 722	23.65	Gated	Radial	8	14.63	12.72	Natural Rocks	66.46
JULIUS	Leichhardt	1976	239.16	213.9	1:210,000 (AEP for PMPOF)	44 625	14 590	219.5	Ungated	—	—	—	—	Stilling Basin	66.5

Table 2.3
(B) Data on Australian Large Spillways

NAME	RIVER	YEAR	MAX RESERVOIR FLOOD ELEV (m)	MAX TAILWATER FLOOD ELEV (m)	DESIGN FLOOD CRITERION	PEAK INFLOW (m³/s)	SPLW DESIGN FLOW (m³/s)	WIDTH OF SPLW EXIT (m)	SPILLWAY TYPE	GATES TYPE	NUMBER	W (m)	H (m)	TYPE OF ENERGY DISSIPATION DEVICE	SPECIFIC DISCH. FLOW (m³/s/m)
BOONDOOMA	Boyne	1982	300.29	256.3	1:250,000 (AEP for PMPDF)	14 570	13 420	115.0	Ungated	—	—	—	—	Stilling Basin	116.7
OPTHALMIA	Fortescue	1982					13 000		Ungated	—	—	—	—		
WIVENHOE	Brisbane	1985					12 000		Gated plus Fuseplug						
GLENBAWN	Hunter	1958	286.10		PMF	22 291	11 115	50.0	Ungated & fuseplug	—	—	—	—	Natural Stilling Basin	223.3
KEEPIT DAM	Namoi	1960	333.53		PMF	55 301	10 489	89.6	Gated & fuseplug	Radial	6	14.9	11.2	Bucket and dentated sill, deep roller	98.3
LESLIE DAM	Sandy Ck	1986	477.36	458.4	1:1,660,000 (AEP for PMPDF)	18 668	3 920	92.0	Gated (radial gates)	Radial	7	12.74	6.64	Roller Bucket	42.6

Table 2.4
Data on Venezuelan Large Spillways

NAME	RIVER	YEAR	MAX RESERVOIR FLOOD ELEV (m)	MAX TAILWATER FLOOD ELEV (m)	DESIGN FLOOD CRITERION	PEAK INFLOW (m³/s)	SPLW DESIGN FLOW (m³/s)	WIDTH OF SPLW EXIT (m)	SPILLWAY TYPE	GATES Type	No.	W (m)	H (m)	TYPE OF ENERGY DISSIPATION DEVICE	SPECIFIC DISCL. FLOW (m³/s/m)
SIMON BOLIVAR (GURI)	Caroni	1986	271.60	148.0	PMF	48 100	28 750	137.16	Chute	Radial	9	15.24	21.66	Flip bucket	209.63
ANTONIO JOSE DE SUCRE (MACAGUA)	Caroni	1997	54.50	34.0	PMF	30 000	30 000	285.00	Crest overflow	Radial	12	22.80	15.60	Straight to River bed	105.26
FRANCISCO DE MIRANDA (CARUACHI)	Caroni	2005	92.40	57.0	PMF	30 000	30 000	161.16	Crest Overflow	Radial	9	15.24	21.66	Flip bucket & plunging pool	186.15
MANUEL PIAR (TOCOMA)	Caroni	UC	127.50	97.0	PMF	28 750	28 750	153.16	Crest Overflow	Radial	9	15.24	21.66	Flip bucket & plunging pool	187.71

Table 2.5
Data on Mexican Large Spillways

NAME	RIVER	YEAR	MAX RESERVOIR FLOOD ELEV (m)	DESIGN FLOOD CRITERION	PEAK INFLOW (m³/s)	SPILLWAY DESIGN FLOW (m³/s)	WIDTH OF SPLW EXIT (m)	SPILLWAY TYPE	GATES TYPE	NUMBER	W (m)	H (m)	TYPE OF ENERGY DISSIPATION DEVICE	SPECIFIC DISCHARGE OF EXIT FLOW (m³/s/m)
LA AMISTAD	BRAVO	1969	347,59	PMF	54 000	43 700	243,84	Crest overflow	Radial	16	15.24	16.43	Flip bucket and plunge pool	179.00
HUITES	FUERTE	1995	290,00	10 000-yr	30 000	22 450	62,00	Crest overflow	Radial	2	15.50	21.00	Flip bucket and plunge pool	362.00
CERRO DE ORO - main	STO DOMINGO	1988	72,80	10 000-yr	25 980	6 000		3 Tunnel splws	Radial	6	5.90	15.20	Flip bucket and plunge pool	113.00
CERRO DE ORO - aux									Radial	3	5.90	12.80		
MALPASO - main	GRIJALVA	1964	188,00	Regional Maxima	21 750	11 100	45.0	Crest overflow	Radial	3	15.00	15.00	Stilling basin	247.00
MALPASO - aux						10 650	60.0	Crest overflow	Radial	4	15.00	18.70	Flip bucket and plunge pool	178.00
EL INFERNILLO - main	BALSAS	1963	176.40	Creager Form	38 800	13 800	66.78	3 Tunnel splws					Flip bucket and plunge pool	207.00
EL INFERNILLO - aux			183.20	10 000-yr	41 600	10 500								157.00
FALCON	BRAVO	1953	95.77	PMF	20 000	13 000	91.44	Crest overflow	Radial	6	15.24	15.24	Stilling basin	142.00

3. HIGH HEAD SPILLWAYS – THE CHALLENGE OF ENERGY DISSIPATION AND SCOUR CONTROL DOWNSTREAM[*]

3.1. INTRODUCTION

The safety of dams during flood events has to be ensured by an appropriate capacity of the releasing structures. In the case of high head spillways, besides of well-known hydraulic structures problems due to high velocity flows and cavitation risk, one main issue is the challenge of energy dissipation and scour control downstream (Schleiss, 2002). High velocity jets can occur which are guided by the releasing structures into the tail-water at a certain distance from the dam. At the zone of impact of these high-energy jets, the riverbed will be scoured. Since scour due to plunging jets can reach considerable depth even in rocky riverbeds, instability of the valley slopes has to be feared, which can endanger in some cases the foundation and the abutment of the dam itself. Such scour problems occur especially at dams were the spillways are combined with the dam structure itself and, consequently, the impact zone of the high-velocity falling jets is relatively close to the dam. This is typically the case with concrete dams, where high velocity falling jets can be created by gated or ungated crest spillways (arch dams only), chute spillways followed by a ski jump and orifice spillways as well as high-capacity bottom outlets. Severe scour conditions occur especially in the case of high concrete arch dams in narrow valleys with high flood discharges.

Such a typical spillway arrangement is shown in Figure 3.1 and Figure 3.2 at the example of Khersan III Dam Project in Iran. Flood handling at the 175 m high double arch dam will be provided by three separate spillway facilities:

- a two-bay chute flip bucket spillway on the right abutment with an ogee crest at El. 1404.5 m controlled by 11.5 m wide by 13.5 m high radial gates (capacity of 4240 m³/s at PMF flood level El. 1426.3 m);

- an uncontrolled crest spillway with ogee crest divided in 6 bays; two 13.5 m and two 19.5 m wide bays at El. 1418 m and two 12.5 m wide bays at El. 1421 m (total capacity of 3360 m³/s at PMF flood level El. 1426.3 m);

- two bottom outlets with centerline at El. 1330 m and 1345 m with service gate openings, 3 m wide by 4 m high (total capacity of 395 m³/s at PMF flood level El. 1426.3

(*) This Chapter corresponds to an updated and enhanced version of Schleiss (2002)

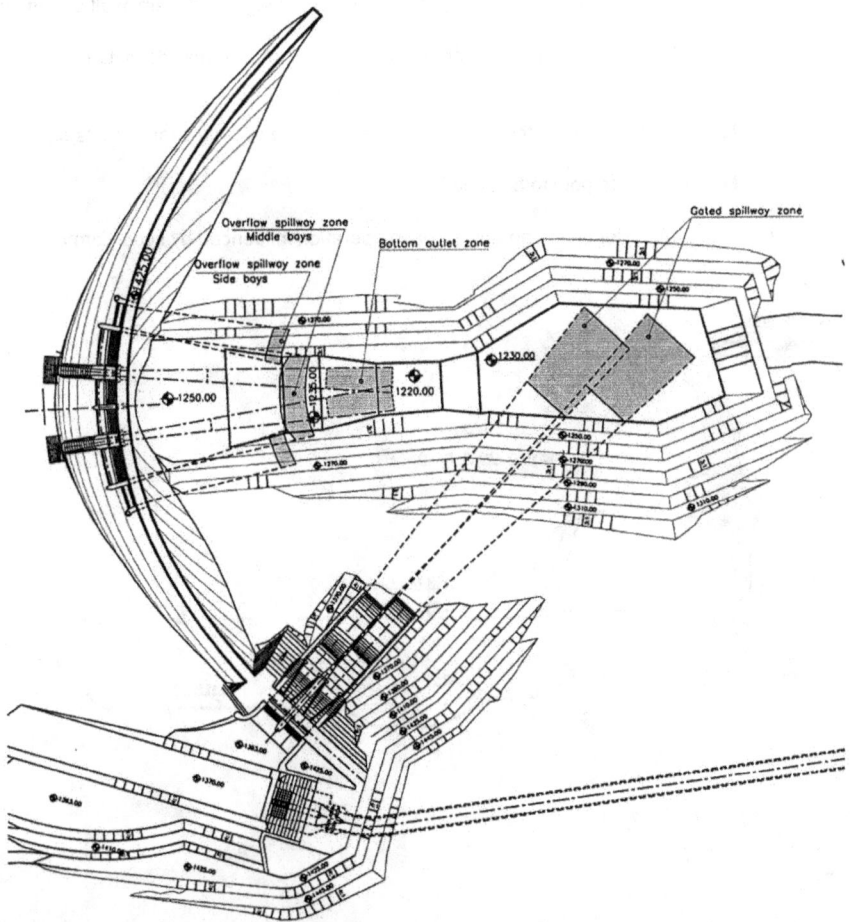

Fig. 3.1
Khersan III Dam Project in Iran. Layout of arch dam with spillway structures and its jet impact zones.

In today's spillway design of dams there is a tendency of increasing the unit discharge of the high velocity jet leaving the appurtenant structures. In gated chute flip bucket spillway, specific discharges in the range of 200 to 300 m³/s/m are not rare anymore, since cavitation risk in chutes can be mitigated by the help of bottom aerators. Uncontrolled crest spillways for arch dams are designed nowadays for specific discharges up to 70 m³/s/m and up to 120 m³/s/m by installing gates on the crest. With the latest high pressure gate technology, low level orifice spillways can evacuate specific discharges in the range of 300 to 400 m³/s/m.

This trend is also confirmed by many high dam projects in China with large discharge flows and built in narrow valleys. Special experiences on high gravity dams, high arch dams and high rockfill dams have been reported by Gao *et al.* (2011).

Besides the hydraulic design questions of the water release structures itself, the challenge regarding energy dissipation and scour control, must answer the following questions:

- What will be the evolution and extent of scour downstream of the dam at the jet impact zone?

- Are the stability of the valley slopes and the foundation of the dam itself endangered?

- Is a tail-pond dam required to create a water cushion and how does it affect the scour depth?

- Is a pre-excavation of the rocky riverbed required and what should be its shape

- Has the plunge pool to be lined?

- Is the tail-water level and powerhouse operation influenced by scour formation?

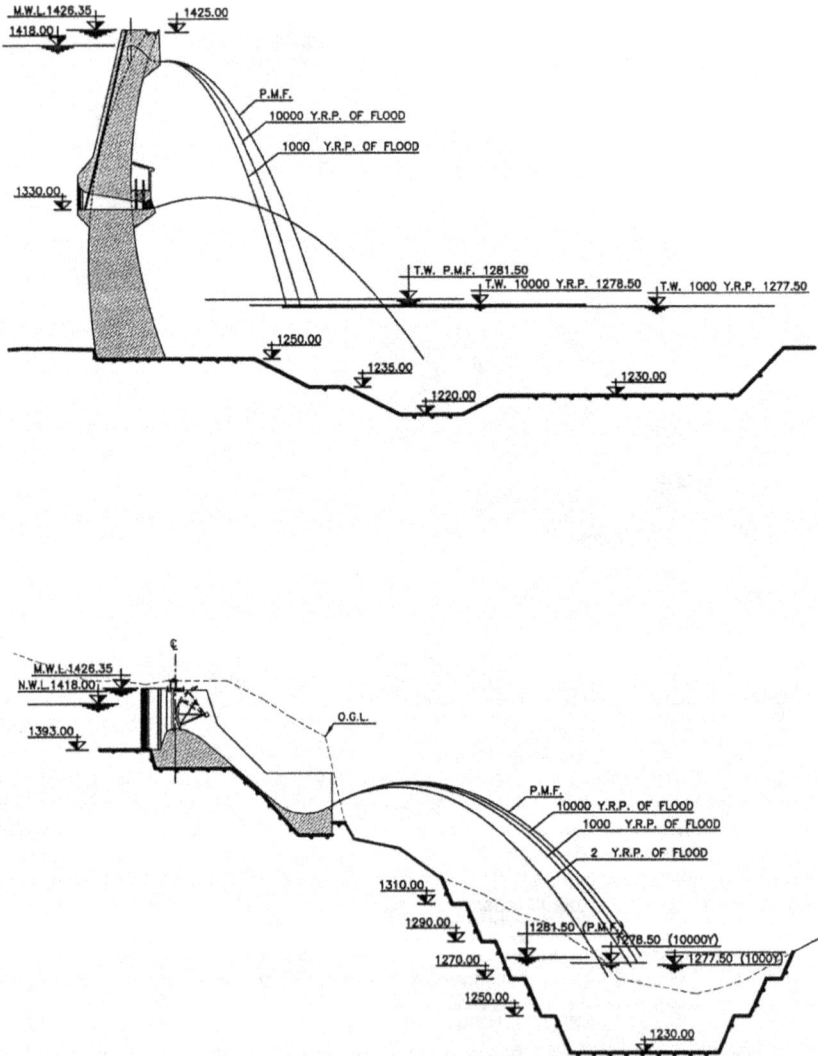

Fig. 3.2
Khersan III Dam Project in Iran. Longitudinal profiles through crest spillway and gated side spillway and plunge pools.

3.2. THE SCOUR PROCESS

3.2.1. The physical processes

Scouring is a complex problem and has been studied since longtime. As illustrated in Figure 3.3, scour can be described by a series of physical processes as (Bollaert, 2002; Mason, 2011):

a. free falling jet behavior in the air and aerated jet impingement;

b. plunging jet behavior and turbulent flow in the plunge pool;

c. pressure fluctuation at the water-rock interface;

d. propagation of dynamic water pressures into rock joints;

e. hydrodynamic fracturing of closed end rock joints and splitting of rock in rock blocks;

f. ejection of the so formed rock blocks by dynamic uplift into the plunge pool and entrain and circulate the excavated rock around the pool and vertically over the height of the pool;

g. break-up and degrading of the rock blocks by the ball milling effect of the turbulent flow in the plunge pool;

h. ejecting the rocks downstream to a point where they cannot roll back into the pool and thus formation of a downstream mound;

i. displacement downstream of the scoured materials by the flow transport in the river.

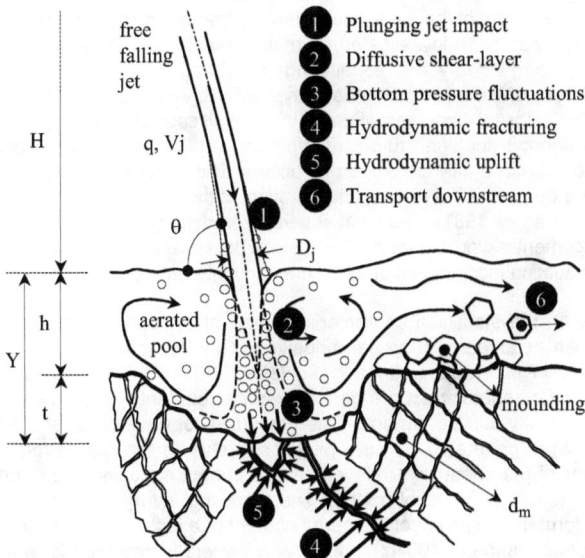

Fig. 3.3
Main parameters and physical-mechanical processes involved in the scouring phenomenon (Bollaert, 2002)

3.2.2. Jet behavior in the air

In evaluating the scour caused by free jets, it is at first necessary to predict the jet trajectory so that the location of the jet impingement in the plunge pool and the zone of the scour hole are known (Whittaker & Schleiss, 1984). The behavior of an ideal jet can be easily assessed by using ballistic equations. Nevertheless, at prototype jets effects as air drag, disintegration of the jet in the air and initial flow aeration in long chutes have to be considered. A number of researchers have developed equations to predict the jet trajectory, i.e. the jet travel length accounting for these effects (Gun'ko et al., 1965; Kamenev, 1966; Kawakami, 1973; Zvorykin et al., 1975; Taraimovich, 1980; Martins, 1977).

The spread of the jet during fall is also an issue and has been addressed on an empirical basis by Taraimovich, 1980 and U.S.B.R. 1978. More recently, the lateral spread has been related to its initial turbulence intensity (Ervine & Falvey, 1987; Ervine et al., 1997). Typical angles of jet spread are 3 to 4 % for roughly turbulent jets. Turbulence intensity for free over-fall jets is less than 3 %, for flip bucket jets between 3 and 5 % and for orifice jets between 3 and 8 % (Bollaert, 2002). Based on experimental results and an extensive literature survey, jet issuance parameters relevant for engineering practice can be found in Manso et al. (2008).

3.2.3. Jet behavior in the plunge pool and pressure fluctuations

As the jet plunges into the pool, a considerable amount of air is entrained, corresponding to an air concentration of 40 to 60% at typical jet velocities of 30 m/s at impingement (Bollaert 2002). Several expressions defining the air content at the point of impact of the jet in the plunge pool have been developed by scale model tests. Some of them can be reasonably extended towards prototype velocities (Van de Sande, 1973; Bin, 1984; Ervine et al., 1980; Ervine 1998).

The flow conditions in the plunge pool can be characterized by a high-velocity, two-phase turbulent shear layer flow and a macro-turbulent flow outside of this zone (Bollaert, 2002; Manso, 2066). The shear layer produces severe pressure fluctuations at the water-rock interface and is highly influenced by aeration. Existing theories on two-dimensional vertical jet diffusion in a semi-infinite or bounded medium define the outer limits of the water-rock interface that is directly subjected to these pressures. The diffusion of 2D-jet has been investigated by many researchers (Bollaert & Schleiss, 2003). The concept of a jet of uniform velocity field penetrating into a stagnant fluid is based on the progressive growing of the thickness of the related boundary layer by exchange of momentum. In this shear layer the total cross section of the jet increases whereas the non-viscous core of the jet decreases. The inner angle of diffusion is about 8° for highly turbulent impinging jets. When the plunge pool is deep enough, the core jet disappears, and a fully developed jet occurs. The angle of diffusion of the jet through the pool depends on the degree of turbulence and aeration of the jet at impact and can be estimated at around 15° (Ervine & Falvey, 1987). The most severe hydrodynamic action on the plunge pool bottom occurs in the impingement region, where the hydrostatic pressure of the free jet region is progressively transformed into fluctuating high stagnation pressures and into an important bottom shear stress.

Knowledge on the statistical characteristics and of the spatial distribution of the pressure fluctuations has been acquired by physical modelling of hydraulic jumps and plunging jets (Tos & Bowers, 1988; Ballio et al., 1992; Bollaert & Schleiss, 2003, 2005; Manso, 2006). The pressure patterns generated by core jet impact are generally quite constant with high values in the core and much lower values directly outside the core. The pressures for developed jet impact are completely different from pressures generated by core jet impact. Due to the high turbulence level and the two-phase character of the shear layer, developed jet impact conditions can generate much more severe dynamic pressures at the pool bottom. Therefore, not every water cushion has a retarding effect on the scour formation. The spectral energy content of a high velocity plunging jet extends over a much wider frequency range (> 100Hz) than what is generally assumed for macro-turbulent flows in plunge pools (up to 25 Hz).

Systematic experimental investigations in plunge pools with different lateral confinements showed that the pool geometry influences not only plunging jet diffusion and air entrainment in the pool, but also impact pressures at the water-rock interface and inside the fissured rock mass (Manso et al., 2009). Favourable plunge pool geometries may limit the pressure fluctuations at water rock interface and therefore the scour potential is reduced (Bollaert et al., 2012).

3.2.4. Propagation of dynamic water pressures into rock joints, hydrodynamic fracturing and dynamic uplift

The transfer of pool bottom pressures into rock joints results in transient flow that is governed by the propagation of pressure waves. For closed-end rock joints, as encountered in a partially jointed rock mass, the reflection and superposition of pressure waves generate a hydrodynamic loading at the tip of the joint. If the corresponding stresses at the tip of the joint exceed the fracture toughness and the initial compressive stresses in the rock, the rock will crack and the joint can further grow. In the case of open-end rock joints in a fully jointed rock mass, the pressure waves inside the joints create a significant dynamic uplift force on the rock blocks. This dynamic uplift force will break up the remaining rock bridges in the joints by fatigue and, if high enough, eject the so formed rock blocks from the rock mass into the macro-turbulent plunge pool flow (Manso et al., 2007).

The presence of such pressure waves in rock joints and significant amplifications due to resonance phenomena could be observed and measured with an experimental study at near prototype scale (Bollaert, 2002; Bollaert & Schleiss, 2003, 2005; Manso, 2006; Federspiel, 2011; Duarte et al., 2012).

3.2.5. Ball milling effect of the turbulent flow in the plunge pool and formation of a downstream mound

Once the rock blocks are formed and ejected by the dynamic uplift from the surrounding rock mass into the pool, they can be taken up by the macro-turbulent eddies. If the block is too big to be transported by the flow, it will be broken-up after some time by the ball milling effect of the eddies in the plunge pool. Having attained the required minimum size, the rock blocks will be displaced downstream by the flow and deposited on the mound or carried away by sediment transport in the river. The mound may limit the depth of scour but also raise the tail-water level. Exceeding a critical level, the increased tail-water may interfere with the operation of bottom outlets or reduce the net head of the powerhouse, if it is located upstream of the mound. If the mound does limit the depth of the scour, the scour is considered to have attained a so-called dynamic limit. However, if the mound is removed and the scour proceeds to a maximum possible extent, it is considered to have attained the ultimate static limit (Eggenberger & Müller, 1944).

3.3. SCOUR EVALUATION METHODS

3.3.1. General overview

The existing scour evaluation methods can be grouped as follows (Bollaert & Schleiss, 2003):

- empirical approaches based on laboratory and field observations;

- analytical-empirical methods combining laboratory and field observations with some physics;

- approaches based on extreme values of fluctuating pressures at the plunge pool bottom;

- techniques based on time-mean and instantaneous pressure differences and accounting for rock characteristics;

- scour model based on fully transient water pressures in rock joints.

The most common methods used for scour evaluation due to falling, high velocity jets are illustrated in Figure 3.4 for the considered physical parameters involved in the scour process which can be related to the three phases water, rock and air. Time evolution is a further parameter.

3.3.2. Empirical formulae

A large number of empirical equations have been developed for predicting the scour from plunging jets. These empirical formulae were mostly derived from hydraulic model tests in the laboratory but also from prototype observation and are widely used in practice for design purposes. Some expressions are of general applicability, others are specific to free over-fall jets, ski-jumps or orifice spillways. A comprehensive overview and comparison of most of the known formula have been given by Whittaker & Schleiss (1984), Mason & Arumugam (1985) and Bollaert (2002).

The complex scouring process is reduced by the empirical formulae to a few parameters. The ultimate total scour depth Y measured from tail-water level is thought to be a function of:

- the specific discharge q (discharge per unit width of jet);

- the fall height H;

- the tailwater depth h (measured from initial riverbed level);

- the characteristic sediment size or rock block diameter d.

Most of the formulas are written in the form

$$Y = t + h = K \cdot \frac{H^y \cdot q^x \cdot h^w}{g^v \cdot d_m{}^z} \tag{1}$$

where t is the scour depth below the initial bed level and K a constant.

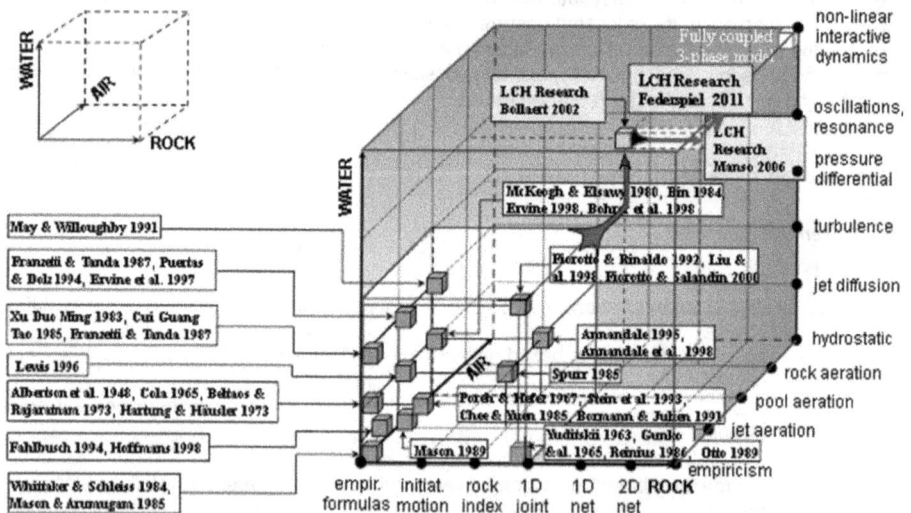

Fig. 3.4
Summary of existing scour evaluation methods considering physical parameters involved in the scour process, which can be related to the three phases: water, rock and air (completed according Bollaert, 2002)

Mason & Arumugam (1985) applied this form of formula to a large number of scour data, 26 sets from prototypes and 47 from model tests. Following further work, Mason (1989) suggested the following best fit exponents and constants for both model and prototype conditions:

$K = (6.42 - 3.10H^{0.10})$

$v = 0.30$

$w = 0.15$

$x = 0.60$

$y = 0.05$

$z = 0.10$

This was shown to give the most consistent and accurate prediction for the 46 model test data set and a general upper bound for the prototype data set. It was also shown to dimensionally balance which was a pleasing indicator of probable validity (Mason, 1989).

According to the data sets, the formula is applicable for hydraulic model fall heights H between 0.325 m and 2.15 m and between 15.82 m and 109 m for prototypes in the case of free over-fall jet, ski-jump or orifice spillways. The use of the mean particle size d_m gave better results than the use of the d_{90} particle size. For prototype rock, an equivalent particle size $d_m = 0.25$ m is recommended in the above formula.

3.3.3. Semi-empirical equations

As already mentioned, semi-empirical methods are combining laboratory and field observations with some physics as:

- initiation of motion of the bed material by shear stress;

- energy conservation equations;

- geomechanical characteristics;

- angle of impingement of the jet;

- steady-state two-dimensional jet diffusion theory;

- aeration effects.

A detailed overview of these semi-empirical equations and methods can be found in Whittaker & Schleiss (1984) and Bollaert (2002). The hydrodynamic process of scour is often derived from the two-dimensional jet diffusion theory. The geomechanical behavior of the rock mass is considered by the shear-stress based initiation of motion concept for non-cohesive granular materials or by an index that defines the resistance of the rock mass against erosion. Both hydrodynamic and geomechanical characteristics are for example combined in Spurr's (1985) and Annandale's (1995, 2006) erodibility index methods for rocks and in the momentum conservation equations established by Fahlbush (1994) and Hoffmans (1998) for non-cohesive material.

3.3.4. Approaches based on extreme values of fluctuating pressures at the plunge pool bottom

At the plunge pool bottom, fluctuating, dynamic pressures occur due to the direct jet core impact in the case of relatively small water cushion. For high plunge pool depth (higher than 4 to

6 times the thickness of the incoming jet) a turbulent shear flow or developed jet impact according to the two-dimensional jet diffusion theory is created. These two types of jet impact generate completely different pressure patterns as already mentioned.

The dynamic pressures at the plunge pool bottom can penetrate into fissures of the underlying rock mass. The approaches based on the extreme pool bottom pressures assume that the maximum pressures occurring at the water-rock interface are transferred through the joints underneath the rock blocks. These maximum pressures underneath the rock blocks combined with the minimum pressures at the plunge pool bottom create a net uplift pressure Δp on the rock blocks (Figure 3.5). The ultimate scour is reached when this net pressure difference Δp on the rock block is not able anymore to eject it. Since the maximum and minimum pressures are not occurring at the same moment, the so defined net uplift pressure represents a physical upper limit of dynamic loading conditions and is therefore rather conservative.

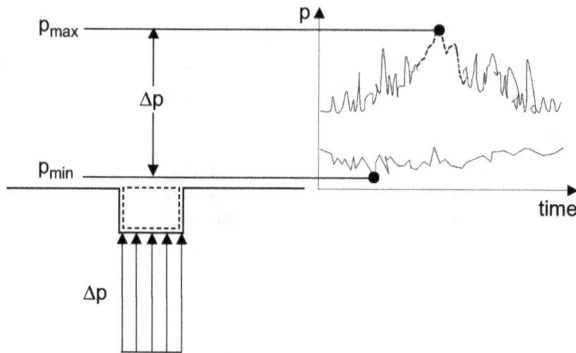

Fig. 3.5
Definition sketch of extreme dynamic pressures at the plunge pool bottom. The maximum and minimum pressures are defined at the center of the block for a long enough time interval (according to Bollaert, 2002).

Studies on pressure fluctuations in plunge pools have mainly been carried out by Ervine et al. (1997), Xu-Duo-Ming (1983) and Franzetti & Tanda (1984, 1987) for circular jet impingement and by Tao et al. (1985), Lopardo (1988), Armengou (1991), May & Willoughby (1991) and Puertas & Dolz (1994) on rectangular jets. These studies give useful information on bottom pressure fluctuations but do not describe their propagation inside the joints of the underlying rock mass. The simultaneous application of extreme minimum and maximum bottom pressures above and underneath rock blocks can result in net pressure differences of up to 7 times the root-mean-square value or up to 1.5 - 1.75 times the incoming kinetic energy of the jet. Even if this seems to be a conservative design criterion it has to be noted, that violent transient pressure phenomena, which could occur inside the rock joints, are not considered (Bollaert, 2002).

3.3.5. Techniques based on time-mean and instantaneous pressure differences and accounting for rock characteristics

Contrary to the approach presented in the previous section, time-averaged or instantaneous pressure differences occurring at or during a certain time at the surface and underneath the rock blocks, are considered. This means that the fluctuating pressures have not only to be measured at the plunge pool bottom but also inside the rock joints.

Yuditskii (1963) (reported by Gunko et al., 1965) was probably the first who stated that time averaged and pulsating pressures are responsible for rock block uplift in the scour process. He measured the forces on a single rock block on flat plunge pool bottoms due to the impact of a jet produced by a ski-jump spillway in a scale model. Measurements techniques at that time allowed only to obtain time averaged forces. Otto (1989) quantified time-averaged uplift pressures acting on a rock block for plane jets impinging obliquely. Depending on the relative protrusion of the block and the point of jet impact, important surface suction effects occurred, leading to mean uplift pressures of almost the total incoming kinetic energy. Without considering this suction effect, the maximum uplift pressures were still half of the incoming kinetic energy.

The destructive effects of instantaneous pressure differences entering in tiny rock joints was highlighted the first time by Hartung & Häusler (1973). Kirschke (1974) at first performed an analytical and numerical analysis of water hammer propagation in rock joints. Instantaneous pressure differences based on transient flow assumptions were measured and analysed for the first time for the case of concrete slab linings of stilling basins (Fiorotto & Rinaldo, 1992; Bellin & Fiorotto, 1995; Fiorotto & Salandin, 2000; Fiorotto & Caroni, 2007). However, the scale of the model tests and the data acquisition rate did not allow measuring any oscillatory or resonance effects in the joint under the slabs.

Annandale et al. (1998) simulated the erosion of fractured rock by the use of lightweight concrete blocks, placed in a series of two layers on a 45° dip angle. Jet impingement confirmed their theory that the erosion threshold criterion for rock and soil material can be defined by means of a geomechanical index, the so-called Erodibility Index. The erosive power of the water can be related to the erosion resistance of the material.

Experimental and numerical studies focusing on fluctuating net uplift forces on simulated rock blocks were also performed by Liu et al. (1998). A design criterion for rock block uplift was given based on a transient flow model but without considering resonance effects (Liu, 1999).

All the mentioned studies on concrete slabs and rock blocks consider the surface pressure field as a function of space and time. But the pressure field underneath is assumed constant over the surface of the element and equal to the pressure at the entrance of the joints i.e. at the surface. Therefore, fully transient flow conditions in the joints such as pressure wave reflections and amplifications are neglected by these extreme pressure techniques.

3.3.6. Scour model based on fully transient water pressures in rock joints

Bollaert (2002) measured for the first time the transient pressures in rock joints due to high-velocity jet impact with systematic laboratory tests with and experimental set-up reproducing near prototype conditions and reproduced them in a numerical model. New phenomena could be observed and explained as the reflection and superposition of pressure waves, resonance pressures and quasi-instantaneous air release and resolution due to pressure drops in the joint.

The analysis revealed that the pressure wave velocity is highly influenced by the presence of free air bubbles in the joints. These bubbles can be transported by flow from the plunge pool into the joint but also be released from the water during sudden pressure drop below atmospheric pressure as also experimentally shown by Manso (2006).

In open-end joints instantaneous net uplift pressures of 0.8 to 1.6 times the incoming kinetic energy of the impacting jet has been measured. This is significantly higher than any previous assumptions in literature (see also section 3.3.5) and underlines that transient pressure effects in rock joints are a key physical process for scour formation.

Based on the experimental results and the numerical simulation, a new model for the evaluation of the ultimate scour depth has been developed, the so-called Comprehensive Scour Method (CSM), which represents a comprehensive assessment of the two physical processes: hydrodynamic fracturing of closed-end rock joints and dynamic uplift of rock blocks (Bollaert, 2002, 2004; Bollaert & Schleiss, 2005). All relevant processes as the characteristics of the free-falling jet (velocity and diameter at issuance, initial jet turbulence intensity), the pressure fluctuations at the plunge pool bottom and the hydrodynamic loading inside rock joints are dealt with and compared with the resistance of the rock against crack propagation considering fatigue.

The CSM model was further enhanced by considering geometry of plunge pool and lateral confinement of plunging jet (Manso, 2006) as well as the dynamic response of a rock block due to fluid-structure interaction (Federspiel, 2011; Asadollahi et al., 2011). In a recent research the influence of jet aeration was investigated and implemented in the CSM model (Duarte et al., 2012, 2013).

The application of the CSM model needs knowledge on rock parameters as number, spacing, direction and persistency of fracture sets, in-situ stresses in the rock mass, fracture toughness and unconfined compressive strength. These parameters are normally assessed during the geological and geotechnical reconnaissance campaign for any dam foundation project and are therefore available. Nevertheless, uncertainties have to be addressed by sensibility analysis and engineering judgment.

Li and Liu (2010) presented a model of 1D discrete fracture network (DFN) based on the Monte-Carlo method. Using a wave celerity of 1000 m/s the transient water pressures in the fracture network are calculated under the assumptions that the effect of the drag term on pressure transfer and the influence of air entrained can be neglected. With the help of a stability criterion it is possible to assess if the blocks of the fractured rock can be dislodged layer by layer by the transient water pressure. However, resonance effects and vibration of the blocks are neglected.

3.4.   DIFFICULTIES ENCOUNTERED WHEN ESTIMATING SCOUR DEPTH

3.4.1.   Which is the appropriate formula or theory?

In the feasibility design stage of a dam project normally the easily applicable empirical and semi-empirical formulae are used to get a first idea of the expected ultimate scour depth occurring downstream of the spillway structures during the project lifetime. Most of the formulae have been developed for a specific case such as a ski-jump, a free crest over-fall or an orifice spillway, while some others are of general applicability. Therefore, a careful selection of the appropriate equations should be made for each project. Nevertheless, even after this selection, the results of the remaining formulae show very often a wide scatter. In such a case, the engineering decisions can be based on a statistical analysis of the results by using, for example, the average of all formulae or the positive standard deviation in a more conservative way. This analysis is carried out for a certain spillway discharge with the corresponding tail-water level by varying the characteristic rock block size (if considered) according to expected geological conditions.

Sometimes prototype scour measures of an existing dam, situated in similar geological conditions than the foreseen new project, are available. Then the formula predicting best the scour depth can be identified and be applied to the new project.

3.4.2.   How model tests should be performed and interpreted?

If free jets impinging on rock underlying a plunge pool have to be modeled in a laboratory, three difficulties may arise (Whittaker & Schleiss, 1984):

- the appropriate choice of a material that will behave dynamically in the model as fissured rock does in the prototype;

- the grain size effects;

- the aeration effects.

In most models the disintegration process, that means the hydrodynamic fracturing of closed end rock joints and splitting of rock in rock blocks, is assumed to have taken place. Thus, only the ejection of the rock blocks into the macro-turbulent plunge pool flow and the transport of the material out from the scour hole are modelled. Reasonable results may be obtained if fissured rock is modelled by appropriately shaped concrete elements (Martins, 1973), but their regular pattern and size does not fully represent a rock mass with several intersecting fracture sets. Nevertheless, crushed rock having at least a grain size distribution similar to the expected rock blocks, should be used instead of round river gravel. In any case, model tests cannot simulate the break-up of the rock blocks by the ball milling effect of the turbulent flow in the plunge pool. Therefore, in the model a mound is formed which is higher and more stable than in the prototype. As a result, the prototype scour depth is underestimated. This can be compensated to some extent by choosing the material carefully including downscaling. Normally good predictive results for scour depth can be obtained by using non-cohesive material, but the extent of the scour may be not correct because steep and near vertical slopes cannot be reproduced. Therefore, the use of slightly cohesive material by adding cement, clay caulk, chalk powder, paraffin wax etc. to the crushed gravel is proposed (Johnson, 1977; Gerodetti, 1982; Quintela & Da Cruz, 1982; Mason, 1984).

It is known that for grain sizes smaller than 2 to 5 mm, the ultimate scour depth becomes constant (Veronese, 1937; Mirtskhulava et al., 1967; Machado, 1982). For an acceptable scale of a comprehensive dam model of 1:50 to 1:70, the smallest prototype rock blocks, which can be reproduced, are in the range of 0.1 to 0.35 m.

Finally, air entrainment cannot be scaled appropriately in comprehensive models unless using an unpractical large scale (in the order of 1:10). Air entrainment has a highly random character, which influences the scour process considerably (see section 3.2). Mason (1989) studied systematically the effect of jet aeration on scour in a mobile gravel bed with a specially designed apparatus in the laboratory for head drops of maximum 2 m. Duarte et al. (2012, 2013) studied systematically the effect of jet aeration on the dynamic response of a rock block impacted by high velocity jet in an experimental set-up reproducing near prototype conditions (Duarte et al., 2015 et 2016).

Through the use of hybrid modelling the behaviour of rock scour can be correctly represented. This was successfully done for defining mitigation measures for the scour of Kariba dam (Noret et al., 2013; Duarte et al., 2011). Several historical scour geometries were reproduced in a physical model with a fixed mortar surface equipped with pressure sensors. Four different spillway operation dynamic pressures were measured at the water rock interface. These recorded dynamic pressures were then introduced in the comprehensive scour method (CSM) of Bollaert & Schleiss (2005), which can predict how deep rock blocks can be formed by fatigue and ejected into the plunge pool by the dynamic pressures.

3.4.3. How to analyse prototype observations properly?

When analysing prototype observations on scour depth in order to derive an equation for similar conditions, the following questions have to be answered:

- What was the duration of the operation of the spillway for different specific discharges (discharge-duration curve)? (An example of discharge-duration curve of a spillway is given in Figure 3.6).

- Which was the prevailing, specific discharge, which has formed the scour depth?

- Was the duration of this specific discharge long enough to create ultimate scour depth?

Since in practice it is very often difficult to answer precisely to these questions, probably significant uncertainties have been introduced into the existing formulas derived from prototype observations. This may also explain the large scatter when predicting scour for other prototypes.

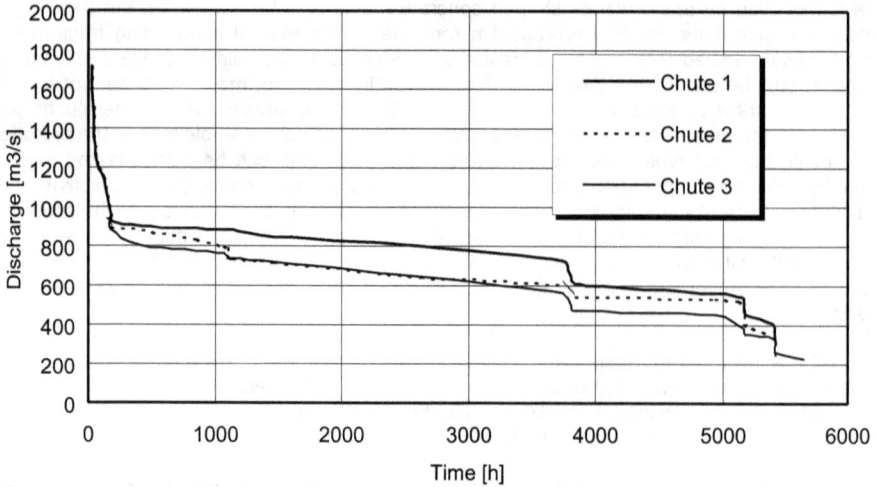

Fig. 3.6
Discharge-duration curve of the chute spillway of Karun I Dam in Iran for the period between March 1980 and July 1988 (width per chute 15 m) (Schleiss, 2002).

3.4.4. Can ultimate scour depth be achieved during spillway operation period and what is the scour rate?

In principle, the scour depth estimated by use of empirical and semi-empirical formula will occur only for a long duration of spillway operation, after steady conditions in the scour hole are achieved. This will happen only after a minimum duration of spillway operation, mainly depending on the quality and jointing of the rock mass. Therefore, the specific discharge which will have a sufficiently long duration to form the ultimate scour during the technical life of the dam has to be known. Higher and therefore rare discharges are not able to create ultimate scour.

Since plunge pool scour t + h is known to develop at an exponential rate with time T, the scour rate can estimated with the following relationship (Spurr, 1985):

$$(t+h)(T) = (t+h)_{end}(1 - e^{-aT/T}e) \tag{2}$$

where a is a site-specific constant. The evaluation of T_e (instant at which equilibrium is attained) depends on how rapidly hydro-fracturing and washing out of the material from the scour hole will occur, taking into account the primary and secondary rock characteristics. Primary rock characteristics comprise RQD, joint spacing, uniaxial compressive strength, and angle of jet impact compared to main faults or bedding planes; the secondary characteristics are the hardness and degree of weathering (Spurr, 1985). Knowing the depth of a scour, which occurred during a certain period of operation and estimating the maximum scour depth by one of the formulae, the site specific constant a/T_e can be determined.

As a rough estimation based on some prototype data, ultimate scour is normally attained only after $T_e = 100$ to 300 hours of spillway operation for a certain discharge considered.

It may be concluded, that the ultimate scour depth for a certain specific discharge occurs only if the duration of this discharge is long enough. Scour for a smaller duration can be estimated with an exponential rate relationship.

3.4.5. Which will be the prevailing discharge for scour formation during a flood event?

A flood event and the corresponding discharge curve of the spillway can be characterized by a hydrograph (Figure 3.7). For all discharges of the hydrograph with duration shorter than the instant T_e at which equilibrium is attained, the ultimate scour will not be reached. Knowing the scour rate relationship [Equation (2)], the prevailing discharge which will produce maximum scour depth during the flood event can be determined. The scour is estimated successively for discharges :

$$q_u(T=T_e=T_u), q_1(T_1<T_u), q_2(T_2<T_1<T_u), ..., q_{peak}(T_{peak}<T_i<T_u)$$

The discharge, which gives the deepest scour, is the prevailing discharge.

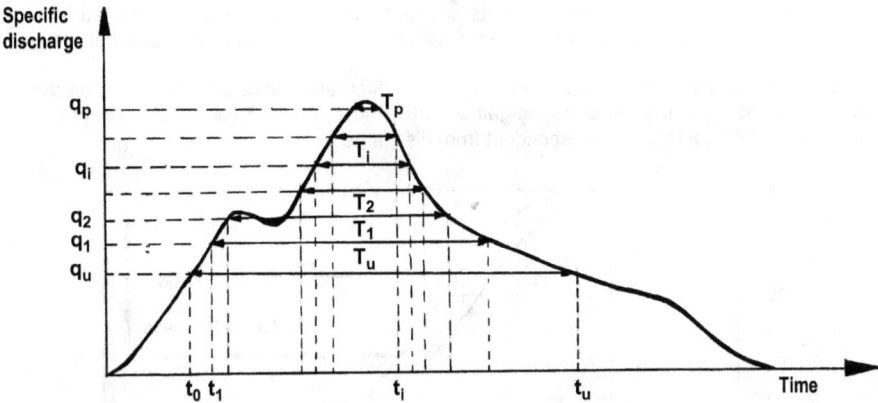

Fig. 3.7
Flood event and the corresponding discharge curve of the spillway showing discharges with a duration shorter than the instant $T_e = T_u$ at which equilibrium of scour and its ultimate depth is attained with the purpose to determine the prevailing discharge (Schleiss, 2002).

It has to be noted that these considerations are valid only for ungated free surface spillways. For gated free surface spillways the discharge may not be directly related to the reservoir inflow but be prescribed by operation rules. When lowering the reservoir level during floods, outflow discharges are higher than reservoir inflow. This can also be the case for pressurized orifice spillways

3.5. SPILLWAY DESIGN DISCHARGE AND SCOUR EVALUATION

Spillways are designed for the so-called design or project flood, typically a 1000-year flood, and checked for the so-called safety check flood, typically to a flood between a 10 000-year flood and PMF. The question arises for what flood the scour depth has to be evaluated, and the constructive scour control measures have to be based on.

As discussed in section 3.4.4, the ultimate scour depth will occur only after steady conditions in the scour hole are achieved, which will happen only after certain duration of spillway operation. Therefore, it is very conservative to base the estimation of the scour depth or the design of mitigation measures on low frequency floods (PMF or 10 000-year flood). It will be very unlikely that during the technical life of the dam, these rare floods can produce ultimate scour depth.

Therefore, for each flood event with a certain return period, the prevailing discharge and the maximum scour depth has first to be determined according to section 3.4.5. Furthermore, it has to be decided, for which flood return period the maximum scour depth during the technical life of the dam, will be estimated. The probability of the occurrence of a flood with a given return period during the useful life of a dam is as follows:

$$r = 1-(1-1/n)^m \qquad (3)$$

where r is the risk or the probability of occurrence, n the return period of the flood (years) and m the useful life of the dam.

In Figure 3.8 the probability of occurrence of floods for different useful lifetime of the dam is illustrated. It can be seen, that the probability of the occurrence of a 200-year flood during 200 years of operation is 63 %, whereas for a 1000-years flood is only 20 %.

It seems reasonable to choose a serviceability design discharge with a probability of occurrence of about 50 % during the useful lifetime of a dam to assess scour and protection measures. Higher design discharges with lower probability of occurrence are too conservative for this purpose. Nevertheless the stability of the dam itself including its abutments and foundation should not be endangered up to the safety check flood of the dam. Under safety check flood conditions damage is acceptable, but failure not.

It has to be noted, that in the case of gated, surface and orifice spillways, high discharges can be released at any time by opening the gates. Furthermore, for low-level outlet spillways the core impact velocity of the jet is nearly independent from discharge.

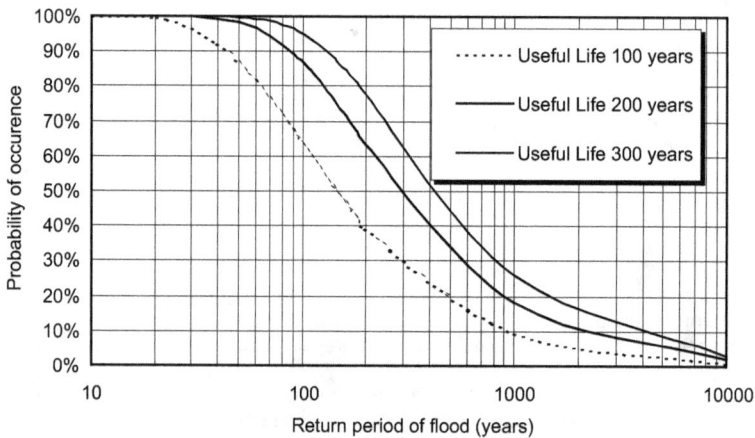

Fig. 3.8
Pre-excavation of the plunge pool for a scour depth, which would be formed by a 50- to 100-year flood, and slope stabilization measures in case of ultimate scour formation in order to avoid abutment instabilities (Schleiss, 2002).

3.6.   MEASURES FOR SCOUR CONTROL

3.6.1. Overview

To avoid scour damage, three active options may be considered:

- avoid scour formation completely;

- design water release structures such that the scour occurs far from dam foundation and abutments;

- limit the scour extent.

Since structures for scour control are rather expensive, only the two latter are normally economically feasible (Ramos, 1982). Besides elongating as much as possible the impact zone of the jets by an appropriate design of the water release structures, the extent of the scour can be influenced by the following measures:

- limitation of the specific spillway discharge;

- forced aeration and splitting of jets leaving spillway structures;

- increasing tailwater depth by a tailpond dam;

- pre-excavation of the plunge pool.

The location of the scour depends on the selected type of spillway and its design.

To avoid scour completely, structural measures as lined plunge pools are required. Besides the active options, scour damage can be prevented also by passive measures, for example by protecting dam abutments with anchors against instability due to scour formation.

3.6.2. Limitation of the specific spillway discharge

This measure is mainly important in the case of arch dams and free ogee crest spillways with jet impacts rather close to the dam. The jet can be guided by an appropriate crest lip design for a given specific discharge at a certain distance from the dam toe. If the dam foundation can be endangered by the scour the discharge per unit length of the ogee crest has to be limited. But by reducing the specific discharge, the available velocity at the crest lip and therefore the travel distance of the jet is also reduced.

In the case of gated ski-jump spillways and low-level outlets, the specific discharge depends on the size of the outlet openings. Since the available velocity at the outlet is high enough to divert the jet far away from the dam and its foundation, the limitation of the specific discharge is normally less important than for free crest spillways.

3.6.3. Forced aeration and splitting of jets

In order to split and aerate the jets leaving flip buckets and crest lips, they are often equipped with baffle blocks, splitters and deflectors. Furthermore, high velocity flows in spillway chutes are normally aerated by aeration ramps and slots along the chute. All these measures will increase air entrainment, which will reduce the scouring capacity of the plunging jets. Nevertheless, the amount of air entrained is difficult to estimate. Because of scale effects the efficiency of these measures can only be checked qualitatively by hydraulic model tests.

Martins (1973) suggested a reduction of 25 % of the calculated scour depth in the case of high air entrainment and 10 % for intermediate air entrainment. Mason (1989) proposed an empirical expression considering the volumetric air-to-water ratio β. The proposed empirical equation based on spillway models does not depend on the fall height H, since he used a direct relationship between β and H as developed by Ervine (1976). The empirical formula is accurate for model data and seems to give a reasonable upper bound of scour depth for prototype conditions.

Nevertheless, in the case of prototype scour in fractured rock, air strongly influences water hammer velocity and consequently resonance effects of pressure waves in rock joints. Recent research reveals that forced aeration may reduce pressure fluctuations at the water rock interface but may increase resonance effects due to reduced pressure wave celerity (Duarte et al., 2012, 2013). Thus forced aeration is not always reducing scour potential but may increase under certain conditions the risk of rock block break up and its ejection into the plunge pool.

3.6.4. Increasing tail-water depth by a tail-pond dam

Another way to control scour from jets is to increase tail-water depth by building a tail-pond dam downstream of the jet impact zone. The efficiency of a water cushion is often overestimated (Häusler, 1980). For plunge pool depths smaller than 4 to 6 times the jet diameter, core jet impact (see section 3.2.3) is normally observed at the plunge pool bottom (Bollaert, 2002). The jet core is characterized by a constant velocity and is not influenced by the outer two-phase shear layer condition of the impinging jet. The pressures are also constant with low fluctuations, which have significant spectral energy at very high frequencies (up to several 100 Hz). For tail-water depths larger than 4 to 6 times the jet diameter (or thickness), developed jet impact occur at the plunge pool bottom with a different pressure pattern produced by a turbulent two-phase shear layer. Very significant fluctuations are produced with high spectral content at frequencies up to 100 Hz. Maximum fluctuations have been observed for tailwater depths between 5 to 8 times the jet diameter (Bollaert & Schleiss, 2003). Substantial high values persist up to tailwater depth of 10 to 11 times the jet diameter. From these observations it may be concluded, that water cushions in the range of 5 to 11 times the jet diameter can generate even more severe dynamic pressures at the plunge pool bottom than smaller tail-water depths. Therefore, only water cushions deeper than 11 times the diameter of the jet at impact have a retarding effect on the scour formation.

This tendency was also confirmed by the empirical scour equation of Martins (1973), which gives a maximum scour depth for a certain tail-water depth. Johnson (1967) already found that too small water cushions are even worse than no cushion since the material can be transported more easily out of the plunge pool.

It has to be noted, that increased tail-water level by the help of a tail-pond dam may interfere with bottom outlets.

3.6.5. Pre-excavation of the plunge pool

In principle, pre-excavation increases the tail-water depth and in view of scour control, the same remarks are valid than given in section 3.6.4.

In general, the pre-excavation of the expected scour may be also appropriate for scour control, when the eroded and river transported material can form dangerous deposits downstream, for example near the outlet of the powerhouse. Such deposits could increase the tail-water level and reduce the power production. Such problems normally should not be expected, when the scour is formed about 200 m upstream of the powerhouse outlet.

Pre-excavation of the scour hole is also often considered when instabilities of the valley slopes may be considered. In such cases the excavation has to be stabilized by anchors and other measures. As an alternative, even under such conditions, pre-excavation could be omitted if the valley slopes are stabilized by appropriate measures, in such a way that the slopes remain stable even after ultimate scour formation.

The selection of the design discharge for an excavated geometry of the plunge pool has to be based on considerations similar to the ones discussed in section 3.3.5. In general, it is not economically interesting to excavate deeper than the scour depth, which would be formed by 50- to 100-year floods. If instabilities of the valley slopes are considered possible as a result of deeper scour due to higher spillway discharges, rock anchors and pre-stressed tendons can be used, as already mentioned (Figure 3.9).

Instead of full excavation, pre-splitting of plunge pool rock may by adequate under certain conditions, making the rock more readily erodible in certain places. Pre-splitting can therefore be used to influence the eventual geometry of the plunge pool without the expense of complete excavation. In some cases, of course, such as where scour affects power production, complete excavation may be more appropriate.

The pre-excavated plunge pool geometry has to be based on the expected natural scour geometry. Several authors proposed empirical formula for the estimation of the length (Martins, 1973; Kotoulas, 1967) and the width (Martins, 1973) of the scour hole. Amaniam & Urroz (1993) performed a number of tests on a model scale flip bucket spillway with a gravel bed plunge pool in order to develop equations to describe the geometry of the scour hole created by the jet impinging into the plunge pool. They observed that the performance of the pre-excavated scour holes is better when they are close to the self-excavated hole for the same flow parameters.

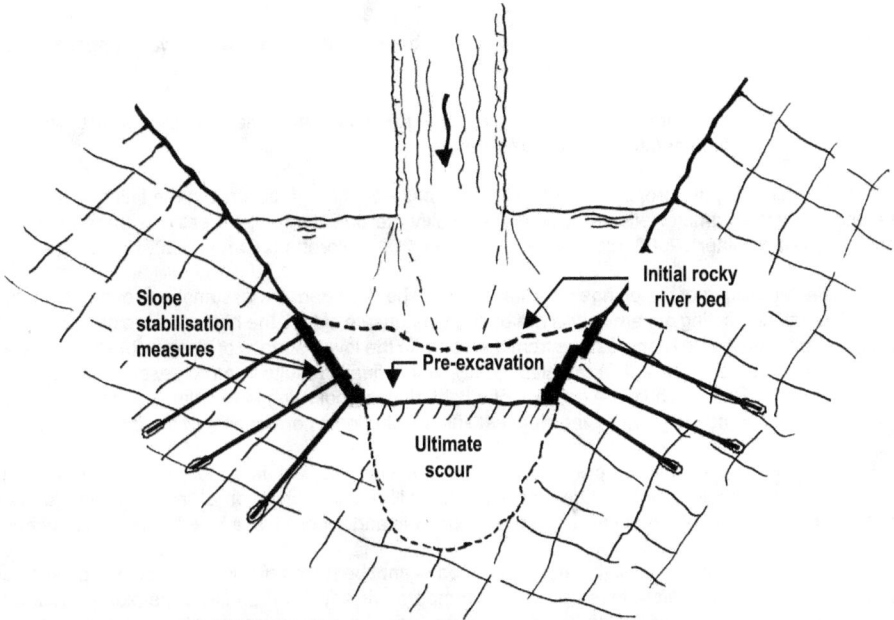

Fig. 3.9
Pre-excavation of the pit for a scour depth that would be formed by a 50- to 100-year lifetime, and slope stabilization measures for ultimate scour formation to avoid bank instabilities (Schleiss, 2002)

For the well-known scour hole case at Kariba dam on Zambezi River, which has reached a depth of more than 80 m below the tail-water level, an innovative solution has been developed by reshaping the existing scour hole with an excavation mainly in downstream direction (Noret *et al.*, 2013). This reduces significantly the dynamic pressures at water rock interface due to jet impact. With hybrid modelling techniques it could be proved that the reshaped scour hole will not progress anymore in future (Duarte *et al.*, 2011).

3.6.6. Concrete lined plunge pools

If absolutely no scour formation in the rock downstream of a dam can is required, the plunge pool has to be reinforced and tightened by a concrete lining. Since the thickness of the lining is limited by construction and economic reasons, normally high tension or pre-stressed rock anchors are required to ensure the lining stability in view of the high dynamic loading. Furthermore, the surface of the lining has to be protected with reinforced (wire mesh, steel fibers) and high tensile concrete having also high resistance against abrasion. Construction joints have to be sealed with efficient water stops. In addition, the stability of the lining against static uplift pressure during dewatering of the plunge pool has to be guaranteed by a drainage system. This can also limit dynamic uplift pressures in case of limited cracking of the lining.

The design of plunge pools linings has to be based on the following sequence of events (Fiorotto & Rinaldo, 1992; Fiorotto & Salandin, 2000):

- pulsating pressures can damage the joint seals between slabs (construction joints);

- through these joint seals, extreme pressure values may propagate from the upper to the lower surface of the slabs;

- instantaneous pressure differentials between the upper and lower surfaces of the slabs can reach high values;

- the resultant force stemming from the pressure differential may exceed the weight of the slab and the anchor's resistance.

Furthermore, the propagation of dynamic pressures through fissures in the lining reveals the presence of water hammer effects, which can amplify the pulsating uplift pressures underneath the concrete slabs (Bollaert, 2002; Melo, et al., 2006; Liu, 2007; Fiorotto & Caroni, 2007).

Since cracking of the plunge pool lining cannot be excluded, the assumptions of an absolutely tight lining and neglecting dynamic uplift are on the very unsafe side. If the high dynamic pressures can propagate through a small, local fissure from the upper to the lower surface of a concrete slab, dynamic pressures from underneath will lift the slab locally, which finally results in a progressive failure of the whole plunge pool lining. Thus, the concept of a tight plunge pool lining is as risky as a chain concept: the system's resistance is given by the weakest link, i.e. the local permeability of the concrete slab.

Furthermore, fluctuating pressures due to the plunging jets into the plunge pool are high compared to the proper weight of the slab and thus will result in vibration of the concrete slab and consequently in the development of cracks in the concrete and in possible fatigue failure of the anchors.

Cracking of a concrete slab before operation cannot be ruled out fully even using sophisticated construction joints and water-stops because of temperature effects when filling the plunge pool with water and of deformation of the underground. Therefore, the design criteria of the plunge pool liner have to take into account the following:

- the load case of dynamic uplift pressures during operation;

- reinforcement in the concrete slab has to be designed for crack width limitation for possible dynamic vibration modes (depending on anchor pattern and stiffness);

- the grouted and pre-stressed rock anchors have to be designed for fatigue.

The drainage system under the plunge pool lining is of very high importance since it increases considerably the safety against dynamic uplift pressures. Nevertheless, since limited cracking of the lining cannot be excluded, as already mentioned, pressure waves can be transferred through the cracks into the drainage system. The response of the drainage system in view of these dynamic pressures with a wide range of frequencies at the entrance of a possible crack has to be controlled by a transient analysis, in order to be sure that no amplification of the pressures in the drainage system occurs (Mahzari et al., 2002). These amplifications of the pressures in the drainage system or any dynamic pressure underneath the slabs have to be considered in the design of the anchors (Mahzari & Schleiss, 2010).

As a conclusion, since plunge pool linings are a risky concept as already mentioned before, a "belt-and-braces" approach for prototype design should be used with the following recommendations:

- slabs are sized generously so that localized instantaneous fluctuations average over larger area and the slabs will generally be reinforced;

- water bars are provided between slabs to prevent dynamic pressure communicating with the underside of the slabs;

- as a further precaution anchor bars are used and sized based on mean pressure differentials across different parts of the apron and on the assumption that some water bars may have failed;

- finally, generously sized under drainage is provided so that any pressure fluctuations feeding through small joints between the slabs can be dissipated in generously sized drainage zones.

3.7. CONCLUSIONS

Although the physical understanding of the scour process has considerably improved during the last 10 to 20 years, the scour evaluation still remains a challenge for dam designers. Scour models are now available which take into account the pressure fluctuations in the plunge pool and the propagation of transient water pressures into the joints of the underlying rock mass such as the Comprehensive Scour Method. It allows modelling the physical process of ejection of rock blocks due to dynamic uplift and the influence of air bubbles in erosion process in the plunge pool.

The rock parameters, which have to be assessed during the reconnaissance campaign for the dam foundation together with the hydrological conditions during the considered lifetime of the dam, influence the evolution of scour with time. The latter results in uncertainties which have to be overcome by engineering judgment.

In order to check and calibrate complex scour models, more detailed prototype data on scour evolution with fully documented discharge records are still needed. These observations are essential for a continuous safety assessment of a dam and to allow predicting the scour evolution.

3.8. REFERENCES

Amanian, N. and G. E. Urroz, (1993). 'Design of pre-excavated scour hole below flip bucket spillways'. *Proceedings of the ASCE International Symposium on Hydraulics*, San Francisco, USA

Annandale, G. W. (1995). 'Erodibility', *Journal of Hydraulic Research*, IAHR, vol. 33, N°4, pp. 471-494.

Annandale, G.W, R. J. Wittler, J. Ruff, and T. M. Lewis. (1998). 'Prototype validation of erodibility index for scour in fractured rock media', *Proceedings of the 1998 International Water Resources Engineering Conference*, Memphis, Tennessee, USA.

Annandale, G.W. (2006). *Scour Technology*, McGraw-Hill, New York, USA.

Armengou, J. (1991). *Disipacion de energia hidraulica a pie de presa en presas boveda*, PhD thesis, Universitat Politechnica de Catalunya, Barcelona, Spain.

Asadollahi, P., F. Tonon, M. Federspiel, M. and A. J. Schleiss, (2011). 'Prediction of rock block stability and scour depth in plunge pools' *Journal of Hydraulic Research*, vol. 49 (5), pp. 750-756.

Ballio, F., S. Franzetti, and M. G. Tanda, (1994). 'Pressure fluctuations induced by turbulent circular jets impinging on a flat plate', *Excerpta*, vol. 7.

Bellin, A. and V. Fiorotto, (1995). 'Direct dynamic force measurement on slabs in spillway stilling basins', *Journal of Hydraulic Engineering*, ASCE, vol. 121, N° HY 10, pp. 686-693.

Beltaos, S. and N. Rajaratnam, (1973). 'Plane turbulent impinging jets', *Journal of Hydraulic Research*, IAHR, vol. 11, N° 1, pp. 29-59.

Bin, A.K. (1984). 'Air entrainment by plunging liquid jets', Proceedings of IAHR Symposium on Scale Effects in Modelling Hydraulic Structures, Esslingen.

Bollaert, E. 2002. 'Transient water pressures in joints and formation of rock scour due to high-velocity jet impact'. *Communication N° 13 of the Laboratory of Hydraulic Constructions (LCH)*, EPFL, Lausanne. (Ed. By A. Schleiss).

Bollaert E. and A. J. Schleiss. (2003a). 'Scour of rock due to the impact of plunging high velocity jets - Part I: A state-of-the-art review'. *Journal of Hydraulic Research*, vol.41, No. 5, pp. 451-464.

Bollaert E. and A. J. Schleiss. (2003b). 'Scour of rock due to the impact of plunging high velocity jets. Part II: Experimental results of dynamic pressures at pool bottoms and in one- and two-dimensional closed end rock joints', *Journal of Hydraulic Research*, vol. 41, No. 5, pp. 465-480, 2003

Bollaert, E. (2004). 'A comprehensive model to evaluate scour formation in plunge pools'. *International Journal of Hydropower & Dams* 1, 94-101.

Bollaert E. and A. J. Schleiss. (2005). 'Physically based model for evaluation of rock scour due to high-velocity jet impact'. *Journal of Hydraulic Engineering* ASCE, March 2005, pp.153-165

Bollaert, E., R. Duarte, M. Pfister, A. J. Schleiss and D. Mazvidza. (2012). 'Physical and numerical model study investigating plunge pool scour at Kariba Dam'. *Proceedings of the 24th Congress of CIGB –ICOLD*, Kyoto, Japan, Q. 94 – R. 17, pp. 241-248

Duarte, R., Bollaert, E., Schleiss, A.J., Pinheiro, A. 2012. Dynamic pressures around a confined block impacted by plunging aerated high-velocity jets. Proceedings of the 2nd European IAHR Congress, 27-29 juin 2012, Munich, Allemagne, réf. B14.

Duarte R., Schleiss A.J., Pinheiro A. (2013) Dynamic pressure distribution around a fixed confined block impacted by plunging and aerated water jets. Proc. of the 35th IAHR World Congress, Chengdu, China, September 8-13, 2013.

Eggenberger, W. & Müller, R. 1944. Experimentelle und theoretische Untersuchungen über das Kokproblem. Mitteilungen Nr. 5 der VAW, ETH Zürich (in German)

Ervine, D.A. 1976. The entrainment of air in water, Water Power and Dam Construction, 28(12), pp. 27-30.

Ervine, D.A. 1998. Air entrainment in hydraulic structures: a review, Proceedings of the Institution of Civil Engineers Wat., Marit. & Energy, Vol. 130, pp. 142-153.

Ervine, D.A. & Falvey, H.R. 1987. Behavior of turbulent jets in the atmosphere and in plunge pools, Proceedings of the Institution of Civil Engineers, Part 2, Vol. 83, pp. 295-314.

Ervine, D.A. & Falvey, H.R. 1988. Aeration in jets and high velocity flows, Conference Proceedings, Model-Proto Correlation of Hydraulic Structures, P. Burgi, 1988, pp. 22-55.

Ervine, D.A.; Falvey, H.R.; Withers, W. 1997. Pressure fluctuations on plunge pool floors, Journal of Hydraulic Research, IAHR, Vol. 35, N°2.

Ervine, D.A.; McKeogh, E.; Elsawy, E.M. 1980. Effect of turbulence intensity on the rate of air entrainment by plunging water jets, Proceedings of the Inst. Civ. Eng., Part 2, pp. 425-445.

Fahlbusch, F. E. (1994). Scour in rock riverbeds downstream of large dams, Hydropower & Dams, pp. 30-32.

Federspiel, M.P.E.A., 2011. Response of an embedded block impacted by high-velocity jets. Communication 47 of Laboratory of Hydraulic Constructions, Lausanne, Switzerland, ISSN 161-1179 (Ed. Schleiss A.).

Federspiel, M., Bollaert, E., Schleiss, A. J. 2011. Dynamic response of a rock block in a plunge pool due to asymmetrical impact of a high-velocity jet. Proc. of 34th IAHR World Congress, 26 June - 1st July 2011, Brisbane, Australia, CD-Rom, ISBN 978-0-85825-868-6, pp. 2404-2411

Fiorotto, V. & Rinaldo, A. 1992. Fluctuating uplift and lining design in spillway stilling basins, Journal of Hydraulic Engineering, ASCE, Vol. 118, HY4.

Fiorotto, V. & Salandin, P. 2000. Design of anchored slabs in spillway stilling basins, Journal of Hydraulic Engineering, ASCE, Vol. 126, N° 7, pp. 502-512.

Fiorotto, V. & Caroni, E. 2007. Discussion of 'Forces on plunge pool slabs: influence of joints location and width' by J. F. Melo, A. N. Pinheiro, and C. M. Ramos, J. Hydraul. Eng., 133(10), 1182-1184.

Franzetti, S. & Tanda, M.G. 1984. Getti deviati a simmetria assiale, Report of Istituto di Idraulica e Costruzioni Idrauliche, Politecnico di Milano.

Franzetti, S. & Tanda, M.G. 1987. Analysis of turbulent pressure fluctuation caused by a circular impinging jet, International Symposium on New Technology in Model Testing in Hydraulic Research, India, pp. 85-91.

Gerodetti, M. 1982. Auskolkung eines felsigen Flussbettes (Modellversuche mit bindigen Materialen zur Simulation des Felsens). Arbeitsheft N° 5, VAW, ETHZ, Zürich (in German)

Gao, J.; Liu, Z. and Guo, J. 2011. Newly achievements on dam hydraulic research in China. In: Valentine, EM (Editor); Apelt, CJ (Editor); Ball, J (Editor); Chanson, H (Editor); Cox, R (Editor); Ettema, R (Editor); Kuczera, G (Editor); Lambert, M (Editor); Melville, BW (Editor); Sargison, JE (Editor). Proceedings of the 34th World Congress of the International Association for Hydro- Environment Research and Engineering (IAHR): 2436-2443.

Gunko, F.G.; Burkov, A.F.; Isachenko, N.B.; Rubinstein, G.L.; Soloviova, A.G. ; Yuditskii, G.A. 1965. Research on the Hydraulic Regime and Local Scour of River Bed Below Spillways of High-Head Dams, 11th Congress of the I.A.H.R., Leningrad.

Hartung, F. & Häusler, E. 1973. Scours, stilling basins and downstream protection under free overfall jets at dams, Proceedings of the 11th Congress on Large Dams, Madrid, pp.39-56.

Häusler, E. 1980. Zum Kolkproblem bei Hochwasser-Entlastungsanlagen an Talsperren mit freiem Überfall. Wasserwirtschaft 3.

Hoffmans, G.J.C.M. 1998. Jet scour in equilibrium phase, Journal of Hydraulic Engineering, ASCE, Vol. 124, N°4, pp. 430-437.

Johnson, G. 1967. The effect of entrained air in the scouring capacity of water jets, Proceedings of the 12th Congress of the I.A.H.R., Vol. 3, Fort Collins.

Johnson, G. 1977. Use of a weakly cohesive material for scale model scour tests in flood spillway design, Proceedings of the 17th Congress of the I.A.H.R., Vol. 4, Baden-Baden.

Kamenev, I.A., 1966. Alcance de jactos livres provenientes de descarregadores. (Trans. N° 487 L.N.E.C.) Gidrotekhnicheskoe Stroitel'stvo N° 3.

Kawakami, K. 1973. A study on the computation of horizontal distance of jet issued from ski-jump spillway. Trans. of the Japanese Society of Civil Engineers, Vol. 5.

Kirschke, D. 1974. Druckstossvorgänge in wassergefüllten Felsklüften, Veröffentlichungen des Inst. Für Boden und Felsmechanik, Univ. Karlsruhe, Heft 61 (in German)

Kotoulas, D. 1967. Das Kolkproblem unter besonderer Berücksichtigung der Faktoren "Zeit" und "Geschiebemischung" im Rahmen der Wildbachverbauung. Schweizerische Anstalt für das Forstliche Versuchswesen, Vol. 43, Heft 1 (in German).

Liu, P.Q. 1999. Mechanism of energy dissipation and hydraulic design for plunge pools downstream of large dams, China Institute of Water Resources and Hydropower Research, Beijing, China.

Liu, P.Q.; Dong, J.R.; Yu, C. 1998. Experimental investigation of fluctuating uplift on rock blocks at the bottom of the scour pool downstream of Three-Gorges spillway, Journal of Hydraulic Research, IAHR, Vol. 36, N°1, pp. 55-68.

Liu, P. Q. and Li, A. H. 2007. Model discussion on pressure fluctuation propagation within lining slab joints in stilling basins. J. Hydraul. Eng., 133(6), 618-624.

Lopardo, R.A. 1988. Stilling basin pressure fluctuations, Conference Proceedings, Model-Prototype Correlation of Hydraulic Structures, P. Burgi, pp. 56 – 73.

Machado, L.I. 1982. O sistema de dissipação de energia proposto para a Barragem de Xingo, Transactions of the International Symposium on the Layout of Dams in Narrow Gorges, ICOLD, Brazil.

Mahzari M; Arefi F.; Schleiss A. 2002. Dynamic response of the drainage system of a cracked plunge pool liner due to free falling jet impact. Proc. of Int. Workshop on Rock Scour due to falling high-velocity jets, Lausanne, Switzerland, 25-28 September, (Ed. Schleiss & Bollaert), pp 227-237 .

Mahzari, M., Schleiss, A. J. 2010 Dynamic analysis of anchored concrete linings of plunge pools loaded by high velocity jet impacts issuing from dam spillways. Dam Engineering, Volume XX Issue 4, pp. 307-327

Martins, R. 1973. Contribution to the knowledge on the scour action of free jets on rocky river beds, Proceedings of the 11th Congress on Large Dams, Madrid, pp. 799-814.

Martins, R. 1977. Cinemática do jacto livre no âmbito das estruturas hidráulicas. Memória N° 486, L.N.E.C., Lisboa.

Manso, P. 2006. The influence of pool geometry and induced flow patterns in rock scour by high-velocity plunging jets. Communication 25, Laboratory of Hydraulic Constructions, Ecole Polytechnique Fédérale de Lausanne (EPFL), Lausanne, Switzerland (Ed. Schleiss A.).

Manso P., Bollaert E., Schleiss A.J. 2008. Evaluation of high-velocity plunging jet-issuing characteristics as a basis for plunge pool analysis. Journal of Hydraulic Research, Volume 46, No. 2, pp. 147-157, 2008

Manso, P. A., Bollaert, E. F. R., Schleiss, A. J. 2009. Influence of plunge pool geometry on high-velocity jet impact pressures and pressure propagation inside fissured rock media. Journal of Hydraulic Engineering, ASCE, Volume 135, Issue 10, pp. 783-792

Mason, P. J. 1984. Erosion of plunge pools downstream of dams due to the action of free-trajectory jets. Proc. Instn Civ. Engrs, Part 1, 76 May, 524-537.

Mason, P. J. & Arumugam, K. 1985. Free jet scour below dams and flip buckets, Journal of Hydraulic Engineering, ASCE, Vol. 111, N° 2, pp. 220-235.

Mason, P. J. 1989. Effects of air entrainment on plunge pool scour. Journal of Hydraulic Engineering, ASCE, Vol. 115, N° 3, pp. 385-399.

Mason P J. 1993. Practical guidelines for the design of flip buckets and plunge pools. International Water Power & Dam Construction, 45 (9/10), September/October 1993.

Mason P J. 2011. Plunge Pool Scour: An Update. Hydropower & Dams, Vol 18, (6), 2011.

May, R.W.P. & Willoughby, I.R. 1991. Impact pressures in plunge pool basins due to vertical falling jets, Report SR 242, HR Wallingford, UK.

Melo, J. F., Pinheiro, A. N. and Ramos, C. M. 2006. Forces on plunge pool slabs: influence of joints location and width." J. Hydraul. Eng., 132(1), 49-60.

Mirtskhulava, T.E.; Dolidze, I.V.; Magomeda, A.V. 1967. Mechanism and computation of local and general scour in non cohesive, cohesive soils and rock beds, Proceedings of the 12th IAHR Congress, Vol. 3, Fort Collins, pp. 169-176.

Noret, Ch., Girard J.-C., Munodawafa, M.C., Mazvidza, D.Z. 2013. Kariba dam on Zambezi river: stabilizing the natural plunge pool. La Houille Blanche, n°1, 2013, p. 34-41.

Otto, B. 1989. Scour potential of highly stressed sheet-jointed rocks under obliquely impinging plane jets, PhD thesis, James Cook University of North Queensland, Townsville.

Puertas, J. & Dolz, J. 1994. Criterios hidraulicos para el diseno de cuencos de disipacion de energia en presas boveda con vertido libre por coronacion, PhD thesis, Politechnical University of Catalunya, Barcelona, Spain.

Quintela, A.C. & Da Cruz, A.A. 1982. Cabora-Bassa dam spillway, conception, hydraulic model studies and prototype behaviour, Transactions of the International Symposium on the Layout of Dams in Narrow Gorges, ICOLD, Brazil.

Ramos, C.M. 1982. Energy dissipation on free jet spillways. Bases for its study in hydraulic models, Transactions of the International Symposium on the Layout of Dams in Narrow Gorges, ICOLD, Rio de Janeiro, Brazil, Vol. 1, pp. 263-268.

Schleiss, A.J. 2002. Scour evaluation in space and time – the challenge of dam designers, Proc. of Int. Workshop on Rock Scour due to falling high-velocity jets, Lausanne, Switzerland, 25-28 September, (Ed. Schleiss & Bollaert), pp 3-22.

Spurr, K. J. W. 1985. Energy approach to estimating scour downstream of a large dam, Water Power & Dam Construction, Vol. 37, N°11, pp. 81-89.

Tao, C.G.; JiYong, L.; Xingrong, L. 1985. Translation from Chinese by de Campos, J.A.P.. Efeito do impacto, no leito do rio, da lâmina descarregada sobre uma barragem-abóbada, Laboratório Nacional de Engenharia Civil, Lisboa.

Toso, J. & Bowers, E.C. (1988). Extreme pressures in hydraulic jump stilling basin. Journal of Hydraulic Engineering, ASCE, Vol. 114, N° HY8, pp. 829-843.

Taraimovich, I.I. 1980. Calculation of local scour in rock foundations by high velocity flows, Hydrotechnical Construction N°8.

U.S.B.R. 1978. Hydraulic design of stilling basins and energy dissipators. Water Resources Technical Publication. Engineering Monograph N° 25, 4th Printing.

Van de Sande, E. & Smith, J.M. 1973. Surface entrainment of air by high velocity water jets, Chem. Engrg. Sci., 28, pp. 1161-1168.

Veronese, A. 1937. Erosion of a bed downstream from an outlet, Colorado A & M College, Fort Collins, United States.

Whittaker, J. & Schleiss, A. 1984. Scour related to energy dissipators for high head structures. Mitteilung Nr. 73, Versuchsanstalt für Wasserbau, Hydrologie und Glaziologie, ETH-Zurich: Zürich.

Xu-Duo-Ming 1983. Pressão no fundo de um canal devido ao choque de um jacto plano, e suas características de fluctuação, Translation from Chinese by J.A. Pinto de Campos, Lisboa.

Yuditski, G.A. 1963. Actual pressure on the channel bottom below ski-jump spillways, Izvestiya Vsesoyuznogo Nauchno – Issledovatel – Skogo Instituta Gidrotekhiki, Vol. 67, pp. 231-240.

Zvorykin, K.A., Kouznetsov, N.V., Akhmedov, T.K. 1975. Scour from rock bed by a jet spilling from a deflecting bucket of an overflow dam. 16th Congress of the IAHR, Vol.2, São Paulo.

4.   STEPPED SPILLWAYS

4.1.   INTRODUCTION

The use of stepped chutes in dams and channels is not novel; it was common in ancient hydraulic constructions, especially those constructed with masonry (Chanson, 1994, 2002). The introduction of the Roller Compacted Concrete (RCC) technique on dams has revitalized the use and the study of stepped spillways. This modern period and the ancient period have in common that the stepped surface came as an output of the construction technique. The intense research that has been carried out in the modern time makes the difference between these periods. As a consequence of this, knowledge on hydraulic performance, cavitation risk, durability and energy dissipation has significantly improved.

Stepped spillways have two practical advantages: the first is that they fit very well with the RCC construction procedures, where formworks are placed vertically to provide workspace, resulting in vertical or stepped faces; and the second is that a higher energy dissipation is achieved due to the steps' influence on the flow, allowing to construct smaller energy dissipation structures.

The high energy dissipation rate that is produced by these chutes is now opening a new period, in which stepped spillways are designed regardless of the construction technique.

The main uses of stepped spillways are on RCC gravity dams (Fig. 4.1), where the downstream stepped face is used as the spillway; and on embankment dams with limited hydrological safety, for which an overtopping protection may be placed armouring the downstream shoulder (Fig. 4.2).

Fig. 4.1
La Breña II Dam (Spain). Gravity RCC dam (ACUAES)

Fig. 4.2
Yellow River No. 14 Dam (USA). Overtopping RCC Spillway. (Golder Associates)

Another recent use of the stepped spillways is in combination with embankment dams, where stepped chutes excavated along the abutments are being constructed (Fig. 4.3). These may be lined or unlined, and which may have a variable geometry (step height and chute slope) according to the prevailing topography (Baumann *et al.*, 2006).

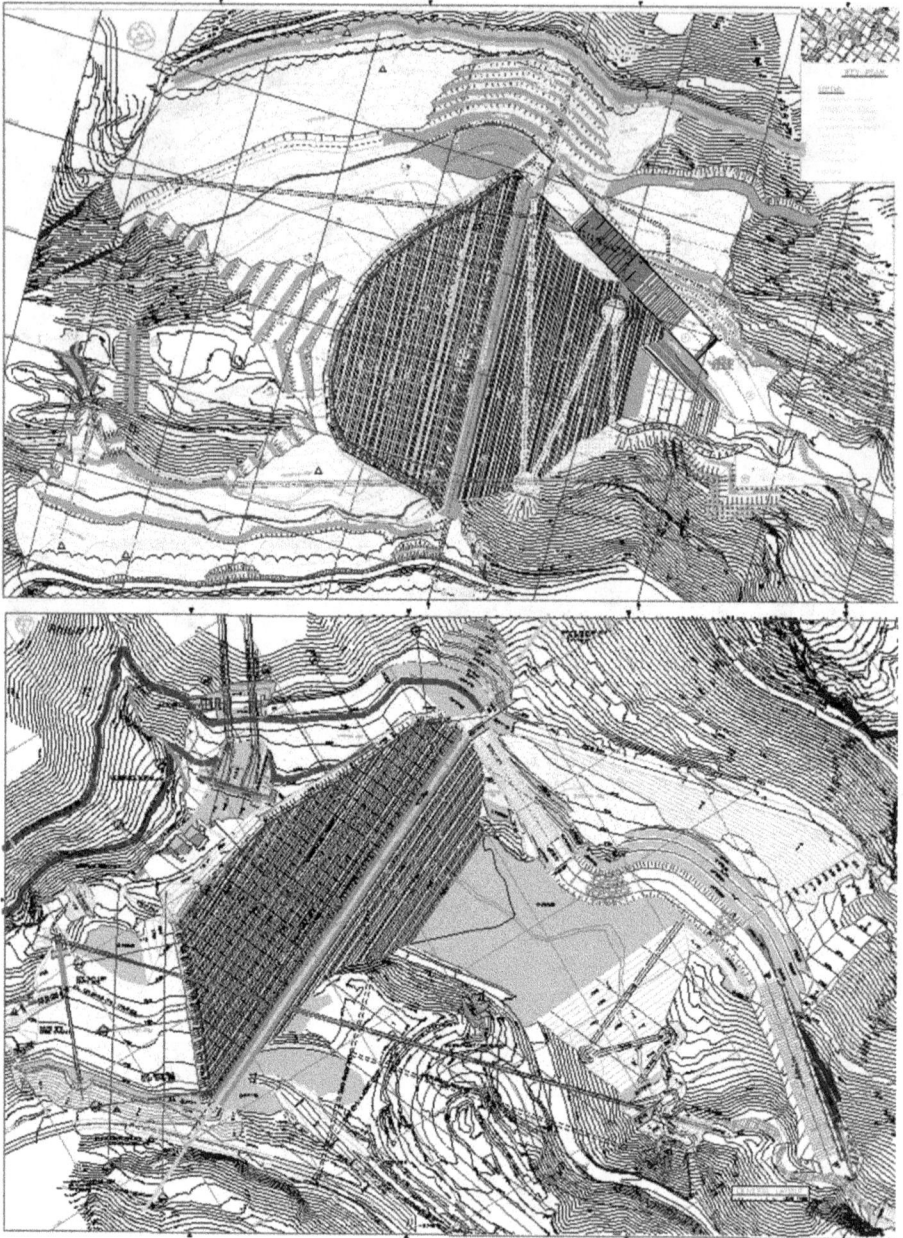

Fig. 4.3
Siah Bishe pumped storage power plant (Iran). Lower Dam (top) and Upper Dam (bottom). Stepped spillways along embankment dam abutments. (Courtesy of Prof. A.J. Schleiss)

There are also many older dams, which are still in use, where stepped masonry spillways were used in conjunction with embankment dams. These still require regular maintenance and assessment to ensure their continued successful use.

This chapter describes the main applications of stepped spillways (section 4.2), their hydraulics and expected performance (section 4.3) and the aspects that require specific attention during the design phase (section 4.4).

4.2. MAIN USES OF STEPPED SPILLWAYS

4.2.1. Stepped spillways on RCC gravity dams

One of the main uses of the stepped chutes is on RCC gravity dams. The RCC construction technique has proved itself as a time and cost-effective solution for developing large hydraulic projects. At the initial stage of the RCC technique many designs were conservative, with conventional vibrated concrete faces and smooth chutes used. Nowadays, most RCC gravity dams have stepped downstream faces as a result of the construction methods. The use of stepped chutes increased worldwide as it complements the construction method and because they provide higher energy dissipation.

Research on their hydraulic performance and experience gained on existing prototypes has contributed to consolidate the use of this type of spillways. According to the studies and the experience, the unit design flow has been increasing throughout the years (Table 4.1). Earlier designs were restricted to relatively small unit discharges (up to 10 m³/s/m) (Sorensen, 1985), because of the uncertainty about the spillway performance and the risk of cavitation, but nowadays higher discharges are being considered. Boes (2012) indicates that designs for discharges below 30 m³/s/m can be considered conventional, while those exceeding that value require special attention.

The layout of these spillways is similar to that of smooth ones on typical concrete gravity dams: control structure, spillway channel and termination structure. However, the design of each component has to be adapted to the hydraulic performance of the stepped chute: the crest-to-chute transition has to be designed carefully to avoid flow deflection and spray, the sidewall height has to account for the flow bulking; the risk of cavitation has to be analysed and aeration provided if needed; and the terminal structure has to account for the air-water flow characteristics and the energy dissipation along the chute. These peculiarities are discussed further in section 4.4.

4.2.2. Overflow RCC stepped spillway on embankment dams

The overtopping protection of embankment dams with a RCC overlay is another important application of stepped spillways. They are typically used for existing dams with limited hydrological safety, where overflow spillways are designed for supplementing the service spillway to discharge large floods (Berga, 1995; PCA, 2002; FEMA, 2014; Toledo et al., 2015). In small dams (<15 m high) this alternative is cost-effective when compared to other measures, such as: enlarging of the existing spillway, constructing a new spillway separated from the dam body, or heightening the dam (Hansen, 2003; Bass et al., 2012). The statistics of 109 overflow RCC spillways in the USA (FEMA, 2014) show that these range in height from 5 to 20 m, and that the maximum height up to date is 35 m; and regarding the design unit discharge the average is below 3 m³/s/m and the maximum around 10 m³/s/m.

Table 4.1
Evolution of the design unit discharge of stepped spillways in gravity dams.
(Adapted and updated from Chanson (2002) and Matos (2003))

Dam (Country)	Year	Design unit discharge (m³/s/m)	Observations
Monskvile (USA)	1986	9.3	
De Mist Kraal (South Africa)	1986	10.3	
Upper Stillwater (USA)	1987	11.4	
Les Olivettes (France)	1987	6.6	
Wolwedans (South Africa)	1990	12.4	
La Puebla de Cazalla (Spain)	1991	9.0	
Shuidong (China)	1994	100	Unconventional stepped spillway. Use of flaring piers to dissipate energy before the spillage over the stepped face.
Rambla del Boquerón (Spain)	1997	17.8	
Dona Francisca (Brazil)	2001	32	Maximum unit discharge.
Dachaosan (China)	2002	165	Unconventional stepped spillway. Use of flaring piers to dissipate energy before the spillage over the stepped face.
Pedrógão (Portugal)	2005	39,9	
La Breña II (Spain)	2008	22.3	Use of rounded edges in the crest-to-chute transition to reduce spray.
Boguchany (Russia)	2012	44	Unit discharge capacity for the Full Supply Level. Use of a large step (3.6 m) at the end of the piers to improve aeration.
Enlarged Cotter (Australia)	2013	48	Design unit discharge of the primary spillway. Use of a deflector -with air-inlets- to provide aeration.
De Hoop (South Africa)	2014	40	Maximum unit discharge. Use of triangular protrusions on the top third to improve aeration.
El Zapotillo (Mexico)	Under const.	38	Use of a large step to provide aeration.

This type of spillway consists of a concrete slab that armours the downstream slope and makes overflowing possible. The RCC technique is often used for constructing these slabs. RCC is typically placed in horizontal layers, and the downstream face, which constitutes the spillway, becomes stepped.

Although these spillways are also constructed with RCC, the design and construction conditions are different than those of gravity dams. Hence, the size and the finishing of the steps vary. The usual step height is 0.30 m (1 lift) and the facing may be formed or unformed. When unformed, the step face is not compacted, and it has to be considered as a sacrificial concrete. Additionally, the hydraulic performance of the unformed protections would differ from the formed stepped ones and less energy dissipation would occur (FEMA, 2014). The lift width is determined by construction and structural requirements. Special care has to be taken with the uplift pressures that could occur beneath the slab, and a drainage system has to be provided. The minimum lift width is around 2.5 m when it is limited by the vibrating roller size. Larger width is required for large unit flows and for flat slopes, where a minimum slab thickness of 0.6 m is recommended (PCA, 2002).

The specific design peculiarities of these spillways, including the crest shape, the terminal structure or the sidewalls design are further discussed in section 4.4.

4.2.3. Stepped spillways along embankment dam abutments

The introduction of the stepped chutes in spillways located along the embankment dam abutments is a recent application that may be used in future projects with the aim of improving energy dissipation. This use benefits from the research about the hydraulic performance of stepped chutes on moderate and flat slopes (Ohtsu et al., 2001; 2004; André, 2004; González and Chanson, 2007; Meireles and Matos, 2009).

A characteristic of these spillways is the necessary adaptation of the chute slopes and the step height to the prevailing topography. In such cases the use of transition channels may be used to keep the step height constant over distances as long as possible (Figure 4.3) (Baumann et al., 2006; Boes et al. 2015).

4.2.4. Stepped masonry spillways

Prior to the era of modern concrete, stepped spillways constructed using masonry often accompanied the construction of earth embankment dams. The slopes of such spillways are typically in the range of 18.4° (3H/1V) to 5.7° (10H/1V). Major failures in the UK of two such spillways, at Boltby Dam in 2005 and Ulley Dam in 2007, led to investigations in their possible causes (Winter et al. 2010). Variations in the hydro-dynamic pressures on the sidewalls were highlighted as a significant vulnerability and the need for continued good maintenance of both the masonry and the mortar was highlighted as one essential aspect for these spillways. This is discussed in section 4.3.7.

4.2.5. Gabion structures and other stepped chutes

Stepped chutes are also found on small gabion dams. This type of dam has some interesting applications, among which are: check dams for solid detention and small weirs for slope correction and erosion prevention. Stepped gabion spillways may be also used on small agricultural reservoirs and dams with small catchment and low unit discharges (up to 3 $m^3/s/m$) (Peyras et al., 1991, 1992; Rice and Kadavy, 1997). A practical further application is the use of porous gabion stepped chutes for noise abatement at municipal sites where low noise is crucial. (Boes and Schmid, 2003).

Another type of stepped chute results from the use of wedge-shaped blocks. Precast blocks are assembled overlapping each other, forming a stepped surface. The blocks benefit from the suction that is produced on the vertical face of the step, to improve its adherence with the foundation, resulting in slimmer pieces. It is a solution which has been used in small dams and agricultural ponds, with unit discharges of up to 4 $m^3/s/m$ (Pravdivets and Bramley, 1989; Hewlett et al., 1997; FEMA, 2014; Morán and Toledo, 2014).

The stepped spillways directly excavated on rock should also be referenced. These features could be found on ancient hydraulic constructions and also in recent works as the spillways along the embankment dam abutments (Baumann *et al.*, 2006; Lutz *et al.*, 2015; Scarella and Pagliara, 2015). The chutes are characterised by larger, irregular steps, up to 2-3 m height or even more, which result from the necessary adaptation to the ground geology and morphology (Chanson, 1994, 2002; Felder and Chanson, 2011). In addition to the study of the hydraulic performance, it is important to check the rock resistance to erosion and scour, as the loss of material may produce changes in geometry and reduce the energy dissipated along the chute, leading to a poor spillway performance.

4.3. HYDRAULICS OF STEPPED SPILLWAYS

The hydraulic characterisation of flow is essential for the spillway design. The hydraulic design of any spillway requires the determination of the flow regime, depth, velocity distribution and residual energy. On stepped spillways it is also important to know where and how the aeration develops and to characterize the pressure field over the steps, in order to prevent cavitation and damage.

4.3.1. Flow regimes

The hydraulics of stepped spillways is driven by several variables with the most important being the critical depth (h_c) (which is a function of the unit discharge (q) and the cross-section), the step height (s) and the chute slope (Φ). The flow could be classified into three different types (Figure 4.4), which basically depend on the ratio between the flow depth and the step height. When this ratio is small (the step size predominates) the flow occurs as *nappe flow*, consisting of a succession of small cascades, with the stream falling from step to step (Essery and Horner, 1978). When the ratio is large (the flow depth is predominant) the flow is called *skimming flow*. This flow is characterised by the formation of a coherent mainstream that skims over the edges of the steps (Essery and Horner, 1978; Sorensen, 1985; Rajaratman, 1990). The intermediate situation, when flow is unstable, partly nappe, partly skimming, is called *transition flow* (Ohtsu and Yasuda, 1997; Chanson and Toombes, 2004).

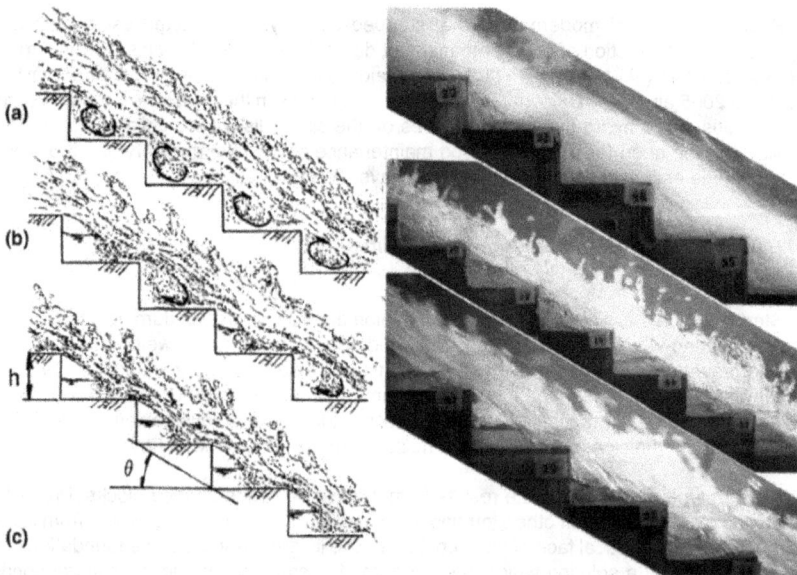

Fig. 4.4
Nappe flow (bottom), transition flow (centre) and skimming flow (top). (Othsu *et al.*, 2001)

Nappe flow develops when the critical depth is small in comparison to the step height for a given chute slope. In ungated spillways this is the flow that develops at the beginning and at the end of the routing process, and in times when the discharge is relatively small. When the flow depth is small in relation to the step size the flow jumps from one step to another. The jet impacts on the horizontal face of the step, causing the formation of a small hydraulic jump, prior to jump onto the next step, so that energy is gradually dissipated on each drop. The hydraulic jump could be fully or partially developed depending on the discharge and the step geometry (Chanson, 1994, 2002). The air pocket bounded by the vertical face of the step and the nappe of the falling jet is a characteristic of this flow type.

Skimming flow develops when the critical depth is larger than about 80% of the step height for a given chute slope (Essery and Horner, 1978). In such conditions, a main compact stream flows over the steps, which constitute a rough bottom. The cavity defined by the stream underside and the step faces, is completely filled with water. The shear stresses between the main flow and the water that fills the step lead to the formation of a recirculating vortex. A continuous momentum transfer feeds the vortex circular movement; this transfer is the main cause of the energy dissipation on stepped spillways.

Ohtsu *et al.* (2004) studied a wide range of slopes, from 5.7º (10H/1V) to 55º (0.7H/1V). They observed that skimming flow may be sub-classified into two categories. For the slopes between 19º (slope 2.9H/1V) and 55º (slope 0,7H/1V) the water surface is parallel to the pseudo-bottom defined by the step edges. This type of flow is the one which develops in most stepped spillways, those located on gravity dams and on many embankment dams. For the mildest slopes, from 5.7º (10H/1V) to 19º (2.9H/1V), the flow surface is not completely parallel to the pseudo-bottom, the horizontal face is so long that the influence of gravity on the flow causes a flow-deflection and an impact before the edge. This effect is reproduced by the water surface which has a wavy pattern.

A detailed description of the different types and subtypes of flow can be found in Othsu *et al.* (2004) and in González and Chanson (2007).

4.3.2. The use of skimming flow for design

As mentioned before, the flow on stepped chutes occurs either as nappe or skimming flow, with a transition regime between them. The determination of the limits and the conditions of each flow has been a traditional research topic. This characterisation is an important issue, since it is advisable to ensure that the design flow is distinctly nappe or skimming, but not transition flow. The transition regime must be avoided for design purposes, given that it is characterized by significant hydrodynamic instabilities, which may lead to high spray and poor spillway performance (Chanson and Toombes, 2004).

Both, the nappe and the skimming flow were studied in depth, but more analysis has been focussed on the skimming flow as it is usually selected for the hydraulic design. Under this condition – the design flood is discharged in the skimming regime– the extreme flood would also be spilled in the skimming regime. For smaller flows, which occur for lower recurrence floods and during the beginning and the end of the routing process, the spillage would be inevitably in nappe flow and possibly in transition flow. As a consequence of that, despite the adjustment of the design parameters for large floods, it is necessary to check whether a good hydraulic performance for smaller discharges takes place as well.

Typically, the onset of the transition flow should occur near a 100-year recurrence interval. This is reasonable since for the more frequent floods the nappe flow regime with high dissipation energy will occur.

The development of either one type of flow or the other depends basically on the ratio between the step height and the critical depth, and also on the chute slope. The onset of the skimming flow may be estimated by:

(Boes and Hager, 2003b) For slopes between 26º<Φ<55º.

$$\frac{h_c}{s} = 0.91 - 0.14 \tan\Phi$$

(Ohtsu et al., 2004) For slopes between 5.7°<Φ<55°.

$$\frac{h_c}{s} = 0.857 \left(\tan \Phi\right)^{-0.1667}$$

(André, 2004) For slopes between 18.6°<Φ<30°.

$$\frac{h_c}{s} = 0.939 \left(\tan \Phi\right)^{-0.364}$$

(Amador et al., 2006) For gravity dams (Φ~51°).

$$\frac{h_c}{s} = 0.854 \left(\tan \Phi\right)^{-0.169}$$

Table 4.2 shows the unit discharge threshold for the formation of skimming flow for typical slopes and step heights, using the referred methods.

Table 4.2
Unit discharge (in m³/s/m) for the onset of skimming flow, under different typical stepped spillway configurations.

Method and scope[*]		Gravity dams			Embankment dams			
		(0,8H/1V) Φ=51.3°			(1,5H/1V) Φ=33.7°	(2H/1V) Φ=26.6°	(3H/1V) Φ=18.4°	(4H/1V) Φ=14.0°
		s=0.9 m	s=1.2 m	s=1.5 m	s=0.3 m	s=0.3 m	s=0.3 m	s=0.3 m
Boes and Hager (2003b)	26°<φ<55°	1.7	2.6	3.6	0.4	0.4	---	---
Ohtsu et al. (2004)	5.7°<φ<55°	2.0	3.1	4.3	0.4	0.5	0.5	0.6
André (2004)	18.6°<φ<30°	---	---	---	---	0.7	0.9	---
Amador (2006)	φ~51°	2.0	3.1	4.3	---	---	---	---

* As defined by authors.

Among the three variables which influence the flow type, two of them (i.e. step height and chute slope) are usually difficult to modify since they are conditioned by other factors such as the construction method, dam type or ground morphology. Therefore, to adjust the flow type, it would be necessary to modify the critical depth, which in turn may be adjusted by changing the crest length.

4.3.3. Location and characteristics of the inception point

Two regions differing in aeration are distinguished for skimming flow (Fig. 4.5). In the upstream part of the spillway, nearby the crest, the flow is non-aerated, this region is also known as black-water or clear-water region. Downstream, typically in the medium and lower parts of the chute, the flow is aerated, this region is also named white-water region. In the non-aerated region, the flow features a compact, transparent look, without bubbles. The approach flow to the spillway crest is laminar. The turbulent boundary layer starts growing from the crest. When the boundary layer reaches the surface, the air-water friction is large enough to cause surface irregularities, creating air pockets that are incepted and rapidly distributed within the flow (Amador et al. 2009). Downstream of this inception point the flow is aerated.

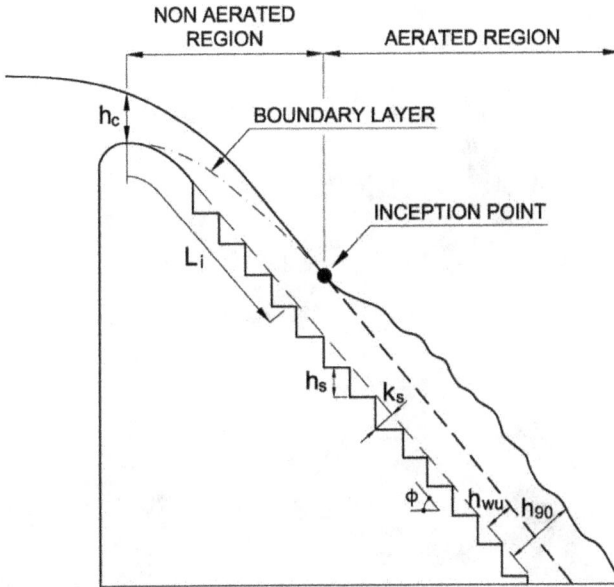

Fig.4.5
Skimming flow development. Non-aerated and aerated regions.
Design variables.

In turn, three different regions may be differentiated within the aerated zone regarding aeration and flow development (Ohtsu et al., 2001; Amador et al. 2006) (Fig. 4.6). In the first, located immediately downstream of the inception point, the air is rapidly distributed by the turbulence effect and the flow is rapidly varied. This region lasts until the aeration process stabilizes. The flow in the second aerated region is gradually varied until quasi-uniform flow is attained, which constitutes the third aerated flow region.

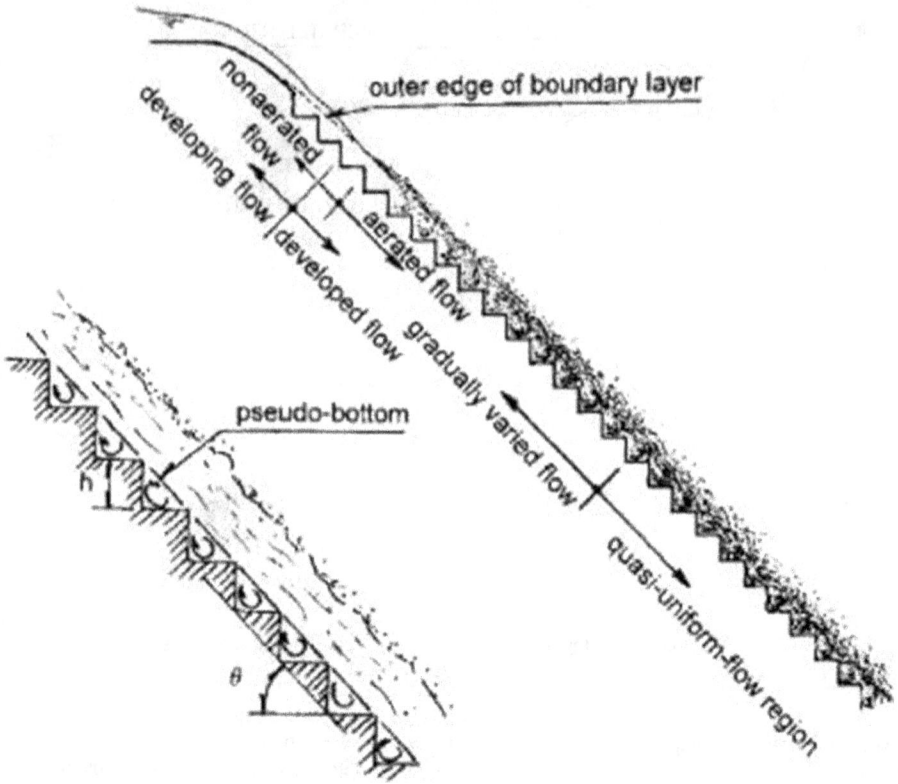

Fig. 4.6
Sub-regions of the aerated zone. (Ohtsu et al., 2001)

As explained above, the inception point defines where the air-entrainment process begins. Upstream of this point the flow is un-aerated. The location of the inception point is important to assess the cavitation risk. The area which is more prone to develop cavitation is around the inception point, as it is the non-aerated zone with the highest flow velocities and therefore the highest sub-atmospheric pressures. Downstream of this area the risk of cavitation is not significant, since the flow is highly aerated.

Some equations which may be used for determining the location of the inception point are shown below.

(Boes and Hager, 2003a) for slopes between $26° < \Phi < 55°$:

$$L_i = \frac{5.90 \, h_c^{6/5}}{(\sin \Phi)^{7/5} \, s^{1/5}}$$

where L_i is the distance from the crest to the inception point.

The analysis of this formula illustrates the influence of the different factors on the non-aerated region. To be precise, larger critical depths (say larger unit flows) lead to larger distances of the non-aerated flow, milder slopes produce larger distances (for instance, embankment dams have larger un-aerated areas than gravity dams), and smaller steps results also in larger distances. The location of the inception point is more sensitive to the critical depth and to the chute slope than to the step height.

(Amador et al., 2009) for gravity dams (Φ~51º):

$$\frac{L_i}{K_s} = 5.982 \, F_*^{0.84}$$

where K_s is the bottom roughness, that is computed as the step height in the stream direction; and F_* is the roughness Froude number:

$$K_s = s \cos \Phi$$

$$F_* = \frac{q}{\sqrt{g \sin \Phi \, K_s^3}}$$

(Meireles et al., 2012) for gravity dams (Φ~53º):

$$\frac{L_i}{K_s} = 6.75 \, F_*^{0.76}$$

(André, 2004) for spillways over embankment dams (18.6º<Φ<30º):

$$\frac{L_i}{K_s} = \frac{8}{\tan \Phi} \, F_{*\Phi}^{0.73}$$

where $F_{*\phi}$ is the roughness Froude number corrected for accounting with the slope effect:

$$F_{*\Phi} = \frac{q}{\sqrt{g \cos \Phi \, K_s^3}}$$

(Meireles and Matos, 2009) for spillways over embankment dams (16º<Φ<26.6º):

$$\frac{L_i}{K_s} = 5.25 \, F_*^{0.95}$$

Hunt and Kadavy (2013) have analysed the influence of the roughness Froude number, noting that for large discharges (F_*>28) the chute has a different performance; in which the effect of the steps is less as the distance increases to the location of the inception point from the crest. These authors proposed the following equations for broad crested stepped spillways over embankment dams (Φ≤26.6º):

$$\frac{L_i}{K_s} = 5.19 \, F_*^{0.89} \qquad 0.1 < F_* < 28$$

$$\frac{L_i}{K_s} = 7.48 \, F_*^{0.78} \qquad 28 < F_* < 10^5$$

Table 4.3 includes L_i values for typical dam configurations (slopes and step heights) and unit discharges, using the referred methods.

The flow depth at the inception point (h_i) has also been analysed in different studies, with some of the formulas proposed are:

(Boes and Hager, 2003a) for slopes between 26º<Φ<55º:

$$\frac{h_i}{s} = 0.40 \, F_{*1}^{0.6}$$

Note that these authors defined the roughness Froude number (F_{*1}) using the step height (s) instead of the bottom roughness (K_s).

$$F_{*1} = \frac{q}{\sqrt{g \sin \Phi s^3}}$$

Table 4.3
Location of the inception point (in m) from the spillway crest, under different typical stepped spillway configurations and for different unit discharges.

Method and scope*		Gravity dams			Embankment dams					
		(0.8H/1V) Φ=51.3° s=1.2 m			(2H/1V) Φ=26.6° s=0.3 m			(3H/1V) Φ=18.4° s=0.3 m		
		q=5 m³/s/m	q=10 m³/s/m	q=20 m³/s/m	q=1 m³/s/m	q=3 m³/s/m	q=5 m³/s/m	q=1 m³/s/m	q=3 m³/s/m	q=5 m³/s/m
Boes and Hager (2003b)	26°<ϕ<55°	11.6	20.3	35.5	9.3	22.3	33.4	---	---	---
Amador et al. (2009)	ϕ~51°	10.6	18.9	33.9	---	---	---	---	---	---
Meireles et al. (2012)	ϕ~53°	11.0	18.6	31.5	---	---	---	---	---	---
André (2004)	18.6°<ϕ<30°	---	---	---	8.3	18.4	26.8	12.1	26.9	39.0
Meireles and Matos (2009) (1)	16°<ϕ<26.6° 1.9<F*<10	---	---	---	4.6	13.1	---	5.3	15.0	---

* As defined by authors.
(1) This formula has been developed with a broad crested weir model and the location of the inception point (L) is referred to the downstream edge of the broad crested weir.

(Amador et al., 2009) for gravity dams (Φ~51°):

$$\frac{h_i}{K_s} = 0.385 \, F_*^{0.58}$$

(Meireles and Matos, 2009) for spillways over embankment dams (16°<Φ<26.6°):

$$\frac{h_i}{K_s} = 0.28 \, F_*^{0.68}$$

After Boes and Hager (2003a) the averaged air concentration at the inception point $\left(\overline{C_i}\right)$ could be determined as:

$$\overline{C_i} = 1.2 \cdot 10^{-3} \left(240° - \Phi\right)$$

4.3.4. Attainment and characteristics of uniform flow

As aforementioned, the aerated region may be divided into three areas depending on the aeration and the flow development (Fig. 4.6). The first, located immediately downstream of the inception point, is where the aeration process develops, and where the flow depth increases significantly due to the highly turbulent aeration process. Downstream of this region the stream is a two-phase flow emulsion of air and water (Boes and Hager, 2003a) where the flow gradually varies towards the becoming uniform flow. The mixture flow depth increases progressively until reaching a constant value under uniform flow conditions. Uniform flow may not be reached in some cases as the chute may not be long enough. Such condition could be written in terms of the chute height (H_{chute}) as:

(Boes and Hager, 2003b) required relative chute height (H_{chute}/h_c) to reach uniform flow:

$$\frac{H_{chute}}{h_c} \sim 24 \, (\sin \Phi)^{2/3}$$

In accordance with this equation, the steeper the slope is, the higher the chute should be to attain uniform flow. For instance, the relative chute height to reach uniform flow on a typical gravity dam (slope 0.8H/1V; Φ= 51.34°) is H_{chute}/h_c = 20.35; while for an embankment dam (slope 3H/1V; Φ= 18.43°) it is 11.14. Hence, spillways over gravity dams require larger chute heights to attain uniform flow than spillways overlying embankment dams, for a given unit discharge.

The equivalent clear water depth of uniform flow (h_{wu}) (i.e. equivalent to the flow depth on a smooth conventional chute) may be determined as follows:

(Boes and Hager, 2003b)

$$\frac{h_{wu}}{h_c} = 0.215 \, (\sin \Phi)^{-1/3}$$

Skimming flow on stepped spillways is highly aerated, resulting in larger depths than those which correspond to a smooth chute. The high turbulence in skimming flow makes it difficult to exactly determine the air-water mixture flow depth. To cope with this limitation the depth where the local air concentration is 90% is considered as the characteristic mixture flow depth (h_{90}) (Chamani and Rajaratnam, 1999). This variable, that considers flow bulking, is often used for the sidewall design, including a safety freeboard to avoid lateral spilling. This is discussed further in section 4.4.3. The characteristic mixture flow depth can be determined with the following methods:

(Chanson, 1994, 2002) Characteristic mixture depth in uniform flow (h_{90u})

$$h_{90u} = h_c \sqrt[3]{\frac{f_e}{8 \, (1-C_e)^3 \, \sin \Phi}}$$

where f_e is the equivalent Darcy friction factor for the air-water mixture and C_e is the equilibrium air concentration for uniform flow:

$$\frac{f_e}{f} = 0.5 \left[1 + \tanh \left(2.5 \frac{0.5 - C_e}{C_e (1 - C_e)} \right) \right]$$

$$C_e = 0.9 \sin \Phi \quad (\text{for } \Phi < 50°)$$

where f is the Darcy friction factor for non-aerated flows, which could be computed as:

$$\frac{1}{f} = 1.42 \, \ln \left(\frac{D_h}{K_s} \right) - 1.25$$

with D_h being the hydraulic diameter $D_h = 4A/P$; where A is the cross-sectional area of flow and P the wetted perimeter.

(Boes and Hager, 2003b) characteristic mixture depth in uniform flow (h_{90u}):

$$\frac{h_{90u}}{s} = 0.5\,F_{*1}^{(0.1\tan\Phi+0.5)}$$

An approximation of the differential equation of the backwater curve by Boes and Minor (2000) could be used for computing the characteristic mixture flow depth (h_{90}) in the gradually varied flow region:

$$h_{90}(x) = 0.55\left(\frac{q^2 s}{g\sin\Phi}\right)^{1/4}\cdot\tanh\left[\frac{\sqrt{gs\sin\Phi}}{3q}(x-L_i)\right]+0.42\left[\frac{q^{10}s^3}{(g\sin\Phi)^5}\right]^{1/18}$$

where x is the distance along the spillway from the crest.

A review into the hydraulics of stepped chutes by the UK Environment Agency (Winter et al., 2010) indicated that the Chanson approach is likely to give the most accurate results for the relatively shallow sloping chutes associated with earth embankment dams, while the Boes and Hager, as well as the Boes and Minor, equations are applicable to the more steep sloping chutes associated with RCC dams and for which they were originally developed.

4.3.5. Energy dissipation

<u>Friction factor</u>

The large energy dissipation rate provided by stepped spillways is one of the main advantages of stepped spillways. A significant part of the initial head is dissipated as a consequence of the momentum transfer to the vortices that form between the steps and the pseudo-bottom. This is fundamental for the design of the dissipation structures; thus several research studies and analyses have been carried out regarding this topic.

Energy dissipation on a chute depends on the friction factor of bottom roughness (f_b). In a stepped chute this is related to the bottom roughness (K_s) and the spacing between the step edges. The factor of friction could be determined analytically using the Darcy-Weisbach formula by making a shape correction to a rectangular open channel and deducting the lateral effect of the smooth sidewalls. Apart from this approach, it should be noted that, adjusting a friction factor formula for stepped spillways is complex due to the highly turbulence in which the two-phase flow develops. Boes and Hager (2003b) proposed the following equation to determine the friction factor bottom roughness for slopes between $19° < \Phi < 55°$.

$$f_b = \left(0.5-0.42\sin(2\Phi)\right)\left(\frac{K_s}{D_h}\right)^{0.2}$$

When $0.1 < K_s/D_h < 1.0$, this equation may be rewritten as:

$$\frac{1}{\sqrt{f_b}} = \frac{1}{\sqrt{0.5-0.42\sin(2\Phi)}}\left[1.0-0.25\log\left(\frac{K_s}{D_h}\right)\right]$$

Another important feature that should be taken into account when studying the friction factor is the influence of aeration. The air bubbles have a lubricant effect, limiting the shear stresses between the mainstream and the recirculating vortices; hence reducing the energy dissipation that is achieved along the chute (Chanson, 1994). Accounting for this effect Boes and Hager (2003b) recommended to compute the hydraulic diameter using the equivalent clear water depth (h_w) instead of the characteristic flow depth (h_{90}). This parameter is named D_{hw}. According to these authors the use of the characteristic flow depth may lead to large friction factors, which in turn may produce an overestimation of the energy dissipation and a non-conservative design of the terminal structure.

Residual energy

The energy dissipation depends on the flow type. Mateos and Elviro (2000) found that the energy dissipation rate becomes significant once the flow is close to attaining the uniform flow. According to these authors for relative chute heights $H_{chute}/h_c < 10$, energy dissipation is similar to that of smooth chutes.

Boes and Hager (2003b) proposed two formulas for calculating the residual energy (H_{res}) depending on the type of flow attained:

(Boes and Hager, 2003b) for $H_{chute}/h_c < 15\text{-}20$ (i.e. uniform flow is not attained):

$$\frac{H_{res}}{H_{max}} = \exp\left[-0.045 \left(\frac{K_s}{D_{hw}}\right)^{0.1} (\sin\Phi)^{-0.8} \frac{H_{chute}}{h_c}\right]$$

where H_{max} represents the maximum head –potential energy of water in the reservoir– which could be estimated as $H_{max} = H_{chute} + 1.5h_c$.

(Boes and Hager, 2003b) for $H_{chute}/h_c \geq 15\text{-}20$ (i.e. uniform flow is attained):

$$\frac{H_{res}}{H_{max}} = \frac{E}{\dfrac{H_{chute}}{h_c} + E}$$

with E equal to:

$$E = \left(\frac{f_b}{8\sin\Phi}\right)^{1/3} \cos\Phi + \frac{\alpha}{2}\left(\frac{f_b}{8\sin\Phi}\right)^{-2/3}$$

where α is the kinetic energy corrector coefficient, which could be assumed as $\alpha = 1.1$.

According to André (2004) the residual energy at the base of a moderate sloped chute ($\Phi = 30°$) is given by:

$$\frac{H_{res}}{H_{max}} = \frac{1.5\tan\Phi}{10}\exp\left(25.26\frac{h_c}{N_s \cdot s}\right)$$

where N_s is the number of steps.

Table 4.4 summarizes the energy dissipation potential for typical dam configurations (slopes and step heights), dam heights and unit discharges.

Table 4.4
Normalized residual energy (H_{res}/H_{max}) and energy dissipation rate (%), under different typical stepped spillway configurations and for different unit discharges.

Chute height	Gravity dams $(0.8H/1V)$ $\phi=51.3^0$								
	s=0.9 m			s=1.2 m			s=1.5 m		
	q=5 m³/s/m	q=10 m³/s/m	q=20 m³/s/m	q=5 m³/s/m	q=10 m³/s/m	q=20 m³/s/m	q=5 m³/s/m	q=10 m³/s/m	q=20 m³/s/m
H_{chute}=30 m	0.32 68%	0.52 48%	0.68 32%	0.32 68%	0.51 49%	0.68 32%	0.31 69%	0.51 49%	0.67 33%
H_{chute}=50 m	0.22 78%	0.33 67%	0.52 48%	0.22 78%	0.32 68%	0.51 49%	0.21 79%	0.31 69%	0.50 50%
H_{chute}=100 m	0.13 87%	0.19 81%	0.28 72%	0.12 88%	0.19 81%	0.28 72%	0.12 88%	0.18 82%	0.28 72%
H_{chute}=150 m	0.09 91%	0.14 86%	0.21 79%	0.08 92%	0.13 87%	0.21 79%	0.08 92%	0.13 87%	0.20 80%

Method: Boes and Hager (2003b)

The grey cells indicate that the uniform flow is attained at the chute end.

Chute height	Embankment dams s=0.3 m							
	(3H/1V) $\phi=18.4^0$			(2H/1V) $\phi=26.6^0$			(3H/1V) $\phi=26.6^0$	
	q=1 m³/s/m	q=3 m³/s/m	q=5 m³/s/m	q=1 m³/s/m	q=3 m³/s/m	q=5 m³/s/m	q=0,20 m³/s/m	q=0,30 m³/s/m
H_{chute}=15 m	0.09 91%	0.19 81%	0.35 65%	0.14 86%	0.26 74%	0.45 55%	0.10 90%	0.11 89%
H_{chute}=30 m	0.05 95%	0.10 90%	0.14 86%	0.07 93%	0.15 85%	0.21 79%	0.09 91%	0.09 91%
H_{chute}=50 m	0.03 97%	0.06 94%	0.09 81%	0.04 96%	0.10 90%	0.14 86%	0.08 92%	0.08 92%
Method	Boes and Hager (2003b)			Boes and Hager (2003b)			André (2004)	

In the calculations done with the Boes and Hager equations (2003b), the grey cells indicate that the uniform flow is attained at the chute end.

Energy dissipation could be enhanced introducing end-sills, blocks and protrusions. André *et al.* (2003; 2004) studied the effect of end-sills and blocks on stepped spillways and concluded that the latter can increase energy dissipation by 5 to 8% and also reduce the risk of cavitation at the beginning of the chute. Wright (2010) has investigated the use of triangular protrusions, obtaining similar conclusions. This type of block, which has been used in De Hoop Dam in South Africa, has been shown to be an effective measure to increase the friction factor, and to improve aeration and energy dissipation.

4.3.6. Pressure field on the steps

Another important issue of stepped spillways hydraulics is the description of the hydrodynamic pressure field. It is important to identify the areas with cavitation risk. The works by André *et al.* (2004), Amador *et al.* (2006), Sánchez-Juny *et al.* (2007) and Amador *et al.* (2009) describe in detail the development and fluctuations of the pressure field on the horizontal and vertical faces of the steps, and provide equations for their computation on gravity dams, both on the horizontal and vertical faces. These two regions which have to be distinguished to describe the pressure field are the external zone located near the step edge, which is influenced by the main stream; and the internal part located around the inside corner, which is most influenced by the recirculating vortices.

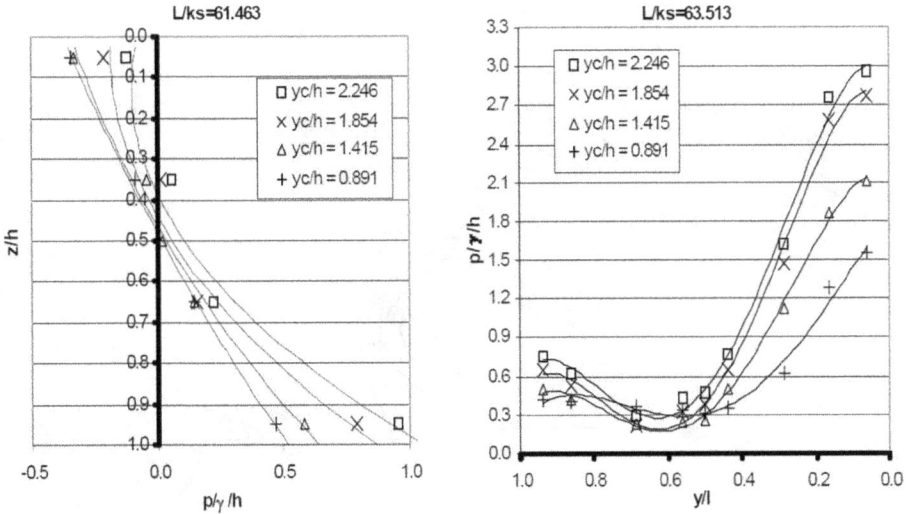

Fig.4.7
Profiles of mean pressure on the vertical and the horizontal faces of the steps. (Sánchez-Juny et al. 2007)

On the horizontal face the maximum pressure is located on the external zone, due to the impact of the main flow, which is influenced by gravitational action, resulting in a pressure decrease towards the interior of the step (Fig. 4.7). On the vertical face the minimum pressure is located on the external zone, due to the drag effect caused by the mainstream. Pressure has a positive gradient to the interior corner, where its value is similar to that of the horizontal face (Fig. 4.7). Fluctuation on both faces is very high, being more pronounced near the step edges and for higher discharges.

Regarding the evolution of the pressure field along the spillway, researchers agree that the most delicate area is that located in the non-aerated region, while the aerated region is not prone to cavitation damage (Zhang et al., 2012). The minimum and maximum pressures were recorded around the inception point; downstream of this zone the presence of air bubbles in the flow produces a cushion effect that reduces the pressure and its fluctuation (Amador et al. 2005). Research regarding the design of aerators at the beginning of stepped spillways is ongoing.

It has to be noted that pressure measurements performed on scaled stepped spillway scale models do not show high frequency pressures as they occur on prototype spillways. Therefore model results should be transferred to prototype spillways with care and only if having similar slopes because of scale effects. Scale effects could also impact other variables such as the air concentration and, in turn, influence the friction factor and the residual energy estimation. Boes (2000) suggested a minimum Froude scale between 1:10 and 1:15 for hydraulic models to minimise these scale effects.

4.3.7. Pressure field on the sidewalls

As mentioned in section 4.2.4, the collapse of two old stepped masonry spillways in the UK, at Boltby Dam in 2005 and at Ulley Dam in 2007, led to a research programme into hydrodynamic effects on stepped chutes (Winter et al., 2010; Winter, 2010). These studies revealed that at high skimming flows the centres of the horizontal vortices at each step can generate high negative pressures which in turn act directly on the associated local zones of the sidewall. Moreover, this will occur adjacent to other wall zones subject to high positive pressures due to step impact (Fig. 4.8). Poor maintenance and cracks in the mortar can, therefore, lead to the high back-pressurisation of masonry in zones subject to external negative pressures. This results in blocks being "plucked" from the wall, a local turbulence increase and, in extreme cases, complete wall collapse.

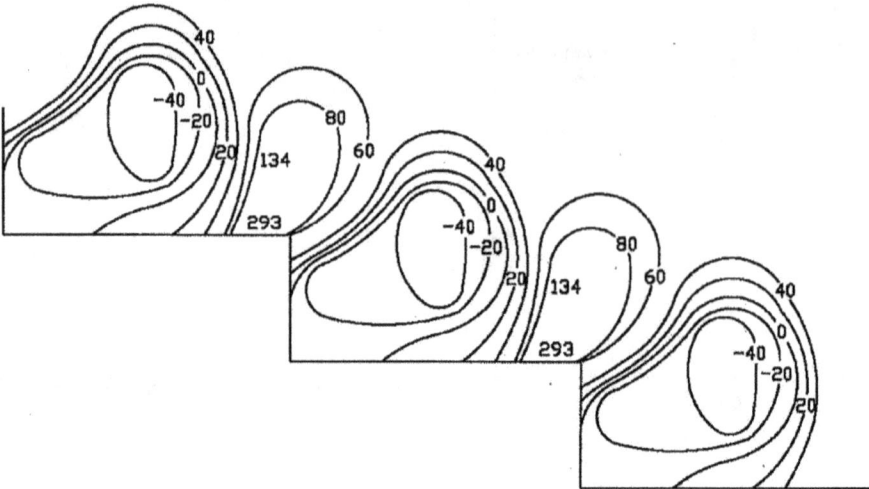

Fig.4.8
Hydrodynamic pressure variation on the sidewall of a typical stepped chute under skimming flow conditions (Winter et al., 2010)

4.4. DESIGN FEATURES

4.4.1. Step height

There is an optimal relation between the step height and the energy dissipation that is around $s/h_c \geq 0.3$ (Tozzi, 1992) and $s/h_c \geq 0.25$ (Othsu et al. 2004). However, it is important to remark that the step height is mainly determined by the existing construction techniques. On RCC stepped spillways the step height is a multiple of the concrete layer thickness. The standard thickness is 0.3 m. Initially smaller thicknesses were used on the pioneering RCC dams, but nowadays 0.3 m is predominant. For example: 23 out of the 27 existing RCC dams in Spain were constructed using 0.3 m thick layers (de Cea et al., 2012). In gravity dams it is usual to lift the formwork every four layers, resulting in steps heights of 1.2 m, whereas, in embankment overtopping RCC protection the most usual step height is 0.3 m. This is due to the fact that smaller steps adapt better to milder slopes (otherwise the step length would be too large, and concrete is wasted) and because the steps are not always formed. As the step heights are largely influenced by the construction technology, layer thickness and step height that are common nowadays may slightly change in future depending on further RCC development.

New developments where the stepped slope does not come as a direct output of the construction method, such as in conventional gravity dams, like the Boguchany Dam in Russia (Bellendir et al., 2012), or in side spillways of embankment dams like the Siah Bishe Lower and Upper Dams in Iran (Baumann et al., 2006), step heights are not determined by the construction technique and may be more flexible to be adjusted for maximising energy dissipation.

Some authors indicate that the step height may be adjusted during the design phase to ensure that design flow is discharged distinctively as skimming flow. It should be emphasised that the step height is not the most adequate parameter to do so, since it is highly dependent on the construction method. As discussed in section 4.3.2, the most flexible parameter is the critical depth, which could be modified by means of changing the crest length (if possible).

4.4.2. Crest shape and transition to chute

Most gravity dam spillways have an ogee crest (as WES or Creager). This crest shape fits very well with the typical triangular cross section and slopes of gravity dams. The transition between the crest and the spillway chute should be designed in such a way to prevent flow deflection and impact on the initial steps, which later results in spray and bad performance for small discharges, and cavitation risk for the larger flow. The work carried out by CEDEX (Madrid) largely focussed on that design issue. Mateos and Elviro (1995; 2000) proposed a stepped transition which follows the crest profile, by means of steps of increasing height and length (Fig. 4.9).

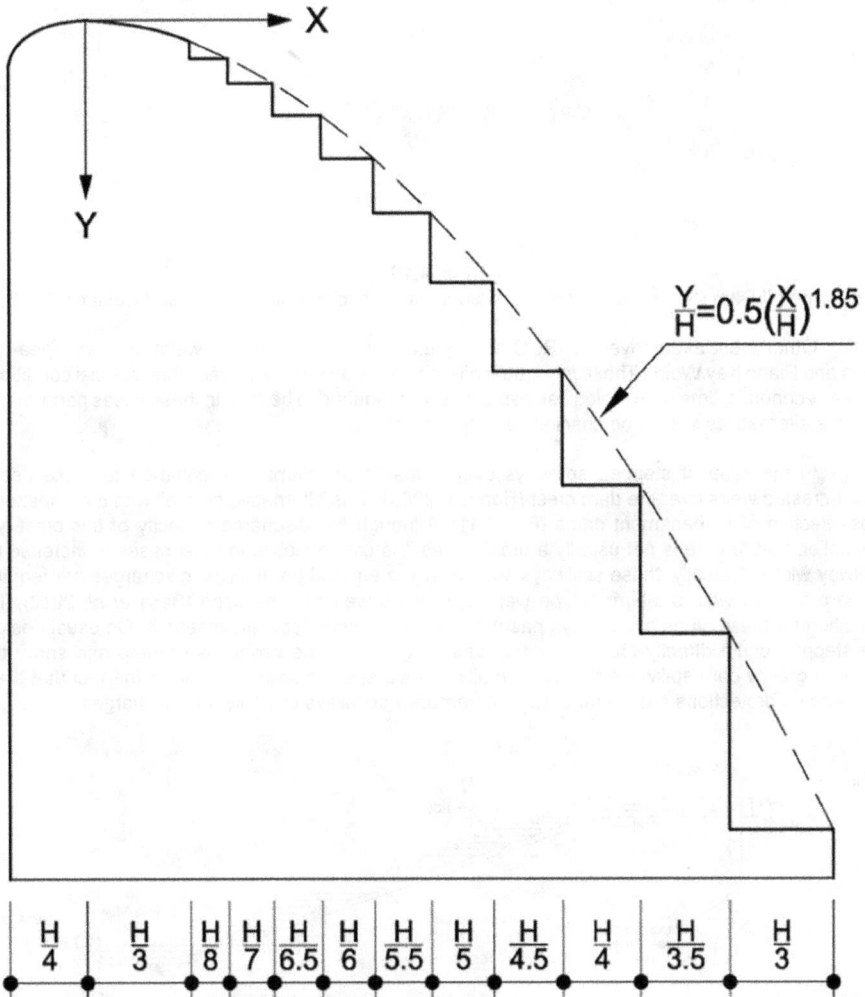

In the figure: $\frac{Y}{H} = 0.5\left(\frac{X}{H}\right)^{1.85}$

Step dimensions along the base: $\frac{H}{4}$, $\frac{H}{3}$, $\frac{H}{8}$, $\frac{H}{7}$, $\frac{H}{6.5}$, $\frac{H}{6}$, $\frac{H}{5.5}$, $\frac{H}{5}$, $\frac{H}{4.5}$, $\frac{H}{4}$, $\frac{H}{3.5}$, $\frac{H}{3}$

Fig.4.9
CEDEX's proposal of crest-to-slope transition, with H being the design head on spillway crest
(Mateos and Elviro, 1995)

Another alternative that has been proposed to reduce impact and spray is modifying the impact angle (Pfister *et al.*, 2006a). When impact occurs, the flow is deflected with an angle that is approximately symmetrical to the impacting flow. Thus, chamfering the edge of the first steps reduces spray. This action may also be reproduced by means of an insert or rounding the steps edges, as has been done on the first three steps of La Breña II Dam (Fig. 4.10) (Elviro and Balairón, 2008).

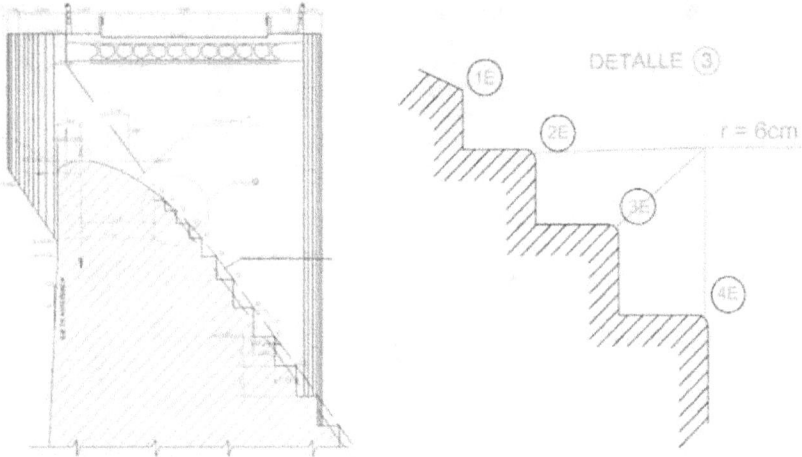

Fig. 4.10
La Breña II Dam (Spain). Use of rounded edges for reducing spray. (Elviro and Balairón, 2008)

Other crest alternatives for RCC gravity dams are broad-crested weirs, precast ogee-type weirs and Piano Key Weirs. Those infrequent alternatives may be considered when special conditions, such as economic, time or hydrological restrictions, are required to be met. In these cases performance of those alternatives should be checked with specific studies.

In the case of stepped spillways over embankment dams it is common to make use of broad-crested weirs over the dam crest (Hansen, 2003). This alternative fits well with the trapezoidal cross section of embankment dams (Fig. 4.11). Although the discharge capacity of this crest-type is smaller, that feature is not usually a problem, as it is cost-effective in most cases to increase the spillway width. Typically, these spillways span over the entire dam. If larger discharges are required a sharp crested weir, a labyrinth-type weir or an ogee weir may be used (Bass *et al.* 2012). The transition for these types of spillways has not been an object of specific research. On usual designs the stepped chute directly starts after the crest (Fig. 4.11). The problems of jump and spray that arise on gravity dam spillways do not normally receive specific attention, due to the fact that these overtopping protections are in many cases emergency spillways with low unit discharges.

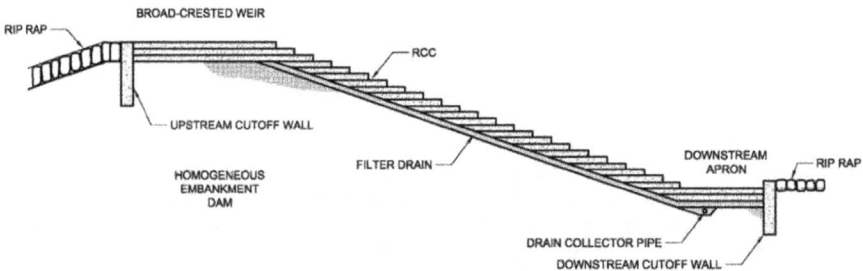

Fig. 4.11
Typical cross-section of an embankment dam with RCC overtopping spillway. (Adapted from Bass *et al.* (2012))

Gated stepped spillways are not common since they usually imply larger unit discharges, thus many gated spillways on RCC dams have smooth chutes. Specific design features such as enlarging the smooth part to fully cover the non-aerated flow region (Amador et al., 2006) or providing aeration (Guo et al. 2003) should be considered on gated spillways.

Dachaosan Dam (China) spillway is a gated stepped spillway which has been designed for a very large unit discharge (165 m³/s/m). The upstream part of this spillway is smooth and the air entrainment is ensured by means of flaring piers in conjunction with a small flip bucket. The flaring piers concentrate the flow, turning it into a vertical position, achieving a high lateral aeration due to friction with air. Dachaosan Dam has discharged up to 6.173 m³/s, a unit discharge of 93 m³/s/m, showing good performance and no significant damages reported (Shen, 2003). This type of energy dissipater is an interesting alternative for large projects that has been successfully used in other high head/velocity spillways in China.

4.4.3. Cavitation risk and aeration measures

The zone prone to cavitation on stepped spillways is that located on the non-aerated flow region. The most delicate area is the surroundings of the inception point, since the aeration process is not fully developed, and velocity is larger than upstream. Boes and Hager (2003a) establish an upper limit for the velocity at the inception point of 20 m/s, based on the air-concentration requirements to avoid cavitation near the inception point. Amador et al. (2005) determined a lower velocity threshold upon the inception point of 15 m/s, based on the analysis of the pressure field. Frizell et al. (2013) indicate that no cavitation problems or severe damage were reported on stepped spillways to date; hence those limits may be increased with the support of future research (use of low-ambient pressure chambers to induce cavitation) and also with more data analysed from prototypes. It is however advisable to include measures to prevent cavitation when the referred limits are to be surpassed and also for unit discharges over 30 m³/s/m (Boes, 2012).

The best measure to counter cavitation is providing aeration. Research that has been carried out at VAW of ETH Zurich focussed on the effect of step aerators and deflectors (Pfister et al., 2006b; Schiess-Zamora et al., 2008). The step aerator is an air-supply conduit that exits at the first vertical face (first step after the smooth upstream part) (Fig. 4.12). The negative pressures on the upper part of the vertical face contribute to drag the air, which is then mixed within the flow. The use of a step aerator enables to increase unit discharges up to 40 m³/s/m (for usual gravity dam and step heights of s=1.2 m). Larger unit discharges (q>40 m³/s/m) require the introduction of a second aerator, that should be located upstream of the inception point.

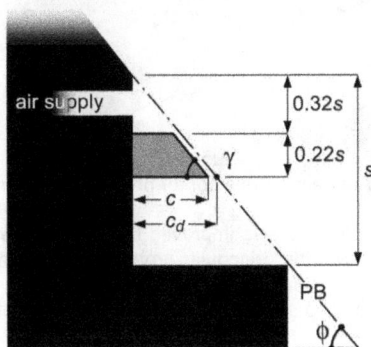

Fig. 4.12
Step aerator for providing aeration. (Schiess-Zamora et al. 2008)

A more effective and brisk measure for providing aeration is the use of deflectors. The deflector is located at the beginning the stepped chute, and it produces a jump over the initial steps creating a large air-cavity under the jet (Pfister et al., 2006b). The air is entrained into the flow when the jet impacts the chute downstream. The deflector has to be complemented with air-supply conduits (Fig. 4.13). The design of these devices should take into account that high spray would be produced around the impact point. A higher step designed with the same aim of creating an air-cavity under the flow, has been also employed to enhance aeration as an alternative to deflectors.

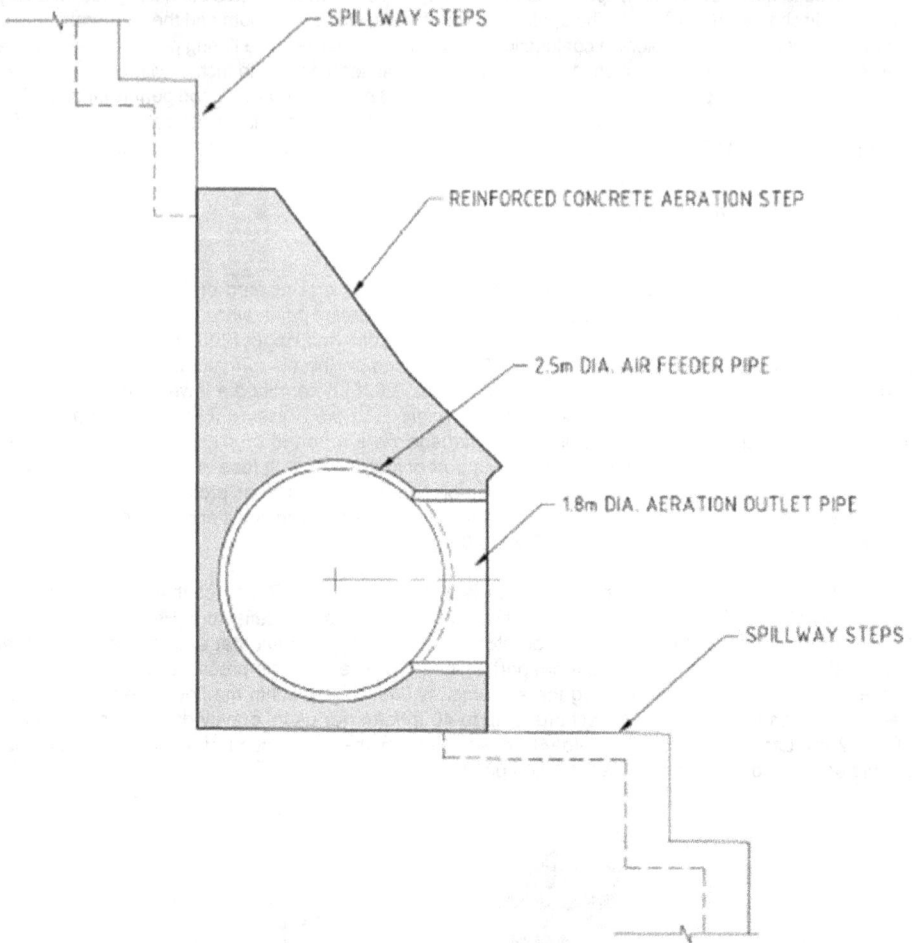

Fig. 4.13
Enlarged Cotter Dam (Australia). Deflector for providing aeration. (Willey *et al.*, 2010)

Furthermore, aeration could be also improved by using splitters and protrusions. The work carried out by Wright (2010) shows that triangular protrusions located on the upstream part of the spillway help to shorten the length of inception and enhances aeration. Thus, for an upper velocity limit of 20 m/s at the inception point a spillway fitted with triangular protrusions will be effective up to 40 m³/s/m.

4.4.4. Sidewall design and behaviour without sidewalls

The sidewall design should take into account the bulking effect of the aerated flow. The flow depth of the two-phase fluid is larger than that of the equivalent discharge over a smooth chute. Hence, in assessing the required wall height a safety factor over and above the calculated height is usually applied depending on the nature of the surrounding topography. Special care is needed if the spillway overflow could affect an adjacent earth embankment dam.

For stepped spillways with parallel sidewalls, the required sidewall height perpendicular to the pseudo-bottom (h_t) could be calculated applying a freeboard coefficient (η) to the characteristic flow depth:

$$h_t = \eta \cdot h_{90}$$

Boes and Hager (2003b) propose the use of η =1.2 for gravity dams and η=1.5 for embankment dams.

The calculation of the sidewall height in converging spillways requires a specific approach (Hunt et al., 2012), since there is a flow concentration near the wall. Those types of converging spillways can be found on narrow valleys and, most commonly, on embankment dams where the spillway overflow protection may also spill over the dam abutments constituting a converging stepped channel.

The performance without sidewalls is a novel approach. Is characterized by the lateral expansion of flow, which results in a lower unit discharges at the dam toe. It could be a cost-effective alternative in cases with low unit discharge and good foundation. The studies that were carried out at UPC (Barcelona) show that unit discharge reduction with such arrangement is between 50% and 70% (Estrella et al., 2012).

4.4.5. *Terminal structure*

One of the main advantages of the stepped spillways is the energy dissipation that could be achieved along the chute. This renders the design of smaller terminal structures compared to that of a similar smooth chute possible, which results in cost and construction time savings.

In stepped spillways on RCC gravity dams the use of a stilling basin as a terminal structure is common. The sizing of the stilling basin is based on the determination of the conjugate depth. For smooth spillways the velocity and the depth at the chute toe are used for calculating the conjugate depth. For stepped spillways the equivalent clear water depth and equivalent terminal velocity should be determined at the chute toe upon the base of the residual energy (Boes and Hager, 2003b).

Precaution has to be taken when using standard stilling basins, as these were developed for smooth chutes. Particular care has to be taken for those which used baffle blocks and sills to reduce the pool length. Work by Frizell et al. (2009) regarding the performance of USBR-Type III stilling basins (Peterka, 1978) shows that the change in the velocity distribution in a stepped chute reduce the effect of the baffles, with the result being that similar or even larger pool length would be required.

In the case of overflowing spillways for embankment dams constructed with RCC, both the unit discharges (average of 3 m^3/s/m) and the dam heights (average of 15 m) are smaller than those of the RCC gravity dams, hence the required terminal structures are simpler. In many cases a downstream apron is sufficient to prevent erosion, and stilling basins are only used for the larger protections.

4.5.   NOTATION

The following symbols are used in this chapter:

A   Cross-sectional area of flow

C_e Equilibrium air concentration for uniform flow

$\overline{C_i}$ Averaged air concentration at the inception point

D_h	Hydraulic diameter ($D_H = 4A / P$)
D_{hw}	Hydraulic diameter computed using the equivalent clear water depth
E	Parameter used to determine the residual energy when uniform flow is attained
F_*	Roughness Froude number ($F_* = q / (g \cdot \sin \Phi \cdot K_s^3)^{1/2}$)
F_{*1}	Roughness Froude number ($F_{*\Phi} = q / (g \cdot \cos \Phi \cdot s^3)^{1/2}$)
$F_{*\Phi}$	Roughness Froude number corrected for the slope effect ($F_{*\Phi} = q / (g \cdot \cos \Phi \cdot K_s^3)^{1/2}$)
f	Darcy friction factor for non-aerated flows
f_b	Friction factor of bottom roughness
f_e	Equivalent Darcy friction factor for the air-water mixture
g	Gravitational acceleration
H_{chute}	Chute height
H_{max}	Maximum head –potential energy of water at the reservoir–
H_{res}	Residual energy
h_c	Critical depth
h_i	Flow depth at the inception point
h_t	Sidewall height perpendicular to pseudo-bottom
h_w	Equivalent clear-water depth
h_{wu}	Equivalent clear-water depth on uniform flow
h_{90}	Characteristic mixture flow depth
h_{90u}	Characteristic mixture flow depth on uniform flow
K_s	Bottom roughness ($K_s = s \cdot \cos \Phi$)
L_i	Streamwise length from spillway crest to inception point
N_s	Number of steps
P	Wetted perimeter
q	Unit discharge
s	Step height
x	Streamwise distance along the spillway from the crest
α	Kinetic energy corrector coefficient
η	Freeboard coefficient
Φ	Chute slope

4.6. REFERENCES

Amador A., Sánchez-Juny M. and Dolz J. (2005). "Discussion of 'Two phase flow characteristics of stepped spillways' by R.M. Boes and W.H. Hager" *Journal of Hydraulic Engineering*, 131(5):421-423.

Amador A., Sánchez-Juny M. and Dolz J. (2006). "Diseño hidráulico de aliviaderos escalonados en presas de HCR" *Ingeniería del Agua*, 13(4):289-302. (in Spanish)

Amador A., Sánchez-Juny M. and Dolz J. (2009). "Developing flow region and pressure fluctuations on steeply sloping stepped spillways" *Journal of Hydraulic Engineering*, 135(12):1092-1100.

André S. (2004). "High velocity aerated flows on stepped chutes with macro-roughness elements" *Communications du Laboratoire de Constructions Hydrauliques No. 20*, Ed. Schleiss A., EPFL, Lausanne (Switzerland).

André S., Manso P., Schleiss A.J. and Boillat J.L. (2003). "Hydraulic and stability criteria for the rehabilitation of appurtenant spillway structures by alternative macro-roughness concrete linnings" *Proc. 21st ICOLD Congress*, Montreal (Canada), Q.82, R.6.

André S., Matos J., Boillat J.L. and Schleiss A.J. (2004). "Energy dissipation and hydrodynamic forces of aerated flow over macro-roughness linings for overtopped embankment dams" *Proc. Intl. Conference on Hydraulics of dams & River structures*, Tehran (Iran), 189-196.

Bass R.P., Fitzgerald T. and Hansen K.D. (2012). "Lesson learned - More than 100 RCC overtopping spillways in the United States" *Proc. 6th Intl. Symposium on Roller Compacted Concrete Dams*, Zaragoza (Spain).

Baumann A., Arefi F. and Scheiss A. (2006). "Design of two stepped spillways for a pumped storage scheme in Iran" *Proc. HYDRO 2006 Conference*, Porto Carras (Greece), CD-ROM.

Bellendir E.N., Volynchikov A.N. and Sudolskiy G.N. (2012). "Boguchany HPP additional spillway: Necessity of construction and peculiar features of design" *Proc. 24th ICOLD Congress*, Kyoto (Japan), Q.94, R.12.

Berga L. (1995). "Hydrologic safety of existing embankment dams and RCC for overtopping protection" *Proc. 2nd Intl. Symposium on Roller Compacted Concrete Dams*, Santander (Spain), 639-652.

Boes R.M. (2000). "Scale effects in modelling two-phase stepped spillway flow", *Proc. Intl. Workshop Hydraulics of stepped spillways*, VAW-ETH Zurich, Eds. Minor, H.E., and Hager, W.H., Balkema, Rotterdam (The Netherlands), 53-60.

Boes R.M. (2012). "Guidelines on the design and hydraulic characteristics of stepped spillways" *Proc. 24th ICOLD Congress*, Kyoto (Japan), Q.94, R.15.

Boes R.M. and Hager W.H. (2003a). "Two-phase flow characteristics of stepped spillways" *Journal of Hydraulic Engineering*, 129(9):661-670.

Boes R.M. and Hager W.H. (2003b). "Hydraulic design of stepped spillways" *Journal of Hydraulic Engineering*, 129(9):671-679.

Boes R.M., Lutz N. and Lais A. (2015). "Upgrading spillway capacity at large, non-overtoppable embankment dams", *Proc. 25th ICOLD Congress*, Stavanger (Norway), Q.97, R.23.

Boes R.M. and Minor H.E. (2000). "Guidelines to the hydraulic design of stepped spillways" *Proc. Intl. Workshop Hydraulics of stepped spillways*, VAW-ETH Zurich, Eds. Minor, H.E., and Hager, W.H., Balkema, Rotterdam (The Netherlands), 163-170.

Boes R.M. and Schimd H. (2003). "Weir rehabilitation using gabions as a noise abatement option" *Proc. HYDRO 2003 Conference*, Cavtat (Croatia), CD-ROM.

Chanson H. (1994). Hydraulic design of stepped cascades, channels, weirs and spillways, Pergamon, Oxford (UK).

Chanson H. (2002), *The hydraulics of stepped chutes and spillways*, Balkema, Lisse (The Netherlands).

Chanson H. and Toombes L. (2004). "Hydraulics of stepped chutes: the transition flow" *Journal of Hydraulic Research*, 42(1):43-54.

Chamani M.R. and Rajaratmam N. (1999). "Characteristics of skimming flow over stepped spillways" *Journal of Hydraulic Engineering*, 125(4):361-368.

de Cea J.C., Ibáñez de Aldecoa R., Polimón J., Berga L. and Yagüe J. (2012). "30 years constructing RCC dams in Spain" *Proc. 6th Intl. Symposium on Roller Compacted Concrete Dams*, Zaragoza (Spain).

Elviro V. and Balairón L. (2008). "Recrecimiento de la presa de La Breña. Estudio en modelo reducido del aliviadero escalonado" *Proc. VIII Jornadas Españolas de presas*, SPANCOLD, Córdoba (Spain). (in Spanish)

Essery I.T.S. and Horner M.W. (1978). "The hydraulic design of stepped spillways" *CIRIA-Report 33*, London (UK).

Estrella S., Sánchez-Juny M., Pomares J., Dolz J., Ibáñez de Aldecoa R., Domínguez M., Rodríguez J. and Balairón L. (2012). "Recent trends in stepped spillways design: behaviour without sidewalls" *Proc. 24th ICOLD Congress*, Kyoto (Japan), Q.94, R.28.

Felder S. and Chanson H. (2011). "Energy dissipation down a stepped spillway with nonuniform step heights" *Journal of Hydraulic Engineering*, 137(11):1543-1548.

FEMA (2014). Technical Manual: Overtopping protection for dams FEMA P-1015, Federal Emergency Management Agency (USA).

Frizell K.W., Kubitschek J.P. and Matos J. (2009). "Stilling basin performance for stepped spillways of mild to steep slopes - Type III basins" *33rd IAHR Congress*, Vancouver (Canada).

Frizell K.W., Renna F.M. and Matos J. (2013). "Cavitation potential of flow on stepped spillways" *Journal of Hydraulic Engineering*, 139(6):630-636.

González C.A. and Chanson H. (2007). "Diseño hidráulico de vertedores escalonados con pendientes moderadas: Metodología basada en un estudio experimental" *Ingeniería Hidráulica en México*, 22(2):5-20. (in Spanish)

Guo J., Liu Z., Liu J. and Lu Y. (2003). "Field observation on the RCC stepped spillways with the flaring pier gate on the Dachaoshan project." *Proc. 30th IAHR Biennial Congress*, Eds. Ganoulis J. and Prinos P., Thessaloniki (Greece), 473-478.

Hansen K.D. (2003). "RCC use in dam rehabilitation projects" *Proc. 4th Intl. Symposium on Roller Compacted Concrete Dams*, Madrid, Eds. Berga *et al.*, Balkema, Rotterdam (The Netherlands), 79-89.

Hewlett H.W.M., Baker R., May R.W.P. and Pravdivets Y. (1997). "Design of stepped-block spillways" *CIRIA-Special publication 142*, London (UK).

Hunt S.L, Temple D.M., Abt S.R., Kadavy K.C. and Hanson G. (2012). "Converging stepped spillways: simplified momentum analysis approach" *Journal of Hydraulic Engineering*, 138(9):796-902.

Hunt S.L and Kadavy K.C. (2013). "Inception point for embankment dams stepped spillways" *Journal of Hydraulic Engineering*, 139(1):60-64.

Lutz N., Lucas J., Lais A. and Boes R.M. (2015). "Stepped chute of Tränsglet Dam: Physical model study" *Journal of Applied Water Engineering and Research*, 3(2):166-176.

Mateos C. and Elviro V. (1995). "Stepped spillways. Design for the transition between the spillway crest and the steps" *Proc. XXVI IAHR Congress. HYDRA 2000*, Thomas Telford, London (UK), 260-265.

Mateos C. and Elviro V. (2000). "Stepped spillway studies at CEDEX" *Proc. Intl. Workshop Hydraulics of stepped spillways*, VAW-ETH Zurich, Eds. Minor, H.E., and Hager, W.H., Balkema, Rotterdam (The Netherlands), 87-94.

Matos J. (2003). "Roller compacted concrete and stepped spillways: from new dams to dam rehabiltation" *Proc. Intl. Congress on Dam Rehabilitation and Maintenance*, Madrid, Eds. Llanos J.A. *et al.*, Balkema, Lisse (The Netherlands), 553-559.

Meireles I. and Matos J. (2009). "Skimming flow in the nonaerated region of stepped spillways over embankment dams" *Journal of Hydraulic Engineering*, 135(8):685-689.

Meireles I., Renna F., Matos J. and Bombardelli F. (2012). "Skimming, nonaerated flow on stepped spillways over roller compacted concrete dams" *Journal of Hydraulic Engineering*, 138(10):870-877.

Morán R. and Toledo M.A. (2014) "Design and construction of the Barriga Dam spillway through an improved wedge-shaped block technology" *Canadian Journal of Civil Engineering*, 41(10):924-927.

Ohtsu I. and Yasuda Y. (1997). "Characteristics of flow conditions on stepped channels" *Proc. 27th IARH Congress*, San Francisco (USA), 583-588.

Ohtsu I., Yasuda Y. and Takahashi M. (2001). "Discussion of 'Onset of the skimming flow on stepped spillways' by M.R. Chamani and N. Rajaratman" *Journal of Hydraulic Engineering*, 127(6):522-524.

Ohtsu I., Yasuda Y. and Takahashi M. (2004). "Flow characteristics of skimming flow in stepped channels" *Journal of Hydraulic Engineering*, 130(9):860-869.

PCA (2002). *Design manual for RCC spillways and overtopping protection*, prepared by URS Greiner Woodward Clyde, Portland Cement Association, Illinois (USA).

Peterka A.J. (1978). "Hydraulic design of stilling basins and energy dissipators" *Engineering monograph No. 25*, Bureau of Reclamation, Colorado (USA).

Peyras L., Royet P. and Dégoutte G. (1991). "Ecoulements et dissipation sur les déservoirs en gradins de gabions" *La Hoille Blanche*, 46(1) :37-47. (in French)

Peyras L., Royet P. and Dégoutte G. (1992). "Flow and energy dissipation over stepped gabion weirs" *Journal of Hydraulic Engineering*, 118(5):707-717.

Pfister M., Hager W.H. and Minor H.E. (2006a). "Stepped chutes: pre-aeration and spray reduction" *Intl. Journal of Multiphase Flow*, 32(2):269-284.

Pfister M., Hager W.H. and Minor H.E. (2006b). "Bottom aeration of stepped spillways" *Journal of Hydraulic Engineering*, 132(8):850-853.

Pravdivets Y.P. and Bramley M.E. (1989). "Stepped protection blocks for dam spillways" *Water Power and Dam Construction*, 41(7):60-66.

Rajaratman N. (1990). "Skimming flow in stepped spillways" *Journal of Hydraulic Engineering*, 116(4):587-591.

Rice C.E. and Kadavy K.C. (1997). "Physical model study of the proposed spillway for Cedar Run Site 6, Fauquier County, Virginia" *Applied Engineering in Agriculture*, 13(6):723-729.

Sánchez-Juny M., Blade E. and Dolz J. (2007). "Pressures on stepped spillways" *Journal of Hydraulic Research*, 45(4):505-511.

Scarella M. and Pagliara S. (2015). "A challenging solution for Zarema May Day Dam spillway design and model tests", *Proc. 25th ICOLD Congress*, Stavanger (Norway), Q.97, R.37.

Schiess-Zamora A., Pfister M., Hager W.H. and Minor H.E. (2008). "Hydraulic performance of step aerator" *Journal of Hydraulic Engineering*, 134(2):127-134.

Shen C. (2003). "RCC dams in China" *Proc. 4th Intl. Symposium on Roller Compacted Concrete Dams*, Madrid, Eds. Berga *et al.*, Balkema, Rotterdam (The Netherlands), 15-25.

Sorensen R.M. (1985). "Stepped spillway hydraulic model investigation" *Journal of Hydraulic Engineering*, 111(12):1461-1472.

Toledo M.A., Morán R. and Oñate E. (Eds.) (2015), *Dam protections against overtopping and accidental leakage*, CRC press, Leiden (The Netherlands).

Tozzi M.J. (1992). *Caracterização/comportamento de escoramentos em vertedouros com paramento em degraus*, PhD thesis, Universidade de São Paulo, São Paulo (Brazil). (in Portuguese)

Willey J., Ewing T., Lesleighter E. and Dymke J. (2010). "Refinement of hydraulic design of a complex stepped spillway arrangement through numerical and physical modelling" *Proc. ASIA 2010 Conference*, Sarawak (Malaysia).

Winter C., Mason P.J., Baker R. and Ferguson A. (2010). *Guidance for the design and maintenance of stepped masonry spillways*, UK Environment Agency Project SC080015, Bristol (UK).

Winter C. (2010). "Research into the hydrodynamic forces and pressures acting within stepped masonry spillways" *Dams and Reservoirs*, 20(1):16-26.

Wright H.J. (2010). "Improved energy dissipation on stepped spillways with the addition of triangular protrusions" *Proc. 78th ICOLD Annual Meeting*. Hanoi (Vietnam).

Zhang. J., Chen J. and Wang Y. (2012). "Experimental study on time-averaged pressures in stepped spillways" *Journal of Hydraulic Research*, 50(1):236-240.

5.   LABYRINTH SPILLWAYS

5.1.   INTRODUCTION

Labyrinth spillways can convey higher discharges at a given pool elevation than a linear overflow structure of comparable width. The spillway layout can be adapted to site-specific requirements to significantly increase spillway capacity, improve flood control, and offer additional reservoir storage. These '3-D' spillways also provide passive-control operation, reliability, energy dissipation, flow aeration, and desirable aesthetics. The hydraulic optimisation of these spillways is commonly moderated by project economics and constructability. They are particularly well suited for applications where it is advantageous to minimize the width of the spillway or overall size of the spillway footprint.

An example of a labyrinth spillway is shown in Figure 5.1. The spillway has a discharge capacity of 15 000 m³/s and features aeration vents and bridge piers.

Fig. 5.1
View of the Labyrinth Spillway of Maguga Dam in Swaziland (Photo courtesy of Aurecon).

One of the first studies of labyrinth weirs may perhaps be credited to Gentilini, (1941). The precursor to the first published design method by Hay and Taylor in 1970 was a systematic study performed by Kozák and Sváb, (1961). Numerous physical model studies and ensuing design methods have been published since that time. Labyrinth weir design publications of note have been produced at the U.S Bureau of Reclamation (Denver, Colorado, USA), the Laboratório Nacional de Engenharia Civil (Lisbon, Portugal), and the Utah Water Research Laboratory (Logan, Utah, USA).

Falvey, (2003) summarised many of the contributions to labyrinth weir hydraulics up to the turn of the century. Crookston, (2010) expanded the labyrinth weir research knowledge base with a more comprehensive design method and by investigating the influences of crest shape, nappe behaviour, configuration (in-channel vs. reservoir application), arched weir geometries, and size-scale effects on labyrinth weir hydraulics.

5.2. GENERAL DESCRIPTION

Labyrinth weirs have been placed in assorted channel and reservoir applications and have featured a wide variety of weir geometries and crest shapes. Labyrinth spillways commonly feature multiple trapezoidal-shaped cycles placed in a linear fashion at a single crest elevation, as shown in Figure 5.2. However, arced spillway configurations have been constructed in reservoir applications, such as Avon Dam (Australia, Darvas 1971), Kizilcapinar Dam (Turkey, Yildiz and Uzecek, 1996), Maguga (Swaziland, Van Wyk *et al.* 2006), María Cristina Dam (Spain, Page *et al.* 2007) and Weatherford Dam (USA, Tullis, 1992). Furthermore, the U.S. Army Corp of Engineers is designing a large arced labyrinth spillway at Isabella Dam in California, USA. In general, labyrinth spillways are symmetrical with some type of chute or energy dissipation basin located downstream. In addition, some labyrinth weir spillways have incorporated multiple crest elevations and various appurtenant structures such as foot and vehicle bridges, piers, base-flow notches, cool-water releases, and low-level outlets.

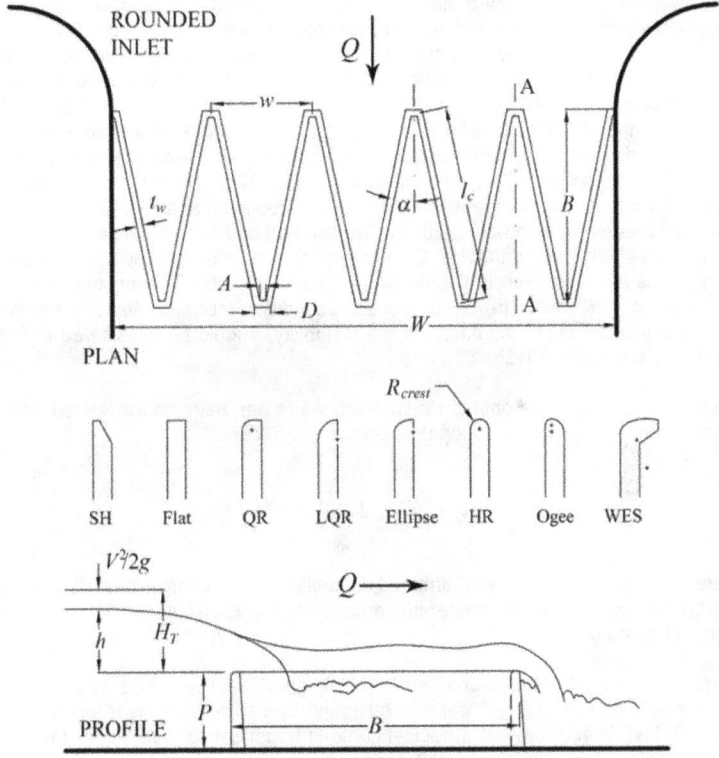

Fig. 5.2
Definition sketch for Labyrinth Spillway geometry (Crookston, 2010).

Crookston, (2010) provides a comprehensive overview of the hydraulic design of labyrinth spillways. It presents a thorough history of labyrinth weir research and design, including prior design methods, numerous case studies, and a comprehensive bibliography of labyrinth weir research contributions. The results of Crookston, (2010) have been summarised and published in Crookston and Tullis, (2013a, b) and Crookston and Tullis, (2012a, b, c). These technical articles comprise the principal references for the following discussion regarding the hydraulic design of labyrinth spillways and shall be generally referred to as the Crookston and Tullis Design Method.

5.3. DISCHARGE CAPACITY

Total discharge over a labyrinth spillway is a function of the flow conditions and weir geometry, expressed as:

$$Q = f(g, v, H_T, H_d, L_c, \alpha, A_c, l_c, P, B, w, t_w, crest, approach, nappe) \tag{5.1}$$

where Q = total discharge; g = acceleration constant of gravity; v = kinematic viscosity of water; H_T = total upstream head measured relative to the weir crest, where $H_T = V^2/2g + h$; V = average cross-sectional velocity of flow; h = piezometric head upstream of the weir (relative to the weir crest elevation); H_d = downstream tailwater total head; L_c = total centreline length of the weir crest, where $L_c = 2N(A_c + l_c)$; N = number of labyrinth cycles; A_c = centreline length of the apex; l_c = the centreline length of the sidewall; α = sidewall angle; P = weir height; B = cycle depth (longitudinal length); w = cycle width; t_w = wall thickness at the crest. Crookston and Tullis, (2013a) documented the influence of crest shape on discharge efficiency. The upstream topography and approaching flow field, as well as the abutment geometries and location relative to the weir placement (projecting, flush, etc.), also influence discharge efficiency (Crookston and Tullis, 2012b). The influence of weir height and cycle width on labyrinth hydraulics are also discussed in Crookston et al. (2012) with recent physical modelling based research by Seamons, (2014). The collision of adjacent nappes (near the upstream apex) influences labyrinth weir hydraulics; the occurrence and effects of local submergence wakes, and standing waves are discussed in Crookston and Tullis, (2012c). The influence of apex width is presented in Seamons, (2014). The nappe aeration condition (clinging, aerated, partially aerated, drowned) and behaviour (nappe vibration, nappe instability) should be considered in labyrinth weir design (Crookston and Tullis, 2013b).

The head-discharge relationship for labyrinth weirs has been characterized using Equation (5.2), a common form of the weir equation (Henderson, 1966):

$$Q = \frac{2}{3} C_{d(\alpha°)} L_c \sqrt{2g} H_T^{3/2} \tag{5.2}$$

where $C_{d(\alpha°)}$ = experimentally determined dimensionless discharge coefficient that is sidewall-angle-specific, and accounts for parameter influences noted in Equation (5.1) that are not expressly accounted for in Equation (5.2).

Figures 5.3 and 5.4 present Crookston, (2010) graphical $C_{d(\alpha°)}$ data as a function of α and dimensionless headwater ratio (H_T/P) for quarter-round and half-round crest shapes, respectively. Crookston and Tullis, (2013a) provide empirical curve-fit equations for quarter- and half-round $C_{d(\alpha°)}$ data. Linear interpolation may be appropriate for estimating $C_{d(\alpha°)}$ for other values of α not tested by Crookston, (2010).

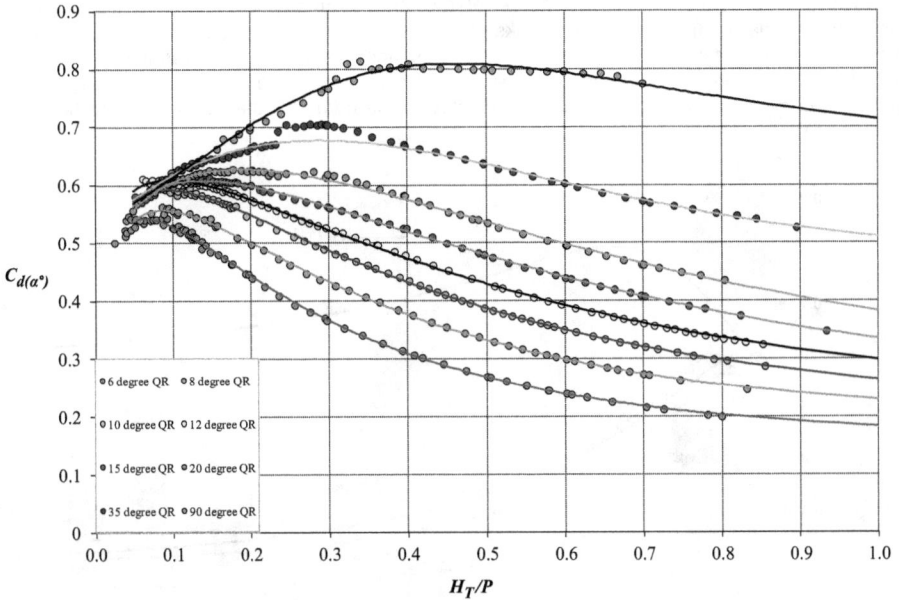

Fig. 5.3
$C_{d(\alpha°)}$ *versus* H_T/P for quarter-round trapezoidal labyrinth spillways

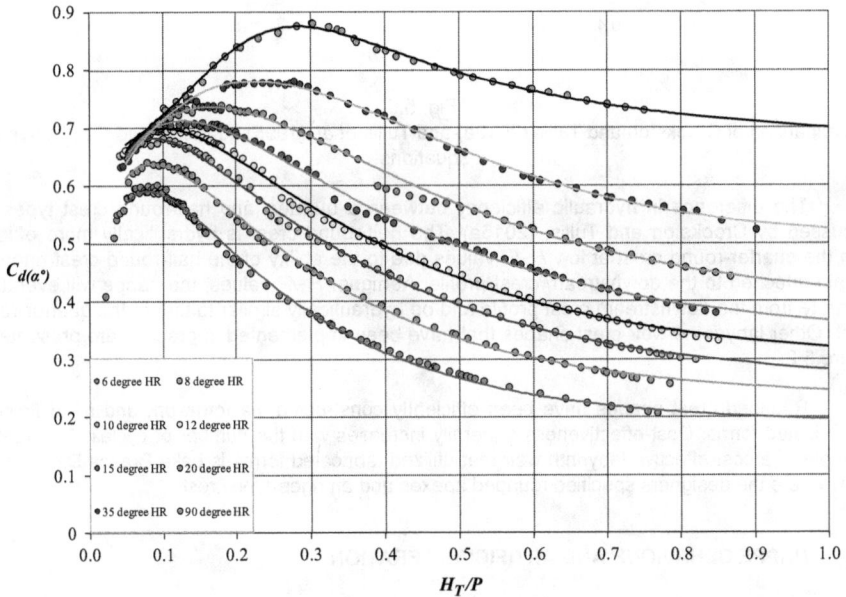

Fig. 5.4
$C_{d(\alpha°)}$ *versus* H_T/P for half-round trapezoidal labyrinth spillways.

The $C_{d(\alpha°)}$ curve-fit equations presented by Crookston and Tullis, (2013a) are based on experimental data limited $H_T/P \leq 1.0$. The specific form of the curve-fit equations, however, was selected to facilitate $C_{d(\alpha°)}$ extrapolation beyond H_T/P=1.0 (see Figure 5.5). The α=15° equation was validated for $H_T/P \leq 2.1$ by Crookston *et al.* (2012) through physical and numerical modelling, increasing the design flexibility of the Crookston and Tullis Method. In contrast, the polynomial curve fits, used by Tullis *et al.* (1995) (quarter-round crest shape) are limited to the range of experimental data from which they were derived ($H_T/P < 0.9$) and cannot be used for extrapolation (i.e., $C_{d(\alpha°)}$ values become negative with increasing H_T/P due to the nature of the polynomial functions, see Figure 5.5).

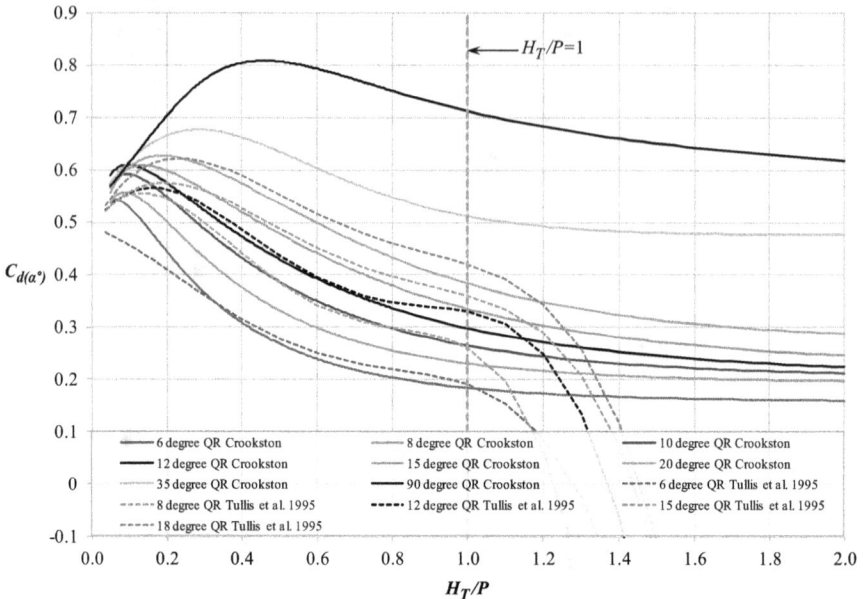

Fig. 5.5
Comparison of Crookston and Tullis (2013a) and Tullis *et al.* (1995) quarter-round $C_{d(\alpha°)}$ curve-fit equations.

The difference in hydraulic efficiency between a quarter- and half-round crest types are discussed by Crookston and Tullis, (2013a). The half-round crest is hydraulically more efficient than the quarter-round crest at low H_T/P values due to the ability of the half-round crest nappe to remain attached to the downstream crest profile. At higher H_T/P values, the nappe will eventually separate from the downstream crest profile and be hydraulically similar to that of the quarter-round crest. Other labyrinths weir crest shapes that have been implemented in practice are presented in Figure 5.2.

Rounded crest shapes have been efficiently constructed via formwork and hand finishing or fabricated forms. Cost-effectiveness generally increases with the number of cycles. An excellent example of a cost-effective labyrinth weir that utilized fabricated forms is Lake Brazos Dam (Texas, USA) where the designers specified rounded apexes and an ogee-type crest.

5.4. NAPPE BEHAVIOUR AND ARTIFICIAL AERATION

In addition to discharge, labyrinth spillway nappe behaviour should also be considered in design. Figure 5.6 illustrates the various nappe aeration behaviours conditions, and nappe instability (also termed flow surging) for labyrinth spillways with quarter-round crest shape; similar information for half-round crests is also presented in Crookston and Tullis, (2013b). Not included in Figure 5.6 is nappe vibration, which may also occur (Crookston *et al.* 2014; Anderson, 2014). Four nappe aeration conditions exist for labyrinth weirs: clinging, aerated, partially aerated, and drowned. The influence of artificial aeration, vent pipes, and nappe breakers or flow splitters on discharge and nappe behaviour is also discussed, including information regarding device placement and general nappe breaker shape.

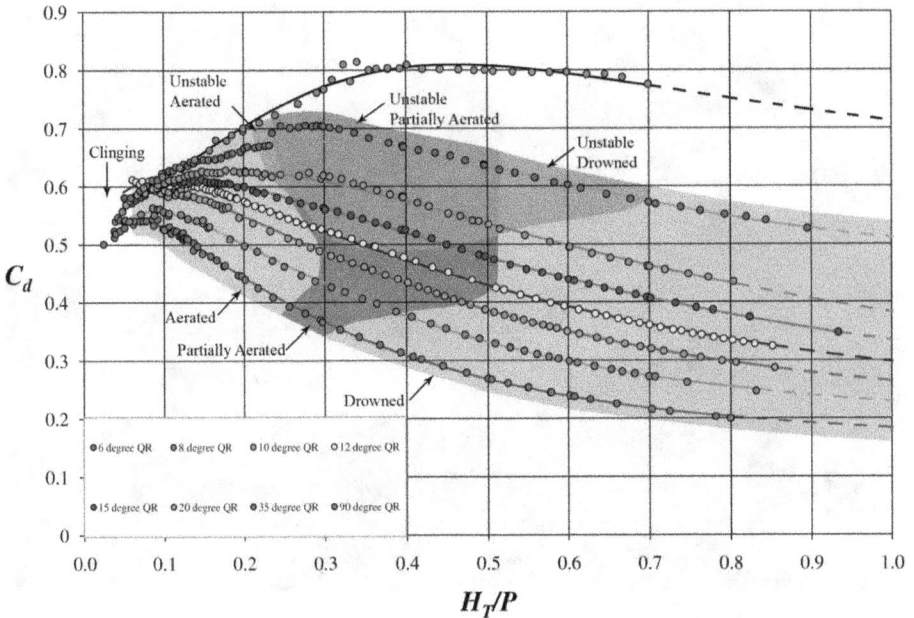

Fig. 5.6
Nappe behaviour for labyrinth spillways with a quarter-round crest.

For some hydraulic conditions, the size of the aeration cavity behind the nappe can fluctuate temporally, resulting in nappe instability. It is a low-frequency nappe trajectory fluctuation phenomena that occurs at moderate H_T/P ratios, produces noise, and is also a function of α and crest shape. Flow splitters may decrease the magnitude of nappe instability but were observed to be insufficient to prevent occurrence in the laboratory (Crookston 2010). Additional information regarding nappe instability is discussed in Crookston and Tullis (2013b).

Separate from nappe instability, nappe vibration (see Figure 5.7) can occur at low-head flow conditions (observed in some prototypes with flow depths above the crest of approximately 15 cm or less) and can develop with linear and non-linear weirs (Crookston et al. 2014). Nappe vibration can result in intense acoustic pressure waves and noise (Casperson, 1995) and has been aptly described as sounding similar to a helicopter. In general, nappe vibration is visually observed as closely spaced horizontal bands that initiate where the nappe departs from the crest. Very thin nappes may undulate or flutter. Also, vibration can be amplified by wind.

The most widely accepted theory for this phenomenon attributes vibrations to shear forces at the interface between the falling water sheet and the surrounding air, known as the Helmholtz mechanism (Helmholtz, 1868). Instability due to said mechanism can be amplified by an enclosed air pocket behind the nappe (Naudascher and Rockwell, 2005); however, a fully vented nappe displaying intense vibrations has been observed by Falvey, (1987), in prototype structures, and in large models at the Utah Water Research Laboratory (Anderson, 2014).

Fig. 5.7
Nappe vibration (photo courtesy of Schnabel Engineering, USA).

The specific influence of crest shape, weir length, fall height, and flow characteristics on nappe vibration for labyrinth spillways is still unclear. Adding roughness elements to the crest surface (Metropolitan Water, Sewerage and Drainage Board, 1980) is an example of one successful method reported in the literature for mitigating nappe vibration Vibrations have also been mitigated using numerous nappe breakers at some minimal interval determined by experimentation.

5.5. CYCLE EFFICIENCY

Per Equation (5.1), Q is proportional to $C_{d(\alpha°)}$, which decreases with decreasing α (see Figures 5.3 and 5.4), and Lc, which increases with decreasing α (for a given cycle width). Cycle efficiency ($\varepsilon' = Cd(\alpha°)Lc\text{-}cycle/w$) is representative of the discharge per cycle (at a given HT/P value) for a given labyrinth weir geometry and can be used to illustrate the net effect of these two opposite-trending parameters on discharge efficiency. ε' data for quarter- and half-round crest shapes and various values of α and HT/P are shown in Figure 5.8 (Crookston and Tullis, 2013a). The ε' data show that the increase in weir length more than compensates for the declining $Cd(\alpha°)$ values as α decreases. This means that the largest discharge per cycle occurs at the smaller α values. The hydraulic benefit of increased ε' of smaller α angles decreases with increasing HT/P. ε' only evaluates the discharge efficiency per cycle and should be yoked with a cost analysis to select the optimal labyrinth spillway geometry as a less efficient spillway that meets design requirements may be a more cost-effective option. ε' comparisons between structures are particularly useful when the structures share common HT/P ratio. The actual discharge per cycle width requires multiplying ε' by $HT^{3/2}$.

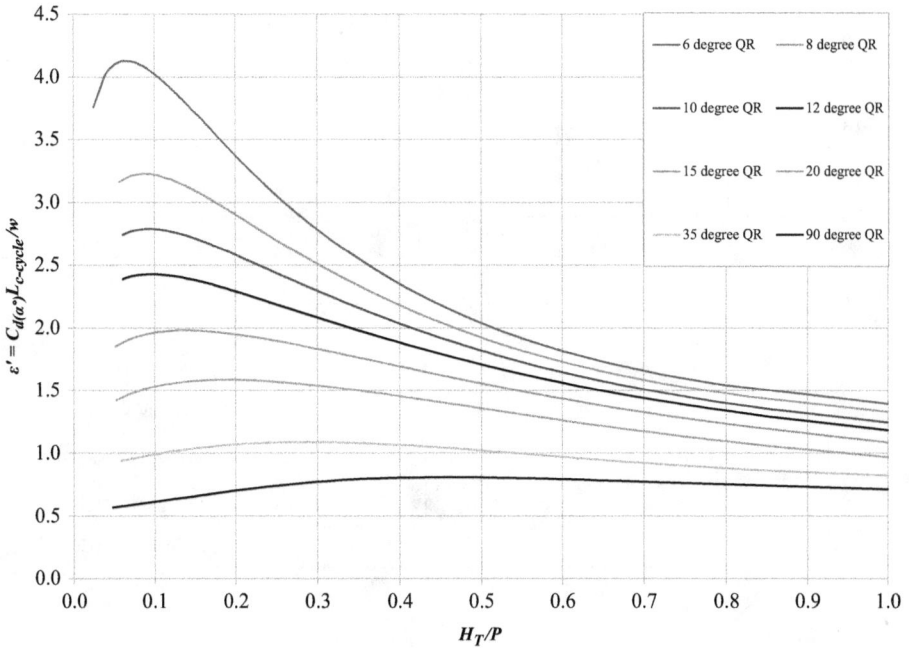

Fig. 5.8
Cycle efficiency for labyrinth spillways with a quarter-round crest.

5.6. TAILWATER SUBMERGENCE

Weir tailwater submergence occurs when the elevation of the downstream water surface exceeds the crest elevation. A tailwater submergence condition that does not increase the headwater elevation is referred to as modular submergence. Submergence levels higher than the modular submergence limit will increase the upstream water elevation for a give discharge. Local submergence differs from tailwater submergence in that it is independent of the downstream tailwater condition. A labyrinth weir dimensionless submerged head-discharge relationship is presented in Figure 5.9 (Tullis *et al.* 2007). In Figure 5.9, H^* and H_d are respectively the total upstream and downstream heads under submerged labyrinth weir flow conditions. H_T is the free-flow upstream total head for the same discharge (Q represents the independent variable in this analysis). The modular submergence range corresponds to $H^*/H_T=1.0$. The data in Figure 5.9 can be used to determine H^* knowing H_T and H_d. Similar head-discharge data for submerged piano key weirs is presented by Dabling and Tullis, (2012).

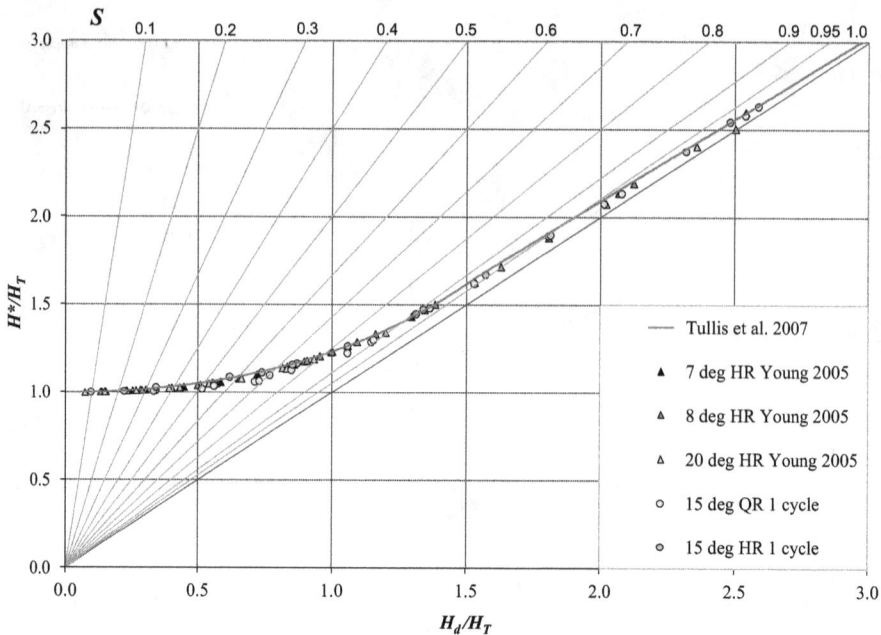

Fig. 5.9
Tailwater submergence dimensionless head relationship for labyrinth spillways (adapted from Tullis *et al.* 2007).

5.7. CYCLE WIDTH RATIO

A minimum cycle width ratio (*w/P*) greater than 2 has been suggested in previously proposed design methods; this recommendation has been based upon supporting data and discharge efficiency. However, increasing *P* may improve the approach flow conditions and therefore improve discharge capacity. In addition, reductions in *w* and hence, *w/P*, typically result in reduced construction costs for a given design flow and therefore increase spillway effectiveness, despite reduced flow efficiency (Crookston *et al.* 2013, Paxson *et al.* 2013). Furthermore, the Tullis *et al.* (1995) Design Method included $C_{d(\alpha°)}$ data for *w/P* ratios less than 2 despite suggesting ratios from 3 to 4. In light of this, research performed by Crookston, (2010) explored the influence of *w* and *P* on discharge, as identical *w/P* ratios may exist for different value combinations of *w* and *P*. General guidance regarding the influence of *w* and *P* on discharge of labyrinth spillways is presented in Crookston *et al.* (2013) and Seamons, (2014).

5.8. WEIR PLACEMENT AND ABUTMENT EFFECTS ON HYDRAULIC EFFICIENCY

Research performed at the USBR for labyrinth prototypes provided some insight regarding the effect of labyrinth weir placement in reservoir applications (Houston, 1983). Additional applied research regarding the effects of spillway abutments for labyrinth spillways with a half-round crest is presented in Crookston and Tullis, (2012b). The various in-channel and reservoir application specific geometric configurations tested are presented in Figure 5.10.

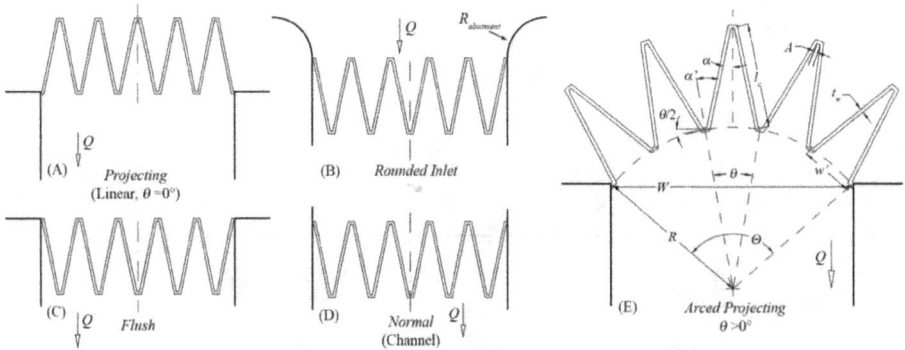

Fig. 5.10

Crookston and Tullis, (2012b) reservoir geometries: (A) *Projecting*, (B) *Rounded Inlet*, (C) *Flush*, (D) *Normal*, and (E) *Arched*.

A comparison of $C_{(\alpha°)}$ for four reservoir geometries [Figure 5.10 (A), (B), (C), and (E)] relative to an In-Channel [Figure 5.10 (D)] application (α=12°) is presented in Figure 5.11 (Crookston and Tullis 2012b). Based on the specific geometries tested, the *Flush* configuration produced the largest hydraulic efficiency reduction, relative to the in-channel application, due to flow separation at the abutments. Adding rounded abutments [Figure 5.10 (B)] improved the hydraulic efficiency; however, both the *Rounded Inlet* and *Projecting* configurations were approximately 3% to 7% less efficient than the in-channel configuration. Additional information regarding the hydraulic performance of these configurations is presented in Crookston and Tullis, (2012b).

A projecting labyrinth spillway with a cycle angle θ = 10° [see Figure 5.10 (E)] exceeded the discharge efficiency of the *In-channel* configuration by approximately 5% to 11% and was the most efficient geometry tested. An *Arched* labyrinth weir configuration increases the discharge capacity for reservoir applications by orienting the cycles toward the approach flow. The *Arched* configuration also facilitates a narrower downstream channel or chute width, potentially reducing construction costs. Crookston and Tullis, (2012a) present standardized nomenclature for arched labyrinth spillways and discuss geometrically comparable layout designs. Several *Arched* configurations were tested and found to be approximately 5 to 30% more efficient than in-channel linear configurations (H_T/P dependent). The hydraulic advantage of the *Arched* labyrinth weir diminishes with increasing H_T/P (increased influence of local submergence increase).

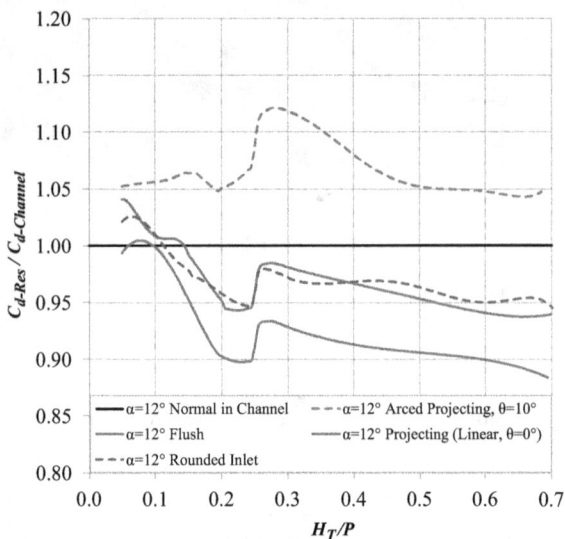

Fig. 5.11
Discharge efficiency for reservoir-specific labyrinth spillways, relative to in-channel application
(α=12°)

5.9.   ENERGY DISSIPATION AND RESIDUAL ENERGY

Labyrinth spillways are highly effective at energy dissipation. Lopes *et al.* (2006, 2008) present design information regarding relative residual energy (H_1) downstream of labyrinth weirs. Relative residual energy increases as H_T/P increases (i.e., less relative energy dissipation); labyrinth weirs produce a smaller residual energy than vertical drops (regardless of α). Formulae proposed by Lopes *et al.* (2006, 2008) are recommended for estimating residual energy downstream of labyrinth weirs.

5.10.   DEBRIS

Debris is often transported to hydraulic structures during flood events (Pfister *et al.* 2013). A survey regarding the debris handling of 75 labyrinth spillways in the USA and Portugal was recently completed (Crookston *et al.* 2015). The results indicate that in general, the debris handling performance of labyrinth spillways is acceptable with few reports regarding debris maintenance. For example, storm events in forested catchments may result in flows laden with woody debris. Depending upon the spillway hydraulics and site conditions, this type of debris may pass over the labyrinth weir during floods and require little maintenance. Conversely, conditions may result in large accumulations of woody debris that raise safety concerns. Therefore, the accumulation of debris on labyrinth weirs should not be overlooked for situations where there is the potential for debris to reduce spillway capacity. Research on driftwood for PKW spillways provides unique insights that are generally applicable to labyrinth weirs (Pfister *et al.* 2013).

5.11. HYDRAULIC DESIGN AND ANALYSIS

The recommended procedure for designing a labyrinth weir is presented as Table 5.1 (Crookston and Tullis, 2013a). This format was first introduced by Tullis *et al.* (1995), continued by Falvey, (2003), and has been refined and expanded by Crookston and Tullis. The top section of the design table features user-defined hydraulic conditions or requirements for the labyrinth weir. The design flow rate (Q_{design}) typically represents a flood discharge (e.g., 100-yr storm, PMF, etc.) estimated from a hydrologic analysis. In practice, Q_{design} is commonly estimated using computer programs such as HEC-HMS, its predecessor HEC-1, or other suitable methodologies. H represents the maximum allowable reservoir pool elevation. H_T may be assumed to be the difference in the reservoir pool elevation and the crest elevation, H_{crest} (negligible velocity head). However, it may be appropriate to consider the upstream flow conditions and any minor losses when estimating H_T. H_d is the total downstream head measured relative to the spillway crest and may be determined by a backwater curve. User-defined labyrinth weir geometric parameters are presented in the second section; the use a nappe aeration device can be specified if desired. The third section computes the labyrinth weir hydraulic performance data and determines additional weir geometric parameters. Although the Crookston and Tullis Design Method only provides $C_{d(\alpha°)}$ for quarter- and half-round crest shapes, the method easily accommodates experimentally determined $C_{d(\alpha°)}$ for site-specific conditions or other crest shapes. $C_{d(90°)}$ and the corresponding linear weir crest length (same crest shape) required to match the Q_{design} and H_T requirements are also reported for comparison. The last section of the design method includes the submerged head-discharge relationships developed by Tullis *et al.* (2007). Table 5.1 can be expanded to include a complete head-discharge relationship for the specified labyrinth weir geometry. This design method can also be used to estimate the head-discharge relationship for existing labyrinth spillways. Such a procedure, which also adapts easily to a spreadsheet program format, is outlined in Crookston and Tullis, (2013a); the labyrinth weir geometric parameters are specified rather than calculated. The effects of tailwater submergence may be determined by solving for Q or H^* as discussed by Tullis *et al.* (2007).

The design method and support data are limited to the geometries (Crookston, 2010) and hydraulic conditions tested (e.g., $0.05 \leq H_T/P \leq 0.9$). These results may be conservatively applied (with sound engineering judgment) to other labyrinth weir geometries and flow conditions (design verification with a hydraulic model study is recommended). Linear interpolation is recommended to determine $C_{d(\alpha°)}$ for α values other than those presented. Based on the available support data from the current study, the design method (Table 5.1) may be used for $H_T/P \leq 2.0$, as the experimental results are well behaved and the $\alpha=15°$ was tested up to $H_T/P = 2.1$; the resulting $C_{d(\alpha°)}$ data agreed well with the curve-fit equation proposed by Crookston and Tullis (2013a).

Table 5.1
Labyrinth weir geometry calculation template

Parameter	Symbol	Value	Units	Notes
Hydraulic Conditions – Input Data				
Design Flow	Q_{design}	1,500.00	(m³/s)	$g = 9.81$ m/s²
Design Flow Water Surface Elevation	H	1,680.00	(m)	
Approach Channel Elevation	H_{apron}	1,674.00	(m)	
Crest Elevation	H_{crest}	1,678.00	(m)	
Unsubmerged Total Upstream Head	H_T	2.00	(m)	Piezometric Head + Velocity Head - Losses
Downstream Total Head	H_d	0.50	(m)	
Labyrinth Weir Geometry – Input Data				
Angle of Side Legs	α	12	(°)	$\alpha \sim 6° - 35°$
Number of Cycles	N	9	-	whole or half cycles
Crest Height	P	4.00	(m)	$P \sim 1.0 H_T$
Thickness of Weir Wall at the Crest	t_w	0.50	(m)	$t_w \sim P/8$
Inside Apex Width	A	0.50	(m)	$A \sim t_a$
Crest Shape	Crest Shape	Quarter	-	Quarter- or Half-Round
Aeration Device (Nappe Breakers, Vents)	-	Breakers	-	Breakers, Vents, or None
Calculated Data				
Headwater Ratio	H_T/P	0.50	-	Data for $H_T/P<1.0$, extrapolation for $H_T/P \leq 2.0$
Labyrinth Weir Discharge Crest Coefficient	$C_{d(a°)}$	0.429	-	$C_{d(a°)} = f(H_T/P, \alpha, Crest\ Shape)$
Total Centerline Length of Weir	L_c	418.28	(m)	$L_c = 3/2 Q_{design}/[(C_{d(a°)}H_T^{3/2})(2g)^{1/2}]$
Centerline Length of Sidewall	l_c	2.33	(m)	$l_c = (B-t_w)/\cos(\alpha)$
Outside Apex Width	D	1.30	(m)	$D = A+2t_a\tan(45-\alpha/2)$
Cycle Width	w	11.10	(m)	$w = 2l_c\sin(\alpha)+A+D$
Width of Labyrinth (Normal to Flow)	W	99.87	(m)	$W = Nw$
Length of Apron (Parallel to Flow)	B	22.35	(m)	$B = [L_c/(2N)-(A+D)/2]\cos(\alpha)+t_w$
Magnification Ratio	M	4.19	-	$M = L_c/(wN)$
Cycle Width Ratio	w/P	2.77	-	Normally $2 \leq w/P \leq 4$
Relative Thickness Ratio	P/t_w	8.0	-	~8
Apex Ratio	A/w	0.05	-	<0.08
Cycle Efficiency	ε'	1.80	-	$\varepsilon' = C_{d(a°)}M$
Efficacy	ε	2.23	-	$\varepsilon = C_{d(a°)}M/C_{d(90°)}$
# of Nappe Breakers or Vents	-	9	-	Breaker on ds Apex, 1 Vent per Sidewall
Linear Weir Discharge Coefficient	$C_{d(90°)}$	0.808	-	$C_{d(90°)} = f(H_T/P, \alpha, Crest\ Shape)$
Length of Linear Weir for same Flow	$L_{c(90°)}$	222.33	(m)	$L_{c(90°)} = 3/2 Q_{design}/[(C_{d(90°)}H_T^{3/2})(2g)^{1/2}]$
Submergence (Tullis et al. 2007)				
Downstream/Upstream Ratio of Unsubmerged Head	H_d/H_T	0.25	(m)	
Submerged Head Discharge Ratio	H^*/H_T	1.013	-	Piecewise function Tullis et al. (2007)
Submerged Upstream Total Head	H^*	2.025	(m)	
Submergence Level	S	0.247	-	$S = H_d/H^*$
Submerged Weir Discharge Coefficient	C_{d-sub}	0.421	-	$C_{d(a°)}(H_T/H^*)^{3/2}$

┤Design limited to extent of experimental data; designs that exceed these limits may warrant a physical model study

5.12. PHYSICAL AND NUMERICAL (CFD) MODELLING

Further discussions on the use of Physical and CFD modelling in spillway design are provided in Chapter 9.

5.13. SELECTED PROTOTYPE LABYRINTH WEIR SPILLWAY EXAMPLES

The bibliography included herein contains references for many labyrinth spillway case studies; a summary of these structures is presented in Table 5.2.

Table 5.2
Labyrinth Spillway Prototypes

Name	Location	Q_{design}	H_t	N	Source
		(m^3/s)	(m)	( )	
Accord Pond Dam	USA	-	0.46	2	Crookston et al. (2015)
Agua Branca	Portugal	124	1.65	2	Quintela et al. (2000)
Alfaiates	Portugal	99	1.60	1	Quintela et al. (2000)
Alijó	Portugal	52	1.23	1	Magalhães & Lorena (1989)
Alloway Lake Dam	USA	-	-	5	Crookston et al. (2015)
Antelope Creek Channel	USA	-	-	3	Crookston et al. (2015)
Arcossó	Portugal	85	1.25	1	Quintela et al. (2000)
Avon	Australia	1790	2.80	10	Darvas (1971)
Bartletts Ferry	USA	5920	2.19	20.5	Mayer (1980)
Belia	Zaire	400	2.00	2	Magalhães & Lorena (1989)
Beni Bahdel	Algeria	1000	0.50	20	Afshar (1988)
Berg	South Africa	270	2.50	2	Fourie (1999)
Boardman	USA	387	1.77	2	Babb (1976), Lux (1985)
Bospoort 1	South Africa	2465	4.20	7.5	ARQ (2008)
Bospoort 2	South Africa	1613	4.20	2.5	ARQ (2008)
Boyde Lake	USA	1209	0.59	59	Brinker (2005)
Calde	Portugal	21	0.60	1	Quintela et al. (2000)
Capital City Country Club	USA	-	2.44	1	Crookston et al. (2015)
Carty	USA	387	1.80	2	Afshar (1988)
Cloe Dam	USA	-	-	7	Crookston et al. (2015)
Castelletto-Nerv. Canal	Italy	25	0.12	24	Magalhães & Lorena (1989)
Cimia	Italy	1100	1.5	4	Lux & Hinchliff (1985)
Concourse Lake Dam	USA	-	1.07	2	Crookston et al. (2015)
Crystal Bridges	USA	-	-	2	Crookston et al. (2015)
Dog River	USA	1572	2.74	8	Savage et al. (2004)
DRA Detention Structure	USA	-	1.22	4	Schnabel (2013)
Dungo	Angola	576	2.4	4	Magalhães & Lorena (1989)
Eikenhof	South Africa	190	3.5	2	ARQ (1998)
Elmendorf Lake	USA	-	-	6	Crookston et al. (2015)
Estancia	Venezuela	661	3.01	1	Magalhães & Lorena (1989)
Forestport	USA	76	1.02	2	Lux (1989)
Fort Miller	USA	-	-	45	Crookston et al. (2015)
Garland Canal	USA	25.5	0.37	3	Lux & Hinchliff (1985)
Gema	Portugal	115	1.12	2	Magalhães & Lorena (1989)
Glen Park	USA	-	-	25	Crookston et al. (2015)
Grahamstown	Australia	628	1.4	8	Barker et al (2001)
Greystone	USA	-	1.45	1	Schnabel (2013)
Hard Labor Creek	USA	-	3.66	4	Schnabel (2013)
Harrezza	Algeria	350	1.9	3	Lux (1989)
Hollis G Lathem Reservoir	USA	-	2.44	3	Crookston et al. (2015)
Huntington Hills	USA	-	0.58	3	Crookston et al. (2015)

Indian Run	USA	-	2.44	2.5	Crookston et al. (2015)
Infulene Canal	Mozambique	60	1.00	3	Magalhães & Lorena (1989)
Juturnaiba	Brazil	862	0.70	4	Afshar (1988)
Kauffman	USA	128	5.0	5	Crookston et al. (2015)
Keddara	Algeria	250	2.46	2	Magalhães & Lorena (1989)
King Falls	USA	-	-	6	Crookston et al. (2015)
Kizilcapinar	Turkey	2270	4.6	5	Yildiz (1996)
Lake Brazos	USA	24609	-	24	Tullis and Young (2005)
Lake Natalie	USA	-	2.44	4	Crookston et al. (2015)
Lake Paupacken	USA	-	-	2	Crookston et al. (2015)
Lake Sovereign	USA	-	1.22	2	Crookston et al. (2015)
Lake Townsend	USA	3483	4.57	7	Tullis & Crookston (2008)
Lake Upchurch	USA	-	1.52	6	Crookston et al. (2015)
Leaser Lake	USA	289	3.26	2	Crookston et al. (2015)
Linville Land Harbor	USA	693	2.61	4	Schnabel (2013)
Little Blue Run	USA	-	-	10	Crookston et al. (2015)
Lyman Run	USA	-	-	8	Crookston et al. (2015)
Maguga	Swaziland	15000	8.25	9	Van Wyk et al (2006)
Midmar	South Africa	3052	4.15	10	ARQ (2002)
María Cristina Dam	Spain	5444	-	7	Page et al. (2007)
Meacham Grove	USA	-	-	4	Crookston et al. (2015)
Mercer	USA	239	1.83	4	CH2M-Hill (1976)
Navet Pumped Storage	Trinidad	481	1.68	10	Phelps (1974)
New London	USA	-	-	4	Crookston et al. (2015)
Ohau C Canal	New Zealand	540	1.08	12	Walsh (1980)
Opossum	USA	-	-	4	Crookston et al. (2015)
Pacoti	Brazil	3400	2.72	15	Magalhães & Lorena (1989)
Pine Run	USA	-	1.40	4	Crookston et al. (2015)
Pisão	Portugal	50	1.00	1	Quintela et al. (2000)
Pye Lake	USA	-	0.34	3	Schnabel (2013)
Quincy	USA	552	2.13	4	Magalhães & Lorena (1989)
Rapp Run Flood Control	USA	-	-	6	Crookston et al. (2015)
Rocklands-Berg	South Africa	-	-	2	Fourie (1999)
Roy F. Varner Reservoir	USA	-	2.13	8	Crookston et al. (2015)
São Domingos	Portugal	160	1.84	2	Magalhães & Lorena (1989)
Sam Rayburn Lake	USA	-	-	16	USACE (1991)
Santa Justa	Portugal	285	1.35	2	Magalhães & Lorena (1989)
Sarioglan	Turkey	490.7	1.06	7	Yildiz (1996)
Sarno	Algeria	360	1.5	8	Afshar (1988)
South River No 29	USA	-	3.05	3	Schnabel (2013)
Standley Lake	USA	1539	1.98	13	Tullis (1993)
Teja	Portugal	61	1.05	1	Quintela et al. (2000)
Upper Dam - Rangeley	USA	-	1.19	4	Schnabel (2013)
Ute	USA	15570	5.79	14	Houston (1982)
Weatherford	USA	-	-	4	Tullis (1992)
Woronora	Australia	1020	1.36	11	Darvas (1971)

The plan and profiles of a few of these installations are shown on the following pages to illustrate the varied configurations that have been used.

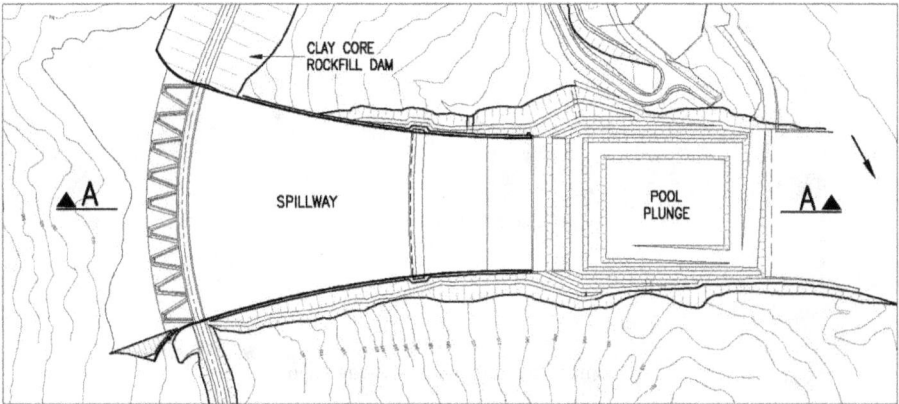

Fig. 5.12
Maguga Dam Labyrinth Spillway Layout

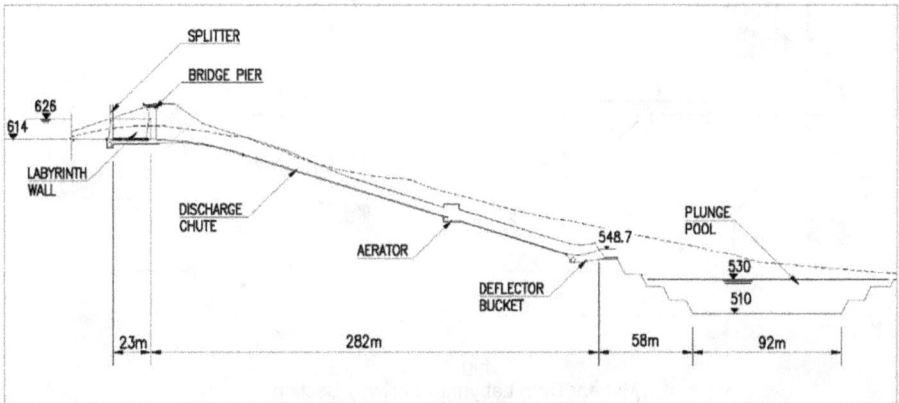

Fig. 5.13
Maguga Dam Labyrinth Spillway Longitudinal Section

A bridge had to be constructed over the labyrinth spillway. In order to accommodate the road alignment over the spillway and the embankment, the labyrinth was curved to a radius of 300 m. The discharge chute tapers from 181 m wide at the labyrinth to 100 m wide at the deflector bucket lip. In order to make the transition as gradual as possible, curved sidewalls with a radius of 876 m was fitted tangentially to both the labyrinth and the deflector bucket. This resulted in the individual flows from each cycle to be directed down the discharge chute in such a way that no cross waves developed.

- **MIDMAR DAM, SOUTH AFRICA** (Information courtesy of ARQ, South Africa)

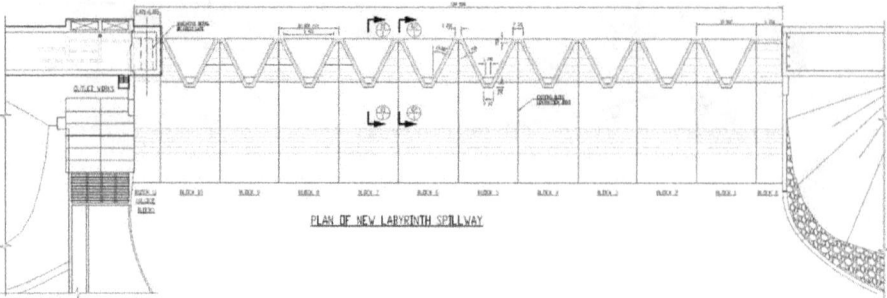

PLAN OF NEW LABYRINTH SPILLWAY

Fig. 5.14
Midmar Dam Labyrinth Spillway Layout

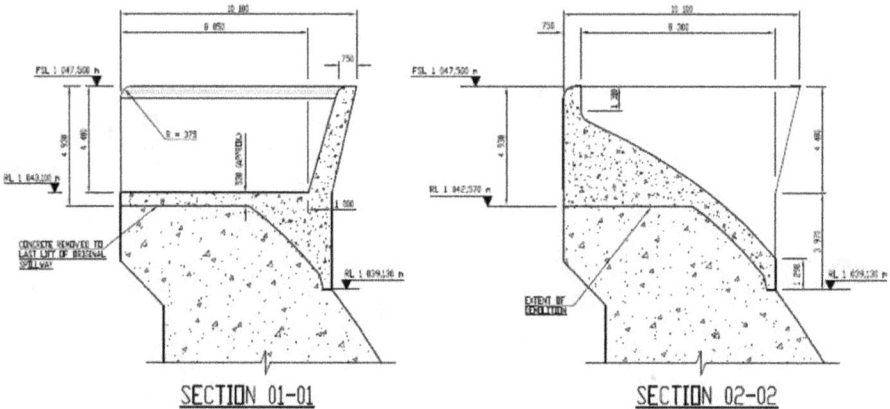

SECTION 01-01   SECTION 02-02

Fig. 5.15
Midmar Dam Labyrinth Spillway Sections

This structure was incorporated on the crest of an existing mass concrete gravity dam originally intended to be raised through the addition of radial gates. It is believed to be the first implementation of a labyrinth in this configuration in the world. The bridge, piers and trunnion points, as well as a portion of the ogee crest were demolished to make way for the new labyrinth. Additional mass was provided on the downstream between the labyrinth bays to assist with overall stability.

The primary motivation for the deviation from the original design was the reliability of a fixed labyrinth and its capacity to safely attenuate the Design and Safety Evaluation Floods.

The labyrinth design was based on the bucket depth and the structure was model tested using three full cycles at 1:27 scale to confirm theoretical capacities.

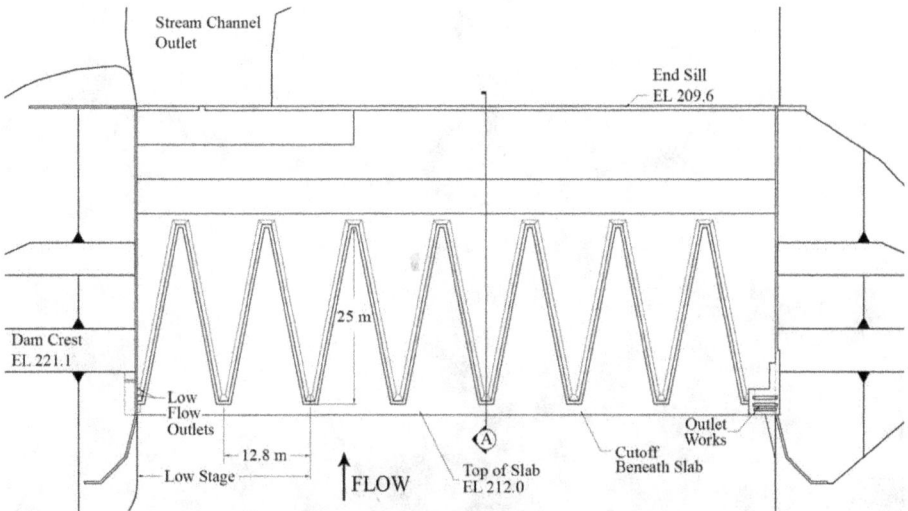

Fig. 5.16
Lake Townsend Labyrinth Spillway Layout

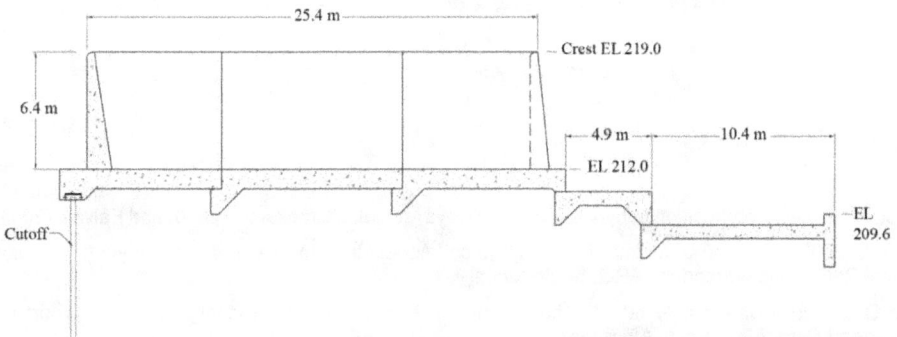

Fig. 5.17
Lake Townsend Labyrinth Spillway Section

The labyrinth spillway (6.1 m high) at Lake Townsend Dam replaced an 84 m wide gated concrete spillway. The 7-cycle labyrinth features a staged crest with 2-cycles at normal pool and 5 cycles comprising the high stage. It is also subject to tailwater submergence for large flows. The design was tested with physical modelling performed at the Utah Water Research Laboratory and CFD modelling performed at Idaho State University (Paxson et al. 2007). The spillway design flow is about 3,483 m³/s; the embankment dam includes overtopping protection to allow passage of the spillway design flood (SDF) plus flows resulting from the failures of three upstream dams.

Fig. 5.18
Lake Townsend Labyrinth Spillway

5.14. REFERENCES

For additional references on Labyrinth Weir Spillways, see also Crookston, (2010) and Falvey, (2003)

ARQ, (2002) - Raising of Midmar Dam – Physical model study for the proposed labyrinth spillway. Final Report and Addendum, ARQ, South Africa.

ARQ, (2008) - Dam safety rehabilitation programme. Second draft discharge capacity report for Bospoort Dam. ARQ, South Africa.

Barker, M.B., R.M. Holroyde and T. Qui (2001) - Grahamstown Dam Stage 2. Augmentation selection and design of a labyrinth spillway and baffle chute. ANCOLD Bulletin, Issue No. 118, August 2001.

Crookston, B. and B. Tullis (2008) - "Labyrinth weirs." In: S. Pagliara (ed.) *Hydraulic Structures*, 2nd IJREWHS on Hydraulic Structures, Edizioni Plus, University of Pisa, Italy.

Crookston B. and B. Tullis (2010) - "Hydraulic performance of labyrinth weirs." In: *3rd International Junior Researcher and Engineer Workshop on Hydraulic Structures*, Edinburgh, Scotland (currently unpublished).

Falvey, Henry T. (2003) - *Hydraulic design of labyrinth weirs*. ASCE Press, 1801 Alexander Bell Drive, Reston, Virginia, USA.

Fourie, S. (1999) - Revised design report for Rocklands-Berg Dam enlargement and upgrading of spillway. Report No 3690/5972, October 1999.

Lopes, R., J. Matos and J.F. Melo (2006) - "Discharge capacity and residual energy of labyrinth weirs." *International Junior Researcher and Engineer Workshop on Hydraulic Structures*, J. Matos and H. Chanson (Eds), Report CH61/06, Div. of Civil Eng., The University of Queensland, Brisbane, Australia.

Lopes, R., J. Matos and J.F. Melo (2008) - "Characteristic depths and energy dissipation downstream of a labyrinth weir." *Hydraulic Structures*, 2nd IJREW on Hydraulic Structures, S. Pagliara (ed.), Edizioni Plus, University of Pisa, Italy.

Lux III, F.L. and D. Hinchcliff (1985) - "Design and Construction of Labyrinth Spillways", *Transactions of the Fifteenth International Congress on Large Dams*, Vol. 4, Q. 59, R. 15, pp. 249-274, International Commission on Large Dams, Paris, France.

Paxson, G., D. Campbell and J. Monroe (2011) - "Evolving Approaches and Considerations for Labyrinth Spillways". *21st Century Dam Design – Advances and Adaptations*, 31st Annual USSD Conference, San Diego, California, April 11 – 15, 2011.

Savage, B., K. Frizell and J. Crowder (2001) - *Brains versus Brawn: The Changing World of Hydraulic Model Studies.* United States Bureau of Reclamation, www.usbr.gov/pmts/hydraulics_lab/pubs/PAP/PAP-0933.pdf.

Tullis, J.P., N. Amanian and D. Waldron (1995) - "Design of Labyrinth Spillways", *Journal of Hydraulic Engineering*, ASCE Volume 121, No 3, March 1995.

Van Wyk, D., A. Officer, W. Schwartz, G. Goodey and A. Rooseboom (2006) - "Design and Model Testing of a Labyrinth Spillway for Maguga Dam", *Transactions of the Twenty Second Congress on Large Dams*, Q.84 – R.54, ICOLD, Barcelona.

Amanian, N. (1987) - *Performance and design of labyrinth spillways.* M.S. Thesis, Utah State University, Logan, Utah, USA.

Anderson, A.A. (2014) - *Causes and Countermeasures for nappe oscillation: An experimental approach.* M.S. Thesis. Utah State University, Logan, Utah, USA.

Cassidy, J.J., C.A. Gardner and R.T. Peacock (1983) - "Labyrinth-crest spillway – planning, design, and construction." *Proceedings of the International Conference of the Hydraulic Aspects of Flood and Flood Control,* London, England.

Cordero-Page, D., V. García and C. Nonot (2007) - "Aliviaderos en laberinto. Presa de María Cristina." *Ingeneiería Civil,* 146, 5-20 (in Spanish)

Crookston, B.M. (2010) - *Labyrinth weirs.* Ph.D. Dissertation. Utah State University, Logan, Utah, USA.

Crookston, B.M., G.S. Paxson and B.M. Savage (2012) - *It can be done! Labyrinth weir design guidance for high headwater and low cycle width ratios.* Proc. of the 2012 ASDSO Annual Conference, Denver, Colorado. CD-ROM.

Crookston, B.M., A.A. Anderson, L. Shearin-Feimster and B.P. Tullis (2014) - "Mitigation investigation of flow-induced vibrations at a rehabilitated spillway." *5th International Symposium on Hydraulic Structures*, Brisbane, Australia.

Crookston, B.M., G.S. Paxson, B.M. Savage and B.P. Tullis (April 2013) - "Increasing hydraulic design flexibility of labyrinth spillways." *ICOLD 2013 International Symposium.* Seattle, Wash.Crookston, B.M and B.P. Tullis (2012a) - "Arced labyrinth weirs." *Journal of Hydraulic Engineering, ASCE,* 138(6). 555-562.

Crookston, B.M and B.P. Tullis (2012b) - Discharge efficiency of reservoir-application-specific labyrinth weirs. *J. Irrig. Drain. Engr., ASCE,* 138(6). 773-776.

Crookston, B.M and B.P. Tullis (2012c) - "Labyrinth weirs: Nappe interference and local submergence." *J. Irrig. Drain. Engr., ASCE,* 138(8), 757-765.

Crookston, B.M and B.P. Tullis (2013a) - "Hydraulic design and analysis of labyrinth weirs. Part 1: Discharge relationships." *J. Irrig. Drain. Engr., ASCE,* 139(5), 363-370.

Crookston, B.M and B.P. Tullis (2013b) - "Hydraulic design and analysis of labyrinth weirs. Part 2: Nappe aeration, instability, and vibration." *J. Irrig. Drain. Engr., ASCE,* 139(5), 371-377.

Dabling, M.R. and B.M. Crookston (2012) - "Staged and Notched Labyrinth Weir Hydraulics." *4th International Junior Researcher and Engineer Workshop on Hydr. Structures.* B. Tullis and H. Chanson (eds.).

Dabling, M.R., B.P. Tullis and B.M. Crookston (2013) - "Staged labyrinth weir hydraulics." *J. Irrig. Drain. Eng.,* posted ahead of print May 22, 2013.

Darvas, L. (1971) - "Discussion of performance and design of labyrinth weirs, by Hay and Taylor." *J. of Hydr. Engrg., ASCE,* 97(80), 1246-1251.

Easterling, D., K. Kunkei and X. Yin (2013) - "Observed increases in probable maximum precipitation over global land areas." *Proceedings of 25th Conference on Climate Variability and Change,* Austin, TX, USA.

Falvey, H. and Trielle (1995) - "Hydraulics and design of fusegates." *J. of Hydr. Engrg., ASCE,* 121(7), 512-518.

Frizell, K. (2003) - *Dog River Dam Hydraulic Model Study Results.* USBR Water Resources Research Laboratory Report, Denver, Colo.

Gentillini, B. (1940) - *Stramazzi con cresta a planta obliqua e a zig-zag.* Memorie e Studi del Instituto di Idraulica e Construzioni Idrauliche del Regil Politecnico di Milano, No. 48 (in Italian).

Hay, N. and G. Taylor (1970) - "Performance and design of labyrinth weirs." *J. of Hydr. Engrg., ASCE,* 96(11), 2337-2357.

Henderson, F.M. (1966) - *Open Channel Flow.* MacMillan Company, New York, USA.

Hinchliff, D. and K. Houston (1984) - *Hydraulic design and application of labyrinth spillways.* Proc. of 4th Annual USCOLD Lecture. Dam safety and Rehabilitation, Bureau of Reclamation U.S. Dept. of the Interior, Washington, DC, USA.

Houston, K. (1982) - *Hydraulic model study of Ute Dam labyrinth spillway.* Report No. GR-82-7, US Bureau of Reclamation, Denver, Colo., USA

Houston, K. (1983) - *Hydraulic model study of Hyrum Dam auxiliary labyrinth spillway.* Report No. GR-82-13, US Bureau of Reclamation, Denver, Colo., USA.

Laugier, F. (2007) - "Design and construction of the first piano-key weir (PKW) spillway at the Goulours Dam." *Hydropower and Dams,* Vol 12, Issue 5.

Lempérière, F. and A. Ouamane (2003) - "The PK Weir: a new cost-effective solution for spillways." *Hydropower & Dams,* Vol. 8, Issue 5.

Magalhães, A. and M. Lorena (1989) - *Hydraulic design of labyrinth weirs.* Report No. 736, National Laboratory of Civil Engineering, Lisbon, Portugal.

Melo, J., C. Ramos and A. Magalhães (2002) - "Descarregadores com soleira em labirinto de um ciclo em canais convergentes. Determinação da capacidade de vazão." *Proc. 6° Congresso da Água,* Porto, Portugal. (in Portuguese).

Paxson, G., D. Campbell and J. Monroe (2011) - "Evolving design approaches and considerations for labyrinth spillways." *Proc. 31st Annual USSD Conf.,* San Diego, CA, USA. CD-ROM.

Paxson, G., J. Monroe, B.M. Crookston and D. Campbell, D. (2013) - "Balancing site considerations with hydraulic efficiency for labyrinth spillways." *ICOLD 2013 International Symposium.* Seattle, Washington, USA.

Pfister, M., A. Schleiss and B. Tullis (2013) - "Effect of driftwood on hydraulic head of Piano Key weirs." *Labyrinth and Piano Key Weirs II.*

Quintela, A., A. Pinheiro, J. Afonso and M. Cordeiro (2000) - "Gated spillways and free-flow spillways with long crests. Portugese dams experience." *20th ICOLD Q79-R12,* Bejing, China. 171-89.

Savage, B., K. Frizell and J. Crowder (2004) - "Brian versus brawn: The changing world of hydraulic model studies." *Proc. of the ASDSO Annual Conference,* Denver, CO, USA, CD-ROM.

Seamons, T. R. (2014) - Labyrinth weirs: A look into geometric variation and its effect on efficiency and design method predictions. M.S. Thesis. Utah State University, Logan, Utah. USA

Taylor, G. (1969) - *The performance of labyrinth weirs.* PhD Thesis, University of Nottingham, Nottingham, England.

Tullis, B. and B.M. Crookston (2008) - *Lake Townsend Dam spillway hydraulic model study report.* Utah Water Research Laboratory, Logan, Utah, USA.

Tullis, B., J. Young and M. Chandler (2007) - "Head-discharge relationships for submerged labyrinth weirs." *J. of Hydr. Engrg., ASCE,* 133(3), 248-254.

Tullis, J.P. (1992) - *Weatherford spillway model study.* Hydraulic Report No. 311, Utah Water Research Laboratory, Logan, Utah, USA.

Tullis, J.P. (1993) - *Standley Lake service spillway model study.* Hydraulic Report No. 341, Utah Water Research Laboratory, Logan, Utah, USA.

Tullis, J.P., N. Amanian and D. Waldron (1995) - "Design of labyrinth weir spillways." *J.of Hydr. Engrg., ASCE,* 121(3), 247-255.

Waldron, D. (1994) - *Design of labyrinth spillway.* M.S. Thesis, Utah State University, Logan, Utah, USA.

6.   PKW SPILLWAYS

6.1.   GENERAL DESCRIPTION

The recently developed Piano Key weir (PKW) spillway, which is an innovative structure that can convey very high specific discharges, is a variation of the traditional labyrinth weirs, originally devised to circumvent some drawbacks of the latter (Barcouda *et al.* 2006). Using a rectangular layout and ramped floors which create overhanging or cantilevered apexes, the PKW is structurally simple and efficient, can be placed on existing or new gravity dam crest sections and multiplies significantly the discharge capacity compared to a standard weir of similar width (Ouamane and Lempérière, 2003).

Fig.6.1
View of the PKW spillway of Gloriettes Dam in France during construction (Photo courtesy of EDF)

The succession of inclined apexes turned alternatively in upstream and in downstream direction gives the name Piano Key weir. Compared to a rectangular labyrinth weir with a common crest layout (in plan form), the PKW spillway has the main advantage that it can be more easily installed at sites featuring limited foundation space (e.g., crest of a gravity dam). In addition, the ramped floors reduce the vertical wall height and thus the volume of reinforcing steel required in concrete. These are the reasons why PKW spillways are an efficient and economical solution for the increase of the flood releasing capacity at existing dams.

Fig. 6.2
View of the PKW spillway of Malarce Dam in France during spillage (Photo courtesy of EDF)

PKW spillways have been first proposed by Hydrocoop in collaboration with the Hydraulic Laboratory of Electricité de France (France), the Roorkee University (India) and the Biskra University (Algeria) (Ouamane and Lempérière, 2003). Since its invention, several works have been carried out all over the world to understand its hydraulic behaviour, optimise its design and objectify its advantages and drawbacks (see for instance Blanc and Lempérière, 2001; Barcouda et al., 2006; Ouamane and Lempérière, 2006; Truong Chi et al., 2006; Machiels et al., 2011a & 2014; Leite Ribeiro et al., 2012a & b; Machiels, 2012; Anderson and Tullis, 2012 & 2013). Studies are still ongoing, for instance with a trapezoidal layout (Cicéro et al., 2013a).

The first PKW spillway was installed by EDF in 2006 at Goulours dam in France (Laugier, 2007). Since then, PKW spillways have been used to increase the flood discharge capacity of five other EDF dams, namely St. Marc (2008), Etroit (2009), Gloriettes (2010), Malarce (2012) and Charmine (2014), or as new overflow structure, (Escouloubre (2011)). Lessons learned from the design of these first PKW spillways can be found in Laugier et al. (2013 and 2017a). Other PKW are presently in operation in Vietnam (Ho Ta Khanh et al. 2017); Sri Lanka (Jayatillake and Perera, 2013 and 2017); Switzerland (Eichenberger, 2013) and Scotland (Ackers et al. 2013). New PKW spillways are under study or construction in Vietnam (Ho Ta Khan et al. 2017); France (Laugier et al. 2017b; Erpicum et al. 2011b; Bail et al. 2013); Algeria (Erpicum et al. 2012); South Africa (Botha et al., 2013) and India (Das Singhal and Sharma, 2011). These works are part of the rehabilitation of existing dams (to increase discharge capacity) or new projects, with PKW spillways built in the river (diversion weir), on the top of a gravity dam, or on a reservoir bank.

So far, almost 30 PKW prototype spillways have been built in 8 countries for the last 10 years. They include a wide range of specific discharge capacities with water heads varying from 30 cm to more than 9 m and structural heights from 1 m to 10 m. Furthermore, total discharge can be as small as dozens of m^3/s to almost 10 000 m^3/s.

Most of the information available so far on PKW spillways has been published in three books (Erpicum *et al.* 2011a, 2013a and 2017a), edited following three specialised workshops held in Belgium (2011), France (2013) and Vietnam (2017). A world register of PKW spillways with a world map of projects has also been established and is available on a dedicated website (http://www. pk-weirs.ulg.ac.be/?q=content/world-register-pkw).

6.2. GEOMETRY AND TYPES

The PKW geometry may appear complex. It involves indeed a large set of parameters. In order to unify the notations, a nomenclature specific to the structure has been developed (Pralong *et al.* 2011a).

Following this nomenclature, the "PKW-unit" can be defined as the basic structure of the weir. It is made of two side-walls, an inlet and two half-outlet keys (Figures 6.3a and 6.3b). The main geometric parameters of the structure are:

- the heights of the inlet and outlet keys P_i and P_o, (very often Pi=Po called "P");

- their widths W_i and W_o, the unit width W_u;

- the number of PKW-units N_u;

- the longitudinal crest length B_h, the lengths B_o and B_i of the up- and downstream overhangs, the base length B_b; and

- the wall thickness T_s.

Subscripts *i*, *o* and *s* refer respectively to the inlet key, the outlet key and the side wall. W_u is equal to $W_i + W_o + 2T_s$ and the total width W of the weir is equal to N_u times W_u. The PKW-unit developed crest length L_u of a PKW-unit is equal to $W_u + 2B_h$ and the total developed crest length L of the weir is equal to N_u times L_u. Parapet walls (vertical extensions of the crest) may be added to the weir. Their height is referred to as P_p.

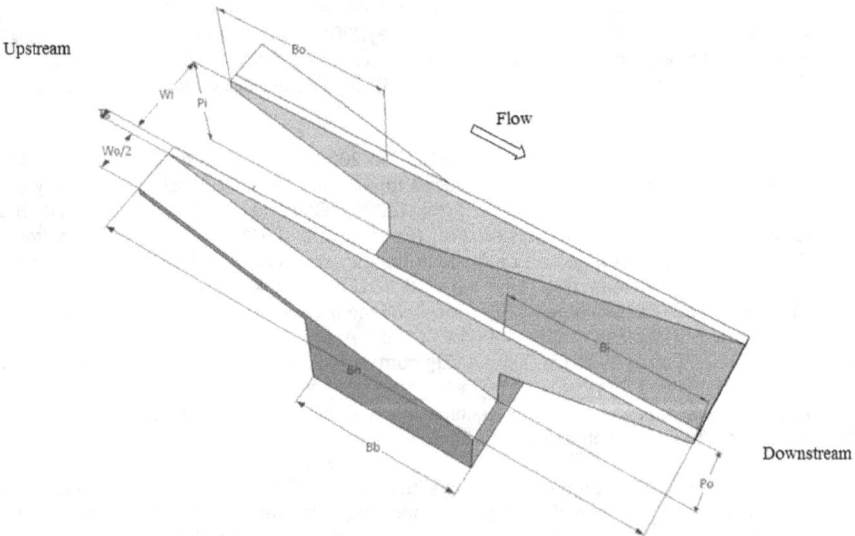

Fig. 6.3
(a) Sketch of a PKW unit and main notations (Erpicum *et al.* 2013b)

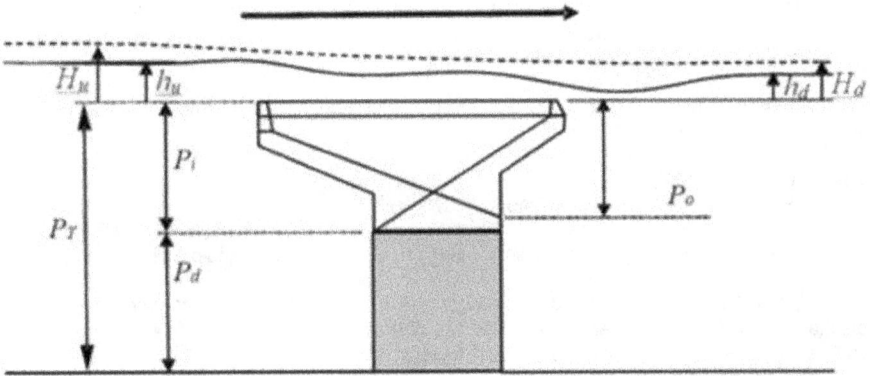

Fig. 6.3
(b) Typical PKW cross section (adapted from Pralong *et al.* 2011a)

Depending on which PKW apexes have overhangs (upstream and/or downstream), PKW spillways have been classified in 4 types (Truong *et al.*, 2006): type-A with symmetric overhangs; type-B with a single upstream overhang; type-C with a single downstream overhang; and type-D without overhang (Figures 6.4).

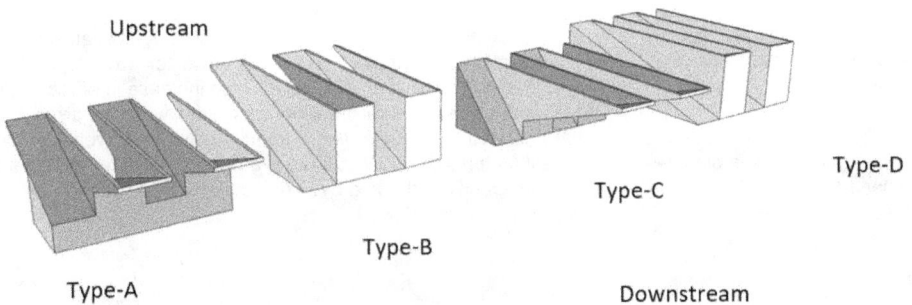

Fig. 6.4
Types of PK weir (adapted from Erpicum *et al.* 2013b)

6.3.   DISCHARGE CAPACITY

The PKW spillway is a free surface weir and its discharge Q_P is thus proportional to the upstream head H as

$$Q_p = \propto \sqrt{2gH^3} \qquad (6.1)$$

As summarised by Leite Ribeiro *et al.* (2012a), two approaches may be chosen to derive the proportion factor, which represents the effect of the crest length and shape.

Referring to the developed crest length L, the discharge coefficient $C_{P,L}$ is closely related to the crest shape. Eq. 6.1 writes as (Leite Ribeiro *et al.* 2012b)

$$Q_p = C_{p,L} L \sqrt{2gH^3} \qquad (6.2)$$

In this approach, L varies with the head as the effective crest length decreases with increasing heads because of local submergence of the upstream apex. $C_{P,L}$ also varies with the head as it includes both frontal and side weirs effects.

Referring to the width of the weir W, Eq. 6.1 writes (Ouamane & Lempérière, 2006; Machiels et al. 2011a):

$$Q_p = C_{p,W} W \sqrt{2gH^3}$$

(6.3)

with a discharge coefficient $C_{P,W}$ accounting for both crest shape and developed length effects.

Whatever the approach used to model the PKW discharge, it is of common use to look at its discharge capacity by comparison with the one of a standard linear weir of same width (Q_S). The discharge increase ratio r is defined by Leite Ribeiro et al. (2012a & b) as

$$r = \frac{Q_P}{Q_S}$$

(6.4)

Considering Eq. 6.3, Eq. 6.4 writes

$$r = \frac{C_{P,W}}{C_S}$$

(6.5)

where C_S is the discharge coefficient of the standard linear weir.

Whatever its geometry a PKW is much more efficient than an ogee crested weir of same width, especially for low heads (Figure 6.5). This explains why most of the existing PKW spillways have been designed for a maximum H/P ratio lower than 1 (Pfister et al. 2012). This high increase in efficiency is due to the developed length of the crest which equals several times the weir width, while the discharge coefficient is close to the one of a sharp crested weir. Regarding a traditional labyrinth weir with the same cycle shape in plan view (same crest footprint), a PKW spillway is around 10% more efficient for a head H equal to its height P_i, as shown in a study performed by Anderson and Tullis, (2012).

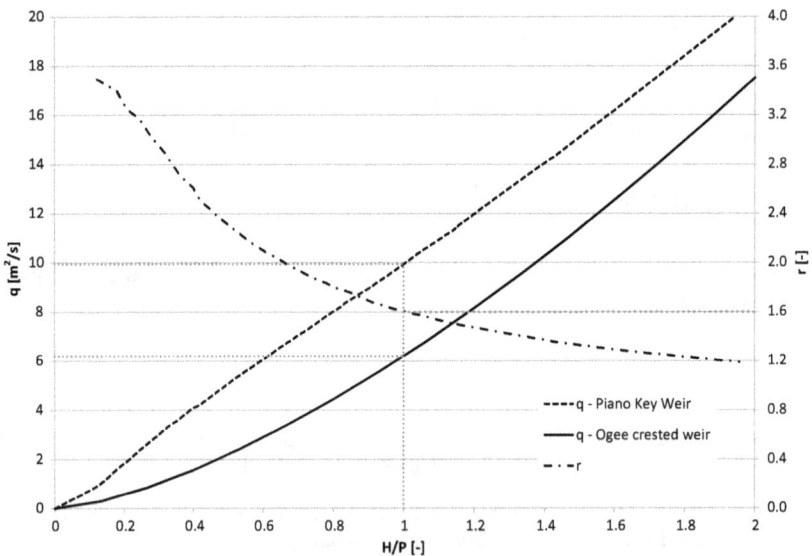

Fig. 6.5

Discharge per unit width W of a PKW compared to an ogee crested weir. PKW: $P=P_i=P_o=2$ m and $L/W=5$; ogee crested weir: design head=2 m, $C_S=0.494$ (Erpicum et al. 2013b)

6.4. MAIN GEOMETRICAL PARAMETERS

6.4.1. Main geometrical parameters

The main geometrical parameters governing PKW discharge capacity are the following:

- Ratio "L/W". This "labyrinth" ratio represents the developed length. It is a major factor for small heads.

- PKW height "P". High P value improves the flow approach and distribution within PKWs inlet keys. It also reduces submergence effects in outlet keys.

- Key width ratio "Wi/Wo". High Wi/Wo value improves flow approach and distribution within PKWs inlet keys. On the other hand, high Wi/Wo value increases submergence effects in outlet keys.

- Sidewall thickness T_s. Thicker sidewalls significantly reduce the available space on the dam crest. This structural factor can affect up to 10-15% PKW specific discharge capacity

6.4.2. Discussion

The PKW discharge efficiency results from the cumulative effects of: (i) three differents types of overflow (linear weir flow over the inlet key apex, linear weir flow over the outlet key apex, and lateral weir flow over the sidewalls), (ii) the developed length of the crest and (iii) the upstream head. Clinging, leaping or springing nappes may happen, depending upon the geometrical parameters of the structure and the head (Machiels et al. 2011a). The greater the head, the smaller the PKW's efficiency increase compared to a linear weir, according to the progressive reduction of the effective developed length and the saturation of the outlet.

Ouamane and Lempérière, (2006) and Leite Ribeiro et al. (2012a) showed that the crest length magnification ratio L/W is the main parameter controlling the discharge capacity. A value of 5 seems to be a reasonable compromise between weir efficiency and structure complexity (Lempérière, 2009; Lempérière et al. 2011), while L/W ratio of existing PKW ranges from 4 to 8 (Pfister et al. 2012). In a further step, Machiels et al. (2011a and 2012b) identify the key height P, the key width ratio W_i/W_o and the overhangs positions ratio B_o/B_i as the main geometric parameters influencing the PKW hydraulic efficiency for a given L/W ratio.

These studies showed that it is of major importance to increase all the geometrical ratios which increase the inlet cross section, as this last one can be seen as the "engine" of the PKW. Increasing the inlet cross section decreases the flow velocity along the lateral crest and thus increases its efficiency. The maximum value of the parameters is reached when the release capacity of the outlet key is affected. Indeed, the outlet key can be seen as the "brake" of the PKW. Too small an outlet key cross section and slope increase the free surface level over the lateral crest elevation (local submergence) and thus significantly limit the weir efficiency.

A PKW design with a height ratio P/W_u equal to 1.3, a key width ratio W_i/W_o equal to 1.25 and an overhangs lengths ratio B_o/B_i equal to 3 was found by Machiels (2012) to provide the highest discharge capacity when the L/W ratio is equal to 5. This finding is consistent with the findings of Leite Ribeiro et al. (2012a), Anderson and Tullis (2013) and Lempérière et al. (2011).

However, the Machiels' study also highlights the importance of the technical and economic criteria in the definition of an optimal PKW design. A high PKW (P/W_u = 1.3) is more effective from a hydraulic point of view and should thus be considered for new dam projects, shorter PKW designs $|(P/W_u \approx 0.5)$, though less hydraulically efficient, would be more practical for dam rehabilitation projects. For the later, W_i/W_o and B_i/B_o ratios equal to 1 are relevant (Erpicum et al. 2014).

Each PKW's design will be a compromise of the aforementioned parameters in order to find an appropriate answer to the specific project's features. In any case, for a given available width, the PKW has (i) a great efficiency at low relative heads (H/P_i), and (ii) can provide a discharge capacity 2 to 5 times greater than an ogee crest using the same width.

6.5.   TAILWATER SUBMERGENCE

Weir tailwater submergence occurs when the elevation of the downstream water surface exceeds the crest elevation and increases the upstream water elevation for a give discharge. For most applications, PKW submergence would only be a consideration for in-channel or in–river structures and not for top-of-dam applications.

Similarly to labyrinth weirs, PKW tailwater submergence has been studied mainly using physical modelling in channel configurations (Belaabed and Ouamane, 2011; Cicéro and Delisle, 2013b; Dabling and Tullis, 2012). It appears that the PKW behaviour regarding submergence is very sensitive to the geometry and type (Figure 6.6), and deserves particular care to be predicted.

Dabling and Tullis, (2012) concluded that for relatively low levels of submergence, the PKW requires less upstream head relative to the labyrinth weir to pass a given discharge. This increase in efficiency was smaller than 6%, and this trend reversed at higher submergence levels.

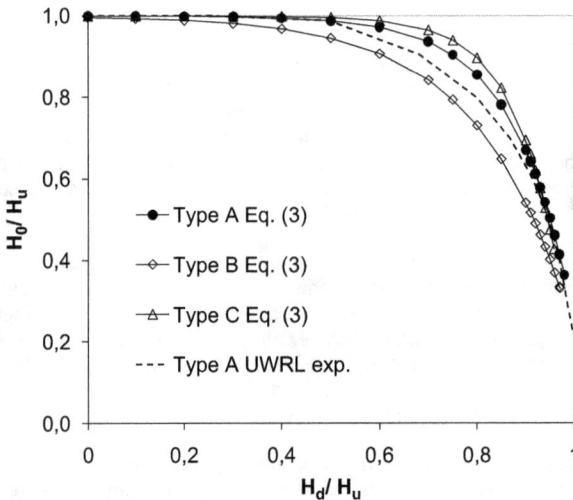

Fig. 6.6
Comparison of the sensitivity to submergence of three PKW different types (Cicéro and Delisle, 2013b)

6.6.   FLOATING DEBRIS

Debris is often transported to hydraulic structures during flood events (Pfister et al. 2013b). The accumulation of debris on PKW should be considered in situations where there is the potential for debris to reduce spillway capacity. For example, storm events in forested catchments may result in flows laden with woody debris. Depending upon the spillway hydraulics and site conditions, this type of debris may pass over the weir during floods and require little maintenance. Conversely, conditions may result in large accumulations of woody debris that reduce spillway capacity (-25% may be observed) and raise safety concerns.

A systematic laboratory study conducted by Pfister *et al.* (2013b) to evaluate the interaction between various PKW geometries and woody debris types and sizes indicated that floating debris blockage probability is highly influenced by trunk diameter and upstream head. The effects of debris accumulation on the upstream head varied with the value of the debris-free reference upstream head condition. For small upstream reference head values, the cumulative debris tests indicated a relative increase of the debris-associated upstream head of approximately 70%; larger upstream reference head values produced upstream head increases limited to approximately 20%. In addition, as per most free flow spillways, most debris are naturally flushed away when the water head increases.

6.7.   AERATION AND ENERGY DISSIPATION

The observed downstream nappes on existing PKW spillway prototypes, mostly equipped with aeration pipes below the inlets apex, are usually well-aerated and not subjected to clinging or vibration, even for very low heads (Figures 6.7 and 6.8).

It is not clear yet whether aeration devices are required or not for PKW spillways. Some prototypes are not equipped with aeration device and apparently spilled many times without any problem. This has to be confirmed.

By default and to be conservative, many PKW prototypes were equipped with aeration devices if their associated costs were minor. This is normally the case for low heads PKW spillways. For high heads, basic design methods might lead to large aeration pipes with relevant cost impact. Aeration design criteria used were based on a typical ratio between air flow and water flow and on-air speed.

Very few measurements of air entrainment in aeration pipes have been done on prototypes or adequately sized hydraulic models. On Malarce spillway (France) in 2014, EDF equipped the prototype aeration pipes with air entrainment measurement devices (Pinchard *et al.* 2013). This PKW spillway spilled a dozen of times between 2014 and 2017, and preliminary data were recently published by Vermeulen, (2017) including a more refined method to design aeration pipe. The newly developed method allows a significant reduction in the required aeration device.

Fig. 6.7
Flow over the Malarce Dam PKW spillway (France) with an upstream head of a few cm. (Photo courtesy of EDF)

Fig. 6.8
Flow over the Escouloubre PKW (France) with an upstream head of a few cm. (Photo courtesy of ULg-HECE)

Laboratory tests have been performed with PKW upstream of stepped spillways (Ho Ta Khanh *et al.* 2011b; Silvestri *et al.* 2013a 2013b). They showed that the flow downstream of a PKW is always well aerated, i.e., that the inception point is located immediately at the weir toe (Figure 6.9).

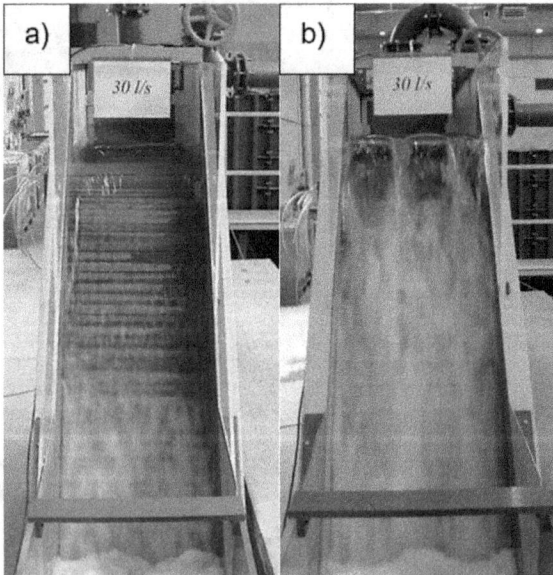

Fig. 6.9
Modification in the inception point location along a stepped spillway with (a) a standard ogee-crested weir, and (b) a PKW (q = 0,06 m²/s) (Adapted from Silvestri *et al.* 2013a)

The possible high specific discharge downstream of a PKW spillway implies that great care has to be provided for the design of the downstream energy dissipation structure, in particular when the weir is located on the crest of a high dam.

The energy dissipation solutions already designed and build downstream of PKW prototype spillway cannot be generalised and have all been found in an innovative way using physical modelling (Figures 6.10 and 6.11) (Bieri *et al.* 2011; Erpicum *et al.* 2011b; Leite Ribeiro *et al.* 2011).

Fig. 6.10
Low slope stepped spillway channel downstream of the Gloriette Dam PKW (Photo courtesy of EDF)

Fig. 6.11
Smooth spillway and inclined flip bucket downstream of the Saint Marc Dam PKW spillway
(Photo courtesy of EDF)

In the scope of the Gloriette Dam project, Bieri *et al.* (2011), showed that PKW combined with stepped chutes may lead to pronounced downstream energy dissipation.

6.8. HYDRAULIC DESIGN EQUATIONS FOR A-TYPE PIANO KEY WEIRS AND DESIGN METHOD

Hydrocoop proposed a very simple preliminary method to specify the discharge per unit width q under an upstream head H for an A-type PKW with a maximum height of the labyrinth walls equal to P_m (Lempérière, 2015). This simplified formula (Eqn. 6.6) applies only in the range of parameters specified by the author.

$$q = 4.3 \, HP_m^{0.5} \qquad\qquad (6.6)$$

With q in m²/s, H and P_m in m (beware not get confused between Pm and P).

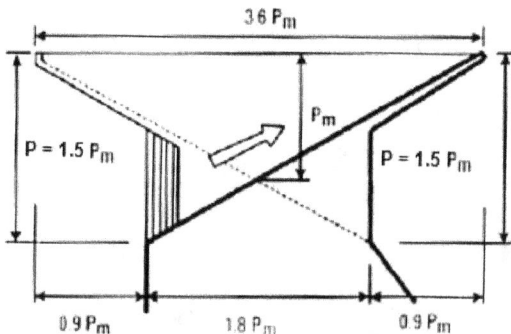

Figure 6.12
Typical cross section of PKW applicable to Lempérière simplified formula. Adapted from Lempérière (2015).

In the meantime, comprehensive and systematic model test series have been conducted in several laboratories (Schleiss, 2011). Based on such tests series, three general hydraulic capacity equations for Type-A PKWs have been proposed and validated (Kabiri-Samani and Javaheri, 2012; Leite Ribeiro *et al.* 2012a; Machiel *et al.* 2014). It is important to notice that these experimentally obtained equations must be used only within the limits of the parameter range specified by their authors, as clearly demonstrated by Pfister *et al.* (2012). It is worth noticing that these limitations should strictly be respected as small discrepancies may result in significant changes in the PKW discharge coefficient.

Pfister & Schleiss (2013a) compared these three hydraulic design formulae for a hypothetical symmetrical Type-A PKW, placed on a Roller Compacted Concrete (RCC) gravity dam with a W = 100 m wide chute spillway on its downstream face, a dam height of P_d = 30 m below the PKW foundation. A design discharge of Q_D = 2 500 m³/s was considered, resulting in a specific discharge of 25 m/s/m as commonly used on stepped spillways.

The PKW spillway geometry had a total streamwise length B = 8.00 m, $P = P_i = P_o$ = 5.00 m as vertical height, T_S = 0.35 m as wall thickness, R = 0 m (without parapet walls), W_i = 1.80 m as inlet key width, W_o = 1.50 m as outlet key width, and $B_i = B_o$ = 2.00 m as overhang lengths. The following characteristics result from this PKW geometry:

- cycle width $W_u = W_i + W_o + 2T_S$ = 4.00 m;

- number of cycles $N = W/W_u$ = 25;

- developed crest length $L = W + (2NB)$ = 500 m;

- and L/W = 5.00, B/P = 1.60, W_i/W_o = 1.20, $B_i/B = B_o/B$ = 0.25, $S_i = S_o$ = 0.83.

Since the tests of the three hydraulic design formulas mentioned before, were performed with different crest shapes, the results were normalized to a broad-crested weir. The resulting discharge capacity curves are shown in Figure 6.13 within their application limits of H/P. As it can be seen, the rating curves of the three PKW studies are similar. In general, the empiric equation of Kabiri-Samani and Javaheri (2012) predicts the highest discharge capacity. Additionally, the rating curve of a linear standard crest profile (ogee) is also shown in Figure. 6.14, derived from Vischer and Hager (1999) considering a design head of H_D = 5.00 m for Q_D.

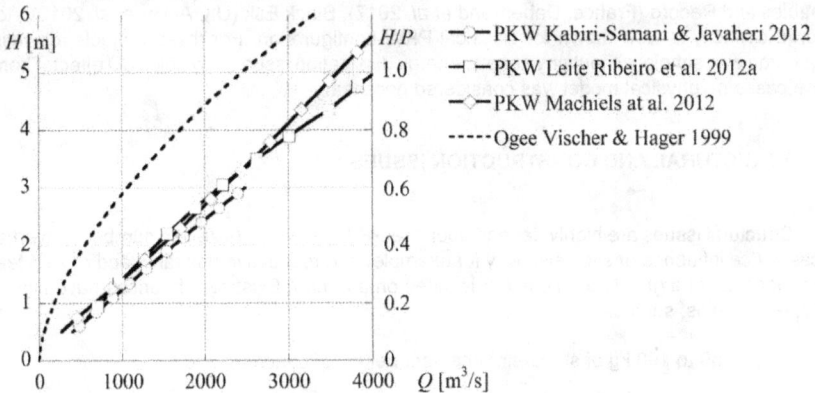

Fig. 6.13
Comparisons of the three PKW Type-A discharge capacity equations and the rating curve of an ogee crest weir (Pfister and Schleiss, 2013a)

For practical application it may be concluded that the most appropriate PKW hydraulic design formula is dependent upon the application range and the prototype characteristics. The crest shape can serve as a second criterion for the formula selection. If several formulas are within the application range of the prototype, the most conservative result may be considered, in order to be on the safe side.

In addition, Machiels *et al.* (2011b & 2012a) proposed a simple preliminary design method to select a PKW geometry from existing rating curves and specific project constraints (maximum discharge, reservoir level, available width, etc.).

6.9. PHYSICAL AND NUMERICAL (CFD) MODELLING

The recent development of PKW spillways leads to intensive comparison between 3D numerical modeling and laboratory hydraulic tests on numerous geometries. These works clearly show that the use of (commercial) 3D software is very promising to predict the discharge capacity of a PKW geometry with a good accuracy (+/-10 %) (Pralong *et al.* 2011b; Lefebvre *et al.* 2013).

Besides, the freeware WOLF1D PKW (available on http://www.pk-weirs.ulg.ac.be) has been developed by the HECE research group of the University of Liege, based on the large-scale physical tests of Machiels (2012). From a PKW geometry, this freeware is able to compute very quickly the flow over half a unit of the weir for a given range of discharge upstream of the structure. The results of the computation are the weir head / discharge curve (+/- 10% accuracy), the water level upstream of the weir depending on the discharge and the distribution of water depth and discharge along both the inlet and the outlet.

Nevertheless, CFD modelling cannot reproduce all the flow characteristics such as instabilities for instance, the effects of complex approach flow conditions or specific abutment geometry as well as the consequences of blocking by floating debris. Furthermore, downstream flow and energy dissipation are critical issues for the overall spillway efficiency which cannot be assessed accurately with numerical models.

Therefore, the final geometry of a PKW together with the energy dissipation system should be tested on a physical model, which is the only way to reproduce properly the complete flow features with the required level of precision. It is worthwhile mentioning that most of large PKW prototypes have been validated and optimised by physical modelling.

However, a couple of recent PKW spillway were designed without physical modelling: Campauleil and Record (France, Dabertrand *et al.* 2017), Black Esk (UK, Acker *et al.* 2013). Actually discharge capacity is well known for classical PKW configuration. For these projects, downstream configuration was simple without any serious energy dissipation issue or complex 3D effect. Therefore, in these cases no physical model was considered necessary.

6.10. STRUCTURAL AND CONSTRUCTION ISSUES

Structural issues are highly depend upon the PKW geometry (size and number of overhangs), load cases (ice influence or site seismicity for example) and respective unit rates and project features (such as location at a gravity dam crest or isolated on a bank.). Existing references have highlighted some specific ratios, such as:

- 50 to 100 kg of steel reinforcement per m^3 of concrete; and

- 60 to 130 m^2 formwork per m of width.

It is common practice to add concrete or anchors in order to reach the required stability factors of safety. Thermal load cases (one face is under the sun, the other one protected by the water) or ice load can easily become controlling sizing parameters. Particular attention should also be given to construction features, such as pre-fabrication (Figure 6.14) or use of steel instead of concrete, as on many prototypes built up to now.

Fig. 6.14
Esk Reservoir spillway – PKW spillway made of prefabricated concrete elements (Courtesy of Black and Veatch Ltd)

6.11. ENGINEERING

So far, PKW spillways have primarily been designed and constructed as flood-routing spillways in the following three configurations:

1. narrow valley where only a limited space is available for installing a new structure;

2. on the crest of gravity dam, where the PKW is easily implemented due to its small footprint, allowing maximisation of the active storage capacity of the reservoir for a given Maximum Water Level. Generally, PKW is combined with surface or bottom outlet gates for the fine tuning of the upstream water level regulation and the flushing of the reservoir;

3. on top of long barrages (gated structure dam) in flat areas. In such cases the PKW, which may be associated with gates, whose number is then smaller than in a traditional alternative, minimises the inundated areas during floods by lowering the Maximum Water Level as compared to a solution with a linear weir.

According to the actual hydraulic knowledge, future design practice may be cost orientated. For example, ongoing research is exploring: (i) standard hydraulic profiles (allowing savings on the design and the construction), and (ii) best compromise between PKW, gates and fuse devices (in order to delay the tilting of the first fuse element).

The Van Phong barrage in Vietnam, 475-m long and 7- m high on the riverbed, with a 15 400 m³/s design flood, is a good example of the third configuration (Figure 6.15) mentioned above. The initial design included 28 radial gates with 15-m width and 7.5-m height. The final design, for the same Maximum Water Level, included only 10 radial gates (and 8 would have been probably enough) in the central part and 60 PKW units with a total length of 302 m on each side. Compared to a labyrinth weir solution, the PKW offered a smaller footprint on the rather deep bedrock, which required less excavation and concrete volume. The main advantages of the final alternative were: (i) the investment and maintenance cost savings, (ii) a higher safety level in operation, and (iii) a better integration with the environment. Taking into account the recent non-functioning of several gates during floods, many Vietnamese engineers think that a safe solution is now to combine, as much as possible, gates and free overflow crest spillway. In this scope, the PKW is often a good alternative, particularly when the available crest length is limited.

Fig. 6.15
The Van Phong barrage in Vietnam. (Courtesy of M. Ho Ta Khanh)

A new project with similar configuration is now under design for the Xuân Minh barrage.

6.12. ONGOING RESEARCH

Ongoing research and design progress are orientated towards new developments of the PKW geometry (side wall angle narrowing the inlet key and widening the outlet one; new hybrid configurations using rectangular labyrinth and PKW) and the analysis of downstream flow feature (aeration and energy dissipation). Research also concerns structural and construction aspects such as the use of steel or combined steel/concrete structures as well as pre-fabrication.

More recent research concerns the application of PKW or labyrinth weirs for river to replace regulating gates. Pfister (2017) studied the toe scour formation at PKW. Belzner (2017) led a general study on PKW and labyrinth weirs regarding downstream water influence, driftwood passage and sediment transportation over weirs. First results are promising and might allow to partially replace river gates by labyrinth weirs or PKW.

It has to be noted that the passage of sediment over PKW was first studied by Das Singhal *et al.* (2011) for the Sawra Kuddu project which was constructed in 2013. The passage of sediment was studied with a physical model. Paper shows that PKW type C (no overhang upstream) are slightly more efficient and allow passing most of sediments over the weir. Those important results were necessary to validate the feasibility of PKW for this run-off river scheme with high rates of sediments to pass.

6.13. PROTOTYPE PIANO KEY WEIR EXAMPLES

The bibliography included in the end of the chapter contains references for many PKW prototypes. A summary of these structures is presented in following table. Almost 30 PKW projects were finalized in less than 10 years in 8 different countries. This is a very quick development regarding typical required time for hydraulic projects.

Table 6.1
Piano Key Weir Spillways Prototypes

Name	Country	Configuration	Q_{design} m³/s	P m	H_{design} m	Date of completion	Source
Beaufort	France	Dam crest	70	1.30	0,70	2013	http://www.pk-weirs.ulg.ac.be/
Black Esk	Scotland	Bellmouth crest	183	2.10	0.97	2013	Ackers *et al.* 2013
Campauleil	France	Dam crest	120	5.35	0.90	2014	Laugier *et al.* 2017a
Charmines	France	Dam crest	300	4.38	1.00	2015	Valley *et al.* 2017a
Da Gang3	Vietnam	Dam crest	7300	5.50	5.00	2016	Ho Ta Khanh *et al.* 2017
Dak Mi 4B	Vietnam	Dam crest	500	3.75	2.00	2013	Ho Ta Khanh *et al.* 2017
Dak Mi 3	Vietnam	Dam crest	6550	5.00	3.50	2013	Ho Ta Khanh *et al.* 2017
Dak Rong 3	Vietnam	Dam crest	6550	5.00	3.50	2013	Ho Ta Khanh *et al.* 2017
Emmenau	Switzerland	River	4.35	1.20	0.45	2013	Eichenberger, 2013
Escouloubre	France	Bank	13	1.80	0.65	2011	Erpicum *et al.* 2013c
Etroit	France	Dam crest	82	5.30	0.95	2009	Laugier *et al.* 2011
Gage	France	Right bank	455	6.00	1.75	2017	Laugier *et al.* 2017b
Giritale	Sri Lanka	River		2.40	0.45	2013	Jayatillake and Perera. 2013
Gloriettes	France	Right bank	90	3.00	0.80	2010	Bieri *et al.* 2011
Gouillet	France	Right bank	18	1.20	0.35	2011	http://www.pk-weirs.ulg.ac.be/
Goulours	France	Right bank	68	3.10	0.95	2006	Laugier, 2007
Hazelmere	South Africa	Dam crest	4300	9.00	3.23	2017	Booyse *et al.* 2017
Loombah	Australia	Bank	416	2.50	2,5	2013	http://www.pk-weirs.ulg.ac.be/
Malarce	France	Dam crest	570	4.40	1.50	2012	Pinchard *et al.* 2013
Oule	France	Right bank	72	1.00	1.00	2017	http://www.pk-weirs.ulg.ac.be/
Ouljet Mellegue*	France	Right bank	3800	6.80	5.00	*	http://www.pk-weirs.ulg.ac.be/
Rambawa Tank	Sri Lanka	Dam Crest	28	1.00	0,35	2015	Jayatillake H.M. & al. 2017
Rattling Lake	Canada	Dam Crest				2011	www.cda.ca
Rassisse	France	Dam crest	306+ +101	3.8/2.7	1.30	2015	Bail *et al.* 2013
Raviège	France	Dam crest	300	4.67	1.40	2015	Cubaynes *et al.* 2017
Record	France	Dam crest	1350	4.67	3.00	2016	Dabetrand *et al.* 2017
Saint Marc	France	Dam crest	138	4.20	1.35	2008	Laugier *et al.* 2009
Sawra Kuddu	India	Dam crest	2500	10.45		2013	Das Singhal *et al.* 2011
Van Phong	Vietnam	Barrage (River)	8750	5.00	5.20	2015	Ho Ta Khanh *et al.* 2014
Xuan Minh*	Vietnam	Barrage (River)	9700	7.50	9.00	*	Ho Ta Khanh *et al.* 2017

*under construction

The following comments can be made:

- PKW prototypes concern a wide range of specific discharge capacity with head varying from 30 cm to more than 9 m. In parallel, total discharge goes from dozens of m³/s to almost 10 000 m³/s;

- PKW have been used for both rehabilitation and new projects; and

- PKW have been installed on the top of all sorts of dams: concrete gravity, arch dam, earthfill, run-off river.

These points outline the strong flexibility of PKW which can be used in small and large projects, new or rehabilitation projects.

6.14. PKW OR USUAL VERTICAL LABYRINTH WEIR?

Comparing traditional vertical walled labyrinth weir and PKW is a difficult exercise. Some attempts to compare both options are proposed by Blancher, (2011); Anderson, (2013) and Paxson (2014). Some considerations on this subject are listed in continuation.

- Blancher (2011) and Anderson (2013) proposed a hydraulic comparison of comparable structure based on numerical models (Blancher) and physical models (Anderson). Comparison is performed for the same structural height P and the same plan view (equal developed length).

- Hydraulic results show that PKW have a discharge capacity from 10 to 15% higher than conventional labyrinth depending on the water head due to the fact that vertical walls create an obstacle inducing head losses. This can be clearly seen on recirculating flow patterns in front of the vertical wall.

- Paxson (2014) proposed a more global approach trying to include both structural and hydraulic aspects to get an optimised economical solution.

- Vertical wall may be simpler to build, but they require more internal reinforcement. Actually, the average height of a labyrinth wall is P whereas it is P/2 for PKW. Thus average moments in labyrinth structure are much higher than for PKW, and so the reinforcement.

- A labyrinth spillway requires a base slab foundation while a PKW do not require the same base slab thickness according to their geometry.

- Due to their large basement footprint, conventional labyrinth weirs are generally not fit to be built on the top of the crest of concrete dam (gravity or arch).

- Conventional labyrinth might be an option if the labyrinth units are very small (1 or 2 m) or if the weir is to be built on a relatively flat bank. The choice will depend on site access, topography, methods and material of contractors.

6.15. REFERENCES

Ackers, J.C., F.C.J. Bennett, T.A. Scott and G. Karunaratne, (2013) - "Raising of the bellmouth spillway at Black Esk reservoir using Piano Key weirs". *Labyrinth and piano key weirs II - PKW 2013*, CRC Press, London, 235-242

Anderson, R.M. and B.P Tullis (2012) - "Comparison of Piano Key and Rectangular Labyrinth Weir Hydraulics", *J. Hydraulic Eng.* 138, 358-361

Anderson, R.M. and B.P Tullis (2013) - "Piano Key Weir Hydraulics and Labyrinth Weir comparison". *Journal of Irrigation and Drainage Engineering*, 139(3), 246-253.

Bail, A., L. Deroo and J.P Sixdenier (2013) - "Designing a new spillway at Rassisse Dam". *Labyrinth and piano key weirs II - PKW 2013*, CRC Press, London, 169-176.

Barcouda, M., F. Laugier, O. Cazaillet, C. Odeyer, P. Cochet, B.A. Jones, S. Lacroix and J.P. Vigny, J.P. (2006) - "Cost effective increase in storage and safety of most existing dams using fusegates or P.K.Weirs", *Proceedings of the 22nd ICOLD Congress*. (Q84, R78). Barcelona, Spain.

Belaabed, F. and A. Ouamane (2011) - "Contribution to the study of the Piano Key weir submerged by the downstream level". *Labyrinth and Piano Key Weirs – PKW 2011*, CRC Press, Leiden, 89-96.

Belzner F., J. Merkel, M. Gebhardt and C. Thorenz (2017) - "Piano Key and Labyrinth Weirs at German Waterways: recent and future research of the BAW". *Labyrinth and piano key weirs III - PKW 2017*, CRC Press, London

Bieri M., M. Federspiel, J.L. Boillat, B. Houdant, L. Faramond and F. Delorme (2011) - "Energy dissipation downstream of Piano Key Weirs - Case study of Gloriettes Dam (France)", *Labyrinth and piano key weirs-PKW 2011*, CRC press, London, 123-130.

Blanc, P. and F. Lempérière (2001) - "Labyrinth spillways have a promising future", *Hydropower & Dams*, 8(4):129-131.

Blancher B., F. Montarros, and F. Laugier F (2011) "Hydraulic comparison between piano-keys weir and labyrinth spillways". *Labyrinth and Piano Key Weirs – PKW 2011*, CRC Press, Leiden, 89-96.

Booyse D. (2017) - "The raising of Hazelmere Dam by means of Piano Key Weir". *Labyrinth and piano key weirs III - PKW 2017*, CRC Press, London.

Botha A.J., J.P. Fitz, A.J. Moore, F.E. Mulder and N.J. Van Deventer (2013) - "Application of the Piano Key weir spillway in the Republic of South Africa". *Labyrinth and piano key weirs II - PKW 2013*, CRC Press, London, 185-194.

Cicéro, G-M., J-R. Delisle, V. Lefebvre and J. Vermeulen (2013a) - "Experimental and numerical study of the hydraulic performance of a trapezoidal Piano Key weir", *Labyrinth and Piano Key Weirs II – PKW 2013*, CRC Press, Leiden, pp. 265-272.

Cicéro, G-M. and J-R. Delisle (2013b) - "Discharge characteristics of Piano Key weirs under submerged flows". *Labyrinth and Piano Key Weirs II – PKW 2013*, CRC Press, Leiden, pp. 101-109.

Cubaynes M., F. Laugier and V. Nagel (2017) - "Construction of a Piano Key Weir spillway at La Raviège dam". *Labyrinth and Piano Key Weirs III – PKW 2017*, CRC Press

Dabling, M. and B. Tullis (2012) - "Piano Key Weir Submergence in Channel Applications." *J. Hydraul. Eng.*, 138(7), 661–666.

Dabertrand F., J. Vermeulen and B. Blancher (2017) - "Construction of a Piano Key Weirs spillway at Record dam". (2017). *"Labyrinth and Piano Key Weirs III – PKW 2017*, CRC Press

Das Singhal G. and N. Sharma (2011) - "Rehabilitation of Sawara Kuddu Hydroelectric Project - Model studies of Piano Key Weir in India". *Labyrinth and piano key weirs-PKW 2011*, CRC Press, London, 241-250.

Dugué V., F. Hachem, J.L. Boillat, V. Nagel, J.P. Roca and Laugier F. (2011) - "PK Weir and flap gate spillway for the Gage II Dam". *Labyrinth and piano key weirs-PKW 2011*, CRC Press, London, 35-42.

Eichenberger, P. (2013) - "The first commercial Piano Key weir in Switzerland". *Labyrinth and piano key weirs II - PKW 2013*, CRC Press, London, 227-234

Erpicum, S., F. Laugier, J-L. Boillat, M. Pirotton, and B. Reverchon (2011a) - *Labyrinth and piano key weirs – PKW 2011*. Schleiss, A.J, eds. CRC Press, Boca Raton Fl, USA.

Erpicum S., V. Nagel and F. Laugier (2011b) - "Piano Key Weir design study at Raviege dam". *Labyrinth and piano key weirs-PKW 2011*, CRC Press, London, 43-50.

Erpicum, S., O. Machiels, B.J. Dewals, M. Pirotton and P. Archambeau (2012) - "Numerical and physical modelling of Piano Key Weirs", *Proceedings of Asia 2012 Conference*, Chiang Mai, Thailand.

Erpicum, S., F. Laugier, M. Pfister, M. Pirotton and G-M. Cicéro (2013a) - *Labyrinth and piano key weirs II – PKW 2013*. Schleiss, A.J, eds. CRC Press, Boca Raton, Fl, USA

Erpicum, S., O. Machiels, B. Dewals, P. Archambeau and M. Pirotton (2013b) - "Considerations about the optimum design of PKW", *Proc. Intl. Conf. Water Storage and Hydropower Development for Africa (Africa 2013)*, Addis Ababa (Ethiopia), CD 13.04.

Erpicum, S., A. Silvestri, B.J. Dewals, P. Archambeau, M. Pirotton, M. Colombié and L. Faramond (2013c) - "Escouloubre Piano Key weir: prototype versus scale models". *Labyrinth and piano key weirs II - PKW 2013*, CRC press, London

Erpicum, S., P. Archambeau, M. Pirotton and B.J. Dewals (2014) - "Geometric parameters influence on Piano Key Weir hydraulic performances". *5th IAHR International Symposium on Hydraulic Structures*, Brisbane, Australia, (1-8). 25-27 June 2014. doi:10.14264/uql.2014.31

Erpicum, S., F. Laugier, M. Ho Ta Khan and Pfister (2017) - *Labyrinth and piano key weirs III – PKW 2017*. CRC Press, Boca Raton, Fl, USA

Ho Ta Khanh M., D. Sy Quat and D. Xuan Thuy (2011a) - "P.K weirs under design and construction in Vietnam (2010)". *Labyrinth and piano key weirs-PKW 2011*, CRC Press, London, 225-232.

Ho Ta Khanh M., T.C. Hien and T.N. Hai (2011b) - "Main results of the P.K weir model tests in Vietnam (2004 to 2010)", *Labyrinth and piano key weirs-PKW 2011*, CRC press, London, 191-198.

Ho Ta Khanh, M., T. Chi Hien and D. Sy Quat (2012) - "Study and construction of PK Weirs in Vietnam (2004 to 2011)", *Proceedings of Asia 2012 Conference*, Chiang Mai, Thailand.

Ho Ta Khanh, M. (2017) - "History and development of Piano Key weirs in Vietnam from 2004 to 2016". *Labyrinth and piano key weirs III - PKW 2017*, CRC press, London.

Jayatillake H.M. and K.T.N. Perera (2013) - "Design of a Piano-Key Weir for Giritale dam spillway in Sri Lanka". *Labyrinth and piano key weirs II - PKW 2013*, CRC Press, London, 151-158.

Jayatillake H.M. and K.T.N. Perera (2017) - "Adoption of a type-D Piano Key Weir spillway with tapered noses at Rambawa Tank, Sri Lanka". *Labyrinth and piano key weirs III - PKW 2017*, CRC Press.

Kabiri-Samani, A. and A. Javaheri (2012) - "Discharge coefficient for free and submerged flow over Piano Key weirs". *Journal of Hydraulic Research*, 50(1), pp. 114-120.

Laugier F. (2007) - "Design and construction of the first Piano Key Weir (PKW) spillway at the Goulours dam", *International Journal of Hydropower and Dams* 14 (5), 94-101.

Laugier F., A. Lochu C. Gille M. Leite Ribeiro and J-L Boillat (2009) - "Design and construction of a labyrinth PKW spillway at Saint-Marc dam, France", *International Journal of Hydropower and Dams* 16 (5), 100-107.

Laugier F., C. Gille and O. Cazaillet (2011) - "Adaptation of Piano Key Weir (PKW) spillway solution to upgrade l'Etroit dam affected by concrete swelling pathology", *Proceedings of Hydro 2011 conference*.

Laugier F., J. Vermeulen and V. Lefebvre (2013) - "Overview of Piano Key Weirs experience developed at EDF during the past few years". *Labyrinth and piano key weirs II - PKW 2013*, CRC press, London, 213-226.

Laugier F. and J. Vermeulen (2017a) - "Overview of design and construction of 11 Piano Key Weirs spillways developed in France by EDF from 2003 to 2016". *Labyrinth and piano key weirs III - PKW 2017*, CRC press, London.

Laugier F., B. Blancher, S. Bouassida and V. Nagel (2017b) - "Hydrothermal - season based design of a new flood spillway at Gage II dam". In French. *Proceedings of CFBR / SHF conference, Chambery, France.*

Lefebvre, V., J. Vermeulen and B. Blancher (2013) - "Influence of geometrical parameters on PK-weirs discharge with 3D numerical analysis". *Labyrinth and piano key weirs II - PKW 2013*, CRC press, London, 49-56

Leite Ribeiro M., M. Bieri, J-L. Boillat, A.J. Schleiss, F. Delorme and F. Laugier (2009) - "Hydraulic capacity improvement of existing spillways - Design of Piano Key Weirs", in *Proceedings of 23rd ICOLD Congress*, Brasilia, Brazil, Q.90, R.43.

Leite Ribeiro M., J-L. Boillat, A. Schleiss and F. Laugier (2011) - "Coupled spillway devices and energy dissipation system at St-Marc Dam (France)", *Labyrinth and piano key weirs-PKW 2011*, CRC press, London, 113-121.

Leite Ribeiro, M., M. Pfister, A.J. Schleiss and J-L.Boillat (2012a) - "Hydraulic design of A-type Piano Key Weirs", *Journal of Hydraulic Research*, 50(4):400-408.

Leite Ribeiro, M., M. Bieri, J-L. Boillat, A.J. Schleiss, G. Singhal and N. Sharma (2012b) - "Discharge capacity of Piano Key Weirs", *Journal of Hydraulic Engineering*, 138:199-

Lempérière, F., J-P. Vigny and A. Ouamane, A. (2011) - "General comments on Labyrinths and Piano Key Weirs: the past and present", *Labyrinth and Piano Key Weirs – PKW 2011*, CRC Press, Leiden, 17-24.

Lempérière F. and A. Ouamane (2015) - "Increasing the discharge capacity of free-flow spillway fivefold", *Hydropower & Dams*, issue 65):80-82.

Loisel, P.E., P. Valley and F. Laugier, F. (2013) - "Hydraulic physical model of Piano Key weirs as additional flood spillways on the Charmine dam". *Labyrinth and piano key weirs II - PKW 2013*, CRC Press, London, 195-202.

Machiels, O., S. Erpicum, B. Dewals, P. Archambeau and M. Pirotton (2011a) - « Experimental observation of flow characteristics over a Piano Key Weir". *J. Hydraulic Res.* 49(3), 359-366.

Machiels O., S. Erpicum, P. Archambeau, B.J. Dewals and M. Pirotton M. (2011b) - "Piano Key Weir preliminary design method - Application to a new dam project", *Labyrinth and piano key weirs-PKW 2011*, CRC Press, London, 199-206.

Machiels, O. (2012) - *Experimental study of the hydraulic behaviour of Piano Key Weirs*, PhD thesis. HECE research unit, University of Liège, Belgium. http://hdl.handle.net/2268/128006.

Machiels O., S. Erpicum, P. Archambeau, B.J. Dewals and M. Pirotton (2012a) - "Method for the preliminary design of Piano Key weirs". *La Houille Blanche*, 4-5, 14-18

Machiels, O., S. Erpicum, P. Archambeau, B.J. Dewals and M. Pirotton (2012b) - "Parapet wall effect on Piano Key Weirs efficiency". *Journal of Irrigation and Drainage Engineering*, 139(6), 506-511.

Machiels, O., M. Pirotton, P. Archambeau, B.J. Dewals and S. Erpicum (2014) - "Experimental parametric study and design of Piano Key Weirs". *Journal of Hydraulic Research*, 52(3), 326-335

Ouamane, A. and F. Lempérière (2003) - "The piano keys weir: a new cost-effective solution for spillways", *Hydropower & Dams*, 10(5):144-149.

Ouamane, A. and F. Lempérière (2006) - "Design of a new economic shape of weir". *Dams and Reservoirs, Societies and Environment in the 21st Century*, Berga et al. (eds), Taylor & Francis, London: 463-470.

Paxson G.S., B.P. Tullis and D.J. Hertel (2013) - "Comparison of Piano Key Weirs with labyrinth and gated spillways: hydraulics, cost, constructability and operations". *Labyrinth and piano key weirs II - PKW 2013*, CRC Press, London

Paxson, G. and F. Laugier (2014) - Labyrinth and Piano Key Weirs, Perspectives and Case Histories from the USA and France). *Proceedings of ASDSO national Conference*

Pfister, M., S. Erpicum, O. Machiels, A. Schleiss and M. Pirotton (2012) - "Discharge coefficient for free and submerged flow over Piano Key weirs - Discussion", *Journal of Hydraulic Research*, 50(6):642-645.

Pfister, M. and A.J. Schleiss (2013a) - "Comparison of hydraulic design equations for A-type Piano Key weirs". Proc. Intl. Conf. Water Storage and Hydropower Development for Africa (Africa 2013), Addis Ababa (Ethiopia), CD 13.05.

Pfister, M., D. Capobianco, B. Tullis and A. Schleiss (2013b) - "Debris-Blocking Sensitivity of Piano Key Weirs under Reservoir-Type Approach Flow." *J. Hydraul. Eng.*, 139(11), 1134–1141.

Pinchard, T., J-L.Farges, J-M. Boutet, A. Lochu and F. Laugier (2013) - "Spillway capacity upgrade at Malarce dam: construction of an additional piano key weir spillway". *Labyrinth and piano key weirs II - PKW 2013*, CRC Press, London, 243-252

Pralong, J., J. Vermeulen, B. Blancher, F. Laugier, S. Erpicum, O. Machiels, M. Pirotton, J-L. Boillat, M. Leite Ribeiro and A. Schleiss (2011a) - "A naming convention for the Piano Key Weirs geometrical parameters". *Labyrinth and piano key weirs - PKW 2011*, CRC press, London, 271-278.

Pralong J., F. Montarros, B. Blancher and F. Laugier (2011b) - "A sensitivity analysis of Piano Key Weirs geometrical parameters based on 3D numerical modelling", *Labyrinth and piano key weirs-PKW 2011*, CRC Press, London, 133-139.

Schleiss, A.J. (2011) - "From labyrinth to piano key weirs: A historical review". *Labyrinth and piano key weirs-PKW 2011*, CRC Press, London, 3-15.

Silvestri, A., S. Erpicum, P. Archambeau, B. Dewals and M.Pirotton (2013a) - "Stepped spillway downstream of a Piano Key weir – Critical length for uniform flow". *International Worskhop on hydraulic structures*. Bundesanstalt für Wasserbau, Karlsruhe, Germany, 99-107.

Silvestri, A., P. Archambeau, M. Pirotton, B. Dewals and S. Erpicum (2013b) - "Comparative analysis of the energy dissipation on a stepped spillway downstream of a Piano Key weir". *Labyrinth and piano key weirs II - PKW 2013*, CRC Press, London, 111-120.

Truong Chi, H., S. Huynh Thanh and M. Ho Ta Khanh (2006) - "Results of some piano keys weir hydraulic model tests in Vietnam", *Proceedings of the 22nd ICOLD Congress*. (Q87, R39). Barcelona, Spain.

Valley P. and B. Blancher (2017) - "Construction and testing of two Piano Key Weirs at Charmines dam". *Labyrinth and piano key weirs III - PKW 2017*, CRC Press, London

Vermeulen, J., F. Laugier, L. Faramondand and C. Gille (2011) - "Lessons learnt from design and construction of EDF first Piano Key Weirs". *Labyrinth and piano key weirs - PKW 2011*, CRC press, London, 215-224

Vermeulen J., C. Lassus and T.C. Pinchard (2017) - "Design of a Piano Key Weir aeration network". *Labyrinth and piano key weirs III - PKW 2017*, CRC press, London

Vischer, D. and W.H. Hager (1999) - *Dam Hydraulics*. Wiley, Chichester UK.

7.   TUNNEL, SHAFT AND VORTEX SPILLWAYS

7.1.   INTRODUCTION

This chapter considers different kinds of tunnel spillways, such as high-level, low-level (or low outlet), vortex, shaft and orifice tunnels. The general arrangement strategies, the control structure, the conveyance structure and the terminal structure, as well as operation issues, risk analysis and considerations, are discussed. Some new developments, innovations and applications on tunnel spillway design for high dams are specially given, with emphasis in what have been achieved in the past 20 to 30 years. Risk analyses both on cavitation risk control and emergency operation are discussed. Some good and valuable cases studies are given as references.

7.2.   GENERAL ARRANGEMENT STRATEGIES FOR HIGH HEAD SPILLWAYS

7.2.1.   Location, type and size

A tunnel spillway can be arranged on one side or on both sides of dam abutments. This last case will happen when a dam is built in a narrow valley with large floods and where surface discharge facilities are not sufficient to discharge the floods. The selection and design of tunnel spillways will depend on the type of dam, the total discharge capacity, the river diversion scheme and on operation and management requirements. Table 7-1, at the end of this chapter, gives the characteristics of some typical large tunnel spillways in the world which are under design, construction and/or operation.

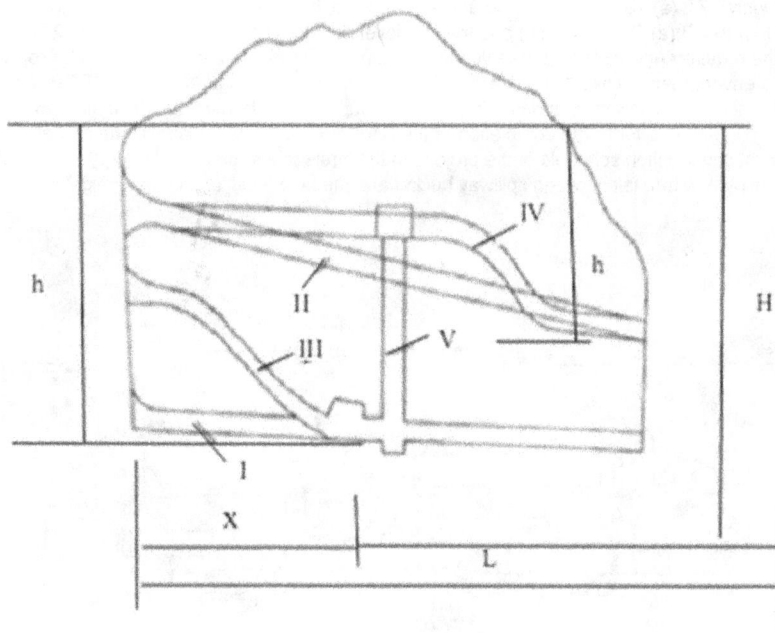

Fig. 7.1
Classification of tunnel spillway layout by longitudinal sectional arrangement

Figure 7.1 depicts a classification of five typical tunnel layouts, based on longitudinal section arrangements. Tunnels can also be classified by types of energy dissipation, as shown in Figure 7.2.

The flow energy dissipation in Type I to Type IV tunnels happens outside the tunnels. High speed flow carries the huge amount of energy to the river reach downstream. On the other hand, the flow energy in Type V tunnel is dissipated inside the tunnel and the flow reaches the river with small velocities.

The layouts of Type I and II tunnel spillways are relatively simple. The tunnel is almost straight and the elevation difference between inlet and outlet, in most cases, is not too large, allowing its application in both high and low dams, depending on the operation purpose of flow discharge; see Figure 7.3.

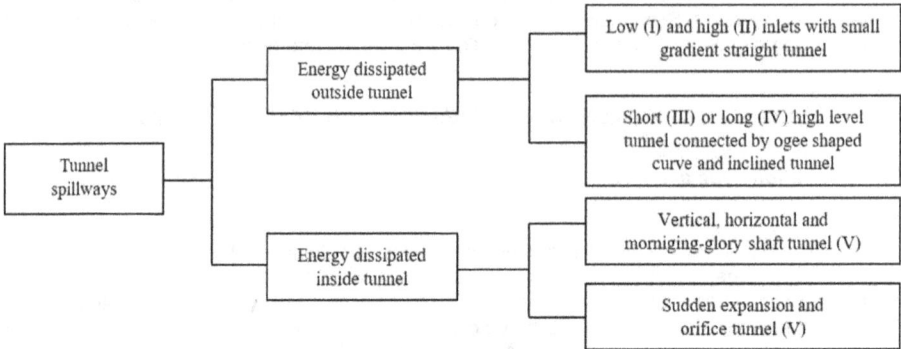

Fig. 7.2
Classification of tunnel spillway layout by type of energy dissipation

Figures 7.3 (a), (b) and(c) show examples of layouts for types I and II tunnels. The low-level outlet tunnel in Figure 7.3 (a) (Moore, 1989) and the mid-level outlet in Figure 7.3(b) (Shakirov, 2013) are built to meet the requirement of starting reservoir impounding and creating a discharge which provided dam safety and environmental flow. In the Brazilian Irapé Dam, shown in Figures 7.3 (c) (CBDB, 2009) there are two surface tunnel spillways and one intermediate outlet. The intermediate outlet allowed the start of reservoir impounding prior to the completion of the dam embankment, which was fundamental to meet the very tight construction schedule of the project. In this project, a large part of energy is dissipated in a plunge pool by flow free fall between spillway bucket and the river level, as the dam height is over 200 m.

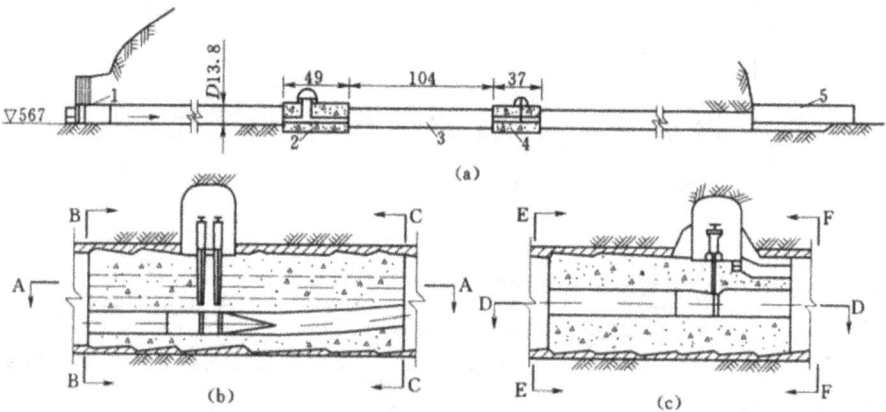

Fig. 7.3
(a) Low-level outlet tunnel spillway in Mica Dam, Canada, with sudden expansion

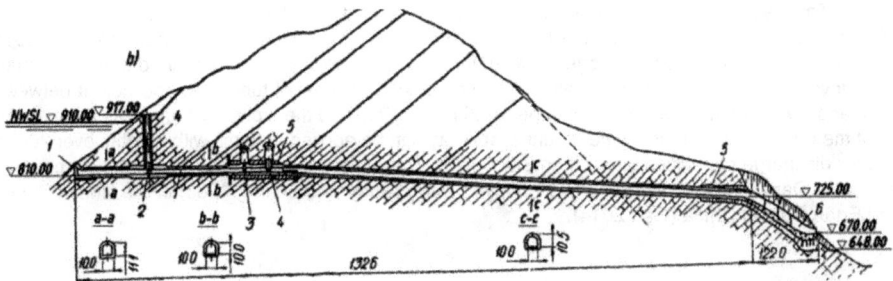

Fig. 7.3
(b) Mid-lever outlet tunnel spillways in Nurek Dam in Tajikstan

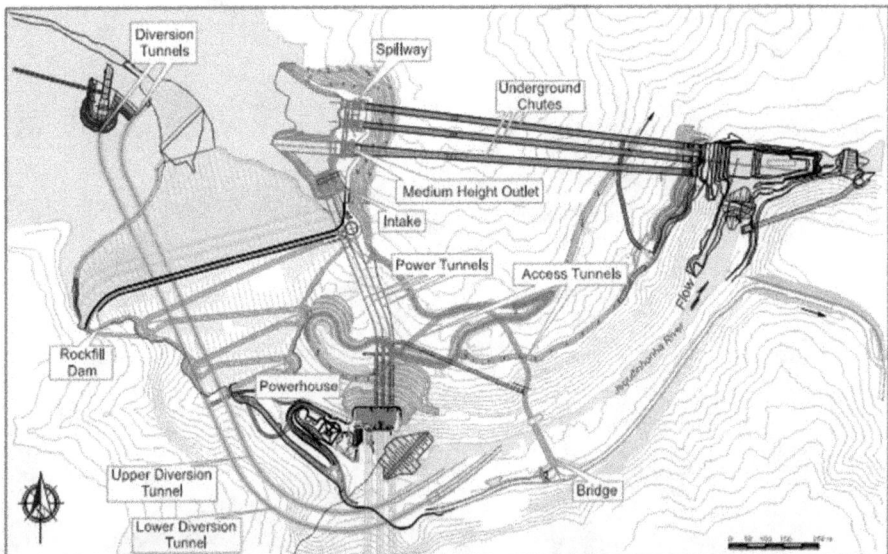

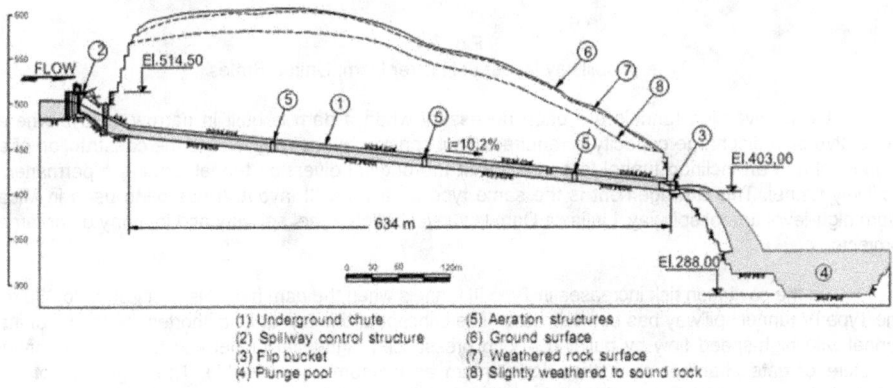

(1) Underground chute   (5) Aeration structures
(2) Spillway control structure (6) Ground surface
(3) Flip bucket   (7) Weathered rock surface
(4) Plunge pool   (8) Slightly weathered to sound rock

Fig. 7.3
(c) Tunnel spillway in Irapé Dam in Brazil

The layout of Type III tunnel spillway is widely applied in high dam schemes since it was first adopted in the USA Hoover Dam in the 1930's; see Figure 7.4. A short high-level tunnel, less than 100 m long, is built and connected to an ogee shaped intake section, followed by an inclined tunnel and then to an inversed curve tunnel and finally to a small gradient low-level tunnel. As the height between high- and low-level tunnels can be as large as 80 m to 100 m and the cross-section area of tunnel can be of the order of 150 m², this type of tunnel spillway can be applied in dams with heights over 200 m and for discharge capacities larger than 3 000 m³/s. The typical applications are the tunnel spillways in Hoover Dam (Falvey, 1990) and Glen Canyon Dam, (Burgi and Eckley, 1988) in the United States and Ertan Dam in China (Gao 2014).

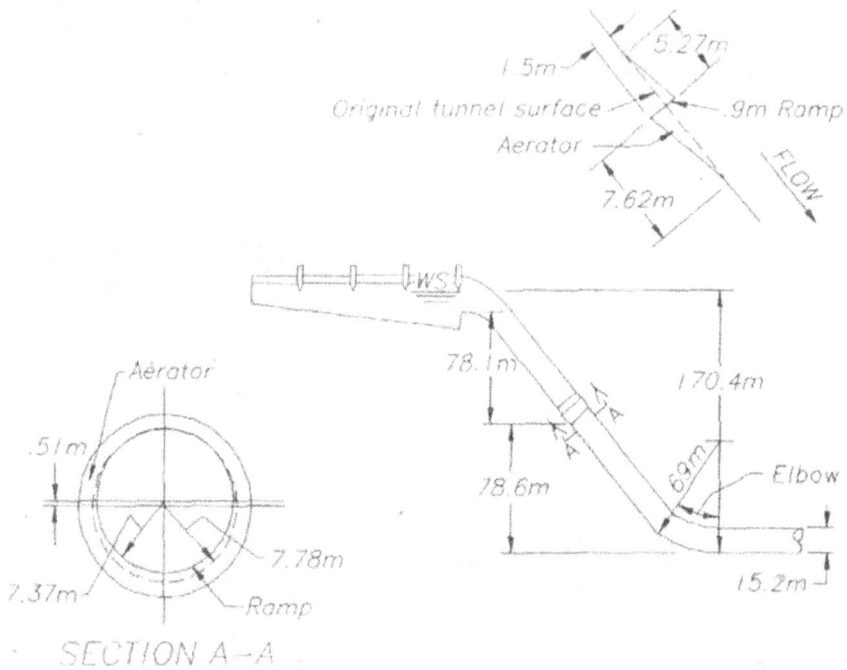

Fig. 7.4
Type III spillway layout at Hoover Dam, United States

Large diversion tunnels are often necessary when a dam is built in narrow valleys where large diversion discharge capacity is required. Cost considerations may lead to the construction of a high-level and an inclined tunnel to connect with the original diversion tunnel forming a permanent spillway tunnel. This arrangement is the same type as a Type III layout. It has been used in Mica Dam high-level tunnel spillway, Liujiaxia Dam low-level outlet tunnel spillway and in many other large projects.

As the cavitation risk increases in Type III tunnels when the dam height is over 250 m to 300 m, the Type IV tunnel spillway has been devised. The concept of this layout is to shorten the length of the tunnel with high-speed flow by building a long pressurised high-level tunnel and moving the control structure of gate chamber as further downstream as economically possible. The alignment of the pressurised tunnel can be changed both in vertical and horizontal directions as convenient for the design.

This kind of tunnel spillway has been applied recently in several Chinese large projects, as for example, in the four large tunnel spillways designed for the high arch Xiluodu Dam. The dam height is 285 m and the discharge capacity of each tunnel is 4 000 m³/s, (Figure 7.5). The high-level pressurised tunnel is 550-m long with a diameter of 17 m, and there are two 90° vertical bends before the service gate chamber. The length of last part of the straight pressurised tunnel has to be long enough to readjust the pressure until the pressure distribution is close to uniformity before the flow goes into the transition section. The length with high-speed flow in the last part of tunnel is only about 20% to 30% of total length. The cross section in open flow tunnel part is 14×12.5m².

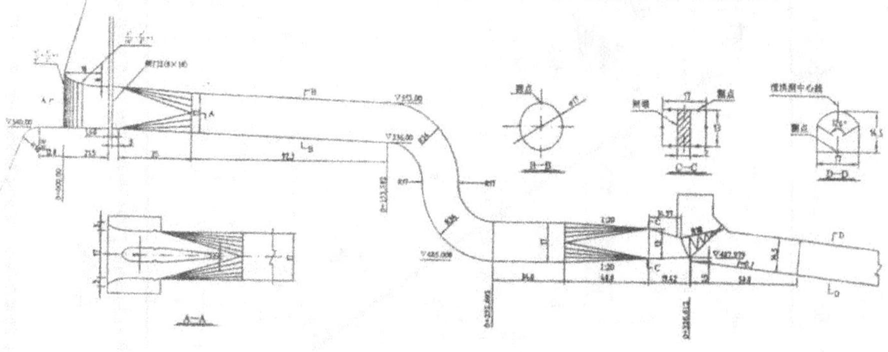

Fig. 7.5
Layout of Type IV tunnel spillway in Xiluodu Arch Dam (China)

The other way of transforming a diversion tunnel into a spillway tunnel is the Type V configuration. It can be designed as a vortex shaft tunnel spillway or an orifice tunnel spillway. The distinct feature of this type of tunnel spillway is to dissipate energy inside the tunnel. The vortex shaft tunnel can easily change flow direction by vortex action and shorten the length of the connecting tunnel.

The hydraulic study of large-scale vortex shaft spillway was started in the former Soviet Union, then in Russia and followed by China, Switzerland and some other countries. Shapai Dam vertical vortex shaft tunnel spillway was the first application of this kind of spillway in China, built in the 1990s, with a working head of about 110 m and discharge capacity of 250 m³/s.

Figure 7.6 show examples of type V tunnel spillway cases with vortex shaft spillways.

Figure7.6(a) (Chen *et al.* 2007) depict the Gongboxia Dam horizontal vortex shaft spillway tunnel which was the second application in China of vortex spillways. It has a working head of about 110 m and discharge capacity of 1 000 m³/s as shown. It was operated in 2006 at the design condition.

Tehri Dam, a 260-m high dam in India, adopted four large scale vortex shaft tunnel spillways, two on each bank: Figure 7.6(b) (Sharma *et al.* 2006). They all used previously built diversion tunnels. The discharge capacity for each tunnel is about 1 800 m³/s to 1 900 m³/s with a total discharging head over 200 m (Fink, 2012 and Shakirov, 2013).

Figures 7.6(c) and (d) show the vertical and inclined vortex shaft spillways designed for Rogun Dam in Tajikistan, with surface and bottom intakes (Shakirov, 2013). These vortex shaft spillways have a total discharge capacity 3 800 m³/s (2 000+1 800 m³/s). The vortex inclined shaft spillway of this project with high level intake, has a total discharge capacity of 4 040 m³/s. This kind of design has the advantage of a wide range of operational water head, especially for the requirement of water release in the initial impounding of the reservoir.

In vortex spillways about 80% of flow energy can be dissipated inside the outlet tunnel without significant dynamic effect to the tunnel lining (Galant *et al.* 1995 and Riquois *et al.* 1967). Water discharge from the tunnel into the river arrives practically with natural river velocities.

Xiaolangdi Dam orifice tunnel spillway, in China, was also built by reconstruction of a diversion tunnel and energy was dissipated by three orifices, as seen Figure 7.7. At Mica Dam, in Canada, low level outlet tunnel spillway uses this kind of energy dissipation by providing a sudden expansion chamber, as shown Figure 7.3 (a).

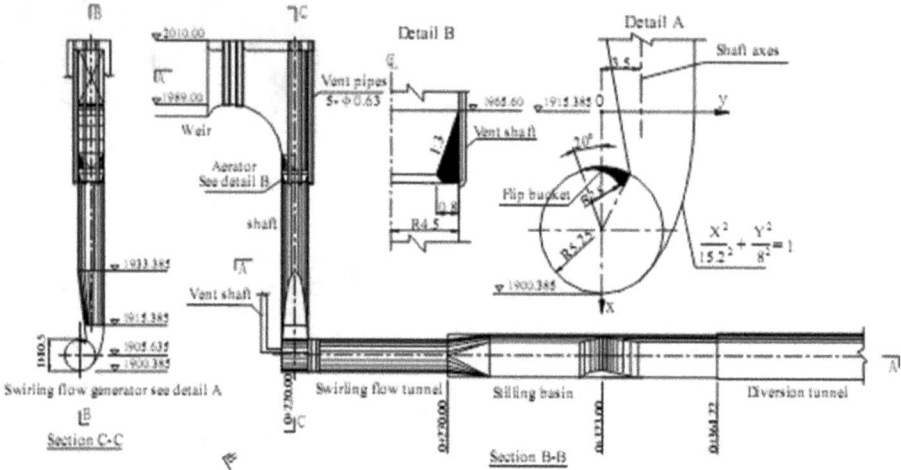

Fig. 7.6
(a) Horizontal vortex tunnel spillway in Gongboxia Dam in China

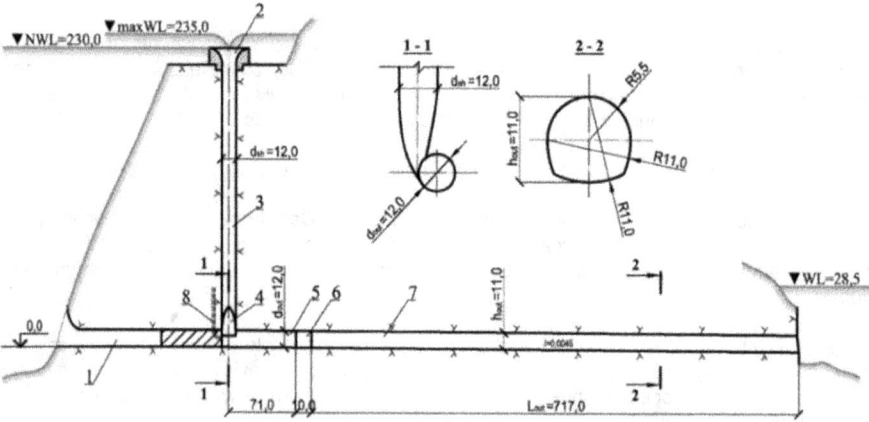

Fig.7.6
(b) Vertical shaft tunnel spillway in Tehri Dam, India

1 – Diversion tunnel.
2 – Surface intake.
3 – Vertical shaft.
4 – Flow swirling unit.
5 – Circle-shaped outlet tunnel section.
6 – Transition section.
7 – Horseshoe-shaped outlet tunnel section.
8 – Deaerator.

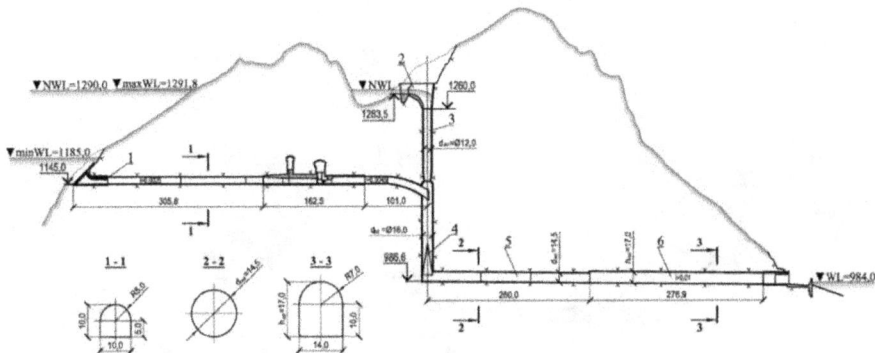

Fig. 7.6

(c) Vortex shaft spillway with surface and bottom intakes, Rogun Dam, Tajikistan

1 – Bottom intake;
2 – Surface intake;
3 – Vertical shaft;
4 – Flow swirling unit;
5 – Circle-shaped outlet tunnel section;
6 – Channel-shaped outlet tunnel section

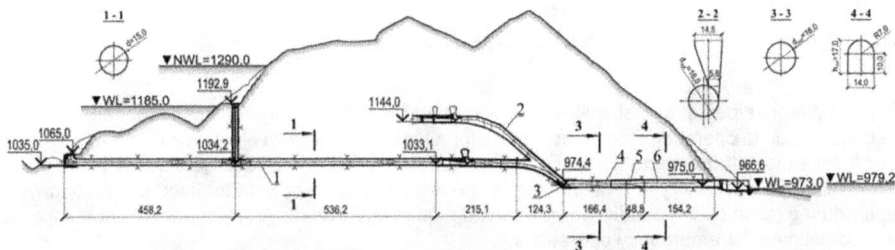

Fig. 7.6

(d) Vortex inclined shaft spillway of Rogun Dam with low- and high-level intakes

1 – Low level tunnel DT-3;
2 – Upper level tunnel;
3 – Vortex device;
4 – Round section tunnel;
5 – Transition section;
6 – Outlet tunnel

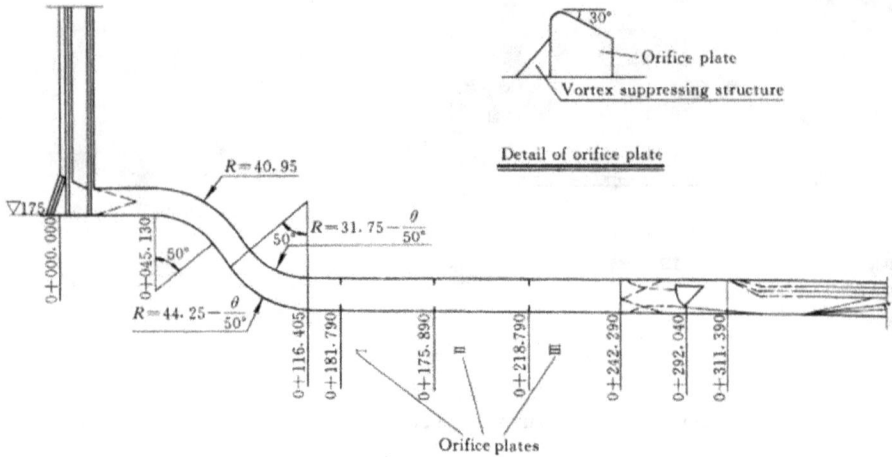

Fig. 7.7
Orifice tunnel spillway in Xiaolangdi Dam, China

The morning-glory shaft spillway has seldom been used in large dam projects. It has a circular inlet connecting to a shaft and low-level horizontal tunnel. The circular inlet can be divided by piers and controlled by gates. The working head and discharge capacity is usually small and this type is not fit to be used where large floods must be discharged.

7.2.2. Single structure vs. different functional components

Different types of tunnel spillway are arranged for different purposes and requirements, from the construction to operation stages, especially for a high dam. High-level tunnels (Type II to Type IV) usually take most of the operation tasks of flood control since they have large discharge capacities. Low-level outlet tunnels (Type I) take primarily the tasks of operating both for flood control and water supply during reservoir impounding, or for lowering the reservoir water level for dam maintenance and also, sometimes, for emergency operations.

Some mid-level or temporary outlet tunnel spillway is sometimes necessary for high dam construction, especially in the case that the dam will is built by stages. A typical example is Rogun Dam in Tajikistan. In this case the mid-level outlet tunnel spillway is connected to the permanent shaft spillway, as shown in Figure 7.6 (c).

7.3. CONTROL STRUCTURE

7.3.1. Surface spillways

For high-level tunnel spillway (Type II to Type IV), the gate chamber is arranged just downstream of the inlet. Two gates are usually adopted: a flat gate for maintenance and a radial gate for the service operation. However, provision of three gates is a more reliable and recommended solution: the first – a flat gate or stop-log (for low head) for maintenance; the second – the emergency gate, and the third – a radial gate for operation service. The working head in this design should be about 40 m to 50 m. There is an open flow after the service gate and throughout the tunnel. The service gate can also be a flat-type gate when the working head is smaller.

The gate chamber is divided into two sections to reduce the load acting on each gate when the working head is high, or discharge capacity is large. This arrangement has been applied in the tunnel spillways of the Hoover Dam and of Xiluodu Dam as the discharge capacities for these two single tunnels are 5 000 m³/s and 4 000 m³/s respectively.

7.3.2. Bottom outlet spillway

The service gate in bottom or low-level outlet tunnel spillway is usually a radial type as the working head can be larger than 60 m to 80 m, reaching even 100 m to 120 m, and the discharge capacity is around several hundred or thousand cubic meters per second. A slide gate can be used when the working head is larger and the discharge is smaller. For example, there are three slide gates built in the concrete plug in the Mica Dam low-level outlet tunnel spillway, as shown in Figure 7.3 (a).

7.4. CONVEYANCE STRUCTURE

7.4.1. Chute channels

The longitudinal layout of a chute channel can be designed in any type of tunnel spillway depicted in Figure 7.1. The gradient of bottom floor is usually less than 10% which is mainly controlled by the transportation capability during the construction. It has been increased to about 12% to 13% as the large and heavy trucks become available. The gradient of tunnel chute in Irape Dam (Figure 7.3 (c)) is 10.2% and the average length of the three tunnels is 634 m. A large gradient of the tunnel can reduce the length of the chute channel.

The cross section of chute channel can either be a circle, a D-shape or a horse-shoe shape. The circle type was used for the tunnel spillways of Hoover Dam, Glen Canyon Dam and some other dams. But the D-shape is more convenient for the excavation and transportation during construction. The concrete lining has to be designed to withstand a high-speed flow. Lining work in D-shape tunnel is much simpler to be built than that in a circular shaped one.

High speed flow in chute channels in Type I to Type III tunnels will necessarily happen. Therefore, the proper handling of such high-speed flow requires cavitation control which is very important in the hydraulic design of the tunnel lining.

7.4.2. Aeration facilities

There are some important cases studies of cavitation damages in tunnel spillways as well as in surface chute spillways. The mitigation of cavitation by aeration was recognised after the cavitation damage in Hoover Dam tunnel, and many studies had been carried out since then. Figure 7.8 shows damage in Glen Canyon Dam left tunnel spillway in 1983 (Burgi and Eckley, 1988 and Falvey, 1990). The 11-m deep "big hole" was found by site investigation.

A group of standard aeration facilities was developed in the 1980s and further on by different countries based on their particular design of tunnels and operation requirements (ICOLD, 1987), (see Figure 7.9). The aeration facilities were widely applied worldwide and cavitation damage in high-speed flow tunnels were greatly reduced, although the mechanism of cavitation mitigation is still under investigation.

In the standard aeration device an abrupt recess is created in the tunnel floor and an air vent is provided, sometimes combining with a small ramp where negative pressure develops. Air is forced to enter the low-pressure zone and aerate the water flow. It is very important to keep the air conduits clear and to ensure the aeration is continuous and sufficient during the tunnel operation. Interval distance between two aerators should be about 120 m to 150 m. Several aerators along the bottom floor of a tunnel may be necessary in a long tunnel spillway.

Fig. 7.8
Cavitation damages at Glen Canyon Dam spillways tunnel

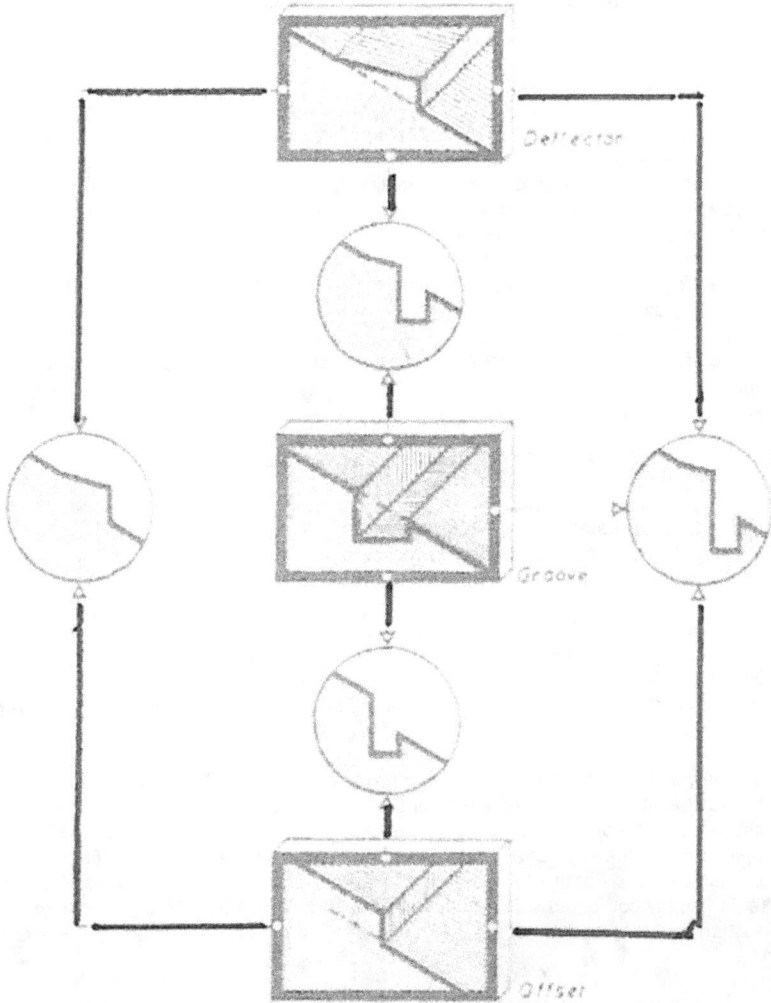

Fig. 7.9
Common aeration devices and their combination (Volkart and Rutschmann, 1984)

A special type of aerator was developed for cases where it is difficult to force air to enter the low-pressure zone in a small gradient floor of a tunnel or where the tunnel cross section has a circular shape. The aerator built along the wetted perimeter is effectively to provide air entrainment for this special case. A half circular aerator has been applied in the tunnel spillways of Hoover Dam, Yellowtail Dam and Glen Canyon Dam, in the USA. A circular aerator can also be applied in shaft tunnel, as for example in Gongboxia Dam (China) shaft tunnel where the physical model study showed that the pressure at the lower part of shaft is quite low. The length of the low-pressure zone measured in the prototype, downstream of the circular aerator device was longer than that measured in the laboratory.

An intermediate aeration tunnel or shaft may be necessary if a tunnel spillway is too long.

7.5. TERMINAL STRUCTURE (ENERGY DISSIPATION FEATURES)

7.5.1. *Stilling basin*

A stilling basin energy dissipater can be applied in a low head tunnel spillway scheme if geological conditions are not good enough - weak rock or soft ground – to allow the use of a flip bucket dissipater. Presence of high velocities and pulsating loads at energy dissipation in stilling basins requires special attention to their structural strength. A good design of a stilling basin is usually based on the results of a physical model study.

Experience in the operation of hydraulic facilities gives many examples of serious damages in stilling basins (Malpaso in Mexico, Tarbella in Pakistan). The most serious damage is characterised by the upward dislodgement of entire bottom slabs, which can be a consequence of the emergence of full hydrodynamic pressure under the slabs or high seepage uplift (see chapter 2, section 2.4.2).

7.5.2. *Flip bucket and plunge pool*

A flip bucket dissipater is more often used in a high head tunnel spillway scheme if geological conditions good, that is, comprise good solid rocks. Jet flow with large erosive power can be sent away from the structures, further downstream in the river channel.

As discussed in chapter 3 (section 3.5.6) of this bulletin, a plunge pool may be necessary and, in some situations, it can be built by pre-excavation. The discharging heads in the tunnels of Figure 7.3 (b) and 7.3 (c) are all over 200 m and the working head between the flip buckets and the water surface downstream are also over 100 m; therefore, the huge flow power must be dissipated in a large water body.

Normally, in plan view, there is a sharp angle between tunnel spillway and river channel, and the bucket design may involve considerable attempts. The bucket can be designed with different shapes and distortions including oblique and slit bucket. The aim of bucket design is to direct the jet flow into the river channel in a proper manner, achieving high efficiency of energy dissipation, reducing the depth of riverbed scour and effects on the river banks. A good design of a bucket is usually combined with physical model study. A slit bucket is very convenient to be applied in a narrow valley and to put the jet flow in longitudinal distribution along the river channel (Guo 2014).

(a) Scour by oblique bucket

(b) Flow pattern of slit bucket

(c) Scour by slit bucket

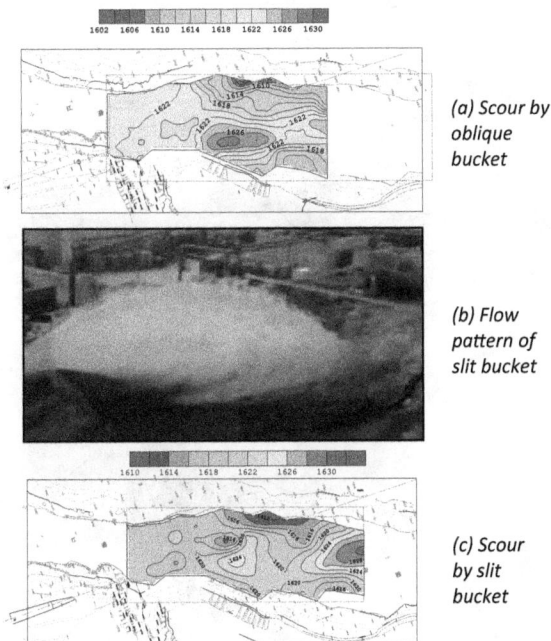

Fig. 7.9
Comparison of energy dissipation between two types of flip bucket
Longitudinal profile of stepped chute

As also discussed in chapter 3 of this bulletin, the configuration and depth of a plunge pool may threaten to reach the dam toe and in narrow sites cause stability loss of the valley abutment slopes.

7.5.3. Energy dissipated inside tunnel

The flow energy can also be dissipated inside the tunnel for shaft tunnel spillway or orifice tunnel spillway, such as in Xiaolangdi Dam shown in Figure 7.6 (e). In vortex shaft spillway tunnels more than 70% to 80% of flow energy can be dissipated inside the tunnel and the velocity in the last horizontal reach of the tunnel can be reduced to less than 20 m/s.

The orifice dissipater is built inside the tunnel by creating a sudden contraction and then an expansion through a chamber or orifice. This was the solution used for Mica Dam low-level outlet tunnel to dissipate energy and it achieved a good operational result. At Xiaolangdi Dam the tunnel spillways were built by reconstructing diversion tunnels with a diameter of 14.5 m, and three orifices in each pressurised tunnel were installed with an interval of 3D and orifice ratios of 0.690, 0.724 and 0.724 respectively. More than 40% of energy is dissipated by the three orifices.

Another example of an energy dissipation arrangement inside the spillway tunnel is the case of the Sayano – Shushenskaya Power Project in Russia as depicted in Fig. 7.10. This rather unusual configuration dissipates the energy of the auxiliary 240-m head spillway in a sequence of 5 underground stilling basins in a stepped chute following the leveled spillway tunnels.

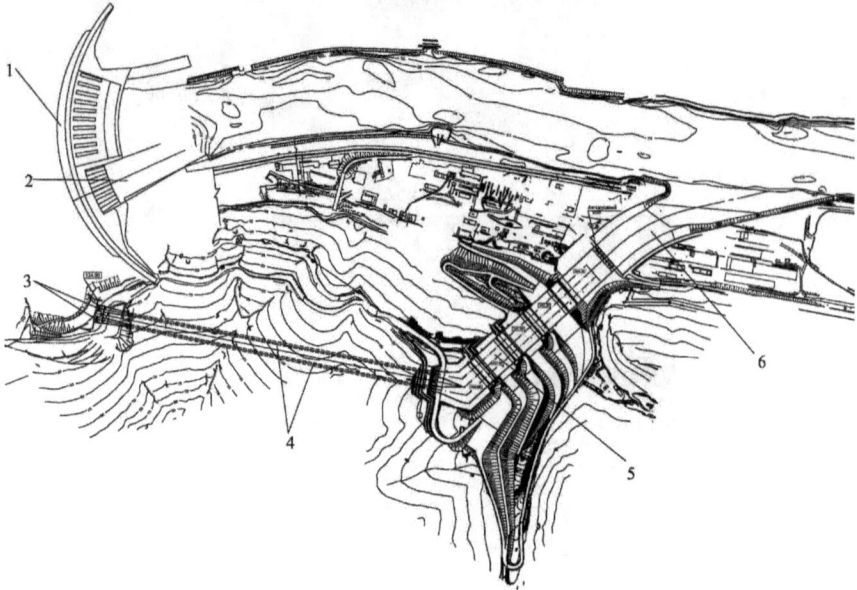

1 – arch-gravity dam;
2 – service spillway;
3 – intake structure of reserve spillway;
4 – tunnels;
5 – stepped chute;
6 – tail race channel

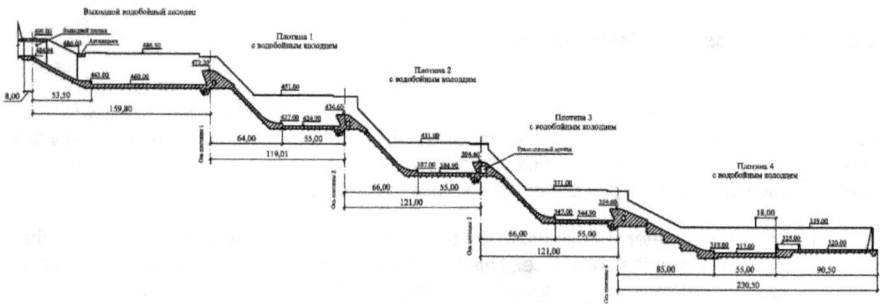

Longitudinal profile of stepped chute

Fig. 7.10
The reserve spillway of Sayano – Shushenskaya HPP (Russia)

7.6.   OPERATIONAL ISSUES

7.6.1. Gate operation

The safety of tunnel spillways is mainly affected by abnormalities in gate operation, ventilation, aeration and energy dissipation during the flood discharge.

Radial gates are usually used for flow control both in high-head and low outlet tunnel spillways to allow flexibility by partially opening to control discharge. The operation of gates must follow strict regulation orders. Any improper or faulty operation may cause serious accidents and damage the gate and/or the tunnel. Small openings, for instance 5% to 10%, should be avoided as vibration may appear at this range of openings according to operation experiences and field observation.

Flat gate is commonly designed mainly for low-level tunnel spillway. Partial openings are not usually allowed as the working head is high. For flat gates it may be necessary to increase the working load on the gate by stages if the head is large.

The electrical and mechanical control system must have the same degree of reliability of the gate as it was demonstrate in the emergency operation experience of Zipingpu Dam (China) low level outlet tunnel spillway during "Wenchuan Earthquake" in 2008. This is necessary to ensure that the gate system is functional in an emergency case.

Ventilation behind gate is also important. The air flow speed must be controlled by selecting proper cross section of air vent because vibration and noise can be induced by a high air flow speed.

7.6.2. Inspection and maintenance

Regular inspections and maintenance of gates, operating equipment, the tunnels proper and aerators must be implemented after each flood season. The importance of this work has been addressed in the General Report of ICOLD Question 79 (Cassidy 2000) as a vital part of assuring spillway operating reliably and safely. Minor and severe damages observed during the inspections must be recorded and repaired.

The inspection by the operator after the initial operation of the tunnel spillway is of utmost importance. The behaviour and performance of all structures have to be checked. These should also be complemented by the analysis of recorded monitoring results.

An underwater survey of the energy dissipation zone may be necessary in the case of a large flood discharge or a long duration of flood discharge operation. This survey will seek to determine the location and shape of scour holes, and will be the basis of an analysis intending to define the need of alterations or provision of additional features in the existing energy dissipation scheme.

7.7.   RISK ANALYSIS AND CONSIDERATIONS

7.7.1. Cavitation risk control

Fig. 7.10 gives the dimensionless geometrical characteristics of Type III and Type IV large scale tunnel spillways (Guo 2006). The dam heights listed are all over 200 m. The three indexes, which are, the relative height (h/H), the relative length (x/L) and the flow velocity, are taken into account to analyse cavitation risks of different kinds of tunnel spillway layouts.

Cavitation damages in Type III tunnels referred to in Figure 7.10 happen in a horizontal reach of the tunnel just downstream of an inverse curve, where 40% to 80% of the water heads prevail. In about 70% to 80% of the length the flow velocity is over 40 m/s. This scheme presents a great risk of cavitation damage. In contrast, the length with high velocity flow in the layout of a Type IV tunnel corresponds to about 20% to 30% of the total tunnel length; therefore, the risk of cavitation damage in these tunnels will be greatly reduced. This kind of arrangement has been applied in several large dams' projects in China with dam heights around 300 m and discharge capacities of individual tunnel over 3 000 m³/s.

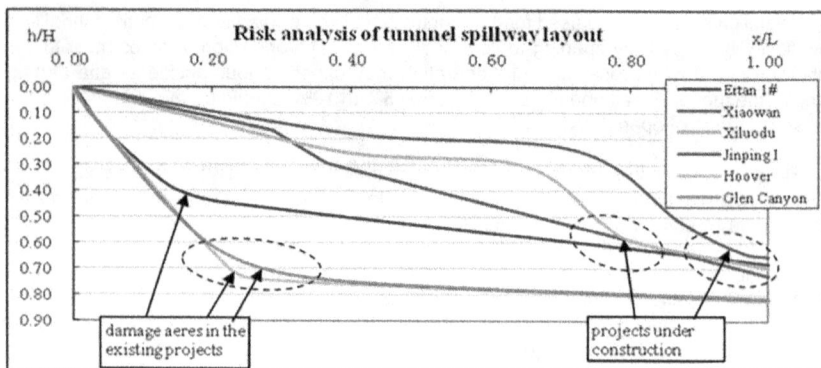

Fig. 7.10
Risk analyses of tunnel spillway layouts

7.7.2. Sediment abrasive action in tunnel linings

Sediment abrasive action of the concrete lining surface is observed when there are tunnel intakes located near the reservoir bottom. Usually, it happens for river diversion tunnels, in the beginning of construction period when the reservoir is too small to decrease river velocity and the flow transportation capacity, so that large quantities of solid river runoff sediment, with fractions up to 100 mm, are discharged by the tunnel during river diversion and construction periods. Such a situation can happen again at the end of a reservoir drawdown period when even larger size particles may reach the spillway intake.

Damages at the flip bucket of the spillway of Grand Coulee Dam (USA) is an example of significant damages of the concrete due to the abrasive action (Keener 1944). Protection against abrasive action can also be provided by the use of special kinds of concrete with hardening the invert part or a steel lining.

7.7.3. Dynamic loads

Dynamic loads in spillway tunnels may be the result, among other causes, of constant change in the modes of operation of the tunnel, the presence of transient processes, or the formation of hydraulic jump in the tunnel in the absence of air supply behind the gate chamber piers leading to formation of vacuum zones, cavitation and separation of the metal liner (Ilyushin 2002)

Significant dynamic loads were the cause of a major accident at auxiliary spillway tunnel of San Esteban Dam (Del Campo A., 1967). The tunnel was designed to operate in free-flow conditions, but at a high downstream water level period, it was flooded. As result a hydraulic jump was formed within the tunnel, which led to the damages of the lining in the area of a rock mass fault and the falling rock caused the obstruction of the spillway tunnel.

7.7.4. Emergency operation

Emergency operation may require that one or more low outlet tunnel be used to lower the reservoir level, especially for earth and rockfill dams, in case of earthquake or a landslide. This was the case of Zipingpu Dam in China, which was subject to an intense earthquake (Case 2, below), which is very instructive of the importance of releasing water to control reservoir water level in emergency situations.

This also emphasises the special role of diversion tunnels and bottom outlets related to dam safety.

7.8. CASES STUDIES

7.8.1. Flood release by shaft tunnel spillway in Tehri Dam (India)

The water level in the Tehri Dam Reservoir rose to El. 882.0 m in September 2009; the right-bank shaft spillways automatically started passing a discharge of 480 m³/s through each of them, which have a maximum discharge capacity of 1 850 m³/s (with MWL at El. 835.0 m). Photo 7.1 shows the operation of the shaft spillways releasing the water-air mix from the deaerating chamber with a jet height exceeding 50 m, demonstrating the effectiveness of operation of the deaerating facilities of the shaft spillway (Sharma *et al.* 2006)

Photo 7.1
- Deaerating jets from Tehri Dam shaft spillway operation in 2009

7.8.2. Emergency operations of low outlet tunnel spillways in Zipingpu Dam after the Wenchuan Earthquake (China).

Zipingpu Dam is located on the upper reach of Minjiang River in China, 9 km upstream of Dujiangyan City and 60 km northwest to Chengdu City. It consists of a 156 m high CFRD, one chute spillway, power generation system, one silt discharge tunnel and two low outlet tunnel spillways. The two low outlet tunnel spillways were built by reconfiguring the diversion tunnels. The main purposes of this project are irrigation and municipal water supply, as well as power generation, flood control, and environment protection. The project was completed in 2006.

A strong earthquake with a magnitude 8.0 happened on 12 May 2008 (so called "Wenchuan Earthquake"). The Zipingpu CFRD is only 17 km away from epicentre of the earthquake (ICOLD Experts, 2009).

The dam had minor damages but power supply was stopped and water supply was terminated. Resuming water and power supply and controlling the reservoir water level were the most important tasks in the emergency management. The first power generating unit was put into operation just 7 minutes after the strong shock and another two units operated right afterwards. The total discharge capacity was 100 m³/s which supplied Dujiangyan City and Chengdu City downstream, serving a population of 20 million people. The silt discharge tunnel operated 24 hours after the earthquake and the discharge rate was increased to 280 m³/s, as shown in Photo 7.2 (a). One low outlet tunnel spillway opened 27 hours after the earthquake. The total discharge flow was up to 850 m³/s, which was close to the reservoir inflow. The total power generation was resumed 128 hours after the earthquake. All tunnel gates were successfully opened 190 hours after the earthquake, and the reservoir water level was effectively controlled from 14 to 20 of May, greatly reducing the risk of dam collapse. The purpose of such operation was to ensure the dam safety allowing inspection of damages in the dam structure and repairing mechanical and electrical equipment; the other purpose was to provide a life rescue waterway in the reservoir area, as shown in Photo 7.2 (b).

An international group of dam experts from ICOLD visited Zipingpu CFRD in April of 2009. Their observations concluded that emergency actions were taken immediately after the earthquake and the previously existing emergency management rules on dam safety have been correctly applied in the "Wenchuan Earthquake" case (ICOLD Experts, 2009).

Photo 2
Emergency actions by Zipingpu CFRD after "Wenchuan Earthquake".
(a) silt sluice tunnel operated 24 hours after shock; (b) temparory jetty on reservoir shoreline to provide a life rescue waterway for reservoir area

Table 7.1
A - Characteristics of typical large scale tunnel spillways in the world (ordered by dam height)

Name	Country	Dam type	Dam height, H (m)	Type of tunnel	tunnel size (with × height)	Q_{max} (m³/s)	Type of aeration	Completion *	Operation experience
Rogun	Tajikistan	ER	335	V(a)	D14×17	4040	aerator in vortex	u.d.	
Rogun	Tajikistan	ER	335	V	D14×17	1800	aerator in vortex	u.d.	
Rogun	Tajikistan	ER	335	V	D14×17	2000	aerator in vortex	u.d.	
Jinping I	China	AV	305	IV	14×12	1×3651	offset, ramp	u.c.	
Nurek	Tajikistan	ER	300	III	11×10	1×2000	Aerator ds radial gate	1980	
Nurek	Tajikistan	ER	300	II	10×10.5	1×2020	Aerator ds radial gate	1980	
Xiaowan	China	AV	292	IV	13×13.5	1×3811	2-steps offset	2011	
Xiluodu	China	AV	278	IV	14×12.5	4×3860	offset, ramp	u.c.	
Tehri	India	ER	261.5	IV	D=8.5	1×1100	-	2008	2010
Tehri	India	ER	260.5	V	4 horse-shoe. D=11	2×1850, 2×1800	Deaerator at the junction between vertical shaft and horizontal tunnel	2008	2010
Mica	Canada	ER	244	V	-	1×850	concrete plug	1977	good operation
Mica	Canada	ER	244	III	-	-	offset	1977	good operation
Ertan	China	AV	240	III	13×13.5	2×3700	offset, ramp, 3D later	1999	damaged in 2001
Sayano-Shushenshkaya	Russia	AV	240	II	2 - 10×12	2×1900	In stepped chute	2011	good operation
Chirkeyskaya	Russia	AV	232	III	11.2×12.6	2900	-	1978	good operation

Table 7.1
B - Characteristics of typical large scale tunnel spillways in the world (ordered by dam height)

Name	Country	Dam type	Dam height, H (m)	Type of tunnel	tunnel size (with × height)	Q_{max} (m³/s)	Type of aeration	Completion*	Operation experience
Hoover	United States	PG	221	III	D15.56	2×5500	circular aerator	1936	damaged in 1941
Glen Canyon	United States	AV	216	III	D12.5	2×3900	circular aerator	1966	damaged in 1983
Irape	Brazil	TE	208	II	10×11.4	2×2000	offset	2005	good operation
Irape	Brazil	TE	208	II	12×11.4	1×2000	offset	2005	good operation
Longyangxia	China	PG	178	III	5×7	1×1340	offset at gate	1989	damaged in 1989
Charvakskaya	Uzbekistan	ER	168	IV, V	D=9.0, D=11.0	1100, 1200	Downstream steel liner	1976	
Xiaolangdi	China	ER	160	II	10×12, 8×9, 8×9.5	1×2680, 1×1973, 1×1796	offset	2000	good
Xiaolangdi	China	ER	160	V	D14.5	1×1727, 2×1549	offset at gate	2000	good
Yellowtail	United States	AV	160	III	2-D15.56	1×2600	circle aerator	1968	damaged in 1967
Zipingpu	China	CFRD	156	III	horse-shoe, D=10.7	1×1672	U-shape sidewall ramp	2006	good aeration, observed
Liujiaxia	China	PG	147	III	8×12.9	1×2105	aerator	1974	damaged in 1972
Cousar	Iran	PG	140	II	2 - 10×10	2×1900		2006	
Aldeadavila	Spain		139.5	III	D=10.4	2800	-	1962	
Gongboxia	China	CFRD	132.2	V	11×14	1×1090	circular aerator on lower part of shaft	2006	good operation
Shapai	China	RCC VA	132	V	4×5.5	1×242	air vent downstream shaft	2002	good operation in 2008

*: u.c. = under construction; u.d. = under design;
**:n.c. = not clear

Table 7.2
A Classification of Tunnel, Shaft and Vortex Spillways

Type	Level	Alignment	Inlet/Outlet Criterion	Hydraulic Condition	Control Structure	Control gate	Air Supply/Aeration/Air Vent	Energy Dissipation
I(a)	Low	Straight	Low level inlet with small gradient straight tunnel. Free outlet without submergence.	Free gravity flow with very high velocity.	Upstream control at inlet	At upstream radial gate for high discharge. For small discharge it can be vertical slide gate	Must have sufficient air supply over the free surface. Due to very high velocity specific aeration groove/ arrangement has to be made for bottom surface.	Energy dissipation outside tunnel
I(b)	Low	After pressurised low reach, the rest is straight	Low level inlet with small gradient. Inlet for pressurised flow, outlet free flow without submergence.	Upstream flow pressurised by sudden expansion.	By sudden expansion	Upstream	Must have sufficient air supply over the free surface flow.	By sudden expansion
II	High	Straight	High level inlet with small gradient straight tunnel. Free outlet without submergence.	Free gravity flow with high velocity	Upstream control at inlet	At upstream radial gate for high discharge. For small discharge it can be vertical slide gate	Must have sufficient air supply over the free surface. Due to very high velocity specific aeration groove/ arrangement has to be made for tunnel floor.	Energy dissipation outside tunnel
III	High and Low	Straight	Short high level tunnel connected by ogee shaped inclined tunnel, inverse curve tunnel and finally to a small gradient outlet tunnel	Free gravity flow with high velocity	Upstream control at inlet	At upstream radial gate for high discharge. For small discharge it can be vertical slide gate	Must have sufficient air supply over the free surface. Due to very high velocity specific aeration groove/ arrangement has to be made for tunnel floor.	Energy dissipation outside tunnel
IV(a)	High	Alignment of pressurised tunnel can be changed both horizontally and vertically.	High level long pressurised tunnel and short length free flow downstream outlet tunnel	Long length of tunnel under pressurised flow. Short length of tunnel under free flow.	Control at the end of pressurised flow by gate	Radial gate at the end of pressurised tunnel. Releasing flow to free flow tunnel	Because of short length of free tunnel, it may not have much problem in air supply and also cavitation	Energy dissipation outside tunnel
V(a)	High and Low	Vortex shaft changing flow direction to the shorter length of connecting tunnel.	High level intake entry to vortex shaft tunnel and low level outlet by free flow.	Flow by generating vortex in vertical shaft and then continuing in horizontal tunnel and subsequent free flow with energy dissipation arrangement in free flow area.	Control in upstream of vertical vortex shaft tunnel.	Radial/ Vertical Slide gate at upstream of vertical vortex shaft.	Sufficient and proper designed arrangement for vent pipe in vertical and horizontal vortex shaft for removal of air/air-water mix.	Energy dissipation inside the tunnel and stilling basin
V(b)	High and Low	Vortex shaft changing flow direction to the shorter length of connecting tunnel.	High level long pressurised tunnel and low level free flow downstream outlet tunnel	Pressurised flow up to control gate. Inclined shaft tangentially joins the tunnel at low level.	Control in upstream of inclined vortex shaft tunnel.	Radial/ Vertical Slide gate at upstream of inclined vortex shaft	Sufficient and proper designed arrangement for vent pipe in vertical and horizontal vortex shaft for removal of air/air-water mix	Energy dissipation inside the tunnel and stilling basin
V(c)	High and Low	Inlet at high level. Upstream pressurised flow with change of horizontal and vertical alignment. Downstream free flow straight tunnel.	High level intake entry with pressurised flow. Low level out let with free flow.	Pressurised flow till control gate. Downstream tunnel with free flow.	Control downstream of pressurised tunnel.	Radial gate at the end of pressurized tunnel. Releasing flow to free flow tunnel	Aeration and free air supply is not a critical issue but should be there.	Energy dissipation by orifice or number of orifices inside the horizontal pressurised tunnel.

7.9. REFERENCES

CBDB (2009) - *Main Brazilian Dams III - Design, Construction and Performance*, Brazilian Committee on Dams, Rio de Janeiro, Brazil

ICOLD (1987) - *Spillways for Dams*, Bulletin 58, CIGB-ICOLD, Paris, France

Cassidy J. (2000) - "Gated spillways and other controlled release facilities and dam safety", General Report of Q.79, *Transactions of the 20 Congress of the ICOLD*, Vol. IV, Beijing, China, 735-758

Burgi, P.H. and M.S. Eckley (1988) - "Tunnel spillway performance at Glen Canyon Dam", *Proceedings of International Symposium on Hydraulics for High Dams*, Beijing, China, 810-818

Chen, W.X., G.F. Li, S.Z. Xie and K.L. Yang (2007) - "Study on aerators of high head spillway tunnels", Proceedings of the IAHR XXXII Congress. Theme D, Venice, Italy, 748-755

ICOLD Experts (2009) - "China shares experience and lessons from the Wenchuan earthquake, Part III: Comments from Symposium Participants", International Journal on Hydropower & Dams, Vol. 16, Issue 3, pp 111-112

Fink (2012) - Experience in design, construction and initial operation of shaft spillways at the Tehri HEP, Personal communication with Dr. Fink

Shakirov, R. (2013) - *Vortex spillways with energy dissipation inside the tunnel*. Contribution presented to the ICOLD Committee on Hydraulic for Dams.

ASCE (1989) - Civil Engineering Guidelines for Planning and Designing Hydroelectric Developments, edited by Edgar T. Moore Vol.1, New York, USA

Falvey, H. T. (1990) - *Cavitation in Chute and Spillways*, Engineering Monograph No. 42, US Bureau of Reclamation, Denver, USA

Gao, J., Z.P. Liu and J. Guo (2014) - "Energy dissipation and high velocity flow" *Dam Construction in China - A Sixty-Year Review*, Editor in Chief: Jia Jinsheng. China WaterPower Press, Beijing, China

Guo, J. (2013) - "Recent achievements in hydraulic research in China Hydro Power", Vol. 6 of *Comprehensive Renewable Energy*, Editor-in-Chief: Ali Sayigh, Elsevier Ltd. UK: Vol. 6, pp 485-505

Guo, J., D. Zhang, Z.P. Liu and L.Fan (2006) - "Achievements in hydraulic problems in large spillway tunnel with a high head and large discharge flow and its risk analysis" (in Chinese) *Journal of Hydraulic Engineering*, 37(10), pp1193-1198

Pugh, C. A. and T.J. Rhone (1988) - "Cavitation in the Bureau of Reclamation tunnel spillways", *Proceedings of International Symposium on Hydraulics for High Dams*, pp 645-652, Beijing, China.

Sharma, R.K., P.P.S. Mann and R.K. Vishnoi (2006) - Tehri Project Shaft Spillways - Example of an effective solution based on analytical and observational design approaches, *http://www.istt.com/doks/pdf/b09_bristane_06.pdf*

Sen, Siba P. (2013) - *Classification of tunnels*. Contribution presented to the ICOLD Committee on Hydraulic for Dams.

Galant M.A., B.A. Zhivotovskiy, I.S. Novikova, V.B. Rodionov and N.N. Rozanova (1995) - "Special features of swirl-type tunnel spillways and hydraulic conditions of their operation". *Gidrotekhnicheskoye stroitelstvo*, № 9, pp. 16-22.

Novikova I.S., V.B. Rodionov, B.A. Zhivotovskiy and N.N. Rozanova (2002) - *Shaft spillways with swirl-type water diversion*. Proceedings abstract. Volume paper on CD-ROM. St.-Petersburg, Russia

Rodionov V., V. Kupriyanov, S. Paremud, V. Vedosov, V. Vladimirov and A.Tolochinov (2006) - "Construction of auxiliary spillways at existing dams". *Transactions of the ICOLD Twenty-Second Congress on Large Dams,* vol. 3, pp.1257-1268 Barcelona, Spain.

Ilyushin V.F. (2002) - "Lessons of the accident at the diversion tunnel during construction of Rogun HPP". *Hydraulic Structures*, No 4, pp.51-56.

Keener K.B. (1944) - "Spillway erosion of Grand Coulee dam". *Engineering News-Record*, vol. 133, № 2, pp.95-101.

Del Campo A., I. Trincado and I.G. Rosello (1967) - "Some problems in operation of San Esteban dam spillways". *Transactions of the ICOLD Ninth International Congress on Large Dams*, vol. II, R. 33, Istanbul, Turkey.

Volkart, P. And P. Rutschmann (1984) – *Air entrainment devices (Air Slots)*. Report Nr. 72. Mitteilungen der Versucsanstalt für Wasserbau, Hydrologie und Glaziologie, Technischen Hochshule Zurich.

8.   SPECIAL PROBLEMS OF SPILLWAYS IN VERY COLD CLIMATE

8.1.   INTRODUCTION

It is known that snow and ice can partly or fully block spillways and thus reduce the spillway and water conveyance capacity in cold regions. These problems, but also other challenges concerning regulation of gates and special loads and impacts from ice on structures, are discussed in this chapter.

8.2.   REDUCED CAPACITY OF SPILLWAY AND WATER CONVEYANCE

Stable operation of hydraulic structures particularly spillways, water intakes and water conveyance structures must be ensured not only in winter when the ice cover is stable, but also during formation and destruction of the ice cover.

It is convenient to separate the reduced capacity of spillways caused by cold climate, into the following points:

- frazil ice;

- aufeis blocking of the cross section;

- snow plug in the spillway; and

- ice drift.

8.2.1.   Frazil ice

The period of formation of the ice cover is characterised by the drop of air temperature below 0˚C. During this period the surface layer of water is cooled to a temperature of about 0˚C which causes formation of a phenomenon known as "frazil ice" or "sludge", especially in the presence of precipitation in the form of wet snow. The frazil ice is water in transition to another solid state of aggregation i.e. ice, however, not yet having the physical properties of ice. It should be born in mind that the ice density is 917 kg/m^3 and the water density is 1 000 kg/m^3, and therefore the density of frazil ice is an average between these values. This fact explains a number of properties of frazil ice: a viscous mass on the water surface resembling glue. When water is super-cooled flowing frazil ice will stick to structures such as booms, trash racks, baffle beams, slot structures, gate edges and stop logs. This will reduce the discharge capacity. The mass of the structures will also increase and make the operation of gates and hoist mechanisms difficult or even damaging them.

Frazil ice is particularly harmful for water intake structures of hydro power plants, pumping stations and hydropower equipment. Preventive measures to control such harmful actions can be taken by providing the devices described in continuation.

- Ice intakes and ice passes (Figure 8.1) diverting the ice laden flow downstream of the chute. The most efficient technology for frazil ice pass is selected based on the intensity of the frazil ice drift, water discharges and other specific conditions of a particular project. To reduce specific water discharges through ice passes to a minimum and to ensure the maximum entrainment of frazil ice, the best elevation of the overflow sill can be achieved by changing the inclination of a flap gate or the position of a swivel as shown in the figures.

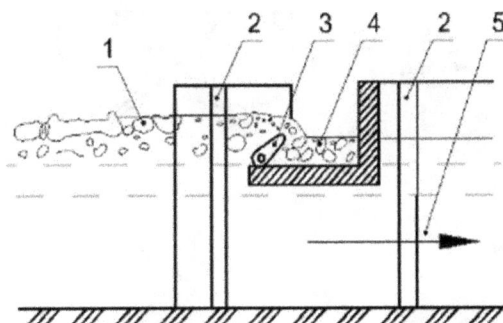

a)   Chute with flap gate

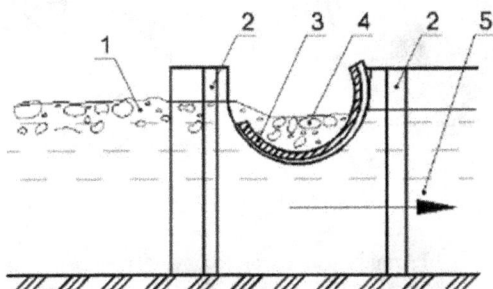

b)   Swivel chute of A. Gostunsky

1 — floating frazil ice; 2— slot; 3— overflow sill with variable elevation through which frazil ice is entrained; 4 — flow with frazil ice in chute; 5— flow free from frazil ice.

Fig. 8.1
Typical arrangement of ice and sludge passes

- Debris deflectors are structures deflecting surface flow streams and objects floating in the water surface, away from the place being protected, such as a water intake (Figure 8.2). The main element of debris deflectors is a deflector baffle or deflector shield. The depth of its submergence has to prevent diving frazil ice under its lower edge. The deflector plane is placed at an accurate angle to the flow so the deflector baffle does not restrict the flow but only deflects it. The adopted depth of the deflector baffle submersion under the water varies from 0.8 m to about 1.5 m, depending on its purpose.

- If the water intake structure is located near the transit portion of the river flow, it can be protected from frazil ice with the aid of a skimmer wall. The principle of its operation is similar to that of the deflector wall. To pass the frazil ice or other non-massive ice formations 1 or 2 gates of spillway openings are made as overflow ones (double-leaf gate). This is also called "skimmer walls".

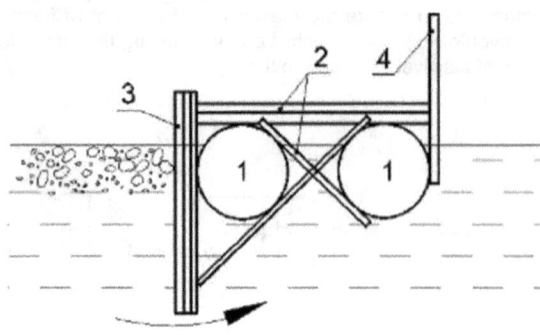

a) Cross section

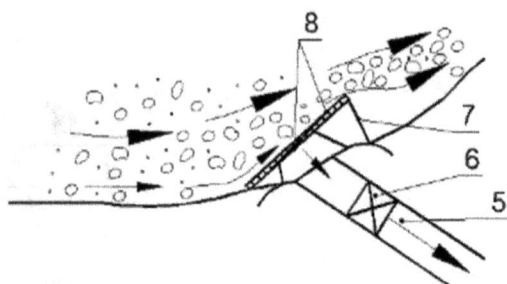

b) Arrangement for protection of river intake - plan

1 —floating elements (steel pipes portions, 600 mm in diameter); 2—bearing structures; 3— slot structure (for example, I-beam) with nested deflector shield; 4 —protection; ,5 — channel; 6 –water intake sluice; 7 — structures holding debris deflector; 8 — links of debris deflector.

Fig. 8.2
Debris deflector

- Arrangement of booms or baffle beams and walls (Figure 8.3a) designed to withstand the impact of frazil- and floating ice.

It is also known that high-flow bubbler systems are used to make a curtain that supress ice growth on upstream part of gates or locks (Tuthill, 2002).

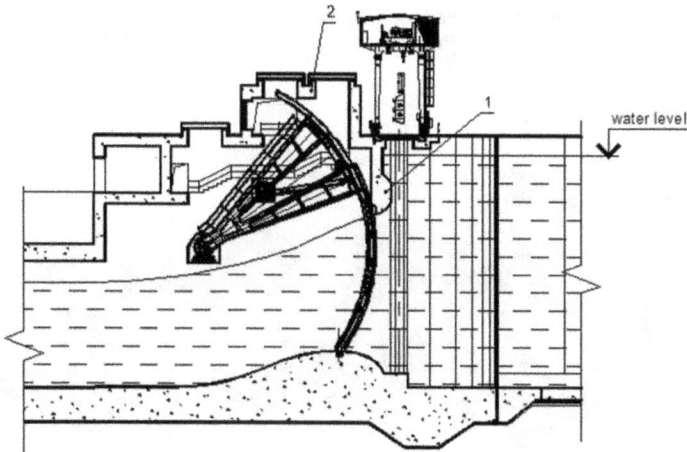

a) Substructure about the gate and boom (baffle-beam);

 1 – Boom installed in a proper way;
 2 – Substructure providing >0°C temperature inside the structure.

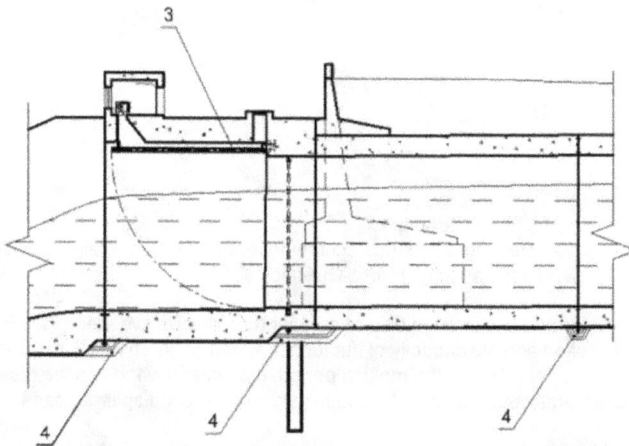

b) Design of thermal shield in the transit section;

 3 – Thermal shield protecting the structure against warm air coming from the downstream side;
 4 – Drainage behind the thermal shield (slab thickness exceeds the frost penetration
 thickness providing stability of the structure and uninterrupted operation of drainage).

Fig. 8.3
Arrangement of booms and baffle beams

8.2.2. *Aufeis blocking*

Aufeis is a term used for ice formed by shallow flow of water which streams over an ice cover. Other terms as *naled* and *icing* are also used for the same phenomenon. Aufeis occurs under natural winter conditions.

The most common situation concerning aufeis in a spillway system is the leakage through gated structures (Figure 8.4). This can block the gate and reduce the capacity significantly.

Fig. 8.4
Pont Arnaud: Gate leakage in winter creating aufeis (ice accumulation) and reducing significantly the spillway capacity (Hydro-Québec).

Another situation causing aufeis is through flow of cold air tunnel spillways. Leakages freeze in and reduce the open cross section and the capacity of the tunnel. Aufeis build-up with 3-m thickness has been observed in spillways tunnels. During the melting period, the water flowing into the reservoir can rise the reservoir level to an unacceptable elevation caused by the reduced spillway capacity.

Description of the build-up and removal of aufeis is based on research by Schohl and Ettema (1986). The aufeis build-up depends on the time, surface area, heat flux, water discharge, slope and different water and ice properties.

The removal of aufeis can be achieved by two processes: melting and mechanical break up. It has been found that if water flows over the aufeis the melting rate depends on the water temperature. If no flowing water occurs, melting will depend on radiation, wind velocity and air temperature. A sudden break-up of the aufeis will result in an ice run and is only possible for local deposits of aufeis. Sokolov (1973) found that a mechanical break-up never removed more than 30-35 % of the ice. According to Ashton (1986) aufeis melting is reported to last for several weeks.

Lia (1998) presented several solutions for reducing the problems with aufeis in spillway tunnels:

- Insulation - A non-structural wall in the outlet (and in some cases also in the inlet) prevents cold air from penetrating the tunnel or at least eliminating the through flow. In addition, insulation of the tunnel walls can be installed.

- Grouting - In order to prevent water from leaking into the tunnel during cold periods, grouting can be used. For the same reason it is recommended to use controlled blasting close to the outlet, inlet and in the area where the spillway weir is to be constructed.

- Improved cross section - Aufeis in a spillway grows when water flows onto the whole tunnel bed and freezes to ice. If a narrow and deep ditch is constructed from the frost-free zone in the tunnel out to the tunnel outlet an ice cover will establish at the free water surface and protect the water flow beneath, as in a natural river, see Figure 8.5.

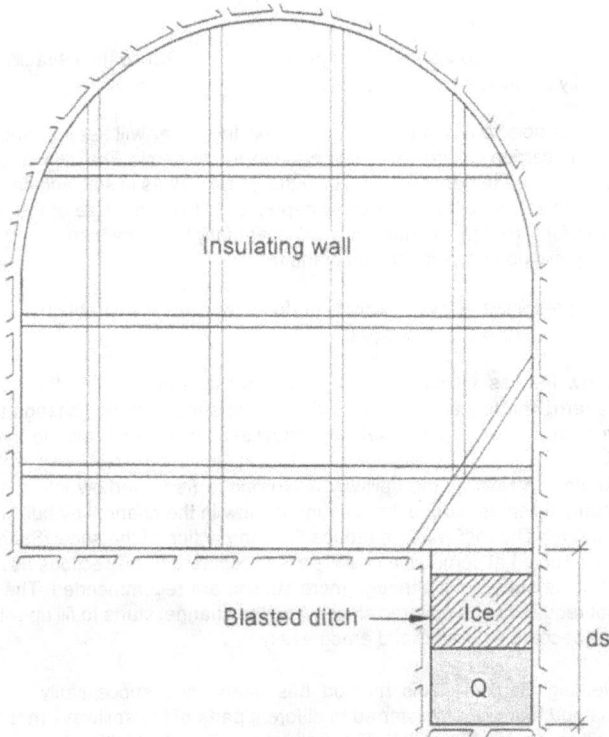

Fig. 8.5
Cross section of the deep ditch with an established ice cover (Lia, 1998)

8.2.3. Snow

Snow is a material that changes behaviour and properties with time. Snow can block the inlet, tunnel/channel and outlet of the spillway, and may prevent the spillway from being inspected and maintained during wintertime.

The average depth of snow cover is dependent on precipitation, temperature and snow drift. For local areas, the topography plays an important role for snow accumulation. The wind removes snow from one area and accumulates it in another. If the wind comes from one main direction (e.g. west) during the winter, the snow will be deposited in hills whose face is in the opposite direction (e.g. east). This is typical in high windy mountains. Although the wind direction changes in consecutive years, the main direction is very often the same. With this in mind, two or three years will be necessary to describe local patterns.

Wind transported snow change properties and becomes more compact when it accumulates in snowdrifts. After some hours a cohesive material develops. Drift snow accumulated in snowdrifts that can last one or several summers before they melt, named a multi-year snowdrifts. Such a snowdrift has blocked a tunnel outlet at a hydropower scheme in Norway for several years.

The snow will form a snow dam in the spillway that will reduce the capacity of the spillway or in the worst case, fully blocking the spillway.

If melt water is able to penetrate the cold snow, the water will seep through the snow dam. Research have been carried out to find the permeability of snow. The rain and melt water flow downwards in the snow, due to the gravity force, in the same way as in soil, and can be expressed by Darcy's law, where the permeability in the snow depends on the grain size of the snow crystals and the density, (Sommerfeld, 1989). The snow dam will eventually break by thermal processes, seepage, floating, overtopping, sliding or a combination of them.

Lia (1998) presented several solutions for reducing the problems with snow in the spillway:

- Snow Fences - This is a convenient tool for manipulating the snow accumulation pattern. The fences do not stop the drifting snow, but they change the wind direction in such a way that the suspended snow accumulates behind the fences.

- Roofs - In Norway the spillway is typically a free overflow into a side channel. It is then efficient to reduce the amount of snow in the channel by building a roof over the channel. The roof will also reduce the compaction of the snow. Such roofs have been constructed at some side channels, see Figure 8.6. Inspections have proved them to be quite successful although more studies are recommended. The entire channel is not required to be covered since when the channel starts to fill up with water the snow will become saturated and erode easily.

- Heating Cables - This method has been used successfully in culverts and the method has been transferred to different parts of the spillways (not reported in tunnel spillways), see Figure 8.7. The method requires electricity supply, but it only has to be switched on in periods when the spillway capacity is required.

- Location of Spillway - During the design of new spillways, the topography and typical wind pattern should be studied before the location of the spillway is set in order to reduce problems with the accumulation of snow.

In the steppe regions of Russia snow fences made with rows of planted trees or local bushes are common. This is an environmental friendly solution to protect structures from being covered and overtopped by snow.

Fig. 8.6
Concrete roof at the side channel at Sønstevatn Project spillway in Norway; view from upstream
(Lia, 1998).

Fig. 8.7
(A) Chute-Allard, Hydro – Quebec
a) and b): Summer 2010: Installation of heating plates for 3 inflatable dams

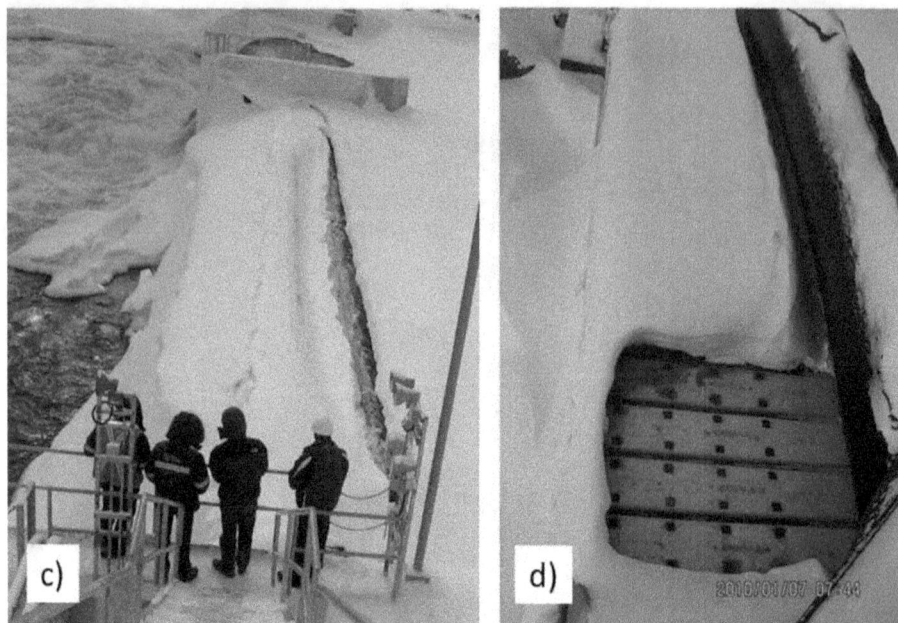

Fig. 8.7
(B) Chute-Allard, Hydro – Quebec
c) Spillway in Winter - Snow accumulation downstream
d) Heating plates experimented during winter

8.2.4. Ice drift

The ice coverage on top of rivers or lakes will break up when the water flow increase. Ice chunks will then be transported by the flow and can cause severe clogging of spillways. The spillway structures must be designed to handle the load from the drifting ice chunks.

8.3. ICE LOAD ON SPILLWAY STRUCTURES

In order to ensure stable operation of spillways, water conveyance structures and water intakes in severe climatic conditions, the impact of ice on the structure, particularly on gates and mechanical equipment should be reduced to a minimum or taken into account. The impacts and loads of ice on structures are discussed by Gosstroy (1989).

Spillway design allowing flow under the gate below the downstream water level should be avoided because it is practically impossible to exclude the impact of ice on the gate from the downstream side.

To minimise the impact of ice on the mechanical equipment (gates) from the upstream side, booms and baffle walls are designed, as depicted in Figures 8.2 and 8.3. Installation of the booms below the ice penetration level should ensure the absence of the direct impact of ice on the gate, i.e. the boom is installed below the depth of frost penetration and below the depth of ice cover. The effect of the boom structure on the discharge capacity has to be considered during design. Booms take up the pressure of ice on the structure and therefore have to be carefully designed to withstand this load. Booms or baffle beams can also serve as protective structures from floating debris.

8.4. OTHER CHALLENGES CONCERNING DIFFERENT PARTS OF THE SPILLWAY SYSTEM

8.4.1. Gated structures

Ice formation in slot structures or behind the gate caused by frozen leakage water should be avoided. This can cause the gate mass to become much larger and may keep the gates completely stuck in the ice or put the hoisting equipment out of service. The following measures can be undertaken to reduce this problem:

- heating slot structures and heat insulation can be provided to prevent the gates to get frozen and ice to be formed at the gate sheathing on the upstream side. Heat insulation layer thickness shall be calculated with the use of thermal physics methods proceeding from thermal conductivity of the material. Heating of slots and the of gate with built-in electrodes helps to remove ice from the structure. For provision of reliable and uninterrupted electric heating the electric circuitry must be doubled;

- when choosing the lifting equipment, the additional weight of heat insulation and possible ice sticking onto the gate structure must be taken into account; and

- the provision of a thermal shield can maintain a microclimate with positive temperatures inside the structure that has a positive effect on the operation of the gates, on the hoisting equipment and on the structure as a whole. Thermal shields can be used at tunnels, discharge pipes, open discharge chutes or behind the spillway structure. Possible leakages from under the gate seals should be collected and diverted to the downstream pool by a drainage system where the outlet must be under the level of frost penetration.

8.4.2. Tunnels and shafts with linings

Frost penetration into tunnel linings and liners must not be allowed. In regions with a wide range of temperature variation (from negative to positive) steel linings of tunnels and shafts experience considerable deformations. The presence of a shaft and a non-gated weir on a control structure complicates the structure protection. Taking into consideration the difference between the atmospheric pressure at the entrance and the exit of a tunnel with a shaft and the natural temperature of the soil below the frost penetration depth, the provision of a thermal shield at the control structure makes it possible to maintain positive temperatures inside the tunnel and the shaft.

8.4.3. Spillway chute

In an open spillway chute, the thickness of the bottom slabs and chute walls must be designed concerning the thermal insulation properties of concrete, not allowing frost penetration under the slabs and behind the walls. If the concrete thickness is not sufficient, heat insulating materials can be added in the structure. Frost penetration into the drainage under the bottom slabs or behind the chute walls should normally not be allowed. Drainage should preferably be located below the frost penetration depth of soil (See Figure 8.3b).

8.4.4. Water mist and icing

In cold climatic regions, floods can start in the midst of the cold season as a result of sharp short-term warming or when average daily temperatures are below 0° C. If the flood reaches the structure at the moment when the thaw is over and the average daily temperatures have become below zero flood passing can cause misting and icing.

If the temperature in the middle of the day is above 0° C while the night temperatures are below 0° C a structure with high aeration (e.g. in the spillway transition section or in an energy dissipater/flip bucket) may cause mist and icing.

Water mist can be transported for quite long distances and may cause icing not only on the spillway structures but also on other nearby structures. The design of structures in severe cold climatic regions should avoid design solutions connected with formation of water mist in the spillway system. A striking example of ice formation over the spillway is the Sayano-Shushenskoye hydropower plant, in Russia, where rapid water discharge through the structures during winter resulted in failure of the hydropower plant and putting it out of service.

8.4.5. Energy dissipation

Designing the energy dissipation structures for severe climatic conditions requires special approaches. The design shall take into account possible freezing – thawing of the structures and ice effect on them. In this connection the energy dissipation arrangements (splitters, toothing, blockwoods etc.) shall be located below natural water level and below depth of frost penetration. This will help the energy dissipaters to perform their functions and protect them against various kinds of defects (crumbling, chips, cracks).

8.5. CONCLUSION

In the design and construction of spillway structures in severe climatic conditions, it should always be kept in mind that the period characterised by negative average daily temperatures (in degrees Celsius), i.e. winter period, can last 6-9 months. The design and construction in permafrost regions are not considered in this paper as it is a separate subject requiring an individual approach. Negative temperatures during construction and operation of spillway structures impose specific requirements to the design of structures and construction material – materials must withstand low temperatures without loosening their properties and, more important, withstand alternating cycles of freezing and thawing without destruction, i.e. to be "frost-resistant".

First of all, this applies to concrete used in structures, elements of rockfill or support, and drainage. Particular attention should be drawn to elements of structures located in the variable area – being alternately in water and in the open air. These areas are the most critical.

When designing spillway structures in cold climatic conditions, thin-walled structures, fully frozen and not ensuring proper thermal insulation of concrete blocks, soil and structures behind them, should be avoided.

In harsh environments steel structures should preserve the geometry at a large temperature range of +40˚C to -40˚C or more.

8.6. REFERENCES

Chapter 8 is mainly based on:

Shakirov, R. (2013) - Some Peculiarities of Operation of Spillways and Water Conveyance Structures in Heavy Climatic Conditions, Contribution to the ICOLD Technical Committee on Hydraulics for Dams.

Lia, L., (1998) - *Snow and ice blocking of tunnels,* The Norwegian University of Science and Technology.

Other references:

Ashton, G. D. (1986) - *River and lake ice engineering*, Water Resources Publications, Littleton Colorado, USA.

Gosstroy (1989) - "Russian Code of Practice in the field of design and construction", *SNiP 2.06.04-82* Loads and effects on hydraulic structures due to waves, ice and water craft*, Moscow, USSR.

Schohl, G. A. and R. Ettema, (1986) - *Naled ice growth*, Iowa Institute of Hydraulic Research, The University of Iowa, Report No. 297, Iowa City, IA, USA.

Sokolov, B. L., (1973) - *Certain features in structures and mechanical breakdown of naleds; their significances in estimates of naled runoff*, Siberian Naleds 1973, Transaction of Naledi Sibiri, Moscow, USSR, Nauka, 1969, pp. 206-226

Sommerfeld, R. A., (1989) - *The Darcy permeability of fine –grained compact snow*, Eastern Snow Conference 1989, pp. 121-128.

Tuthill, A. M. (2002) - *Ice-Affected Components of Locks and Dams*. Technical Report ERDC/CRREL TR-02-4, US Army Corps of Engineers.

9.   COMPOSITE MODELLING OF HYDRAULIC STRUCTURES

9.1.   INTRODUCTION

For many years physical hydraulic modelling has been a standard design tool for hydraulic engineering of dams and their appurtenant structures. Physical modelling is an essential component of spillway design, especially when challenging flow features such as three-dimensional phenomena, multiphase flows, high levels of turbulence, scour and sedimentation issues are present. An engineer's ability to gather and analyse data from physical models continues to improve with the development of more sophisticated electronic instrumentation, faster computers and advanced software. These same advancements in computer technology have enabled the development of more advanced mathematical modelling techniques and several powerful 2-dimensional and 3-dimensional algorithms are commercially available as options to hydraulic engineers.

Physical modelling often utilises numerical modelling as part of the modelling package. For example, it is not uncommon for the design engineer to utilise a numerical model to help develop a proposed design for a hydraulic structure. Numerical modelling can be cost effective and accurate when based upon quantifiable field or physical modelling data. When a numerical model is based upon sound prototype or physical modelling information, the numerical model can be used for years on end as a valuable design and operational tool. Yet numerical modelling is limited when it comes to multi-phase flows (air/water mixtures), highly turbulent flows with vortices and flows where scour and/ or sedimentation are a concern.

The current trend for modelling hydraulic structures is to utilise both a numerical and physical model in parallel during a study (also known as composite or hybrid modelling). Composite modelling is the effective utilisation of both physical and numerical modelling, used either in series or in parallel with one another to solve difficult hydraulic problems.

Although physical modelling is the proven standard for modelling hydraulic structures and has successfully been used for decades, it comes with limitations and constraints. Similarly, numerical modelling, which is relatively new, also carries with it a number of limitations and constraints as well as benefits not found in physical modelling. However, when the two modelling techniques are used together during a hydraulic study, many of the limitations of one technique can be complemented by the other and vice versa. Accordingly, when researchers understand the benefits and limitations of each modelling technique, they can develop a research plan that utilises composite modelling approaches where both modelling techniques are used together to increase the efficiency and effectiveness of the modelling process.

For example, a numerical model may first be utilised to determine a proposed design for the structure. A physical model is then commonly constructed of the proposed design. Data is collected from the physical model and that data is used to calibrate the numerical model to ensure its accuracy. The physical model is then used to optimise hydraulic efficiency and minimise construction costs of the original design. With a calibrated numerical model, it (the numerical model) can then be utilised for many years to check flow conditions or operational procedures even after the physical model has been dismantled. The important thing to remember with this approach is that numerical modellers need real data either from a flood condition through the prototype or from a physical model study. The physical model study is critical to the success of the numerical model. Currently, the trend in modelling is to utilise the benefits of both for design.

9.2.   PHYSICAL MODELLING OVERVIEW

Physical models of hydraulic structures where free-surface flows are present are typically modelled using Froude scaling, which is based on the ratio of inertia to gravity forces. Pressurised flows are predominantly based on Reynolds scaling. The similarity parameter used to operate the model is the Froude number, defined as:

$$Fr = \frac{V}{\sqrt{gy}} \tag{1}$$

where y is a characteristic linear dimension, g is the gravitational acceleration constant, and V is a characteristic velocity. Similarity is achieved by operating the physical model so that the Froude number in the model is the same as the Froude number in the prototype. This ensures that the inertial and gravity forces are scaled properly. The proper length, discharge, velocity, pressures, time and rotational speed for the model in relationship to the prototype can be calculated using Froude relationships.

For a model to properly simulate hydraulic conditions in the prototype it is necessary to maintain geometric, kinematic, and dynamic similitude. Geometric similitude is achieved by carefully building the model to a selected scale ratio with all geometric components scaled appropriately.

The selection of the model scale has a direct impact on its ability to maintain kinematic and dynamic similitude. Physical models are always built as large as possible to minimize the scaling effects attributed to other forces, such as viscous and surface tension forces. When a large physical model is constructed (small scaling ratio), then flow phenomena such as super elevation, flow surging, wave action, hydraulic jumps, vortices, flow separation, local scour and erosion, air entrainment, standing waves, complex and composite roughness, etc., can be accurately modelled. In addition to being able to physically simulate complex flow conditions, the physical model provides a hands-on working simulation that most engineering teams find essential to the design process. Some of the more significant benefits and limitations of physical modelling are listed and summarised below.

9.2.1. Benefits of physical modelling

- As long as the scale ratio is properly selected and the model is constructed as large as possible to avoid potential size scale effects, physical modelling is a trusted and reliable method that will produce accurate results.

- Small changes can be made in the physical model very easily using temporary structures of wood, plastic, mortar, sandbags, etc., so that the resulting hydraulic changes can be seen immediately. It is time consuming to simulate a wall or a raised embankment into the grid of a numerical model, especially if the model is complex and the mesh resolution is fine, but only takes a matter of minutes to see the effect in the physical model.

- Multiple flow rates can be tested on a single model configuration in a very short time. For example, a flood routing or flood hydrograph can be simulated in a matter of minutes. Hydraulic design configurations can experience a hysteresis of flow conditions with changing flow rates. Physical modelling can easily determine the presence and magnitude of such hysteresis effects.

- Specific hydraulic problems are immediately apparent in an operating physical model, and many times, there are multiple problems occurring simultaneously. Problems like vortices, separation zones, scour potential, poor energy dissipation characteristics, wall overtopping, aeration effects or wave action are noted immediately in a physical model.

- Once constructed, a physical model can quickly determine the flow capacity of a flow control structure by increasing supply or inlet flow to the model until the given pool elevation is reached. Depending upon the complexity of the geometry of the control structure and the approach channel, the numerical modelling effort to determine flow capacity can be significantly more time consuming.

- The effects of tailwater or submergence on a hydraulic structure are easily and quickly observed in a physical model by simply varying the height of water in the downstream channel.

- Three dimensional effects, aeration effects, highly turbulent regions, standing waves and super-elevation conditions are easily modelled in a physical model. However, one must be wary of scale effects when evaluating aeration.

- Physical models facilitate "team" engineering. Geotechnical, structural and hydraulic engineers can get immediate feedback and make contributory suggestions to the design of the hydraulic structure as they witness flowing water through the physical model.

- Physical models will allow videotaping or photographing the moving water so that modelled flow conditions can be compared to prototype conditions providing a valuable tool in quality control.

9.2.2. Limitations of physical modelling

- The physical model can be limited by the required structure cost and size, and the time required to build and acquire data. The cost and time are usually proportional to the amount of information that is required and the time it takes to collect the data. It is often difficult to physically model the full extent of the flow region or channel and the approach conditions, due to limited laboratory floor space. Major changes to a physical model are often time/labour intensive and expensive.

- Pressure, flow rate, and velocity measurements taken in the physical model introduce some uncertainties as a function of the accuracy of the instrumentation being used for the measurements. It can be difficult to make detailed pressure measurements in a physical model due to the installation of pressure taps, pressure transducers, and the presence of un-dissolved air in the flow.

- If a physical model is not scaled properly, then the data collected, or the cost of the model is affected. If the model scale is too large (small-sized model), then size scaling effects can influence the accuracy of the results. If the model scale is too small (large-sized model), then the cost of the model will increase. It should be emphasized however, that it is very important to construct any physical model as large as possible to reduce scaling problems.

- It is difficult to vary the boundary roughness in a physical model and often a separate sensitivity study is needed to evaluate the effect of roughness.

- Utilising Froude similitude for the physical model study can be problematic in that full velocity profiles are not developed along the boundary because the model is not operating at the correct Reynolds number when operating at a Froude-scale velocity.

- Project schedule and limited budget usually minimise the number of possible design configurations or modifications that can be tested.

- The physical model has to be dismantled at some point in time, and may then not be available for other design changes resulting from future changes in construction or other constraints.

- There are real hydrodynamic processes which are either poorly or practically impossible to simulate on the physical models. Among these is, for instance, formation of water-air-ice cloud in the spillway tailrace with a high degree of flow aeration (for instance overhang spillways with jet ejection), which is highly dependent on the fluid's surface tension and the laboratory's ambient humidity. Scale models diminish the possibility of this process occurring.

9.3.   NUMERICAL MODEL OVERVIEW

Although there are multiple approaches to numerical modelling such as 1D and 2D methods, the full 3D method commonly referred to as computational fluid dynamics (CFD) is the type of numerical modelling referred to in this chapter. 3D modelling is effective in simulating both subcritical and supercritical flow regimes. The most common CFD programs solve the Reynolds-averaged Navier-Stokes (RANS) equations using a variety of approaches. Most of the leading commercial CFD programs on the market have the ability to track free surface flows commonly associated with hydraulic structures with the Volume-of-Fluid (VOF) method or some variation thereof being the most common algorithm for tracking the free surface interface.

CFD codes solve the RANS equations using a finite volume method applied to a computational domain. The computational domain defines the region of interest as discretised (gridded) into smaller cells. Gridding techniques are usually classified as Boundary (Body) Fitted Coordinates (BFC) or a porosity technique. Solid objects are defined in BFC coordinates by wrapping the grids around the object whereas in porosity methods, the solid is imported into the grid and the solid blocks the fluid flow. Each method has its advantages and disadvantages; BFC tend to be more computationally efficient, especially in resolving boundary layer hydraulics, but grid construction can be time consuming whereas porosity methods are easier to grid but not as computationally efficient. Additionally, the porosity method requires that survey data (site geometry) is imported into the model.

One of the difficulties in numerically solving flow in a hydraulic structure is the presence of a free surface, in which the location must be solved as part of the solution and may be transient in time. This is especially difficult when the water surface is rapidly changing with a high degree of curvature, such as when the flow changes from subcritical flow to supercritical flow as it flows over a spillway crest.

Another common technique used in CFD modelling is to approximate curved surfaces with a straight approximation. It is easy to see that the use of smaller grids and thereby more straight segments provides a better approximation to a curved surface. Therefore, in order to improve the approximation, variable grid spacing can be used with a tighter spacing or smaller grid size located near high curvature areas on the obstacles and in locations where a rapidly varying water surface was expected. The downside to smaller grids is the increase in computational time.

To solve numerically the RANS equations across the gridded domain, there are a variety of numerical approaches that can be used, and each one may have multiple parameters. Primary inputs include defining the boundary conditions, the turbulence model and its parameters, model roughness and the numerical approach for solving the equations (implicit versus explicit, 1st, 2nd or 3rd -order etc.). Some parameters such as the boundary conditions are critical for properly simulating prototype conditions. Other parameters provide more subtle changes. Understanding the effect of the various methods, approaches and parameters is crucial in making sure that the model represents the required features of the true physical flow.

CFD modelling is appropriate for both supercritical and subcritical flow conditions in hydraulic structures, yet it may be limited when conditions are highly turbulent or aerated. As with any engineering application the computed results from CFD can be no more accurate than the data and assumptions on which the computational techniques and methods are based – i.e., accept the values as a decent approximation, but not the absolute truth. A summary of some of the more significant benefits and limitations to numerical modelling are listed and summarised below.

9.3.1. Benefits of numerical modelling

- A numerical model is a good choice for large changes to physical topography or to the location of a control structure.

- Numerical modelling provides the option of testing a configuration without fabricating a physical structure; thus, allowing multiple configurations to be investigated at the same time using multiple computers. Numerical modelling allows for the automated multivariate parametric optimisation of design elements.

- Data produced from numerical modelling results allow for velocities, pressures, turbulence parameters and water surface profiles to be determined at each time step and at any location in the flow domain.

- Numerical modelling helps the researcher see the "big picture", or in other words, where problem areas may be located. Flow patterns such as velocity profiles and vortices can be easily mapped using vector plots, streamline and flow ribbons.

- The numerical model can eliminate the need for physical sectional models [example: section of a spillway constructed at a smaller scale (larger model) so that spillway surface pressures can be evaluated]. The numerical model does a good job of evaluating fluctuating pressures as well as the magnitude of negative pressures along a fluid boundary.

- RANS CFD models are not adequate to simulate pressure fluctuations in two or multiphase flows associated with dynamic eddy structures that are small and therefore smoothed out (e.g. pressure fluctuations present in a hydraulic jump). RANS CFD models are good at surface pressures if the flow is single phase (water) and the flow lines are parallel near the surface and relatively steady with time. Large Eddy Simulation (LES) and Detached Eddy Simulation (DES) CFD models largely eliminate this issue but are computationally expensive and are not (yet) widely used for free surface water flows.

- Significant changes to a modelling configuration are simple to construct on the computer and only have labour and software licensing costs associated with them.

- Numerical modelling costs are often less than a physical model although their continuous use can build up cost to a similar extent.

- Numerical models can be stored and maintained for future use/changes if the software remains available and if the model is compatible with the software updates.

- Animations of the flow, including the free surface can be constructed using a variable of interest such as the velocity or pressure.

9.3.2. Limitations of numerical modelling

- A numerical model benefits from having prototype or physical modelling data for calibration but this is not always available or required.

- Numerical models are limited in the selection of input and operating parameters when there are no prototype data or physical modelling data to use as a basis. Additionally, numerical modellers who have not been properly trained or do not have vast experience in specific applications, introduce the possibility for data errors and software mishandling.

- Numerical models are limited in their application of geometric roughness and the development of the velocity profile at boundaries (law of the wall), especially in high turbulence supercritical flows. Depending on the grid size of the numerical model, the true near-the-wall velocities may not be accurate since the law-of-the-wall approximations are dependent on numerical grid size and other factors.

- Numerical modelling is often constrained by gridding/computational time for models with large spatial and/or time domains. To reduce computational time some CFD codes include hybrid solutions which allow 2D shallow water approximations for large areas where vertical gradients are not critical to be combined with full 3D hydraulic analysis in local regions.

- The number of 3D CFD runs required to evaluate the routing of a hydrograph will be very time consuming for a numerical study. Similarly, simulating the routing of a complete hydrograph can require simulation times that are unrealistic.

- Hysteresis effects between increasing flows of a hydrograph as compared to the decreasing flows of a hydrograph can happen and be significant. Numerical models are limited in their ability to analyse this type of flow relationship.

- It is not possible to videotape or photograph the moving water of a numerical model, so that modelled flow conditions can be compared to prototype conditions as it can be done with a physical model. This type of analysis is extremely valuable as a quality control to ensure that the modelling is accurate. However, it is possible to create visual simulations using numerical output in the form of short video clips that aid in the comparison with prototype conditions.

- A numerical model is limited in its ability to analyse the hydraulic stability and effect of unsteady flow downstream of a control structure.

- Flow conditions like a hydraulic jump can be very sensitive to changes in numerical boundary roughness and there is wide latitude in the application of the theory if prototype or physical model data is not available. Although numerical CFD modelling is appropriate for defining the "macro" characteristics of a hydraulic jump, it is limited in its ability to describe the "micro" characteristics of highly turbulent and highly aerated flows.

- A numerical model is limited by the size of the computational domain both spatially and temporarily. A large physical domain may require a 2D solution rather than a 3D solution, or a hybrid solution as noted above, thereby potentially reducing the accuracy of the solution.

- Numerical models are limited due to the time it takes to construct the numerical grid and run the model when complex flow conditions are being simulated.

- Shallow flows can be computationally intensive, requiring a small grid spacing to accurately simulate a velocity profile. An example would be an extremely low flow over an ogee crest. Fortunately, most design concerns are based on larger flows.

- Complex flow conditions from turbulent flow, re-circulating flow, flow separation, and flow super-elevation around a flow bend may not be accurately calculated in a numerical model due to turbulent modelling limits. Although turbulence models work well for certain documented cases, they may not be appropriate for all flow conditions.

- Incorrect or improperly applied gridding techniques have the potential to introduce numerical errors. Grid convergence testing is necessary to determine the adequacy of the selected grid sizing.

- There is a non-proportional time requirement to simulate and analyse a very small change in a numerical model as compared to the time it takes in a physical model.

9.4. COMPOSITE MODELLING OVERVIEW

Composite modelling is the effective utilization of both physical and numerical modelling, used either in series or in parallel to solve difficult hydraulic problems. Despite weaknesses in the two modelling techniques when they are utilized independently, composite modelling has proven to be extremely effective and efficient. A properly applied composite modelling approach will improve the accuracy of the flow data and measurements by focusing on the benefits of both physical and numerical modelling as they are used together. It should be noted however, that composite modelling will not always reduce overall cost over using one modelling technique or the other. The value of composite modelling is normally in the quality of flow data and the confidence in the proposed hydraulic design and not necessarily in the cost. The composite approach allows for more design options to be evaluated and therefore the design to be optimised.

Composite modelling can be viewed in three components:

1. numerical modelling performed before the physical model is constructed and tested;

2. numerical modelling that is performed at the same time or in parallel with the physical model; and

3. numerical modelling that is performed after the physical model study has been completed.

9.4.1. Numerical modelling performed before the physical model is constructed and tested

- Prior to the construction of a physical model, numerical modelling results will provide information so that the approach geometry (i.e. the physical laboratory head box) of the physical model can be properly aligned with the approach streamlines in the flow, and can be minimized so that the head box that supplies water to the model is not constructed any larger than necessary. This reduces construction costs for the physical model and allows a larger sized model (smaller scale) to be constructed, thereby reducing the Reynolds scale effects especially in the region of interest.

- The numerical model can help reduce the number of physical modelling configurations necessary by performing pre-runs. The numerical model can test design conditions and alternative design concepts before the physical modelling to make decisions about the physical model and to obtain a general acceptance for design.

- The pre-runs from the numerical model will help engineers understand the general hydraulic problems; identifying problematic regions where flow separation, excessive velocities or non-uniform flow may occur.

9.4.2. Numerical modelling that is performed in parallel with the physical model

- One of the most significant benefits of composite modelling is "modelling the model", which means that the exact geometry of the initial or baseline physical model is numerically modelled at a 1:1 scale so that numerical modelling errors can be evaluated and corrected if possible. This quality control effort is effective in reducing or eliminating the uncertainties of the numerical model.

- It is also possible for the calibrated numerical model to then be rescaled to model the prototype to evaluate potential physical modelling scale effects. Again, these scale effects should be minimal for the physical model if the physical model is constructed large enough to minimize the scale effects.

- Once the numeric model has been calibrated or adjusted to the results of the physical model, it can be used to evaluate a myriad of discharges, velocities, water surface profiles and detailed surface pressures and shear stresses that are labour intensive and expensive to perform in the physical model. Adjustments to the numerical model to match the physical model are normally done by adjusting surface roughness, turbulence model parameters and grid cell resolution until the correlation between the two modelling types seems reasonable.

- With a calibrated numerical model, any proposed structural or geometric change that requires major modifications to the physical model can be simulated in the numerical model to evaluate design acceptability as compared to the initial baseline calibrated numerical modelling results. The results of these numerical modelling runs are then carefully compared to determine the appropriate trial or final geometric configuration that should be physically modelled.

- The trial or final physical model configuration is then constructed in order that detailed hydrograph and rating curves can be tested. A physical model can quickly and accurately determine the rating curve of a hydraulic structure, while a numerical model will take a significantly longer time to generate the necessary runs for a rating curve. Again, however, once the noted physical model configuration is constructed, the numerical model can be used to provide an infinite number of data points, streamlines and/or flow ribbons within the flow domain that would take considerable effort to collect in the physical model.

- The physical model can be photographed and the flow conditions and problems associated with the structural design can be videotaped so as the data associated with both the physical and numerical models are reviewed, the recorded flow conditions can be reviewed. These recordings are also extremely valuable in publicizing the design to the general public and to funding agencies.

- During this phase of the composite model, regions in the flow domain in which the numerical model simulations are questionable are documented. This may include areas of highly turbulent flow, areas within the larger model in which extremely small depths and high velocity flows occur, flow conditions that are near critical depth, transient surges or wave oscillations, super-elevation as a result of supercritical flows rounding a tight curve, regions where scour is a potential problem and flow conditions in which the water is highly aerated.

9.4.3. Numerical modelling that is performed after the physical model Study is complete

After every physical model study, it is necessary to dismantle the model so that the space that was utilised for the physical model in the laboratory can be used for another physical model study. However, the numerical model does not take up any space and does not need to be discarded. The benefits of having maintained a calibrated numerical model are:

- the numerical model can be used indefinitely after the physical model is dismantled;

- additional information can be collected within the flow field (pressure and velocities recorded at every cell) if needed after the physical model has been dismantled;

- minor modifications to the geometric design can be made and tested in the calibrated numerical model with confidence that the results will be accurate; and

- the calibrated numerical model can be used for training and operational decisions after the physical model has been dismantled.

9.5.   SUMMARY AND CONCLUSIONS

The use of both a physical and a numerical model together as a composite model has proven to enhance the hydraulic modelling effort, improve modelling accuracy and reduce modelling uncertainty. Composite modelling has been found to be an effective tool for hydraulic structure design.

Composite modelling provides a unique opportunity for researchers and engineers to understand the uncertainties and limitations of both the physical model and the numerical model, since their parallel operation allows for direct comparison and calibration. When complex three-dimensional or turbulent flow conditions or multiphase flows are being modelled, numerical models are often limited in their ability to simulate the flow field in all regions as compared with the physical model. This would include flow conditions with water surface profiles in which transverse super-elevation or non-symmetrical hydraulic jumps occur. This also includes flow conditions where standing waves in flows that are near critical depth are present, where upstream subcritical flows are being merged across a control structure to flows that are rapidly changing downstream of the control structure and where highly turbulent and/or mixed flow regimes occur.

In the past, numerical modelling has been proven to be very effective and accurate when applied to two-dimensional and some three-dimensional flow applications, but many of the hydraulic problems associated with hydraulic structures are complex, highly turbulent, highly aerated and three-dimensional, and numerical modelling can be questionable for these types of flow conditions. Therefore, composite modelling allows for the verification and validation of flow rates, water surface profiles and point velocities within the flow domain and consequently determines specific regions within the numerical model that cannot be simulated appropriately.

Composite modelling can produce a calibrated numerical model to collect very large and detailed amounts of data from the flow domain and produce a model that can be effectively used long after the physical model is dismantled provided that the software remains available and is compatible. Composite modelling provides an opportunity for the numerical model to be calibrated to both the prototype structure at a 1:1 scale and to the physical model at a 1:1 scale. This is all very important to the quality control component of the study.

The use of numerical modelling as an integrated part of the modelling process has proven to be both cost effective and time beneficial for the design hydraulic engineer. Using a numerical model (before the physical model) to define the physical domain of the head pool and the approach flow requirements in the physical model is a valuable component for the modelling process. This pre-modelling provides important information so that the physical model can be constructed as large as possible, thereby minimizing possible scaling effects. This pre-modelling also provides information about how the water supply to the physical model should be designed to ensure appropriate approach flows are simulated.

Additionally, the numerical pre-modelling provides information about potential problems associated with the theoretical design or proposed design changes to the structure, thereby reducing the number of physical modelling configurations necessary during the physical modelling portion of the study. This pre-modelling effort saves time and money once the physical modelling has commenced. With a calibrated numerical model, the numerical model is a valuable tool in noting hydraulic roughness sensitivity and other relative changes that are not so easy to modify in the physical model. As previously mentioned, the numerical model provides great benefits for quality control since it can be scaled to model the physical model directly, thereby producing validation of geometry and operating conditions in the physical model. Finally, the numerical model is a very important tool for maintenance and operation of the hydraulic structure for use after the physical model has been dismantled.

9.6.   REFERENCES

Willey, J., T. Ewing, R. Wark and E Lesleighter, (2012) – "Complementary use of physical and numerical modelling techniques in spillway design refinement", *ICOLD 25th Congress on Large Dams*, Kyoto, Japan.

Erpicum, S., B. J. Dewals, J-M. Vuillot, P. Archambeau and M. Pirotton, (2012) - "Coupling physical and numerical models: example of the Taoussa Project (Mali)", *4th IAHR International Symposium on Hydraulic Structures*, Porto, Portugal.

Rahmeyer, W., S. Barfuss and B. Savage, (2011) - "Composite Modeling of Hydraulic Structures", *Dam Safety 2011*, National Harbor, MD, USA.

Paxson, G., B. Crookston, B. Savage, B. Tullis, and F. Lux III,(2008) - The Hydraulic Design Toolbox-Theory and Modeling for the Lake Townsend Spillway Replacement Project, Assoc. of State Dam Safety Officials (ASDSO), Indian Wells, CA, USA.

Savage, B.M., M.C. Johnson and B. Towler, (2009) - "Hydrodynamic forces on a spillway: can we calculate them?", *Dam Safety 2009*, Hollywood, FL, USA.

10.   ECONOMICS, RISK AND SAFETY IN SPILLWAY DESIGN

10.1.   INTRODUCTION

Spillways are the main safety-assurance feature of any type of dam. For this reason, as a rule, direct cost reduction attempts for these structures will, in many cases, mean increases in the risk and lowering the safety of the overall dam project. In many projects, the direct cost of spillways may represent a reasonable important portion of a project, and its reduction may become a tempting target for designers, but because of their fundamental role in providing safety, the actions or measures intended to obtain savings in spillway design and implementation must be carefully evaluated without losing sight of the role of the spillway. In spite of that, economic and cost optimisations in spillways are possible and desirable. This chapter discusses these aspects.

The importance of a safe spillway cannot be overemphasised as many failures of dams have been caused by improperly designed spillways or by spillways of insufficient capacity. Ample capacity is of paramount importance for earth-fill and rock-fill dams, which are likely to be destroyed if overtopped, whereas concrete dams may be able to withstand moderate overtopping. Usually, an increase in the direct cost of the spillway is not directly proportional to the increase in capacity. Very often, in concrete dam projects, the cost of a spillway of ample capacity will be only moderately higher than that of one with much smaller capacity.

In many dam projects, economic considerations will lead to a design utilising a reservoir surcharge above normal operating level. The most economical combination of surcharge storage and spillway capacity requires flood routing studies and an economic appraisal of the costs of the spillway–dam combinations. However, in conducting these studies, consideration must be given to the minimum size spillway which must be provided for safety. In many cases the information from the inflow flood hydrographs used for defining the spillway design flood (SDF) stresses the flood peak but not the flood volume. Often, such floods may have the highest peak flows but not always the largest volume. When spillways of small capacities in relation to these inflow peaks are considered, precautions must be taken to insure that the spillway capacity will be sufficient to: (i) evacuate reservoir surcharge so that the dam will not be overtopped by a recurring storm; and (ii) prevent the reservoir surcharge from being kept partially full by a prolonged runoff whose peak, although less than the inflow flood peak, exceeds the spillway capacity.

Economics and cost savings in the construction and operation of dams and their flood control features have been the object of various ICOLD Technical Committees and bulletins but are always guided by the fundamental issue of safety assurance. These questions lead to the risks that must be evaluated by the designer who has the duty of balancing the corresponding effects to an acceptable level.

ICOLD Bulletin 152 – *Cost Savings in Dams*, discusses criteria used for the design of dams and appurtenant structures and specifically, for spillways, from the point of view of cost reductions, economics and risks. The concepts discussed included the selection of spillway design floods (SDF), the difficulties of assigning a failure probability to a spillway arrangement and the tendency of design practice to consider only performance and ignore costs.

In recent years innovative approaches have been developed in spillway design to address dam safety concerns at a lower cost. Such include: (i) the use of overtopping protection on embankments to prevent failure during large flood events; (ii) raising of dams with parapet walls to improve flood attenuation and to gain greater storage; and (iii) the application of a risk assessment process to determine the appropriate risk in terms of risk to human life and property posed by dam failure. This latter point is discussed extensively in ICOLD Bulletin 154 – *Dam Safety Management: Operational Phase of the Dam Life Cycle.*

In addition, methodologies and guidelines have been developed to consider the possibility of lowering safety requirements for spillway design criteria when the incremental consequences (or damages) induced by a dam failure become insignificant compared to those generated downstream by natural floods themselves. Such approaches, particularly suitable for adjusting spillway capacity to the level of safety requirements at existing dams, have been implemented in some cases. Cost-benefit analysis methods were used to support the acceptability of not increasing spillway capacity when the cost of spillway upgrading was deemed disproportionate to the net safety improvement gained through the spillway upgrading.

As described in previous chapters of this bulletin, some efficient new spillway alternatives have been developed and implemented improving, cost wise, traditional solutions. This chapter addresses the strategies related to spillway design in dam projects utilising different possibilities, alternatives and methods to gain overall economic advantages. As a rule, no cost reduction measure should be taken to lower the discharging capacity of the spillway system through transfer of part of that capacity to other structures or arrangements.

10.2.   DESIGN FLOOD AND CHECK FLOOD

A survey of existing projects shows that in many modern projects the spillway system is designed for a peak flow value, based on criteria which usually consider the spillway design flood (SDF) as the inflow which must be discharged under normal conditions, with a safety margin provided by an accepted freeboard limit. In addition, a spillway check flood (SCF) is defined as the maximum flood that will not cause the destruction of the dam and beyond which the safety cannot be assured (ICOLD Bulletin 82, 1992). The safety of the project is ultimately based on the spillway system capacity to discharge the SCF. This approach is also the standard criterion in many countries where there is an official recommendation for the design of dams.

In general, the SCF is a flood with a lower probability of occurrence than the conventional SDF. In many cases it is assumed to be the PMF or the theoretical maximum flood that could occur at the dam site. Although the SCF is computed not to overtop the dam, in general, because of its low probability, some degree of damage in the structures is tolerable. As discussed in ICOLD Bulletin 152 (pgs. 133-135) it is possible to reduce the cost of the spillway by increasing the difference between the SCF and the SDF; in other words, by reducing the SDF without diminishing the safety of the project, which is ensured by the capacity of discharging the SCF. This capacity is associated with the possibility of an increase of the elevation of the reservoir water level. Although in many projects this situation corresponds to encroaching into the normal freeboard, raising the dam crest elevation (and computing the corresponding costs) to allow a larger freeboard is also used. The increase in the flood elevation of the reservoir allows a certain degree of flood peak attenuation, which is discussed below.

10.3.   ATTENUATION OF FLOOD PEAK

The provision of an additional volume, above the maximum operational level of the reservoir, to store part of the flood hydrograph allowing a reduction of the maximum flow discharged by the spillway system, is a common strategy for reducing the direct spillway cost without affecting the safety of the dam. However, this is only feasible if the area of the reservoir is large enough to provide a large volume with a limited increase in elevation. The elevation considered in the design, to be reached when the reservoir inflow flood occurs, is generally called the maximum flood level. This elevation is higher than the maximum operational level.

Cost wise, the benefit of reservoir flood attenuation is only effective if the hydrograph coefficient (defined as the ratio of the flood hydrograph volume divided by the product of the reservoir surface area and the height of the dam) is much smaller than unity (Bouvard, 1988).

In any case, a freeboard above the maximum flood level, whether or not coincident with the maximum operational level, should be provided for all dam projects. This provision is intended to protect the dam from waves produced in the reservoir and is normally sized by using standard well established criteria that take into consideration a certain maximum probable wind blowing in the direction of the reservoir's longest fetch with the stored water in the reservoir at the maximum flood level. The probability of a simultaneous occurrence of the maximum flood with the reservoir at its maximum level and the maximum wind producing the maximum wave height, can be assumed as being much lower than the isolated probability of the occurrence of either of them. This has justified the consideration, recommended by some countries (such as China, for example), to consider the reservoir maximum flood level (the SCF level) coincident with, or slightly below, the elevation of the crest of the dam. It should be remembered in this respect, that in a dam where there is a combination of a concrete portion and an embankment portion, it is usual to set the embankment crest elevation about 1m (or more) higher than the concrete crest elevation, to ensure that in the event of an overtopping, the concrete portion would be overtopped first and the risk of overtopping the embankment portion is at least reduced.

This consideration means that the freeboard allowance above the check-flood level will be reduced in relation to the corresponding allowance above the maximum operational level. As mentioned previously, the cost saved for the spillway structure is the result of a smaller SDF to be discharged when the reservoir level is at the maximum operational level. This reduction of the spillway design flood is permitted by the higher water level elevation of the check flood and by the peak attenuation of the flood hydrograph.

10.4. DIVIDING THE SPILLWAY INTO SEPARATE STRUCTURES

One of the challenges to reduce the cost of the spillway for large dams with large flood discharges is to combine different flood discharge outlets. There are many different types of spillways and despite of some types being more adaptable to specific projects or dams, there is the possibility of reducing the investment cost by considering different arrangements or variations of the same type. The economy sought must not be obtained to the detriment of the safety of the project; this means that the design solutions considered must keep the capacity of discharging safely the full project design flood.

Since the frequencies of occurrence of the magnitudes of the flow to be discharged vary, a key strategic approach to the economical design consists in considering the total discharge capacity of the project divided into different type of spillways and designing each one with criteria compatible to its safety and operation expectation. For such cases, clear operating rules must be defined. In concrete dam projects it is possible to have distribution of discharges through chutes, orifices and crest spillways. As shown in Karun III Dam and HPP project, (Figure 10.1) the 20 000 m³/s flood is discharged through a combination of a chute, orifices and crest spillways. Incidentally, as mentioned in chapter 2 of this bulletin, in Karun III Dam, due to deep excavation required at the left bank, a lined plunge pool was provided with a 36 m high tail pond dam which was constructed at the downstream. Here, the chute works as a service spillway, orifices as auxiliary spillway and dam crest as the emergency spillway as shown in the Figure 10.1.

Fig. 10.1
Karun III Dam (Iran): a 205-m high concrete arch dam with service, auxiliary and emergency spillways.

10.4.1. Service spillways

The service spillway is the spillway that provides continuous or frequent regulated (controlled) or unregulated (uncontrolled) releases from a reservoir, without causing damages to the dam, dike or appurtenant structures due to the releases, up to and including its design discharge. When only one spillway is provided, its design discharge must be equal to the project SDF. Service spillways can be classified into basically two types, surface spillway and orifice or tunnel spillways.

The choice as to whether use a surface spillway or an orifice spillway is generally governed by specific site characteristics. In a restricted canyon, it is often difficult to incorporate a surface spillway, and an orifice or tunnel spillway is the only logical choice. An orifice spillway has the disadvantage that discharge through the spillway is a function of the square root of the head ($\sqrt{H}$) available while that of the surface spillway is a function of the head to the power of one and a half ($H^{1.5}$). For this reason, orifice spillways are generally found efficient only on high dams where large heads are available.

Surface service spillways can be gated or ungated. Site characteristics, magnitude of the design flood, estimated risk and cost of different possible alternatives are the main parameters associated with the decision of using either type of spillways. Ungated spillways are without question the safest type of spillway. The ungated spillway is less likely to be obstructed by floating debris, and since there is no equipment to operate, its safe operation is not impaired by possible operator errors. However, ungated service spillways are generally more expensive than gated spillways for a given maximum discharge rate since they will involve long crests and consequently wide chutes or conduit diameters. The contradiction to this general rule on relative cost is given in the ungated labyrinth crest. For cases where the reservoir surface area is relatively large compared with the inflowing flood volume, a labyrinth spillway can provide an economical and safe structure. Ungated service spillways have the distinct advantage that they involve no operating equipment and, thus, require little regular maintenance.

Gated spillways are generally chosen when the site is restricted or the magnitude of the design flood is very large, and it is not physically or economically possible to construct the necessary length of an ungated crest. Control of downstream flooding, or maximization of conservation storage, requires more flexible control than would be provided by an ungated spillway. For a large SDF the total direct investment cost of a gated spillway will generally be smaller than for an ungated spillway of equivalent capacity. Gated service spillway bays should always be designed and constructed with facilities for placing bulkheads upstream of the gate in order that the gate can be serviced in the dry. Examples of problems involving gates which can occur in an emergency situation are depicted in Chapter 2 of this bulletin.

10.4.2. Auxiliary spillways

Where site conditions are favourable, the possibility of gaining overall economy by utilising an auxiliary spillway in conjunction with a service spillway may be considered. In such cases the service spillway will be designed to pass floods likely to occur frequently and the auxiliary spillway set up to operate only after such frequent floods are exceeded.

It must be noted that the concept of auxiliary spillway as a complement of the service spillway so that both have the capacity to discharge the full SDF under normal conditions with a safety margin provided by an accepted freeboard limit, is not unanimously adopted. In many places the auxiliary spillway is considered also an emergency spillway, built to take care of floods larger than the conventional spillway design flood.

Considering the concept of auxiliary spillway as part of the project spillway system complementing the service spillway so that both are required to discharge the full SDF, the safety criteria regarding gate operation (when auxiliary spillway is also gated) such as redundancy of power supply for gate operation, must be the same for all spillway gates, and no cost reduction should be considered for this part of the auxiliary spillway. However, as a rule, because of its less frequent use some degree of structural damage or downstream erosion may be accepted, provided these structural damages or erosion do not affect the structural safety of the project permanent structures and occur in places that can be accessed and repaired after the passage of the flood.

A common solution used to obtain savings in large projects is to reduce or even omit the concrete lining of the outflow channel of the auxiliary spillway. However, the decision to obtain savings by accepting less stringent criteria for the design of outflow channels of auxiliary spillways discharging large floods must be carefully balanced with the possibility of environmental damages such as creating large excavation gradients and landslides. Examples of projects where these solutions were applied are depicted in Chapter 2 of this Bulletin.

One important consideration in designing the service-auxiliary spillway system with less stringent criteria for the auxiliary spillway is the definition of the maximum flood discharge capacity of the service spillway which must be able to discharge all frequent floods. Large projects built with this kind of solution have established the service spillway with a capacity corresponding to return periods varying from 100 to 200 years.

In many cases the auxiliary spillway is located away from the service spillway but there are also cases in which these two structures lie side by side (Figure 10.2). In the first case conditions favourable for the adoption of an auxiliary spillway are the existence of a saddle or depression along the rim of the reservoir leading into a natural waterway, away from the dam or other structure in dam body to allow the construction of a more economical outflow channel without risk of damaging the dam.

Fig. 10.2

Xingó Project (Brazil): service and auxiliary spillway. The auxiliary spillway discharge channel is partially lined. Discharging capacity is 16 500 m³/s each.

10.4.3. Emergency spillways

This is a spillway designed to provide additional protection to the project against overtopping of the dam and is intended to be used under extreme conditions during the occurrence of floods larger than the project design flood or maceration or malfunction of the service or auxiliary spillway. As the name implies, emergency spillways are provided also for additional safety, should emergencies not contemplated by normal design assumptions, arise. Such situations could be the result of a mandatory shut down of the outlet works, a malfunctioning of spillway gates, the clogging of spillway passages by debris or the necessity for bypassing the regular spillway because of damage or failure of some part of that structure. An emergency might arise when flood inflows are handled principally by surcharge storage and a recurring flood develops before the service spillway or the outlet works evacuate a previous flood. Under normal reservoir operation, emergency spillways are never required to function.

In projects where the service spillway alone or in combination with an auxiliary spillway is designed to be apt to discharge the full SDF, the emergency spillway should not be used to justify savings or discharge capacity reductions in the operational spillway system. There are, however, projects in which the functions of auxiliary and emergency spillways are combined, as illustrated by case examples in item 10.5; in these cases, the service spillway is reduced in size and provide an economic benefit for the project investment.

Not all projects have an emergency spillway, but as discussed in ICOLD Bulletin 142, all projects should consider the possibility of occurrence of floods larger than the check-flood.

There are different types of emergencies and ICOLD Bulletin 142 describes the most common types. The operating concept of emergency spillways is a structure built with soil or concrete, which is destroyed or removed when the hydraulic head relative to the reservoir water level elevation, is increased beyond a certain limit. With its destruction or removal, the outgoing flow increases, re-establishing the safety elevation of the water level and preventing the overtopping of the dam.

Therefore, the control crest must be placed at or above the designed maximum elevation of the reservoir water surface. The freeboard requirement for the dam is based on a water level elevation for which the emergency dike or concrete facility is removed. However, in such a case, and depending upon the height of the destroyed embankment (fuse), it may become impossible to re- impound the reservoir up to its full supply level as long as the fuse is not re-built. This may become a real drawback, even if the occurrence of such an event has a very low probability.

10.5.   OPTIMISING THE OVERALL PROJECT ARRANGEMENT

In the formulation of the overall dam project arrangement the analysis of different types of spillway alternatives and different types of arrangements for these structures, for the same safety criteria, may have a significant impact on the overall cost of the project. Different spillway configurations, such as gated versus ungated spillways or a combination of these, and the division of the spillway system in service, auxiliary and eventually emergency spillways, for example, may lead to important cost savings for a dam projects. The following three cases described in continuation are examples of cost-efficient solutions obtained by optimising the project arrangement by using a reduced service spillway and using an additional combination of an auxiliary and emergency spillway to take care of larger floods.

The first example is the Shamil and Nian dams which are 35 m high, earthfill dams with a common reservoir with a storage capacity of 69×10^6 m³. The crest lengths of Shamil and Nian are 320 m and 530 m, respectively. This project is mainly for domestic water supply of the capital of the Hormozgan Province in the south of Iran. In the design of the project the PMF (13 565 m³/s) was selected as a peak inflow to the reservoir. Several design alternatives for flood handling structures were considered to control the PMF. The most economical alternative was found to be a spillway system consisting of a service, an auxiliary and an emergency spillway. This system ensures the safety of the project.

The service spillway is gated and handles floods up to the 150-year flood without assistance from the other spillways. If the flood exceeds this discharge the water will start to spill over the auxiliary spillway, with a 300 m long overflow crest. With both spillways operating the system can safely handle floods up to the 10 000-year flood. If the flood continues to increase a fuse plug used as the emergency spillway will breach and take care of the remaining discharge. The system has a total capacity to handle the PMF without the dam being overtopped.

The probability of the PMF is always very small and where its value is several times larger than the probabilistic computed SDF, a service spillway designed for handling the SDF may be combined with an auxiliary-emergency spillway. One example of this case is the Mnjoli Project in Swaziland (Engels and Sheerman-Chase, 1985). This project had an ungated concrete spillway designed for the 100-yr flood peaking 950 m³/s and a PMF estimated to produce a flood peaking 7,650 m³/s, more than eight times larger than the designed SDF. A fuse-plug spillway was built to discharge the eventual PMF flood, with the unique feature of two levels separated by a dividing wall. One part would overtop when the reservoir reaches the 100-yr flood level and the other, 60 cm higher, would breach for a more severe flood. In January 1984, a tropical cyclone occurred, and the erodible bunds of the fuse-plugs were washed away, allowing discharge of the excess flood to be released by the fuse-plug channel and preventing the overtopping of the dam.

The third case is related to hydroelectric undertaking, the 180-MW Mrica project in Indonesia (Soerachmad, 1988). The Project is located in a tropical mountainous country with a high mean annual rainfall of 3,839 mm and a very rapid flood increase at the project site. The dam is an earth-rock structure, 109-m high. The flood studies concluded for the following return periods:

Return period (year)	Flood discharge (m³/s)
100	4 200
1000	6 100
10000	7 400
PMF	9 300

The designed spillway system includes a gated service spillway, a drawdown outlet and an earth dike fuse-plug emergency spillway. The gated spillway and the drawdown outlet can together discharge the 1000-yr flood of 6 100 m³/s. Floods above the 1000-yr flood and up to the PMF were designed to be discharged by the service spillway and drawdown outlet and by the destroyed dike of the emergency spillway, preventing the overtopping of the dam. In case all gates are damaged and remain closed the emergency spillway can discharge safely the 100-yr flood. This solution was based on economic and risk assessment studies. The project was completed in 1990 and during construction on March 1986 a flood peaking 4 486 m³/s occurred and caused a revision of the hydrological studies and a revision of the spillway features, but not altering the design concept.

10.6. RISK-BASED DESIGN APPROACHES

Flood risk at dams is resulting from the combination of a hydrological hazard, the dam vulnerability or sensitivity, and the consequences in case of flood-induced dam failure (ICOLD 2014 – Bulletin 154). Risk-based design approaches allow considering different design basis criteria depending on the level of consequences downstream (and sometimes also upstream) in case of dam failure, and not only depending on the dam and reservoir characteristics.

Incremental damages methods, and cost-benefit analysis methods within incremental damages approaches, can stand as optimisation tools to find the appropriate design criteria that suit the actual level of flood risk, particularly for existing dams. For existing dams, cost-benefit analysis can be derived in the estimation of the net present value (NPV) of a spillway upgrade project – benefits are expressed as avoided damages and consequences.

Consideration to risk acceptability or tolerability criteria related to different categories of consequences or damages (loss of life, loss of property, loss of economical assets) can be found in Appendix B of ICOLD Bulletin 154, even though this bulletin is not specifically devoted to safe management of floods. Some national committees have been working (Australia, Canada) or are even currently working (France) on these issues and approaches. Reference to the publications of these committees is indicated for more details.

10.7. CONCLUSION

This chapter tried to demonstrate that cost reduction in the design of spillways is possible without impairing the fundamental safety of the project. It has also been shown that formulation of an alternative spillway arrangement in the overall project configuration can lead to significant economy in the project.

10.8. REFERENCES

ICOLD (2011) - *Cost Savings in Dams*, Bulletin 152.

ICOLD (1992) - Selection of design flood- Current methods, Bulletin 82.

ICOLD (2012) - *Safe passage of extreme floods*, Bulletin 142.

Bouvard, M. (1988) - "Design flood and operational flood control", *General Report Question 63, Sixteenth Congress on Large Dams*, San Francisco, USA.

Engels, E. T. and A. Sheerman-Chase (1985) - "Design and operation of a fuse-plug spillway in Swaziland", *Water Power & Dam Construction*, pgs. 26-28, No. 6.

Soerachmad, S. (1988) - "Mrica Dam, Indonesia – Criteria for choice of spillways", R.48, Q.63, *Sixteenth Congress on Large Dams*, San Francisco, USA.

PROGRÈS TECHNIQUES DANS LA CONCEPTION DES ÉVACUATEURS DE CRUES

PROGRÈS ET INNOVATIONS DE 1985 À 2020

L'objectif du bulletin du est de fournir à la communauté des ingénieurs des barrages une information sur les caractéristiques techniques des différents types d'évacuateurs de crue développés et implémentés durant les quatre dernières décennies, ainsi que des caractéristiques hydrauliques associées à ces structures.

Bien que les évacuateurs de crue, en tant que tels, peuvent être construits avec différents types de structures et de dispositions, l'approche de ce bulletin a été de retenir certains types spécifiques de structures – évacuateurs de crue à marche d'escalier, labyrinthes, PKW et évacuateurs de crue en tunnel, et d'analyser des types plus conventionnels d'évacuateurs de crue opérant dans des conditions spéciales – très grands débits, très haute chute, climat très froid.

Dans chaque chapitre, les sujets traités se concentrent sur les principales questions concernant les problématiques et structures spécifiques considérées, dans l'objectif de fournir des informations au concepteur du projet et au propriétaire, pour aider aux choix à opérer dans le cadre du développement du projet.

La récente évolution des technologies en lien avec la conception, la construction et l'exploitation des évacuateurs de crue, s'est concentrée essentiellement sur les sujets économiques et de sécurité. Les évacuateurs de crue sont un élément clé d'un projet de barrage, pour assurer la protection nécessaire du projet vis-à-vis de l'action destructrice des crues. Ceci a mené au développement de différentes approches pour sélectionner et définir le meilleur type d'évacuateur pour chaque projet. Ce bulletin vise à présenter et discuter les résultats les plus récents de cette évolution.

PROGRÈS TECHNIQUES DANS LA CONCEPTION DES ÉVACUATEURS DE CRUES

PROGRÈS ET INNOVATIONS DE 1985 À 2020

B 172

INTERNATIONAL COMMISSION ON LARGE DAMS
COMMISSION INTERNATIONALE DES GRANDS BARRAGES
6 Quai Watier – 78400 Chatou (France)
http://www.icold-cigb.org.

Couverture: Vue en coupe et photographie du barrage d'Ertan – Chine (photo personnelle)

CRC Press/Balkema is an imprint of the Taylor & Francis Group, an informa business

© 2025 ICOLD/CIGB, Paris, France

Typeset by CodeMantra

Published by CRC Press/Balkema
4 Park Square, Milton Park, Abingdon, Oxon, OX14 4RN
and by CRC Press/Balkema
2385 NW Executive Center Drive, Suite 320, Boca Raton FL 33431

**AVERTISSEMENT – EXONÉRATION
DE RESPONSABILITÉ :**

Les informations, analyses et conclusions contenues dans cet ouvrage n'ont pas force de Loi et ne doivent pas être considérées comme un substitut aux réglementations officielles imposées par la Loi. Elles sont uniquement destinées à un public de Professionnels Avertis, seuls aptes à en apprécier et à en déterminer la valeur et la portée.

Malgré tout le soin apporté à la rédaction de cet ouvrage, compte tenu de l'évolution des techniques et de la science, nous ne pouvons en garantir l'exhaustivité.

Nous déclinons expressément toute responsabilité quant à l'interprétation et l'application éventuelles (y compris les dommages éventuels en résultant ou liés) du contenu de cet ouvrage.

En poursuivant la lecture de cet ouvrage, vous acceptez de façon expresse cette condition.

Texte original en anglaise
Traduction par la France et la Suisse
Mise en page par Nathalie Schauner

ISBN: 978-1-041-08378-8 (Pbk)
ISBN: 978-1-003-64875-8 (eBook)

MEMBRES DU COMITÉ
(2009-2016)

Président

    B. P. MACHADO   Brésil

Vice-Président

    J. GUO   Chine

Ancien Présidents du Comité

    A. LEJEUNE   Belgique

    B. PETRY   Pays Bas

Membres

R. WARK		Australie
G. RODRIGUEZ ROCA		Bolivie
Z. MICOVIC	Depuis Fév. 2015	Canada
T. NZAKIMUENA	Jusqu'en Jan. 2015	
J. HODAK	Depuis Jan. 2010	Rép. Tchèque
B. TAQUET	Jusqu'en Déc. 2016	France
F. LAUGIER	Depuis Avr. 2017	
R. HASELSTEINER	Depuis Fév. 2015	Allemagne
H. B. HORLACHER	Jusqu'en Jan. 2015	
SIBA PRASAD SEN		Inde
C. FOULADI		Iran
A. PIETRANGELI	Depuis Oct. 2010	Italie
T. TAKASUKA	Depuis Aout 2015	Japan
T. HINO	Jusqu'en Jul. 2015	
I. KO	Depuis Mar. 2016	Corée
S. LEE	Jusqu'en Fév. 2016	
H. MARENGO	Depuis May 2012	Mexique
R. DE JONG		Pays Bas
H-M. KJELLESVIG		Norvège
A. BASHIR		Pakistan
C. M. RAMOS	Jusqu'en Mai 2010	Portugal
R. SHAKIROV		Russie
H-J. WRIGHT	Depuis Fév. 2015	Afrique du Sud
D. I. VAN WYK	Depuis Mar. 2010 Jusqu'en Jan. 2015	
E. F. SNELL	Jusqu'en Fév. 2010	
K. LAKSIRI		Sri Lanka
A. GRANADOS	Depuis Juin 2012	Espagne
J. YANG	Depuis Mai 2012	Suisse
A. WÖRMAN	Depuis Mar. 2010 Jusqu'en Avril 2012	

A. SCHLEISS		Suisse
P. MASON		Royaume Uni
J. E. LINDELL	Depuis May 2012	États Unis
S. HUI	Jusqu'en Apr. 2012	
A. MARCANO		Venezuela

Membre co-opté

J. E. LINDELL	Jusqu'en Mai 2012	États Unis
A. PINHIERO		Portugal
R. BOES	Depuis Octobre 2013	Suisse

Autres contributeurs au Bulletin

B. CROOKSTON	États Unis
B. TULLIS	États Unis
S. BARFUSS	États Unis
S. ERPICUM	Belgique
L. LIA	Norvège
F. LEMPÉRIÈRE	France
HO TA KHANH	Vietnam
P. MANSO	Suisse

SOMMAIRE

TABLE DES MATIERES

TABLEAUX & FIGURES

FIGURES

REMERCIEMENTS

Ce bulletin a été préparé par les membres du Comité Technique de la CIGB de l'Hydraulique des Barrages Durant la période 2009-2017. Les auteurs et rédacteurs des 10 chapitres sont listés ci-après, et dans certains cas incluent des spécialistes qui ne sont pas formellement membres du Comité, qui a cependant élargi le cercle des contributeurs de manière à pouvoir fournir les dernières informations à jour et ainsi améliorer le contenu du Bulletin.

Chapitre 1 – INTRODUCTION a été rédigé B. P. Machado (Brazil)

Chapitre 2 – DEVERSOIRS DE GRANDE CAPACITE a été rédigé par B.P. Machado (Brazil) et complété par des apports de A. Schleiss (Suisse), S. P. Sen (Inde), R. Wark (Australie), Z. Micovic (Canada), B. Taquet (France), H. Marengo (Mexique) and A. Marcano (Venezuela).

Chapitre 3 – DEVERSOIRS A HAUTE CHUTE ET DISSIPATION D'ENERGIE a été rédigé A. Schleiss (Suisse) avec des contributions de P. Manso (Suisse) et des apports de P. Mason (Royaume Uni).

Chapitre 4 – DEVERSOIRS EN MARCHE D'ESCALIER a été rédigé par A. Granados (Spain) avec des contributions de P. Mason (Royaume Uni), A. Schleiss (Suisse), R. Boes (Suisse) et des apports de H-J Wright (Afrique du Sud).

Chapitre 5 – EVACUATEURS LABYRINTHE a été rédigé par D. van Wyk (Afrique du Sud) avec B. Crookston (États-Unis) et B. Tullis (États-Unis).

Chapitre 6 – EVACUATEURS PKW a été rédigé by B. Taquet (France) avec with S. Erpicum (Belgique) et F. Laugier (France).

Chapitre 7 – EVACUATEURS EN TUNNEL, EN PUITS ET A VORTEX a été rédigé par J. Guo (Chine) avec des contributions importantes de R. Shakirov (Russie) et des apports de S. P. Sen (Indi).

Chapitre 8 – EVACUATEURS DANS LES CLIMATS FROIDS a été rédigé par H-M. Kjellesvig (Norvège) avec des apports de by T. Nzakimuena (Canada), R. Shakirov (Russie) et L. Lia (Norvège).

Chapitre 9 – MODELISATION COMPOSITE DES STRUCTURES HDYRAULIQUES a été rédigé par J. Lindell (USA) avec S. Barfuss (USA), et des apports de R. Wark (Australie) R. Shakirov (Russie) et Z. Micovic (Canada).

Chapitre 10 – ECONOMIE, RISQUE ET SECURITE DANS LA CONCEPTION DES EVACUATEURS DE CRUE a été rédigé par C. Fouladi (Iran) avec des contributions de B. P. Machado (Brésil) et B. Taquet (France), et des apport s de H-J Wright (Afrique du Sud), S. P. Sen (Inde), F. Lempérière (France) et M. Ho Ta Khanh (Vietnam).

Initialement, le Bulletin incluait un chapitre additionnel dédié aux discussions sur les corps flottants et les écoulements chargés en sédiment durant l'utilisation des évacuateurs de crue. Ce chapitre a été déplacé dans un autre bulletin préparé conjointement par le Comité sur l'Hydraulique des Barrages et le Comté sur l'Évaluation de Crues et la Sécurité des Barrages. Le chapitre original a été rédigé par R. Wark (Australie) avec des apports de S. P. Sen (Inde). Cet autre bulletin a été coordonné par R. Wark et s'intitule « *Blocage des évacuateurs de crue, prises d'eau et des vidanges de fond par les corps flottants* ».

L'intégralité du texte anglais du Bulletin a été corrigée pour les corrections typographiques et la cohérence par H-J Wright et B. P. Machado.

La version française a été traduite par un effort commun des comités français et suisses, sous la supervision, les corrections typographiques et la cohérence de F. Laugier.

1. INTRODUCTION

1.1. OBJECTIF ET CADRE DU BULLETIN

L'objectif de ce bulletin est de fournir à la communauté des ingénieurs des barrages une information sur les caractéristiques techniques des différents types d'évacuateurs de crue développés et implémentés durant les quatre dernières décennies, ainsi que des caractéristiques hydrauliques associées à ces structures. L'information présentée dans ce bulletin ne remplace pas le contenu des précédents bulletins de la CIGB sur ce sujet, mais plutôt complète les informations avec des développements techniques récents.

Bien que les évacuateurs de crue, en tant que tels, peuvent être construits avec différents types de structures et de dispositions, l'approche de ce bulletin a été de retenir certains types spécifiques de structures – évacuateurs de crue à marche d'escalier, labyrinthes, PKW et évacuateurs de crue en tunnel, et d'analyser des types plus conventionnels d'évacuateurs de crue opérant dans des conditions spéciales – très grands débits, très haute chute, climat très froid. Ces informations sont complétées avec un chapitre sur l'application des techniques de conception utilisant les modèles numériques et physiques, et un chapitre sur les questions d'économie et de coût. Le sujet très important des corps flottant et du contrôle des sédiments affectant la conception et l'exploitation des évacuateurs de crue, a été reporté sur un autre bulletin de la CIGB, qui inclura les problématiques de sécurité des barrages en crue.

Dans chaque chapitre, les sujets traités se concentrent sur les principales questions concernant les problématiques et structures spécifiques considérées, dans l'objectif de fournir des informations au concepteur du projet et au propriétaire, pour aider aux choix à opérer dans le cadre du développement du projet. Les sujets présentés sont en général basés sur des études hydrauliques expérimentales et physiques récentes et sur des exemples réels de travaux réalisés dans différentes parties du monde.

La récente évolution des technologies en lien avec la conception, la construction et l'exploitation des évacuateurs de crue, s'est concentrée essentiellement sur les sujets économiques et de sécurité. Les évacuateurs de crue sont un élément clé d'un projet de barrage, pour assurer la protection nécessaire du projet vis-à-vis de l'action destructrice des crues. Ceci a mené au développement de différentes approches pour sélectionner et définir le meilleur type d'évacuateur pour chaque projet. Ce bulletin vise à présenter et discuter les résultats les plus récents de cette évolution.

1.2. RELATION AVEC LES PRECEDENTS BULLETINS

La CIGB a produit différents bulletins en relation avec la conception et l'exploitation des évacuateurs de crue. Comme indiqué, le présent bulletin ne remplace pas ces publications, bien qu'il soit reconnu que certains sujets, comme la dissipation d'énergie, mériteraient, un traitement plus ample. Dans tous les cas, la compilation des bulletins suivants constitue un volume significatif d'information technique de grande valeur pour la conception hydraulique des évacuateurs de crue.

N°	BULLETIN	ANNEE
49A	EXPLOITATION DES OUVRAGES HYDRAULIQUES DE BARRAGES	1986
58	EVACUATEURS DE CRUE DE BARRAGES	1987
81	EVACUATEURS – ONDES DE CHOC ET ENTRAINEMENT D'AIR	1992
82	SELECTION OF THE DESIGN FLOOD – CURRENT METHODS	1992
108	COUT DE LA MAITRISE DES CRUES DANS LES BARRAGES	1997
125	BARRAGES ET CRUES – PRINCIPES ET ETUDES DE CAS	2003
130	EVALUATION DU RISQUE DANS LA GESTION DE LA SECURITE DES BARRAGES	2005
142	PASSAGE EN SECURITE DES CRUES EXTREMES	2012
156	GESTION INTEGREE DU RISQUE CRUE	2014

1.3.   SOMMAIRE DU BULLETIN

Une description résumée du sommaire du bulletin est donnée ci-après :

Chapitre 2 – ÉVACUATEURS DE GRANDE CAPACITÉ

Ce chapitre traite de la conception et des dispositions spécifiques des évacuateurs de crue avec des capacités variant de 20 000 m³/s au plus grand évacuateur de crue existant conçu pour 110 000 m³/s. Il inclut des références et des questions relatives aux évacuateurs vannés et non vannés soulevant la problématique de l'utilisation générale de très grandes vannes segment. Il traite aussi les sujets des structures de dissipation d'énergie et du risque intrinsèque associé avec ces structures.

Chapitre 3 – ÉVACUATEUR DE HAUTE CHUTE – LE DÉFI DE LA DISSIPATION D'ÉNERGIE ET DU CONTRÔLE DE L'ÉROSION AVAL

Ce chapitre aborde les problèmes de dissipation d'énergie des jets issus des évacuateurs à haute chute et du développement des processus correspondant d'érosion du rocher aval. Une description physique du processus d'érosion est présentée ainsi que des informations sur les formules semi-empiriques et les méthodes d'évaluation sur base physique, utilisées pour estimer la taille des fosses d'érosion. Les méthodes pour contrôler l'érosion sont également présentées, introduisant la discussion sur la pré-excavation des fosses de dissipation, son dimensionnement et une discussion sur les conditions et critères d'utilisation relatifs à son usage.

Chapitre 4 – ÉVACUATEURS À MARCHE D'ESCALIER

Ce chapitre décrit les principales applications des évacuateurs à marche d'escalier, leur hydraulique et les performances attendues, ainsi que des aspects nécessitant une attention spéciale durant la phase de conception. Le chapitre présente une revue complète de l'hydraulique des évacuateurs à marche d'escalier avec des éléments récents permettant de définir des critères de conception. Des études de cas sont également présentées pour illustrer les performances effectives de structures existantes.

Chapitre 5 – ÉVACUATEURS LABYRINTHE

Des informations pour la conception des évacuateurs labyrinthes sont présentées dans ce chapitre. Une revue complète des différentes géométries de structure et des paramètres hydrauliques correspondants est présentée. Le chapitre inclut des critères de conception hydraulique et une méthodologie de conception est proposée. Il inclut également des références bibliographiques fournissant des détails complémentaires et des études des de cas significatifs.

Chapitre 6 – ÉVACUATEURS PKW

Ce chapitre contient des informations sur les évacuateurs à touche de piano (Piano Key Weir - PKW), une variante de l'évacuateur de crue labyrinthe traditionnel, développée durant la dernière décennie. Il inclut la description de cette structure, ses performances hydrauliques évaluées en laboratoire ainsi que sur des projets réalisés, des critères de dimensionnement hydraulique et des paramètres de conception, des informations sur les problématiques structurelles et de construction, et des cas concrets, ainsi que des variantes dans les dispositions de réalisation. Une bibliographie complète est fournie.

Chapitre 7 – ÉVACUATEURS EN TUNNEL, PUITS ET VORTEX

Ce chapitre considère les différentes sortes d'évacuateurs en tunnel, tels que les tunnels à prise « haute », ou à prise « basse » (ou sortie basse), vortex, puits et tunnels à orifice. Les sujets suivants sont discutés : dispositions générales, structure de contrôle du débit, structure de transfert du débit, structure terminale, problématiques d'exploitation, analyse de risque. Quelques développements récents, innovations et applications aux évacuateurs de crue en tunnel conçus pour des très grands barrages et réalisés ces 20 à 30 dernières années, sont spécialement présentés. Des études de cas illustratifs sont également présentées.

Chapitre 8 – PROBLÈMES SPÉCIFIQUES DES ÉVACUATEURS EN CLIMAT TRÈS FROID

Ce chapitre traite principalement des problèmes sérieux de neige et de blocage par la glace, partiel ou total, des évacuateurs de crue, réduisant ainsi leur capacité d'évacuation des débits dans les régions froides. Cela couvre aussi la question des températures négatives durant la construction et l'exploitation des évacuateurs de crues, et de l'utilisation des matériaux de construction. En complément, d'autres questions concernant la régulation des vannes, et les efforts et impacts spécifiques de la glace sur les structures, sont également évoquées.

Chapitre 9 – MODÉLISATION COMPOSITE DES STRUCTURES HYDRAULIQUES

La modélisation composite est l'utilisation effective et conjointe des modelées physiques et numériques, utilisés soit en série ou en parallèle, pour résoudre des problèmes hydrauliques. Ce chapitre aborde les avantages et limites de la modélisation numérique et physique, dans la pratique de de l'ingénierie hydraulique. Il décrit le processus d'utilisation de modélisation composite / combinée pour la conception des évacuateurs de crue et des vidanges de fond des barrages.

Chapitre 10 – ÉCONOMIE, RISQUE ET SÉCURITÉ DANS LA CONCEPTION DES ÉVACUATEURS

Les évacuateurs de crue constituent la principale assurance de sécurité pour n'importe quel type de barrage. Pour cette raison, la recherche de réduction de coût sur ces structures, pourrait induire le risque de baisser le niveau de sécurité global du projet de barrage. Cependant, pour les mêmes critères de sécurité, l'analyse comparative de différentes options de système d'évacuation des crues, peut avoir un impact significatif sur le coût global du projet. Par ailleurs, des configurations différentes d'évacuateurs de crue, tels que des évacuateurs vannés, versus une combinaison d'évacuateurs vannés et non vannés, par exemple, peut conduire également à des importantes réductions de coûts. Ce chapitre traite ces questions.

2. LES DÉVERSOIRS DE GRANDE CAPACITÉ

2.1. INTRODUCTION

Le but de ce chapitre est de présenter et de discuter des problèmes d'ingénierie associés à la conception, à la construction et aux caractéristiques opérationnelles des déversoirs conçus pour évacuer de très grands débits. Bien qu'il n'y ait pas de différences majeures entre les critères utilisés pour la conception hydraulique des déversoirs de différentes tailles, les aspects constructifs et budgétaires jouent un rôle important dans l'optimisation de la conception des déversoirs de grande capacité. En outre, une attention spécifique doit être portée aux risques opérationnels inhérents à la gestion des gros débits transitant dans la retenue puis étant évacués dans le bief aval du cours d'eau.

Dans le présent bulletin, il est considéré que les déversoirs de grande capacité sont ceux conçus pour évacuer des débits supérieurs à 20 000 m³/s et/ou dont les débits spécifiques sont supérieurs à 130 m³/s/m.

La plupart des projets de barrages actuels et futurs intégrant des déversoirs de grande capacité sont en construction en Asie, en Amérique latine et en Afrique, sur des cours d'eau drainant de très grands bassins versants. Les tableaux disponibles à la fin de ce chapitre présentent les données relatives à un certain nombre de projets, dans différents pays, intégrant de déversoirs de grande capacité. Ces projets ont permis d'accumuler une expérience importante qui constitue la base des discussions présentées ci-après. Par ailleurs, des projets de même ampleur existent également dans d'autres pays.

2.2. PRINCIPAUX ÉLÉMENTS DE DIMENSIONNEMENT DES DÉVERSOIRS DE GRANDE CAPACITÉ

2.2.1. Processus d'altération

Le présent document n'a pas pour objet de discuter des méthodologies permettant de déterminer la crue dimensionnante de l'évacuateur de crue. Cependant il est pertinent de rappeler les principaux éléments généralement utilisés dans le dimensionnement.

La caractérisation de la crue dimensionnante de l'évacuateur de crue s'appuie généralement sur :

- La période de retour (T) de différents débits de pointe.

- Le volume d'eau associé à ces débits de pointe.

- Le CMP ou Crue Maximum Probable, une valeur déterministe supposée être la crue maximale absolue pouvant se produire sur le site du projet

De nombreux grands déversoirs récents ont été conçus pour évacuer la crue décamillénale, correspondant théoriquement au débit naturel du fleuve, dont la probabilité d'occurrence est de 1% sur une durée de vie projetée de 100 ans. En règle générale, ces déversoirs sont capables d'évacuer ce pic de crue avec un niveau de réservoir au niveau maximal autorisé, qui est souvent supérieur au niveau maximal d'exploitation du projet sans toutefois intégrer la revanche du barrage. Si la superficie du réservoir est suffisamment grande, le volume du réservoir créé par cette amplitude de niveau peut contribuer à atténuer l'hydrogramme de la crue, représentant ainsi une certaine marge de sécurité qui est spécifiquement prise en compte dans la stratégie de gestion de la crue.

De manière alternative à la crue décamillénale, la CMP est également largement utilisée pour définir le débit dimensionnant de l'évacuateur de crue. Cependant, en règle générale, pour les grands projets, les deux valeurs de débit sont usuellement calculées ainsi que les conditions d'évacuation correspondantes. Il est courant de définir la crue de conception de l'évacuateur de crue comme étant la crue décamillénale associée au niveau maximal admissible dans le réservoir et de vérifier la capacité de l'évacuateur de crue pour la CMP en autorisant une réduction de la revanche disponible.

Bien que ces critères soient largement utilisés pour la conception de grands déversoirs, il existe un risque inhérent de sous-estimation de la crue à évacuer, en raison de la non-représentativité possible des données de base utilisées pour le calcul des valeurs probabilistes et pour le calcul de la CMP. En effet, ces critères ne prennent pas en compte les simplifications statistiques causées par l'utilisation d'une plage de données limitée pour le calcul d'un évènement de période de retour bien plus long, ni les futures évolutions physiques du site d'étude, tels que les changements climatiques et les modifications des bassins versants affectant le ruissellement. Ces questions sont largement discutées dans le bulletin CIGB n °142 « Passage en sécurité des crues extrêmes » (CIGB, 2012), et méritent une attention particulière lors de la conception de grands déversoirs. La détermination d'intervalles de confiance pour la valeur calculée de la crue et la prise en compte de moyens de décharge supplémentaires tels que des déversoirs d'urgence peuvent représenter une approche prudente.

2.2.2. Emplacement, type et dimensions

Les caractéristiques techniques et le coût d'un évacuateur de crues ne sont pas uniquement liés à ses éléments structurels et équipements intrinsèques, ils sont également fortement dépendants de sa localisation ainsi que d'autres caractéristiques du projet, notamment le type et la hauteur du barrage ou encore l'appréciation du site dans sa globalité. De nombreuses variantes de conception sont envisageables et pour chaque projet il est possible d'optimiser le coût tenant compte des paramètres les plus pertinents : débits à évacuer, exigences opérationnelles ainsi que caractéristiques topographiques et géologiques.

Les réglementations environnementales de nombreux pays exigent qu'un débit réservé, adapté aux conditions environnementales, soit rejeté de manière permanente dans le bief aval du cours d'eau. Cela peut affecter la conception et le fonctionnement de l'évacuateur de crue, en particulier dans les cas où les structures sont construites sur des cours d'eau très larges. La tendance actuelle semble toutefois être en faveur de la localisation de telles installations dans des entités distinctes. Par exemple, le projet de Belo Monte, actuellement en construction au Brésil et court-circuitant un long méandre, aura un déversoir dimensionné pour 62 000 m³/s et devra libérer dans le tronçon court-circuité un débit réservé de 700 m³/s pendant la période sèche et un débit pouvant atteindre 8 000 m³/s pendant la saison des pluies. Au lieu de le faire par le déversoir, ces débits seront relachés par une structure séparée qui comprendra une centrale supplémentaire pour turbiner le débit «environnemental». De nombreux projets dans différents pays suivent ces critères.

Les déversoirs de grande capacité sont logiquement des structures de grandes dimensions, quelle que soit les dimensions des autres éléments du projet. Concernant les barrages en béton, l'emplacement le plus naturel du déversoir se situe en crête ou dans le corps-même du barrage. En général, l'évacuateur de crue sera placé près ou à proximité du lit du cours d'eau afin de faciliter la restitution du débit ; un emplacement qui nécessite d'être optimisé lorsqu'une centrale électrique est prévue au même emplacement. Par exemple dans le projet Itaipu, à la frontière entre le Brésil et le Paraguay (Fig. 2.1), l'évacuateur de crues a été placé en rive droite, toute la largeur du lit étant entièrement occupée par la centrale et ses 18 unités de production.

Fig. 2.1
Aménagement d'Itaipu (frontière Brésil-Paraguay) et son déversoir de 62 200 m³/s

En fonction de la hauteur et du type de barrage – barrage poids en béton ou en remblai, barrage poids-voûte ou barrage voûte - de nombreuses conceptions et configurations sont possibles dont la créativité des ingénieurs du monde entier a fourni d'innombrables exemples.

Pour les barrages situés dans de larges vallées, le déversoir sera normalement construit sur ou dans une portion en béton du barrage, que le reste de la structure soit ou non un remblai. L'aménagement de Tucurui (Fig. 2.2), au Brésil, en est un exemple. Dans tous ces cas, la largeur de la vallée était compatible avec la largeur du déversoir.

Fig. 2.2
Vue de l'aménagement de Tucurui (Brésil) équipé d'un déversoir 110 000 m³/s

Cependant, si le site n'est pas assez large, il peut être difficile de concevoir l'intégralité de l'évacuateur de crue en crête du barrage. Différentes solutions ont été utilisées comme placer les déversoirs complètement à l'extérieur de la structure du barrage ou alors en combinant des déversoirs de surface avec des évacuateurs intermédiaires et de fond, comme dans le cas du barrage Ertan (Fig. 2.3), en Chine. Ce projet comprend sept sorties en crête du barrage rejetant 6 260 m³/s, six sorties à un niveau-intermédiaire d'une capacité de 6 930 m³/s, quatre orifices de fond de 2 084 m³/s et deux tunnels rejetant 7 400 m³/s. La capacité totale atteint alors 22 674 m³/s.

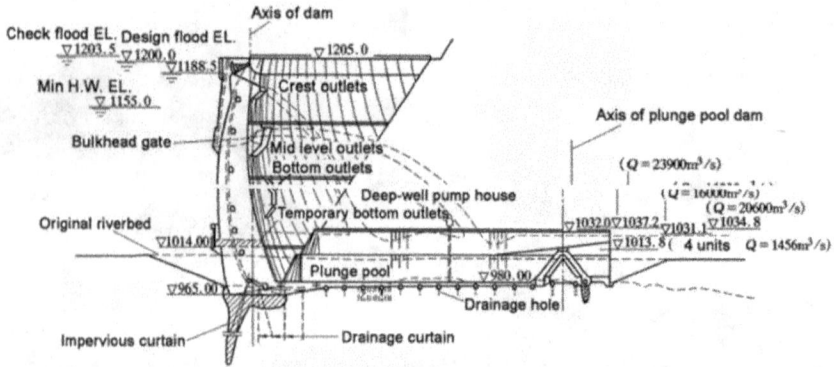

Fig. 2.3
Vue en coupe et photographie du barrage d'Ertan – Chine

2.2.3. Laminage de la crue par le réservoir

L'utilisation d'une partie du volume du réservoir pour stocker une partie du volume de la crue et permettre ainsi une atténuation du pic de crue et donc réduire le débit de conception du déversoir est une pratique courante pour les projets d'évacuateur de crue de grande capacité. Le tableau 2.1 décrit quelques cas dans lesquels cette pratique a été utilisée (comparez les colonnes «Débit de pointe naturel» et «Débit de conception de l'évacuateur de crue»). En pareil cas, le concepteur doit porter une attention particulière à la hauteur atteinte dans le réservoir pour évacuer le débit de pointe naturel.

Toutefois cette pratique augmente la superficie du réservoir et, outre l'impact financier - qu'il convient de comparer au coût du déversoir et du barrage -, l'impact environnemental du projet peut être accru. Pour les grands réservoirs situés dans des zones de faible dénivelé et selon la législation environnementale de chaque pays, cela peut poser des difficultés insurmontables.

2.2.4. Structure unique ou combinant plusieurs éléments

Les déversoirs conçus comme une structure unique devant évacuer des débits très importants seront nécessairement de longues structures et, dans certains cas, il sera économiquement difficile de placer l'ensemble du déversoir dans l'axe du barrage. Dans ces cas, il peut être utile de diviser le déversoir en deux parties ou plus, l'une utilisé de manière fréquente en évacuant les crues courantes (les déversoirs principaux) et les autres (déversoirs auxiliaires) afin d'atteindre la capacité d'évacuation maximale. Même lorsqu'il n'y a pas de problème d'espace, la division du déversoir en éléments séparées peut être une source d'économies en construisant des déversoirs auxiliaires avec des exigences techniques moins strictes, leur utilisation n'étant pas aussi fréquente que la structure principale. Bien entendu, cela implique que, dans tous les cas, la sécurité globale du projet ne soit pas affectée, mais cela peut nécessiter des inspections continues de ses performances. Quelques exemples illustrent cette possibilité.

Pour l'aménagement Itá au Brésil, grâce à deux déversoirs, la capacité totale atteint 52 800 m³/s. Le coursier du déversoir auxiliaire, long de 275 m, n'avait été revêtu de béton que sur les 120 m initiaux (débit de 20 000 m³/s, débit spécifique 234 m³/s/m). La décision originelle de cette optimisation de coût s'appuyait sur l'accessibilité de la zone et la possibilité de réaliser des travaux correctifs quand et si cela s'avérait nécessaire. En fait, les premiers tests effectués sur cette structure ont montré que les risques d'érosion étaient plus élevés que prévu et la dalle de béton a ainsi été étendue de 55 m pour protéger une zone sensible, comme le montre la figure 2.4 (Andrzejewski, 2002).

(a) Protection initiale du courser   (b) Protection étendue après les tests d'érosion

Fig. 2.4
Déversoir auxiliaire de l'aménagement d'Ità (Brésil)

Un autre exemple est le projet Xingó, également au Brésil. Le projet comprend deux déversoirs parallèles d'une capacité combinée de 33 000 m³/s. Le coursier long de 252 m du déversoir auxiliaire, conçue pour un débit maximal de 15 500 m³/s, n'était revêtu d'une dalle en béton que sur 90 m. Ce déversoir a fonctionné dans des conditions d'essai jusqu'à 4 000 m³/s. Une érosion localisée a été observée et remblayée avec du béton. Après les essais, le fonctionnement du coursier a été approuvé, car si des débits plus élevés viennent à provoquer une érosion supplémentaire, un accès aux réparations éventuelles sera disponible (Eigenheer et al., 2002).

Ces deux exemples montrent que l'utilisation d'un évacuateur auxiliaire, fonctionnant peu souvent, et conçu avec une certaine souplesse par rapport aux recommandations techniques peut permettre de réaliser des économies de coût à la construction, mais il doit tout de même garantir un fonctionnement sûr, ce qui dépend d'une inspection continue et d'un accès aisé pour la réparation des dommages éventuels.

L'utilisation d'un déversoir auxiliaire pour le rejet d'une partie de la crue de dimensionnement ne doit pas être confondue avec l'existence d'un déversoir d'urgence. Cette structure est recommandée lorsque la fiabilité des données et études hydrologiques de base, comme indiqué précédemment, est limitée en quantité ou en qualité. Dans ce cas, l'intégration de déversoirs d'urgence ou d'autres types de structures permettant de tamponner ou de dériver le débit en excès peut être incluse dans le projet. Cette question est également abordée dans le Bulletin 142 de la CIGB (2012), évoquant les recommandations concernant les moyens alternatifs de gérer les inondations plus importantes que la crue dimensionnante du déversoir.

2.3. ORGANES DE CONTROLE

2.3.1. Déversoirs de surface

Les déversoirs conçus pour évacuer des débits importants sont généralement des déversoirs de surface, principalement en raison de l'ampleur de la crue. Cependant, des orifices de fond ont été également utilisées, en particulier lorsque la structure était également utilisée pour la dérivation de la rivière et / ou lorsqu'un relargage des sédiments accumulés dans le réservoir était nécessaire.

La plupart des grands déversoirs construits pour des projets hydroélectriques ou d'approvisionnement en eau sont des déversoirs vannés sans quoi la longueur de l'évacuateur de crues ainsi que la réduction de charge utile compromettraient la rentabilité du projet. Cependant, l'utilisation d'évacuateurs vannés a parfois suscité des inquiétudes quant à la possibilité d'une défaillance ou d'un mauvais fonctionnement des vannes dans des circonstances critiques créant des conditions propices à la surverse et à la ruine de l'ouvrage. Bien évidemment les déversoirs libres éliminent ce risque.

Quel que soit l'objectif du projet, il n'est jamais facile de concevoir un seuil libre, aussi bien techniquement qu'économiquement, permettant l'évacuation de grands débits dans un site de projet donné. Néanmoins, cela a été fait dans certains projets où il était possible d'utiliser la plus grande partie de la largeur du site pour l'évacuateur de crue, en localisant la prise d'eau ou d'autres installations du projet dans un endroit n'interférant pas avec l'évacuateur. Là où la possibilité physique existe, l'équilibre entre économie et évaluation des risques est le facteur de décision clé. La figure 2.5 montre une vue du barrage de Burdekin, en Australie, conçu pour un débit de 64 600 m³/s avec une charge maximale de 17 m au-dessus du déversoir. Avec un déversoir vanné, il aurait été possible d'avoir un niveau normal de réservoir 17 m plus élevé que dans la configuration actuelle. Cela montre la limite économique des grands déversoirs non vannés pour les projets hydroélectriques. Le tableau 2.3 présente d'autres exemples de grands déversoirs non vannés australiens.

Fig. 2.5
Le seuil libre d'une capacité de 64 600m³/s du barrage de Burdekin, Australie

La plupart des déversoirs vannés récents conçus pour évacuer des débits importants sont équipées de grandes vannes segments ou vannes Tainter. De grandes vannes levantes verticales peuvent et sont bien sûr utilisées, mais elles ont besoin d'une superstructure très haute pour permettre leur levage et nécessitent également des rainures latérales qui perturbent le flot évacué. Ces vannes fonctionnent habituellement avec des palans à câble car si des vérins hydrauliques étaient utilisés, la hauteur de la superstructure devrait être doublée. Cependant, les vannes levantes verticales suspendues par des câbles sont susceptibles de générer de graves problèmes de vibration lorsque la porte est abaissée dans un flot s'écoulant à grande vitesse.

La configuration des évacuateurs de surface, pour les deux types de vannes, est essentiellement la même que ce soit pour les petits ou les grands ouvrages. L'orientation de l'axe de l'évacuateur de crues par rapport au courant d'approche pour les grands déversoirs, nécessite bien souvent des modélisations hydrauliques afin d'éviter les perturbations hydrauliques et les répartitions inégales.

On utilise parfois une combinaison d'évacuateurs de crue vannés et d'orifices en charge avec des déversoirs d'urgence afin d'éviter ou de réduire au minimum le risque de défaillance des vannes. Ce n'est pas toujours facile à faire pour des projets ayant des débits très importants. Une étude récente portant sur le projet Inga sur le fleuve Congo en Afrique, où la crue de dimensionnement est de 60 000 m³/s, propose la combinaison d'un déversoir vanné conventionnel et d'un déversoir à touche de piano (voir le chapitre 6) pour gérer un aussi grand débit (Lempérière et al., 2012). La combinaison d'évacuateurs de crue en touches de piano et de pertuis de fond a déjà été utilisée pour des projets de plus petits débits.

2.3.2. Vannes segments

L'utilisation de vannes segments pour gérer les grands débits est largement appliquée. De très grandes vannes segments d'une surface supérieure à 440 m² ont déjà été installées et sont en fonctionnement. Des vannes de vingt mètres de large et de plus de vingt mètres de haut ont été utilisées dans les projets d'Itaipu, Tucurui, Estreito, Santo Antonio et Jirau, au Brésil, comme indiqué dans le tableau 2.1. Cela entraînera, bien sûr, des valeurs plus élevées pour le débit spécifique dans le canal de l'évacuateur de crues en aval, mais les conceptions ont permis d'y faire face avec succès.

En règle générale, l'utilisation de vannes de grande taille vise à réduire au maximum la longueur du barrage. Dans de nombreux projets hydroélectriques à très faible hauteur de chute construits sur de très grands fleuves, la puissance de sortie des groupes électrogènes est limitée, ce qui signifie qu'il faut augmenter le nombre de groupes pour répondre à la puissance hydraulique totale disponible. Les crues de conception des déversoirs dans ces projets sont également très importantes, d'où la nécessité d'économiser de l'espace en utilisant de grandes vannes. Par exemple, le projet Jirau sur la rivière Madeira, au Brésil (3 750 MW), compte 50 groupes électrogènes fonctionnant sous une hauteur de chute brute maximale de 15,7 m, et un débit de conception de l'évacuateur de crues de 82 600 m³/s, avec 18 vannes radiales de 20,0 m de large et 22,8 m de haut (figure 2.6). L'évacuateur de crues fait 445,0 m de long et a obligé la centrale à être divisée en deux parties situées sur chaque rive de la rivière.

Dans l'aménagement de Yaciretá-Apipé, entre l'Argentine et le Paraguay, la centrale est composée de 20 unités sur une longueur de 816 m, tandis que la crue de projet de 95 000 m³/s est évacuée par deux structures indépendantes (Fig. 2.7) l'une composée de 18 vannes segments de 15,0 m de largeur sur 19,5 m de hauteur, l'autre avec 16 vannes de 15,0 m de largeur sur 15,5 m de hauteur.

Fig. 2.6
Déversoir de 82 600 m3/s du complexe de Jirau en construction– Rivière Madeira, Brésil

Fig. 2.7
Complexe de Yaciretá-Apipé – Centrale hydroélectrique d'une longueur de 816 m sur la rivière Paraná River jouxtant le déversoir principal d'une capacité de 55 000 m³/s – frontière Argentine / Paraguay

Bien que les vannes segments soient généralement considérées comme des équipements très fiables et sûrs, certains événements ont donné lieu à des défaillances suscitant des préoccupations quant à l'utilisation généralisée de ce type d'équipement. Une analyse générale des vannes a fait l'objet de la question 79 du vingtième Congrès de la CIGB. Le rapport relatif à cette question (Cassidy, 2000) signalait la défaillance des vannes causée par des défauts de conception vis-à-vis des séismes, des vibrations induites par les écoulements et due aux erreurs de fabrication et de montage. Compte tenu de la taille des vannes et de l'importance des débits pour les déversoirs de grande capacité, les concepteurs et les maitres d'ouvrages doivent apporter une attention particulière à ces aspects.

Ce rapport n'a bien sûr pas pour objet de traiter de manière approfondie la question du fonctionnement et de la sécurité des vannes, mais d'attirer l'attention des concepteurs et des maitres d'ouvrage de grands déversoirs sur la nécessité de prendre en compte, dans la conception et le fonctionnement de ces structures, des détails qui, souvent, ne sont pas inclus dans la conception de grands projets.

Pour illustrer le type de problèmes pouvant survenir avec les vannes segments, les cinq cas suivants illustrent la diversité des situations pouvant survenir :

- L'évacuateur de crue du barrage de Folson aux États-Unis : une des 8 vannes segments de l'évacuateur de crue s'est rompue en 1995. La vanne, de 12,8 m de large et 15,2 m de haut, s'est cassée alors qu'elle était en train d'être levée avec le réservoir presque plein (Todd, 1997). Les investigations ont tout d'abord porté sur la possibilité que des vibrations aient affecté la structure de la vanne, mais elles ont finalement conclu que la vanne s'était rompue en raison du frottement du pivot entraînant la défaillance d'un renfort transmettant la poussée de l'eau sur le pivot. Il a été découvert que la force de frottement dans le pivot n'avait pas été prise en compte dans la conception structurelle de la vanne et que la faible fréquence de lubrification et le manque de protection contre les intempéries ont entraîné une corrosion qui a augmenté le frottement avec le temps. La défaillance de la vanne avec le réservoir plein a libéré de manière incontrôlé un débit de 1 132 m³/s.

- L'évacuateur de crue du barrage d'Itaipu, au Brésil (Lima da Silva et al. 2000) : la tige d'un des vérins hydrauliques utilisé pour manœuvrer l'une des vannes s'est brisée en juillet 1994 après 12 ans de bon fonctionnement. Il y a 14 vannes à Itaipu, chacune d'une largeur de 20,0 m et d'une hauteur de 21,4 m, chacune actionnée par deux vérins. La rupture de la tige s'est produite lors de l'abaissement de la vanne lors d'une opération de maintenance programmée sans circulation d'eau. Les investigations sur l'équipement défectueux ont permis de conclure que la défaillance provenait d'une fissure transversale associée à un processus de corrosion et déclenchée par la vibration due au frottement des joints latéraux de la vannes sur les bajoyers en l'absence d'eau. Après la rupture de la tige, le vérin cassé a été jeté et remplacé par un nouveau et l'intégralité des 27 vérins hydrauliques restants ont été démontés, usinés et polis pour éliminer les fissures et la corrosion. Un programme de maintenance détaillé a ensuite été élaboré et mis en œuvre.

- L'évacuateur de crue du barrage de Salto Osório, au Brésil : L'une des 9 vannes segments de ce complexe hydroélectrique, chacune d'une largeur de 15,3 m et d'une hauteur de 20,77 m, a connu un désordre majeur en septembre 2011, après 37 ans d'exploitation en toute sécurité. La manœuvre de la vanne était effectuée par des câbles et des treuils, contrôlés automatiquement par des relais électriques. Une commande de fermeture a été émise mais le treuil n'a pas été arrêté en raison d'une défaillance du relais de fin de course et il a continué à enrouler le câble en sens inverse, ouvrant complètement la vanne. L'enroulement a eu lieu sur plus d'une épaisseur, provoquant un frottement du câble contre le béton, le sectionnant par frottement. La vanne s'est alors fermée violemment par gravité et a ensuite été emportée par le courant, les roulements étant arrachés des appuis en béton. Deux mille mètres cubes par seconde ont été relâchés, heureusement sans conséquence particulière. Le niveau du réservoir a été abaissé, un batardeau installé et une nouvelle vanne installée.

- L'évacuateur de crue du barrage de Shiroro, au Nigéria (Epko et Adegunwa, 2011) : Cet aménagement, mis en service en 1984, comprend un évacuateur de crue composé de 4 vannes segments de 15,0 m de large sur 16,85 m de haut, conçues pour 7 500 m /s. Les portes sont manœuvrées par des treuils hydrauliques. Malgré l'existence de rainures dédiées, aucun batardeau n'était disponible. En 2005 des travaux de maintenance ont été effectuées notamment le remplacement de joints. Cependant, la vanne n°4 ne pouvait pas être ouverte de plus d'un quart de sa course. Les investigations menées ont révélé que la pile supérieure droite s'est déplacée et à réduire la largeur au sommet de 57 mm, à mi-hauteur de 46 mm et à la base de 9 mm. Cela a empêché l'ouverture complète de la vanne et, bien sûr, a limité la capacité d'évacuation du débit. Les rapports indiquent qu'en 2011, la raison de ce problème n'avait pas encore été identifiée, bien qu'il soit probablement lié à l'alcali-réaction de l'agrégat. Aucune solution n'a été proposée ou mise en œuvre pour l'instant.

- L'évacuateur de crue de Tarbela au Pakistan (Khalio Khan et A-Siddiqui, 1994) : L'une des sept vannes du déversoir principal de cet aménagement, chacune large de 15,2 m et haute de 18,6 m, a rompu après 17 ans d'exploitation sans problème. La porte s'est tout d'abord bloquée pendant une opération d'abaissement puis est tombée, cassant deux câbles de levage et endommageant le pont élévateur. L'événement s'est produit lorsque l'opérateur a commencé la fermeture de la vanne située à l'extrémité droite alors que les vannes étaient ouvertes avec l'évacuateur de crues débitant 2 475 m³/s et. Comme indiqué, le moteur de cet appareil de levage a disjoncté et la vanne, après être tombée d'une hauteur indéterminée, s'est coincée en laissant une ouverture de 112 mm à partir du seuil, maintenant un écoulement à grande vitesse. Après une enquête détaillée sur l'accident, il a été conclu que la vanne s'était coincée en raison d'un espacement insuffisant entre les plaques d'étanchéité latérales de la pile et la barre de serrage du joint en caoutchouc de la vanne. Le jeu initial était de 17 mm et, au fil des ans, a été réduit à 2 mm. Les investigations sur la cause de cette réduction ont conclu qu'il s'agissait de la combinaison de la dilatation thermique de la vanne et du glissement du dispositif de maintien du joint en raison du desserrage des boulons au cours des années d'exploitation. La vanne est tombée sous son poids lorsqu'elle a été libérée par les vibrations hydrauliques après le refroidissement de la plaque de peau par les températures plus basses du soir.

En dépit de ces problèmes, le pourcentage de vannes segments ayant subi une défaillance est relativement faible. Le US Bureau of Reclamation, propriétaire du barrage de Folson, a annoncé qu'il possède 314 vannes segments sur ses ouvrages ce qui représente 18 000 vanne-années de fonctionnement et qu'une seule vanne (Folson) a présenté un défaut majeur (USBR, 2011). Le Corps des ingénieurs de l'armée américaine a 90 vannes segments dans divers aménagements et, bien que des incidents techniques et des insuffisances de conception aient été constatés, aucun d'entre elle n'a fait défaut (Ebner et Craig, 2012). Au Brésil, 330 vannes segments ont été installées dans des projets d'évacuateurs de crues d'une capacité supérieure à 20 000 m³/s et, hormis les problèmes survenus sur les aménagements d'Itaipu et Salto Osori comme mentionés plus haut, il y a eu seulement un désordre sur une vanne de l'aménagement de Furnas, liée à la corrosion cristalline des tiges d'acier d'ancrage du pivot. Ce problème n'a eu aucune conséquence sur le fonctionnement de la vanne.

Une caractéristique très importante d'un déversoir contrôlé par des vannes segments est la possibilité d'utiliser des batardeaux dans toutes les passes et, bien sûr, leur disponibilité. Les cas énumérés ci-dessus illustrent le rôle important que jouent les batardeaux pour réduire les risques dus à une défaillance d'une vanne mais ils sont également très importants pour permettre les opérations de maintenance à sec. Comme indiqué dans le rapport général de la question 79 de la CIGB (Cassidy, 2000), l'absence de batardeaux au barrage de Folson a engendré un problème très difficile à résoudre pour corriger la défaillance de la vanne. La réparation d'une vanne du barrage de La Villita au Mexique, endommagée par la rupture soudaine de l'armature en acier, est un autre exemple de problèmes dus à l'absence de disponibilité de batardeaux.

Selon la pratique actuelle pour les grandes portes, les moteurs de porte à servomoteur (vérins hydrauliques) sont plus fiables que les moteurs à câbles ou à chaînes. De nombreuses grandes vannes radiales modernes sont conçues de cette manière, avec deux vérins par vanne. Selon les considérations du projet Itaipu, le fait qu'un seul des cylindres permette de fermer ou d'ouvrir la porte est un critère valable.

Une caractéristique très importante d'un déversoir contrôlé par des vannes segments est la possibilité d'utiliser des batardeaux dans toutes les passes et, bien sûr, leur disponibilité. Les cas énumérés ci-dessus illustrent le rôle important que jouent les batardeaux pour réduire les risques dus à une défaillance d'une vanne mais ils sont également très importants pour permettre les opérations de maintenance à sec. Comme indiqué dans le rapport général de la question 79 de la CIGB (Cassidy, 2000), l'absence de batardeaux au barrage de Folson a engendré un problème très difficile à résoudre pour corriger la défaillance de la vanne. La réparation d'une vanne du barrage de La Villita au Mexique, endommagée par la rupture soudaine de l'armature en acier, est un autre exemple de problèmes dus à l'absence de disponibilité de batardeaux.

Il a également été suggéré, et parfois appliqué, le critère dit N-1, selon lequel le nombre de passages de l'évacuateur de crues est déterminé de manière à ce que la crue totale de conception de l'évacuateur de crues puisse être évacuée par toutes les vannes, à l'exception d'une seule. Cette règle ne fait pas l'objet d'un consensus universel, principalement en raison de son impact économique. Cependant, une approche rationnelle de la question dépendrait de l'importance de la crue entrante et de la vitesse d'augmentation du niveau du réservoir. Pour les très grands réservoirs généralement associés à de grands projets de déversoirs, l'augmentation du niveau du réservoir causée par le blocage d'une vanne est généralement lente et prend un ou plusieurs jours avant d'atteindre un niveau critique. Pendant cette période, il doit être possible d'atteindre la vanne bloquée et de l'ouvrir manuellement. Bien entendu, cela implique que l'accès à la porte endommagée soit disponible à tout moment. Dans ces conditions, il n'est pas nécessaire de prévoir un poste de garde supplémentaire sur l'évacuateur de crues. Cependant, pour les réservoirs dont les petites zones seraient inondées lors d'une crue importante et verraient ainsi leur niveau d'eau augmenter rapidement, l'application du critère N-1 semble être une approche justifiable, à condition qu'aucun autre dispositif d'évacuation d'urgence économiquement alternatif ne soit disponible.

2.3.3. *Évacuateurs de fond et demi-fond*

L'utilisation exclusive d'évacuateurs de fond ou demi-fond pour évacuer des débits très importants est relativement limitée, principalement en raison de leur capacité limitée pour laquelle le débit est proportionnel à la racine carrée de la charge, par rapport aux déversoirs de surface où le débit varie en fonction de la charge à la puissance 1,5. Cela nécessite donc une plus grande charge pour compenser la différence de capacité de débit. Cependant, dans de nombreux projets, un orifice fonctionnant avec une forte charge est néanmoins nécessaire pour permettre un meilleur contrôle du réservoir, ce qui peut s'avérer pratique pour le fonctionnement du projet ainsi que pour le contrôle du passage des crues dans le réservoir.

Les grands orifices de fond sont normalement utilisés lorsque la chasse des sédiments est nécessaire ou lorsque la dérivation de la rivière pendant la construction nécessite un passage à niveau bas qui peut être plus facilement contrôlé lors de la fermeture de la rivière et de la mise en eau du réservoir. L'utilisation exclusive des déversoirs de fond pour l'évacuation des grands débits est limitée aux barrages de petite ou moyenne hauteur, et généralement lorsque la combinaison de la fonction de déversoir avec l'élimination des sédiments ou le contrôle de la dérivation est nécessaire ou pratique.

Cependant, dans de nombreux aménagements comprenant des barrages en béton, une combinaison d'évacuateurs de fond ou demi-fond et de déversoirs de surface est plutôt courante. Le projet Ertan en Chine, mentionné ci-dessus et le projet des Trois Gorges en Chine sont des exemples significatifs de l'utilisation dans des grands fleuves de déversoirs de fond et demi-fond permanents et provisoires (utilisés pour la dérivation des cours d'eau).

La figure 2.8 illustre une coupe transversale du déversoir du barrage des Trois Gorges qui incorpore un système d'évacuation de très grande capacité capable d'évacuer une crue maximale de 102 500 m³/s avec le réservoir rempli jusqu'au niveau de contrôle des crues (élévation. 180,4 m). Comme le montre la figure, il contient deux niveaux d'évacuateurs permanents (surface [7] et demi-fond [8]) en plus de l'orifice de fond de dérivation [9] utilisée pendant la construction puis obstrué.

Le déversoir de surface est formé de 22 vannes levantes de 8 m de large, dont le seuil est à l'altitude 158 m. La sortie de l'orifice de demi-fond ([8], sur la figure) comprend 23 orifices commandées par des vannes segment, calées à l'altitude 90 m, de 7 m de large sur 9 m de haut. La capacité de sortie de l'unique dérivation de fond est de 2 117 m³/s sous une charge de 85 m soit un débit spécifique de 302 m³/s/m. La capacité totale de l'orifice de demi-fond est de 48 691 m³/s, avec un niveau de réservoir au niveau 175 m. C'est l'une des plus grandes installations d'évacuateur de demi-fond et est clairement le résultat de l'adaptation de la structure du projet aux divers besoins de l'entreprise, pendant et après la construction

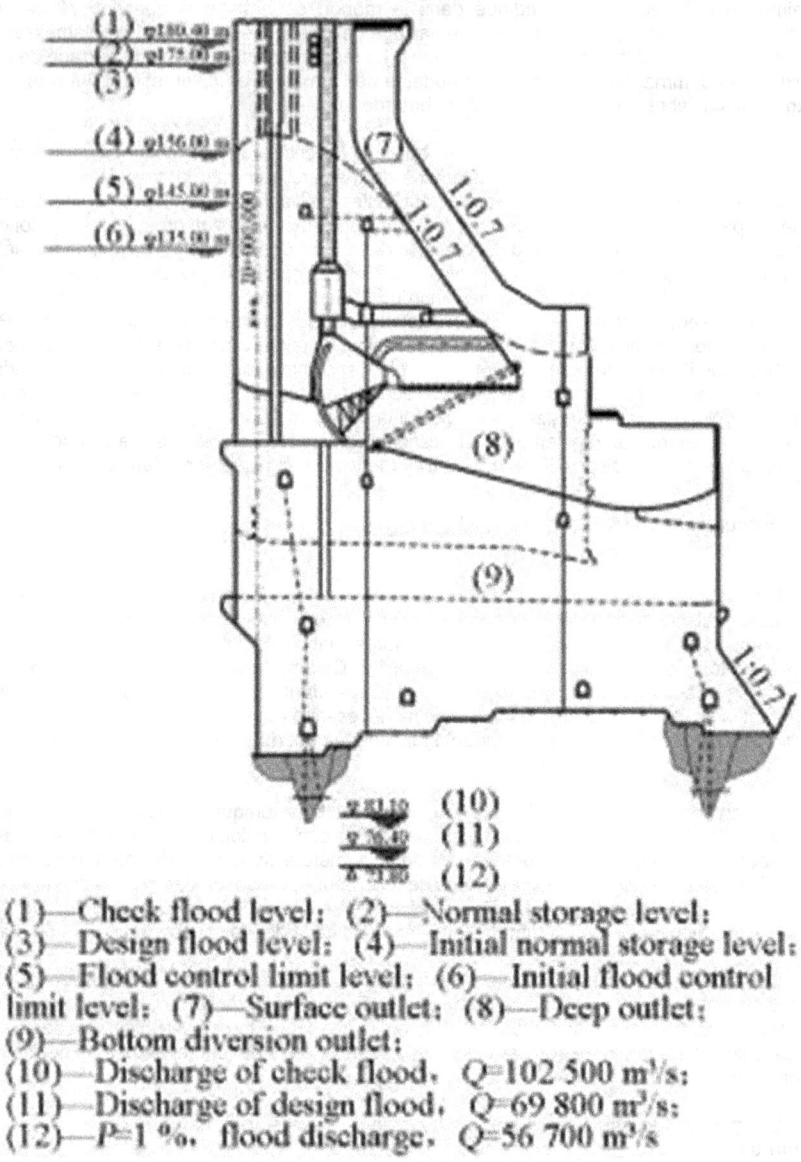

(1)—Check flood level; (2)—Normal storage level;
(3)—Design flood level; (4)—Initial normal storage level;
(5)—Flood control limit level; (6)—Initial flood control
limit level; (7)—Surface outlet; (8)—Deep outlet;
(9)—Bottom diversion outlet;
(10)—Discharge of check flood, Q=102 500 m³/s;
(11)—Discharge of design flood, Q=69 800 m³/s;
(12)—P=1 %, flood discharge, Q=56 700 m³/s

Fig. 2.8
Complexe des Trois Gorges - Coupe transversale du déversoir (Wang et al, 2011)

De très grands déversoirs ont également été construits en Inde. L'évacuateur de crues de l'aménagement de Chamera I conçu pour une capacité de 26 500 m³/s (Fig.2.9) en est un exemple. Il comprend 8 orifices de 10,0 m de large sur 12,8 m de hauteur, contrôlés par des vannes segments et 4 orifices de fond, de 4,0 m de large sur 5,4 m haut, contrôlés par des vannes à glissières.

Fig. 2.9
Orifices et orifices de fond du déversoir de Chamera I Project, en Inde

Un autre grand déversoir de fond est le projet Jupiá, situé sur le fleuve Paraná au Brésil (figure 2.10). Comme l'indique le tableau 2.1, ce projet comprend quatre déversoirs de surface et 37 déversoirs de fond, chacun d'une largeur de 10,00 m et d'une hauteur de 7,61 m, avec une capacité nominale de 44 000 m³/s. Ce type de solution a été retenu pour faciliter le contrôle de la rivière pendant les travaux de construction et la mise en eau du réservoir.

Fig. 2.10
Aménagement de Jupiá – vue de l'évacuateur de crue par orifice de fond de 44 000 m³/s
pendant la construction (1967)

Les passages de sortie de ce projet sont contrôlés par des vannes radiales. Mis en service en 1968, le projet a fait preuve de bonnes performances tout au long de ses 47 années d'exploitation, malgré des problèmes localisés d'érosion, en particulier dans la partie aval de la dalle de fond. Le débit journalier maximum enregistré depuis 1968 s'élève à 28 943 m³/s, mais environ 20 % de ce débit ont été évacués par les quatre déversoirs de surface.

La conception hydraulique des vannes de fond est largement traitée dans la littérature technique, aussi bien en termes généraux qu'au cas par cas. Il convient toutefois de souligner l'importance d'accorder une attention particulière au risque de cavitation et à la nécessité d'assurer une alimentation en air adéquate dans la zone située immédiatement en aval de la vanne de contrôle.

Les évacuations des barrages de fond et intermédiaires peuvent également être assurées par des tunnels. Des références spécifiques à ces ouvrages sont fournies au chapitre 7 de ce bulletin. Dans de nombreux cas, ces exutoires sont construits dans des tunnels qui ont servi à la dérivation de la rivière pendant la phase de construction. Dans ces cas, une attention particulière doit être accordée au calendrier de l'ensemble des travaux, car la transformation d'un élément de construction provisoire en un équipement permanent du projet après son utilisation pour la dérivation de la rivière peut entraîner des retards et des impacts économiques négatifs.

2.4.   OUVRAGES HYDRAULIQUES COMPLÉMENTAIRES

Le retour du débit à la rivière dépend du type et de la hauteur du barrage, de la forme de la vallée et de sa configuration géologique. En général, dans le cas des barrages de hauteur moyenne, les très grands déversoirs qui déversent des débits importants sont des déversoirs de surface, renvoyant le débit dans la rivière par un coursier court et raide ou un coursier plus long. Dans les deux cas, les coursiers se terminent normalement soit sur un saut à ski, si la hauteur du barrage le permet, soit par un bassin à ressaut hydraulique ou un bassin à auge. Les barrages de faible hauteur auront normalement un bassin à ressaut hydraulique immédiatement à l'aval de la crête déversante. En règle générale, les coursiers souterrains sous forme de tunnel, en tant qu'installation d'évacuation unique, ne sont pas adaptés à l'ampleur du débit des grands déversoirs.

Comme discuté ci-dessus, les grands déversoirs sont souvent des structures vannées, ce qui entraîne un débit spécifique assez élevé pour le débit évacué. Ceci influence les critères de conception utilisés pour la dissipation de l'énergie du débit évacué.

2.4.1. Coursiers

Pour les aménagements constitués de hauts barrages en remblai construits dans des vallées de largeur étroite à modérée, les longs coursiers construits sur un des appuis du barrage et se terminant par un saut à ski et se déversant dans une fosse de dissipation, constituent la solution standard pour les déversoirs de petite ou de grande taille. Parmi quelques exemples d'évacuateurs non vannés construits sur des barrages en remblai pour de petits débits, on peut citer les barrages de Crotty en Australie, d'Ahning en Malaisie (Cooke, 1985) et de Tongbai, en Chine (Zhao et Yuyan, 2006).

Lorsque la vallée est suffisamment large, il est possible de concevoir une section du barrage en béton pour accueillir l'évacuateur de crue, avec un coursier court et raide finissant par une cuillère de dissipation, comme pour l'aménagement de Tucuruí (Fig. 2.2). Dans ces conditions, un bassin à ressaut hydraulique peut également être construit, comme pour le barrage déversant en béton de Sardar-Sarovar (Fig.2.11). Ce barrage de 163 m de haut dispose d'un très grand déversoir comprenant un déversoir de service composé de 23 vannes segments en surface, de 16,78 m de largeur par 18,30 m de hauteur, un déversoir auxiliaire contrôlé par 7 vannes segments en surface, de 18,30 m de largeur sur 18,30 m de hauteur et 4 orifices de fond de 2,4 m de largeur par 3,6 m de hauteur. Le barrage avait également 10 vannes de fond de 2,15 m de largeur sur 2,75 m de hauteur utilisées pour la construction puis obturés.

Fig. 2.11
Évacuateur du barrage de Sardar-Sarovar 87 000 m³/s en Inde (piles et vannes segments non installées)

La conception d'un long coursier sur un appui d'un barrage en remblai dépend de l'ampleur du débit ainsi que de la topographie et de la géologie du site. Pour les grands déversoirs, ce schéma nécessite souvent de très grandes excavations qui sont perdues ou qui peuvent être réutilisées comme source de matériau pour le remblai. Cependant, il arrive parfois que le bilan des quantités ou le calendrier de construction du projet ou que la stabilité géotechnique des excavations ne permettent pas la construction d'un très grand déversoir sur un seul appui. Cela peut entraîner la séparation du déversoir en deux ou plusieurs structures distinctes, un déversoir de service et des déversoirs auxiliaires, solution déjà envisagée pour d'autres raisons, comme indiqué précédemment.

Pour ces raisons, l'emplacement et le profil du coursier seront conçus pour minimiser les excavations. En plan, la meilleure solution hydraulique est d'avoir un coursier droit à largeur constante, bien que des coursiers convergents sont également courants. Il est important de placer la cuillère ou le saut à ski au-dessus du niveau d'eau aval maximum et de manière à ce que le jet plongeant n'affecte pas le pied en aval du barrage. Il est souvent pratique, et la plupart du temps nécessaire, de procéder à une modélisation physique hydraulique pour affiner la conception finale de cet ouvrage, et plus particulièrement la géométrie et les performances de la fosse de dissipation.

2.4.2. Bassins à ressaut

Les bassins à ressaut sont nécessairement utilisés pour les barrages de faible hauteur, mais peuvent également être utilisés pour des barrages de plus grande hauteur, comme indiqué ci-dessus. Pour les barrages de hauteur intermédiaire, il n'y a pas de critères évidents permettant de choisir entre un bassin à ressaut et une cuillère alliée à une fosse de dissipation. Les considérations économiques, le contexte géologique local et la pratique habituelle des différents pays définissent généralement le type de solution.

Pour les grands déversoirs avec des débits spécifiques importants, la conception du bassin doit prendre en compte la possibilité d'une érosion aval causée par l'apparition du ressaut en dehors du bassin et le fonctionnement asymétrique des vannes provoquant des courants tourbillonnants entrainant des roches et des débris dans le bassin et pouvant provoquer une abrasion du béton. C'est un problème assez commun des bassins à ressaut. Les dommages survenus dans le bassin à ressaut du déversoir (21 400 m³/s) de l'aménagement de Marimbondo (Carvalho, 2002) au Brésil (figure 2.12) en est un exemple.

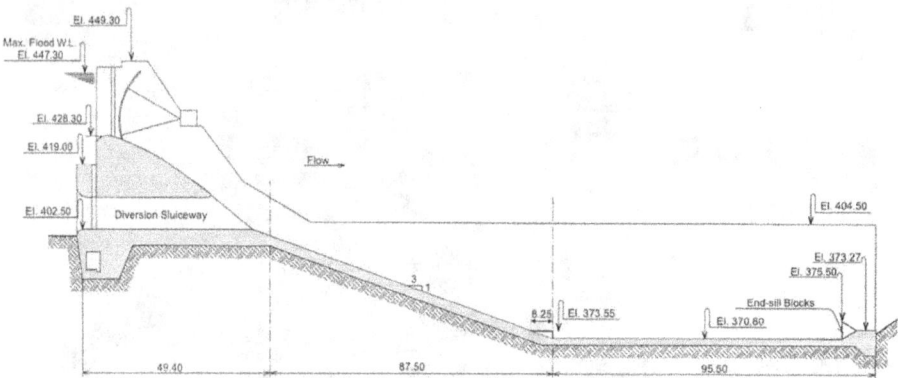

Fig. 2.12
(a) Déversoir du barrage de Marimbondo

Fig. 2.12
(b) – Dommages issus d'un fonctionnement asymétrique du déversoir du barrage de Marimbondo

- Profil en travers du déversoir

- Dépôt de matériaux à cause d'un fonctionnement asymétriques des vannes

- Dégradation du ferraillage du béton

L'utilisation de dents de dissipation, comme celles construites sur l'aménagement de Marimbondo, n'est pas recommandée pour les bassins présentant des débits spécifiques élevés car elles sont susceptibles de provoquer de graves dommages par cavitation. Un tel risque s'est produit sur l'aménagement de Porto Colombia, au Brésil. L'évacuateur de crues de 16 000 m³/s a été construit initialement avec des dents de dissipation placées sur la chute et en sortie de bassin, tel qu'illustré sur la figure 2.14 (a). Après 10 ans d'exploitation, des dommages par cavitation très graves se sont produits à proximité des dents de chute et, dans une moindre mesure, sur les dents placées en sortie. Pour remédier à ces problèmes, les dents ont été enlevées et le seuil d'extrémité arasé, comme le montre également la figure 2.14. Après ces travaux, aucun autre dommage n'a été observé (Carvalho, 2002a).

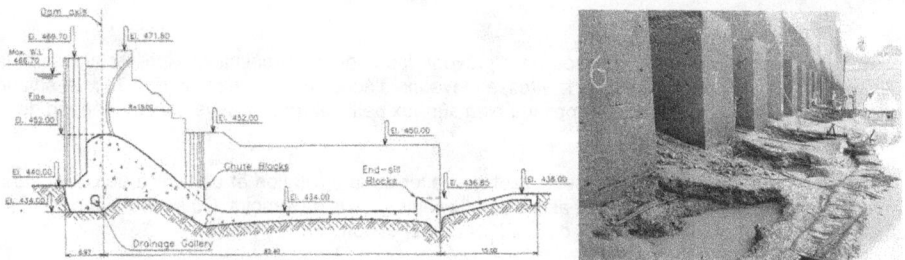

Fig. 2.13
(a) - Déversoir de Porto Colombia : dommages par cavitation et réparations

- Profil en travers du déversoir initial

- Dommages par cavitation des dents de dissipation

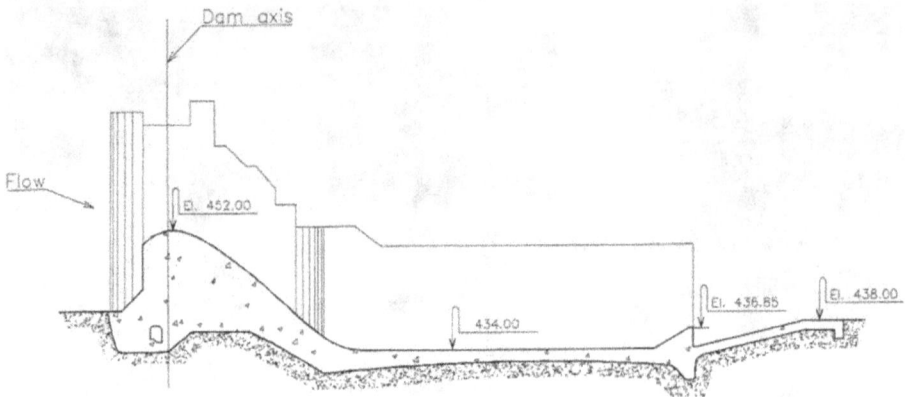

Fig. 2.13
(b) - Déversoir de Porto Colombia

- Profil en travers du déversoir modifié après travaux (enlèvement des dents de dissipation le long de la chute et en sortie de basin)

Outre ce genre de problème, l'intégrité du bassin peut également être affectée par des sous-pressions fluctuantes dont les pics dépassent le poids du radier béton et la résistance des ancrages en acier mis en place pour accrocher l'ouvrage au sol rocheux en place. Les barrages de Malpaso au Mexique (Sanches-Bribiesca et Viscaino, 1973) et de Karnafully au Bangladesh (Bowers et Toso, 1988) ont connus des graves désordres à cause de ces sous-pressions variables. Ces sous-pressions résultent du transfert sous le radier du bassin des pressions dynamiques générées en surface par le ressaut hydraulique, à travers les joints de construction et les tuyaux du système de drainage. Un résumé complet des connaissances actuelles sur ce problème et des recommandations pour la conception sont disponibles dans Bollaert (2009).

Dans de nombreux cas, recherchant une économie, les conceptions ne prennent en compte la formation du ressaut dans le bassin uniquement lors des crues fréquentes, permettant de fait une formation du ressaut en aval lors des évènements plus rares. Bien entendu, cela ne doit être autorisé que si la roche en aval du bassin est considérée comme saine et résistante à l'érosion causée par les évènements les plus rares.

2.4.3. Les aérateurs

Les aérateurs, organes utilisés pour prévenir les risques de cavitation sur le coursier et la cuillère, sont souvent utilisés lorsque la vitesse moyenne d'écoulement atteint environ 30 m/s, soit un indice de cavitation de 0,25. Ceci s'applique bien sûr aux petits et grands déversoirs et dépend de la vitesse et non du débit.

Le présent bulletin n'a pas pour objet de traiter de la cavitation et de l'utilité des aérateurs, cette question n'étant pas strictement associée aux seuls grands déversoirs. On trouvera une analyse complète du sujet dans le Bulletin 81 de la CIGB (1992) et dans Falvey (1990).

2.4.4. Dissipation d'énergie et érosion aval

T La dissipation de l'énergie du débit évacué et le contrôle de l'érosion en aval pouvant être causée par l'énergie résiduelle constituent des problèmes majeurs dans la conception et l'exploitation de grands déversoirs. Cela est particulièrement vrai en raison des valeurs élevées du débit spécifique évacué. Les données présentées dans les tableaux présentés à la fin de ce chapitre et les exemples mentionnés ci-dessus montrent que des débits spécifiques de 150 m³/s/m à 300 m³/s/m. sont assez fréquents pour de grands déversoirs. En général, les débits spécifiques élevés résultent de la nécessité de réduire la longueur vannée, comme indiqué précédemment, ou pour les barrages construits dans des vallées étroites équipés d'orifices de surface et de fond.

Les principaux types de dissipateurs d'énergie sont le bassin à ressaut avec ou sans cuillère, le saut à ski et la fosse de dissipation, ainsi que le jet en chute libre arrivant dans une fosse de dissipation.

Dans les cas où l'évaluation de la qualité de la roche le permet, au lieu d'un bassin à ressaut, il peut simplement être mis en place une dalle de béton menant à un canal excavé dans le rocher. Cette solution a été utilisée au Brésil comme évacuateur auxiliaire pour certains aménagements, comme mentionné précédemment pour les projets d'Itá (Fig. 2.4) et de Xingó, tandis qu'au Canada, il y a bon nombre de projets pour lesquels l'évacuateur principal unique suit cette conception. Il s'agit bien sûr d'une solution économique là où la roche est saine, mais dans tous les cas, des dommages dus à l'érosion peuvent se produire et il est nécessaire d'avoir accès à la zone endommagée et d'avoir un régime d'écoulement laissant des périodes suffisantes pour réaliser les éventuelles réparations. Un exemple intéressant de ce cas est le déversoir de LG-2, une partie du projet de la Baie James au Canada, qui possède un déversoir de 17 500 m^3/s, se déversant dans un coursier en marches d'escalier excavé dans le rocher (figure 2.14). Bien qu'il soit opérationnel depuis septembre 1979, son débit maximal sur 30 ans n'a été que de 3 500 m^3/s (Nzakimuena et Zulfiquar, 1999). Néanmoins, il y a eu un affouillement important immédiatement en aval de la dalle protectrice amont.

Fig. 2.14
Déversoir de LG-2 non revêtu, excavé dans le rocher et dommages
en aval de l'organe de contrôle

Un fort débit spécifique en sortie de bassin de dissipation peut entraîner une érosion en soulevant et en déplaçant des morceaux de roche fracturée, dont certains de très grande taille, créant ainsi des fosses d'érosion. Il est très difficile d'évaluer l'ampleur et l'extension de ce processus d'érosion, différent de celui provoqué par un jet plongeant, également difficile à prévoir mais néanmoins mieux étudié. Des études sur des modèles hydrauliques sont souvent nécessaires pour estimer les fosses d'érosion, cela couplé à une appréciation fine de l'état géologique du site. Il est bien sûr très important d'empêcher l'évolution de la fosse d'affouillement vers les fondations du bassin de dissipation. Dans le cas du déversoir de 30 000 m^3/s de l'aménagement de Macagua, au Venezuela, la dalle de béton placée en aval de l'ouvrage de contrôle a dû être rallongée de 40 m après une érosion profonde à proximité du bassin de dissipation (Marcano, 2009). Dans de nombreux cas, en particulier dans les barrages de faible hauteur, des travaux complémentaires de ce type après la mise en route de l'aménagement peuvent s'avérer difficiles et coûteux, car ils nécessitent de construire des batardeaux dans la zone où la dalle sera étendue. Il est donc prudent, dans ces cas, d'être sécuritaire dans le dimensionnement initial de l'extension de la dalle du bassin de dissipation.

Pour les barrages en béton ou en remblai de plus grande hauteur, un saut à ski en sortie du coursier de l'évacuateur de crue est la règle normale. Le débit est expulsé du tremplin, créant un jet qui heurte le lit de la rivière, créant ainsi une fosse permettant la dissipation d'énergie. La prévision de l'évolution de l'affouillement et le contrôle des dimensions de l''affouillement est l'une des tâches les plus difficiles dans la conception de grands déversoirs avec un grand débit spécifique, en raison des coûts et des risques qui y sont associés. Le chapitre 3 du présent bulletin traite spécifiquement de ce sujet et couvre les divers aspects théoriques et pratiques de cette question.

Parmi les problèmes généralement rencontrés dans les cas de grands déversoirs, il y a la nécessité de disposer d'une fosse de dissipation préalablement excavée si la profondeur du niveau naturel en aval est jugée insuffisante pour dissiper l'énergie et si le projet ne permet pas la construction d'un contre barrage. Pour les grands déversoirs vannés ayant des débits spécifiques élevés, la quantité d'énergie à dissiper est importante.

La détermination de la nécessité d'une fosse de dissipation préalablement excavée, et des dimensions assurant sa pérennité, est basée sur la profondeur estimée et l'extension latérale de l'érosion sous l'action du jet. Comme indiqué également au chapitre 3, cette question a fait l'objet de nombreuses études et recherches, tant au niveau empirique que physique. Le développement de l'affouillement par le jet ne doit pas affecter la fondation des ouvrages ni la stabilité des pentes des versants de la vallée, mais la difficulté d'anticiper la longueur et la largeur de l'érosion et d'établir les dimensions de la fosse, en particulier dans le cas de grands déversoirs avec des valeurs de débit spécifiques élevées, conduisent à une approche conservatrice. Cependant, comme indiqué dans le chapitre suivant de ce bulletin, il est recommandé d'utiliser pour la serviabilité un débit de dimensionnement avec une probabilité d'occurrence d'environ 50 % pendant la durée de vie utile d'un barrage. Néanmoins la stabilité du barrage lui-même avec sa fondation et ses appuis ne doit pas être mise au danger jusqu'à la crue de sécurité du barrage.

À l'heure actuelle, dans la pratique actuelle, la plupart des estimations de la profondeur d'affouillement sont déterminées par des formules empiriques. Le calcul avec ces formules empiriques donne une approximation de la profondeur réelle de la fosse d'affouillement associée à un écoulement donné et fournit des informations sur l'emplacement des exutoires et la conception des fosses de dissipation préalablement excavés.

La Fig. 2.15 (Sucharov et Fiorini, 2002), issue de l'analyse de l'érosion à l'aval du déversoir du barrage d'Itaipu, montre un exemple intéressant illustrant les approximations concernant l'anticipation de la profondeur et de l'extension de l'érosion due à un jet en sortie d'un grand déversoir. Les données sont issues des enquêtes de terrain faites en 1988, six ans après les inondations exceptionnelles de 1982. Pendant toute cette période, l'évacuateur de crues a déversé en continu tout le débit entrant puisque les groupes hydroélectriques étaient toujours en cours d'assemblage. Le débit maximal a été de 40 000 m³/s, soit une période de retour d'environ 500 ans. Le graphique de la Fig. 2.16 présente les données issues du modèle physique et du prototype par rapport aux projections basées sur la formule empirique de Veronese - qui a été utilisée pour prédire la profondeur de l'érosion - avec différentes valeurs du coefficient de formule K et des données provenant d'autres projets.

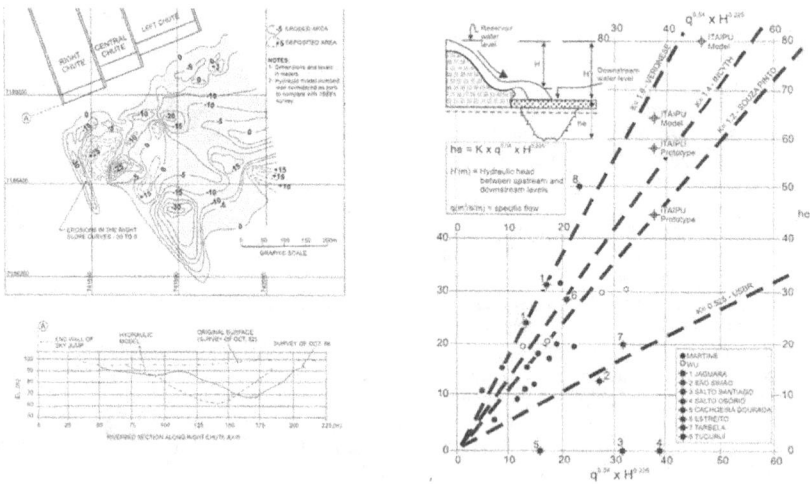

Fig. 2.15
Érosion à l'aval du déversoir d'Itaipu. (Sucharov & Fiorini, 2002)

Un problème important et controversé qui affecte considérablement les coûts et le calendrier d'un aménagement est lié à la nécessité ou non de revêtir la fosse. Si la roche est saine et exempte de fractures, la pratique habituelle a été de laisser la fosse excavée non revêtue. Les fosses de dissipation des aménagements de Tucuruí et d'Itaipu, mentionnés précédemment, ne furent pas revêtues et ont présenté un bon fonctionnement. Toutefois, pour éviter l'évolution de l'affouillement vers les ouvrages, pour certains aménagements il a été jugé nécessaire d'utiliser un revêtement en béton et/ou l'ancrage de la roche. Comme indiqué au chapitre 3 de ce bulletin, les conditions requises pour obtenir un revêtement sans problème sont très strictes, difficiles et coûteuses. Cependant, dans certains projets, les fosses de dissipation ont été totalement recouvertes de dalles de béton ancrées dans le rocher, comme le montre la Fig.2.16 pour l'évacuateur de crues du barrage Karun III (15 000 m³/s) en Iran, afin d'éviter toute conséquence néfaste sur ce grand barrage voûte. Dans ce cas le revêtement de la fosse a nécessité 250 000 m³ de béton armé et 135 000 m de boulonnage, d'ancrage et de goujons dans la roche (IWPC, 2004).

Left: Plunge pool lining under construction
Above: View of the completed works

Fig. 2.16
Fosse de dissipation revêtue en aval du déversoir du barrage de Karun III (15 000 m³/s)

2.5. QUESTIONS OPÉRATIONNELLES ET GESTION DES RISQUES

Les très grands déversoirs sont associés aux grandes rivières et aux grandes inondations. Comme indiqué ci-dessus, ces déversoirs sont normalement vannés et nécessitent par conséquent des règles d'exploitation strictes pour éviter tout incident lié au passage des crues extrêmes. Ceci est bien sûr vrai pour toute taille d'évacuateur vanné, mais l'ampleur des dommages causés par la mauvaise gestion d'un grand déversoir peut avoir des conséquences bien plus graves.

La littérature technique et particulièrement les documents de la CIGB - bulletins, rapports et documents présentés lors de congrès et d'ateliers - ont traité de manière approfondie de ce sujet. Les bulletins 49 (1986) et 142 (2012) et les rapports des Questions 71 de la CIGB (Pinto, 1994) et 79 (Cassidy, 2000) présentent des informations et des résumés complets sur les problèmes opérationnels liés aux déversoirs.

Dans la plupart des cas, et particulièrement pour les aménagements comportant de très grands déversoirs, le Maître d'ouvrage établit un manuel d'exploitation et de maintenance qui comprend des règles et des directives à suivre par les opérateurs dans des situations normales et urgentes. Il est bien sûr très important que ces règles soient appliquées et que les opérateurs soient formés et qualifiés pour mener à bien les activités correspondantes.

Le bon fonctionnement des vannes des grands déversoirs est la question essentielle permettant d'éviter le risque de surverse du barrage et éventuellement de rupture ou de ruine. Certains des problèmes possibles liés aux vannes ont déjà été mentionnés dans ce document, mais en général, les causes de dysfonctionnement peuvent être classées comme suit :

- défaut de l'opérateur de donner à temps l'ordre d'ouvrir les vannes ;

- défaut des vannes à s'ouvrir comme commandé ou défaut des dispositifs d'alimentation en énergie ou de commande nécessaire au bon fonctionnement des vannes ;

- colmatage des baies des vannes par des embâcles de grandes dimensions.

Cependant l'évaluation du risque réel de surverse d'un barrage équipé d'un déversoir vanné est complexe. Par exemple, le concept couramment utilisé de « crue entrante de dimensionnement » ne tient pas de la possibilité d'une «crue opérationnelle» pour laquelle un barrage pourrait surverser et rompre en raison d'une combinaison d'une inondation beaucoup plus petite que la crue de dimensionnement et d'un ou plusieurs défauts opérationnels (par exemple, défaillance d'une vanne). Le nombre de combinaisons d'événements à l'origine d'une telle défaillance est très important et augmente avec la complexité du barrage ou du complexe de barrages. Micovic et al. (2015) ont présenté les résultats de simulations stochastiques d'inondations pour un complexe de trois barrages et réservoirs dans l'Ouest canadien, où l'analyse des risques d'inondation a portée sur les niveaux d'eau dans les réservoirs en lieu et place des débits entrants. Les auteurs ont déduit les probabilités du niveau maximal du réservoir en combinant les probabilités de tous les facteurs impactants, notamment les volumes entrant dans le réservoir, le niveau initial du réservoir, les règles de fonctionnement du (des) réservoir (s) et les défaillances des vannes de sortie. Les résultats ont montré que les barrages étaient beaucoup plus susceptibles de surverser (et par conséquent de rompre) à cause d'une combinaison d'événements individuels plutôt communs plutôt qu'à cause d'une crue extrême telle que la crue de dimensionnement. Micovic et al. (2015) ont également conclu que l'importance de la fiabilité des vannes de l'évacuateur de crue augmente à mesure que la réserve de stockage en supplément diminue. Par exemple, dans le cas d'une petite réserve supplémentaire de stockage, la simulation de la possibilité d'une défaillance des vannes de l'évacuateur lors d'une crue augmente la probabilité de surverse du barrage par cinq par rapport au cas pour lequel toutes les vannes de l'évacuateur seraient exploitables.

Une analyse statistique citée par Hinks et Charles (2004), basée sur Foster et al (2000), indique que 46% des défaillances des barrages en remblai étaient dues à des surverses et que 13% d'entre elles étaient associées à un dysfonctionnement des vannes. Bien qu'il n'y ait aucun chiffre indiquant le pourcentage lié à des erreurs humaines dans la manœuvre des vannes, ce facteur est généralement considéré comme étant très important. Le risque associé aux erreurs humaines est généralement considéré comme pouvant être atténué par la formation des opérateurs et par des règles de fonctionnement simples, claires et non ambiguës. Quoi qu'il en soit, il est recommandé de confier à une équipe de deux personnes la tâche de faire fonctionner les vannes pour des raisons de sécurité du travail (Barker et al, 2006). Indépendamment du type d'erreur, en cas d'inondation, il est évident que des conditions physiques et psychologiques défavorables peuvent affecter le bon raisonnement des opérateurs.

Outre le besoin d'une formation approfondie des opérateurs, il existe des recommandations pratiques pour minimiser les risques d'erreurs et de dysfonctionnements. Parmi celles-ci, il convient d'éviter la surcharge des opérateurs en cas de défaillance des équipements en fonctionnement, de garantir un accès sans obstruction aux vannes et des communications fiables en cas d'urgence. Pour éviter cela, la redondance des équipements, des accès et des communications est obligatoire. Une inspection systématique de ces procédures opérationnelles, dans laquelle les opérateurs ont été inclus, est également très importante (Bister, 2000). Le Bulletin CIGB 154 (2013) décrit de manière plus complète les mesures pratiques permettant de remédier aux défaillances et les mesures de mitigation.

Les exemples mentionnés précédemment dans ce chapitre illustrent le risque de dysfonctionnement du déplacement des vannes ou des instructions de commande en raison de problèmes mécaniques et électriques. La redondance, comme mentionné ci-dessus, et des solutions de conception spécifiques peuvent aider à minimiser ce risque. Cependant, un accès dégagé aux vannes et la fiabilité de l'alimentation électrique pour les manœuvrer même dans des conditions de crues extrêmes, sont très importants, en particulier pour les grands déversoirs vannés.

Il existe des exemples de ruptures complètes de barrages qui auraient pu être évitées si ces précautions avaient été appliquées au moment de l'accident.

Lors de l'accident du barrage Euclides da Cunha sur la rivière Pardo au Brésil, en janvier 1977, un déversoir équipé de 2 vannes de surface conçu pour un débit de 2 040 m³/s et un orifice de fond de 300 m³/s ne fonctionnait pas correctement lors d'une crue soudaine culminant à 3 670 m³/s, ce qui a généré une onde de crue dans le réservoir d'environ 2 000 m³/s. Ceci est dû au fait que l'opérateur avait perdu la communication avec le bureau central du Maitre d'ouvrage auprès duquel il avait l'habitude de recevoir l'ordre de manœuvrer les vannes et craignait de provoquer une inondation en aval affectant certaines populations riveraines. Lorsqu'il a décidé d'ouvrir les vannes, le système de commande était inopérant en raison d'une panne d'alimentation électrique et il ne pouvait pas atteindre l'évacuateur de crue en raison de l'accès inondé. Le déversement a détruit le barrage Euclides da Cunha et le barrage Armando de Salles Oliveira situé plus en aval, sur la rivière Pardo. Après ces événements, le Maitre d'Ouvrage a procédé à une révision générale de toutes ses pratiques d'exploitation du barrage (Siqueira, 1978).

Outre les problèmes électromécaniques, les structures de génie civil doivent également être exploitées et entretenues pour rester en bon état. Les questions qui méritent généralement une attention particulière sont l'action de la cavitation dans les coursiers et les bassins de dissipation, les problèmes hydrauliques liés au fonctionnement asymétrique des vannes, le transport et le dépôt de sédiments dans les bassins de dissipation et l'extension de l'érosion aval vers la fondation des structures. Ces questions sont traitées en détail dans les documents de référence mentionnés dans ce chapitre.

2.6. RÉFÉRENCES

Andrzejewski, R. H. (2002) – "The two spillways of Itá Hydroelectric Powerplant", *Large Brazilian Spillways*, CBDB, Rio de Janeiro, 53-64.

Barker, M., B. Vivian and D. Bowels (2006) – "Gate reliability assessment for a spillway design in Queensland, Australia", *Proceedings of the 26th USSD Conference*, USSD, San Antonio, TX, USA, 1-22.

Bister, D. (2000) – "Practical guidelines for improvement of dam safety during floods", Q.79, R. 29, *Transactions of the 20th Congress on Large Dams*, ICOLD, Beijing, China, Vol. IV, 497-513.

Bollaert, E. F. R. (2009) – "Dynamic uplift of concrete linings: theory and case studies", *Proceedings of the 29th Annual USSD Conference*, USSD, Nashville, Tenn. USA, 149-164.

Bowers, E. and J. Toso (1988) – "Karnafuli Project, model studies of spillway damage", *Journal of Hydraulic Engineering*, ASCE, Vol. 14, No. 5, 469-483.

Carvalho, E. (2002) – "Marimbondo Spillway – Performance and repair of the stilling basin", *Large Brazilian Spillways*, CBDB, Rio de Janeiro, 99-108

Carvalho, E. (2002a) – "Porto Colombia Spillway – Performance and remedial works in the stilling basin", *Large Brazilian Spillways*, CBDB, Rio de Janeiro, 123-132.

Cassidy, J. (2000) – "Gated spillways and other controlled release facilities, and dam safety", General Report Question 79, *Transactions of the 20th Congress on Large Dams*, ICOLD, Beijing, Vol. IV, 735-781

Cooke, J. B. (1985) – *Spillways over embankment dams*, Memo No. 80, Personal Communication.

Ebner, L. and M. Craig (2012) – "Comprehensive spillway tainter gate assessment and identification of interim risk reduction measures", *Proceedings of the 32nd Annual USSD Conference*, USSD, New Orleans, Lo, USA, 1257-1271.

Eigenheer, L. P., A. Vasconcelos, A. Conte and J. A. Souza (2002) – "Design, construction and performance of Xingó Spillway", *Large Brazilian Spillways*, CBDB, Rio de Janeiro, 177-184.

Epko, I. and A. Adegunwa (2011) – "Shiroro Hydroelectric Dam: operation, performance and safety monitoring of the rockfill dam (2004-2010)". *Transactions of the II International Symposium on Rockfill Dams*. CBDB-CHINCOLD, Rio de Janeiro.

Falvey, H. T. (1990) – *Cavitation in chutes and spillways*, US Bureau of Reclamation, Engineering Monograph No. 42, Denver. CO. USA.

Foster M. and M. Spannagle (2000) – "The statistics of embankment dam failures and accidents" *Canadian Geotechnical Journal*, Vol. 37, No.5, 1000-1024.

Hinks, J. L. and J. A. Charles (2004) – "Reservoir management, risk and safety considerations", *Long term benefits and performance of dams*, Tomas Telford, London.

ICOLD (1992) – *Spillways. Shock waves and air entrainment*. ICOLD Bulletin No. 81, Paris,

ICOLD (2012) – *Safe Passage of Extreme Floods,* ICOLD Bulletin No. 142, Paris.

ICOLD (2013) – Dam safety management: operational phase of dam life cycle. ICOLD Bulletin No. 154, Paris

IWPC (2004) – *Karun III Development*, Data sheet on construction works presented at the ICOLD 73rd Annual Meeting, Teheran.

Khaliq-Khan, A. and N. A. Siddiqui (1993) – "Malfunction of a spillway gate at Tarbela after 17 years of normal operation", Q.71, R.27, *Transactions of the 18th Congress on Large Dams*, ICOLD, Durban, South Africa, Vol.IV, 411-428.

Lempérière, F., J-P Vigni and L. Deroo (2012) – "New methods and criteria for designing spillways could reduce risks and costs significantly", *The International Journal on Hydropower & Dams*, Volume 19, Issue 3, 120-128.

Lima da Silva, C.A., R. Garcete, E. da Rosa, E. Fancello and P. Bernardini (2000) – "Failure and repair of hoist rod of a very large radial spillway gate – Itaipu Hydroelectric Powerplant". Q.79, R.43, *Transactions of the 20th Congress on Large Dams*, ICOLD, Beijing, Vol. IV, 709-732

Marcano, A. (2009) – *Lower Caroni developments - Some features of large spillways*. Contribution to the ICOLD Committee on Hydraulic for Dams

Micovic, Z., D. Hartford, M. Schaefer and B. Barker (2015) – "A non-traditional approach to the analysis of flood hazard for dams". *Stochastic Environmental Research and Risk Assessment*, Springer Verlag, Berlin, Germany

Nzakimuena, T. and A. Zulfiquar (1999) - *Rock erosion downstream of the Spillways at Hydro-Quebec Installations, Quebec, Canada – Some case histories*, Contribution presented to the ICOLD Committee on Hydraulic for Dams.

Pinto, N.L.S. (1994) – "Deterioration of spillways and outlet works". General Report Question 71, *Transactions of the 18th Congress on Large Dams*, ICOLD, Durban, South Africa, 1101-1208

Sanchez-Bribiesca, J. L. and A. C. Viscaino (1973) – Turbulence effects on the lining of stilling basins", Q. 41, R. 83, *Transactions of the 11th Congress on Large Dams*, ICOLD, Madrid, Vol.2, 1575-1592.

Siqueira, G. (1978) – "As lições do Pardo" ("The lessons of the Pardo" – in Portuguese), *Atas do XII Seminário Nacional de Grandes Barragens*, CBGB, São Paulo, Brazil, 141-170.

Sucharov, M. and A. Fiorini (2002) – "Itaipu spillway", *Large Brazilian Spillways*, CBDB, 65-78

Todd, R. V. (1997) – "Failure of spillway radial gate at Folson Dam, California", Q.75, R.9, *Transactions of the 19th Congress on Large Dams*, ICOLD, Florence, Vol. IV, 113-126.

USBR (2011) – *Managing water in the West – Gate failures – Best practices*. Powerpoint presentation, US Bureau of Reclamation. Denver, Colo. USA.Wang, X., Xu L. and Liao R. (2011) – "The dam design of the Three Gorges Project", *Engineering Sciences*, Vol. 9, No.3, 57-65.

Zhao, X. and S. Yuyab (2006) – "Exploration into safety and economy of choosing the over dam spillway on lower reservoir's concrete-face rock-fill dam of the Tongbai Pumped Storage power plant", Q.84, R.55, *Transactions of the 22nd Congress on Large Dams*, ICOLD, Barcelona, Spain, Vol. I, 959-967.

Table 2.1
Data on Brazilian Large Spillways

PROJECT NAME	RIVER	YEAR	MAX RESERVOIR FLOOD ELEV (m)	MAX TAIL WATER FLOOD ELEV (m)	DESIGN FLOOD CRITERION	PEAK INFLOW (m³/s)	SPLW DESIGN FLOW (m³/s)	WIDTH OF SPLW EXIT (m)	SPILLWAY TYPE	GATES Type	No.	Width (m)	Height (m)	TYPE OF ENERGY DISSIPATION DEVICE	SPECIFIC DISCH. FLOW (m³/s/m)
TUCURUÍ	Tocantins	1984	75.30	6.80	PMF	114 300	110 000	482.50	Chute	Radial	23	20.00	20.75	Flip bucket & plunging pool	228.00
ST ANTONIO - MAIN	Madeira	2012	72.00	65.25	1:10 000 yr	84 000	70 000	370.00	Crest overflow	Radial	15	20.00	22.00	Hydr jump stilling basin	189.19
ST ANTONIO - AUX	Madeira	2012					14 000	70.00	Crest overflow	Radial	3	20.00	22.00	Hydr jump stilling basin	200.00
JIRAU	Madeira	2012	90.00	74.00	1:10 000 yr	82 600	82 600	445.00	Crest overflow	Radial	18	20.00	22.77	Hydr jump stilling basin	185.62
ITAIPU	Paraná	1982	225.10	142.15	PMF	70 020	62 200	335.00	Chute	Radial	14	20.00	21.14	Flip bucket & plunging pool	185.67
ESTREITO – TOC.	Tocantins	2011	158.00	152.00	1:10 000 yr	62 719	62 719	332.40	Crest overflow	Radial	14	19.10	22.50	Hydr jump stilling basin	188.69
FOZ DO CHAPECÓ	Uruguai	2010	266.60	240.00	PMF	62 190	62 190	343.50	Crest overflow	Radial	15	18.70	20.60	Flip bucket & plunging pool	181.05
PORTO PRIMAVERA	Paraná	1966	259.70	244.60	PMF	62 040	52 800	315.00	Crest overflow	Radial	16	14.96	22.86	Hydr jump stilling basin	167.62
JUPIA - MAIN	Paraná	1968	280.50	270.00	PMF	60 790	44 000	505.00	Bottom outlet	Radial	17	7.61	10.00	Hydr jump stilling basin	87.13
JUPIA - AUX	Paraná	1968					5 000	70.00	Crest overflow	Radial	4	12.80	15.00	Hydr jump stilling basin	83.30
ILHA SOLTEIRA	Paraná	1973	329.00	286.00	PMF	55 230	40 000	355.00	Crest overflow	Radial	19	18.50	21.50	Hydr jump stilling basin	112.68
ITÁ - MAIN	Uruguai	2000	375.00	292.40	PMF	52 800	29 964	131.00	Chute	Radial	6	18.00	21.86	Flip bucket	228.73
ITÁ - AUX	Uruguai	2000					19 976	85.50	Part. Lined Chute	Radial	4	18.00	21.86	Flow over unlined rock	233.64
SALTO CAXIAS	Iguaçu	1999	326.00	268.50	1:10 000 yr	52 400	49 600	231.00	Crest overflow	Radial	14	16.50	20.00	Flip bucket	214.70
LAJEADO	Tocantins	2001	212.30	203.50	1:10 000 yr	49 870	49 870	323.00	Crest overflow	Radial	14	17.00	23.50	Hydr jump stilling basin	154.39
PEIXE ANGICAL	Tocantins	2006	265.21	249.25	1:10 000 yr	42 500	37 044	200.00	Crest overflow	Radial	9	17.00	22.82	Hydr jump stilling basin	185.20
MACHADINHO	Uruguai	2002	485.36	398.00	PMF	39 750	37 674	175.50	Chute	Radial	4	18.00	20.00	Flow over unlined rock	235.81
PAULO AFONSO IV	S. Francisco	1979	253.00	151.00	1:10 000 yr	35 000	10 000	105.00	Chute	Radial	8	11.50	19.60	Flow over concrete slab	95.24
MOXOTÓ	S. Francisco	1974	253.00	230.30	1:10 000 yr	35 000	28 000		Bottom outlet	Radial	20	10.00	8.00		
XINGÓ - MAIN	S. Francisco	1994	139.00	29.70	1:10 000 yr	33 000	16 500	109.00	Chute	Radial	6	14.83	20.76	Flip bucket & plunging pool	151.38
XINGÓ - AUX	S. Francisco	1994					16 500	109.00	Part. lined Chute	Radial	6	14.83	20.76	Flow over unlined rock	151.38
ITAPARICA	S. Francisco	1986	305.40	262.00	1:1 500-yr peak + 1:1 000-yr volume	35 500	26 415	175.00	Crest overflow	Radial	9	15.00	14.70	Roller bucket	150.94
SLT. OSÓRIO - MAIN	Iguaçu	1975	398.00	326.00	1:10 000 yr	28 000	15 000	94.50	Chute	Radial	5	15.30	20.77	Flip bucket & plunging pool	158.73
SLT. OSÓRIO - AUX	Iguaçu	1975					12 000	75.20	Chute	Radial	4	15.30	20.77	Flip bucket & plunging pool	163.93
SALTO SANTIAGO	Iguaçu	1980	509.00	419.00	1:10 000 yr	26 000	24 530	149.50	Chute	Radial	8	15.30	21.57	Flip bucket & plunging pool	164.08
CAPIVARA	Paranapan	1977	336.00	295.00	1:10 000 yr	24 500	17 100	144.40	Chute	Radial	8	15.00	15.56	Flip bucket & deflectors	118.47
BARRA GRANDE	Pelotas	2005	649.17	480.00	PMF	23 840	21 810	111.00	Chute	Radial	6	15.00	20.8	Flip bucket & plunging pool	196.48
ITUMBIARA	Paranaíba	1980	521.20	449.50	PMF	22 100	16 270	118.00	Chute	Radial	6	15.00	18.55	Flip bucket & stilling basin	137.90
MARIMBONDO	Grande	1975	447.36	403.20	1:10 000 yr	21 400	21 400	163.00	Crest overflow	Radial	9	15.00	18.85	Hydr jump stilling basin	131.30
ÁGUA VERMELHA	Grande	1978	386.00	333.80	1:10 000 yr	20 000	20 000	156.00	Crest overflow	Radial	8	15.00	19.60	Flip bucket & plunging pool	128.21

Table 2.2
(A) Data on Chinese Large Spillways

NAME	RIVER	YEAR OF COMPLETION	DESIGN FLOOD CRITERION	PEAK INFLOW (m³/s)	SPILLWAY DESIGN FLOW (m³/s)	SPILLWAY TYPE	GATES TYPE	NUMBER	WIDTH (m)	HEIGTH (m)	TYPE OF ENERGY DISSIPATION DEVICE	SPECIFIC DISCHARGE OF EXIT FLOW (m³/s/m)
Three Gorges - surface	Yangtze	2009		124 300	102 500	Crest orifice	Vertical lift	22	8.0	17.0	Ski-jump flip bucket	134.0
Three Gorges - mid-outlet						Orifice	Vertical lift	2	10.0	12.0		254.0
Three Gorges - deep outlet						Orifice	Radial	23	7.0	9.0		302.0
Longtan - surface	Hongshui	2009		35 500	27 134	Crest	Radial	7	15.0	20.0	Ski-jump flip bucket	223.0
Longtan - deep outlet						Orifice	Vertical lift	2	5.0	8.0		
Ertan - surface	Yalong	1999		23 900	6 260	Crest	Radial	7	11.0	11.5	Ski-jump flip bucket and plunge pool	125.0
Ertan - mid-outlet					6 930	Orifice	Vertical lift	6	6.0	5.0		186.0
Ertan - deep outlet					2 084	Orifice	Vertical lift	4	3.0	5.0		285.0
Ertan - tunnel spillway					7 400	Two tunnels	Vertical lift	2	13.0	13.0		
Daochaoshan - surface	Lancang	2003		23 800	23 800	Crest	Radial	5	14.0	17.8	Flaring pier + stepped spillway	193.6
Daochaoshan - deep outlet						Orifice	Radial	3	7.5	10.0	Ski-jump and flip bucket	267.0
Tianshengqiao I	Nanpan	2000		21 750	21 750	Chute	Radial	5	13.0	20.0	Ski jump and flip bucket	335.0
Wuqiangxi - surface	Yuan	1994	0.1% flood frequency	55 962	40 132	Crest	Radial	9	19.0	23.0	Stilling basin with flaring piers	268.00
Wuqiangxi - middle outlet					3 244	Orifice	Vertical lift	1	9.0	13.0	Stilling basin	287.00
Wuqiangxi - deep outlet					2 915	Orifice	Vertical lift	5	3.5	7.0	Stilling basin	172.00
Panjiakou - surface	Luan	1992		56 200	42 600	Crest	Radial	18	15.0	15.0	Flaring piers and flip bucket	142.00
Panjiakou - deep outlet					3 000	Orifice	Vertical lift	4	4.0	6.0		
Ankang - surface	Han	1995	0.2% design 0.02% check	45 000	37 000	Crest	Radial	5	15.0	17.0	Stilling basin	
Ankang - middle outlet						Orifice	Vertical lift	5	11.0	12.0	Stilling basin	209.3
Ankang - deep outlet						Orifice	Vertical lift	4	5.0	8.0	Stilling basin	
Yantan	Hongshui	1994		33 400	33 400	Crest	Radial	7	15.0	21.0	Flaring piers and stilling basin	308
Yantan - deep outlet						Orifice	Vertical lift	1	5.0	8.0	Flaring piers and stilling basin	210
Geheyan - surface	Qingjiang	1995		27 800	23 458	Crest	Radial	7	12.0	18.2	Flaring piers and stilling basin	231
Geheyan - deep outlet						Orifice	Vertical lift	4	4.5	6.5		225
Geheyan - bottom outlet						Orifice	Vertical lift	3				

Table 2.2
(B) Data on Chinese Large Spillways

NAME	RIVER	YEAR OF COMPLETION	DESIGN FLOOD CRITERION	PEAK INFLOW (m³/s)	SPILLWAY DESIGN FLOW (m³/s)	SPILLWAY TYPE	GATES TYPE	GATES NUMBER	GATES WIDTH (m)	GATES HEIGHT (m)	TYPE OF ENERGY DISSIPATION DEVICE	SPECIFIC DISCHARGE OF EXIT FLOW (m³/s/m)
Manwan - surface	Lancang	1995		20 910	12 025	Crest	Radial	5	13.0	20.0	Ski jump & plunge pool	262
Manwan - low level outlet					2 436	Orifice	Vertical lift	2	5.0	8.0		225
Manwan - tunnel					2 344	Tunnel	Vertical lift	1	12.0	12.0		
Wujiangdu -surface	Wujiang	1983	Design 1:500-yr Check 1:5000-yr	21 350	21 350	Crest		6	13.0	19.0	Ski jump & plunge pool	144
Wujiangdu - middle outlet						Orifice		2	4.0	4.0		201
Wujiangdu - tunnel						Tunnel		2	9.0	10.0		240
Xiluodu - surface	Jinshia			50 311	33 278	Crest orifice		7	12.5	13.5	Ski jump & plunge pool	207
Xiluodu - low level outlets						Orifice		8	6.0	6.7	Ski jump & plunge pool	267
Xiluodu - tunnel					17 600	Tunnel		4	14.0	12.0	Ski jump & plunge pool	283
Xiangjiaba - surface	Jinshia	2015		48 680				12	8.0	26.0	Flaring gate & stilling basin	300
Xiangjiaba - middle outlet								10	6.0	9.5		331
Nuozhadu - chute	Lancang	2012	PMF	31 318	19 814	Chute	Radial	8	15.0	20.0	Ski-jump and plunge pool	162
Nuozhadu - left tunnel					3 211	Tunnel	Vertical lift	2	5.0	8.0	Ski-jump and plunge pool	308
Nuozhadu - right tunnel					3 154	Tunnel	Vertical lift	2	5.0	8.0	Ski-jump and plunge pool	393
Guopitan - surface				26 950				6	12.	13.0		129
Guopitan - middle outlet								7	6.0	8.0		228
Guopitan - tunnel								1	11.0	12.0		254
Xiaowan - surface	Lancang	2009		23 600	20 709	Crest orifice	Radial	5	11.0	15.0	Ski-jump and plunge pool	146
Xiaowan - middle outlet						Orifice	Vertical lift	6	6.0	6.5	Ski-jump and plunge pool	223
Xiaowan - bottom outlet						Orifice	Vertical lift	2			Ski-jump and plunge pool	
Xiaowan - tunnel						Tunnel	Vertical lift	1	15.0	16.5	Ski-jump and plunge pool	238

Table 2.3
(A) Data on Australian Large Spillways

NAME	RIVER	YEAR	MAX RESERVOIR FLOOD ELEV (m)	MAX TAIL WATER FLOOD ELEV (m)	DESIGN FLOOD CRITERION	PEAK INFLOW (m³/s)	SPLW DESIGN FLOW (m³/s)	WIDTH OF SPLW EXIT (m)	SPILLWAY TYPE	GATES TYPE	GATES NUMBER	GATES W (m)	GATES H (m)	TYPE OF ENERGY DISSIPATION DEVICE	SPECIFIC DISCH. FLOW (m³/s/m)
BURDEKIN FALLS	Burdekin	1987	171.9	134.5	1:9,000 (AEP for PMPOF)	57 900	64 600	504.0	Ungated	—	—	—	—	Apron slab with splitter piers	128
WARRAGAMBA	Warragamba	1960	151.2	74	PMF	52 100	42 200	94.5 (Main) 182.0 (Aux)	Gated & fuseplug	Radial Drum	4 1	12.19 27.43	13.3 7.62	Hydraulic Jump (main) Bucket (Aux)	Flip 225 (Main) 100(Aux)
KUNUNURRA DIVERSION	Ord	1963	43.6	42.8	1x10⁶	35 400	35 400		Gated plus auxiliary						118
PARADISE	Burnett River	2005	87.69	79	1:30,000 (AEP for PMPOF)	94 861	33 000	315.0 (Main) 485.0 (Aux)	Ungated	—	—	—	—	Stilling Basin	41
TALLOWA	Shoalhaven	1976	67.0	48	PMPOF	33 000	32 000	352.0	Ungated	—	—	—	—	Roller Bucket	91
BURRINJUCK	Murrumbidgee	1928	380.7		PMF	44 400	29 000		Gated & ungated	Radial	3	2x15.2 1x24.4	4.6	Natural Rocks	124
ORD RIVER	Ord	1972	111.5	48	PMF	200 000	27 750		Main chute plus auxiliary						
FITZROY RIVER BARRAGE	Fitzroy	1970					23 800		Gated						
PINDARI	Severn	1969	527.56		PMF	22 980	21 900	200.0	Ungated	—	—	—	—	Natural Rocks	
HARDING	Harding	1985					21 500		Ungated	—	—	—	—		
FAIRBAIRN	Nogoa	1972	218.59	195.54	1:61,275 (AEP for PMPOF)	27 777	15 580	158.5	Ungated	—	—	—	—	Sill block apron with baffles	98
COPETON	Gwydir	1976	575.4		PMF	37 580	14 800	156.0	Gated & fuseplug	Radial	9	14.6	13.0	Natural Rocks	95
WYANGALA	Lachlan	1971	382.51		PMF	35 520	14 722	23.65	Gated	Radial	8	14.63	12.72	Natural Rocks	66.46
JULIUS	Leichhardt	1976	239.36	213.9	1:210,000 (AEP for PMPOF)	44 625	14 590	219.5	Ungated	—	—	—	—	Stilling Basin	66.5

Table 2.3
(B) Data on Australian Large Spillways

NAME	RIVER	YEAR	MAX RESERVOIR FLOOD ELEV (m)	MAX TAIL WATER FLOOD ELEV (m)	DESIGN FLOOD CRITERION	PEAK INFLOW (m³/s)	SPLW DESIGN FLOW (m³/s)	WIDTH OF SPLW EXIT (m)	SPILLWAY TYPE	GATES TYPE	NUMBER	W (m)	H (m)	TYPE OF ENERGY DISSIPATION DEVICE	SPECIFIC DISCH. FLOW (m³/s/m)
DOONDOOMA	Boyne	1982	300.29	256.3	1:250,000 (AEP for PMPDF)	34 578	13 420	115.0	Ungated	---	---	---	---	Stilling Basin	116.7
OPTHALMIA	Fortescue	1982					13 000		Ungated	---	---	---	---		
WIVENHOE	Brisbane	1985					12 000		Gated plus Fuseplug						
GLENBAWN	Hunter	1958	286.18		PMF	22 291	11 115	58.0	Ungated & fuseplug	---	---	---	---	Natural Stilling Basin	223.3
KEEPIT DAM	Namoi	1960	333.53		PMF	35 381	10 489	89.6	Gated & fuseplug	Radial	6	14.9	11.2	Bucket and dentated sill, deep roller	98.3
LISLIE DAM	Sandy Ck	1986	477.86	458.4	1:1,660,000 (AEP for PMPDF)	18 660	3 920	92.0	Gated (radial gates)	Radial	7	12.74	6.64	Roller Bucket	42.6

Table 2.4
Data on Venezuelan Large Spillways

NAME	RIVER	YEAR	MAX RESERVOIR FLOOD ELEV (m)	MAX TAIL WATER FLOOD ELEV (m)	DESIGN FLOOD CRITERION	PEAK INFLOW (m3/s)	SPLW DESIGN FLOW (m3/s)	WIDTH OF SPLW EXIT (m)	SPILLWAY TYPE	GATES Type	No.	W (m)	H (m)	TYPE OF ENERGY DISSIPATION DEVICE	SPECIFIC DISCH. FLOW (m3/s/m)
SIMON BOLIVAR (GURI)	Caroni	1986	271.60	140.0	PMF	48 100	28 750	137.16	Chute	Radial	9	15.24	21.66	Flip bucket	209.63
ANTONIO JOSE DE SUCRE (MACAGUA)	Caroni	1997	54.50	34.0	PMF	30 000	30 000	285.00	Crest overflow	Radial	12	22.00	15.60	Straight to River bed	105.26
FRANCISCO DE MIRANDA (CARUACHI)	Caroni	2005	92.40	57.0	PMF	30 000	30 000	161.16	Crest Overflow	Radial	9	15.24	21.66	Flip bucket & plunging pool	186.15
MANUEL PIAR (TOCOMA)	Caroni	UC	127.50	97.0	PMF	28 750	28 750	153.16	Crest Overflow	Radial	9	15.24	21.66	Flip bucket & plunging pool	187.71

Table 2.5
Data on Mexican Large Spillways

NAME	RIVER	YEAR	MAX RESERVOIR FLOOD ELEV (m)	DESIGN FLOOD CRITERION	PEAK INFLOW (m³/s)	SPILLWAY DESIGN FLOW (m³/s)	WIDTH OF SPLW EXIT (m)	SPILLWAY TYPE	GATES TYPE	NUMBER	W (m)	H (m)	TYPE OF ENERGY DISSIPATION DEVICE	SPECIFIC DISCHARGE OF EXIT FLOW (m³/s/m)
LA AMISTAD	BRAVO	1969	347,59	PMF	54 000	43 700	243,84	Crest overflow	Radial	16	15.24	16.43	Flip bucket and plunge pool	179.00
HUITES	FUERTE	1995	290,00	10 000-yr	30 000	22 450	62,00	Crest overflow	Radial	2	15.50	21.00	Flip bucket and plunge pool	362.00
CERRO DE ORO - main	STO DOMINGO	1988	72,80	10 000-yr	25 980	6 000		3 Tunnel splws	Radial	6	5.90	15.20	Flip bucket and plunge pool	113.00
CERRO DE ORO - aux									Radial	3	5.90	12.80		
MALPASO - main	GRIJALVA	1964	188,00	Regional Maxima	21 750	11 100	45.0	Crest overflow	Radial	3	15.00	15.00	Stilling basin	247.00
MALPASO - aux						10 650	60.0	Crest overflow	Radial	4	15.00	18.70	Flip bucket and plunge pool	178.00
EL INFERNILLO - main	BALSAS	1963	176.40	Creager Form	38 800	13 800	66.78	3 Tunnel splws					Flip bucket and plunge pool	207.00
EL INFERNILLO - aux			183.20	10 000-yr	41 600	10 500								157.00
FALCON	BRAVO	1953	95.77	PMF	20 000	13 000	91.44	Crest overflow	Radial	6	15.24	15.24	Stilling basin	142.00

3. DÉVERSOIRS À HAUTES CHUTES - LE DÉFI DE LA DISSIPATION D'ÉNERGIE ET DU CONTRÔLE DE L'AFFOUILLEMENT À L'AVAL (*)

3.1. INTRODUCTION

La sécurité des barrages en cas d'inondation doit être assurée par une capacité appropriée des ouvrages d'évacuation des crues. Dans le cas des déversoirs à haute chute, outre les problèmes bien connus dus aux écoulements à grande vitesse et au risque de cavitation, l'un des principaux problèmes est celui de la dissipation d'énergie et du contrôle de l'affouillement à l'aval des barrages (Schleiss, 2002). Il peut se produire des jets à haute vitesse qui sont guidés par les ouvrages d'évacuation dans un bassin aval à une certaine distance du barrage. Dans la zone d'impact de ces jets à haute énergie, le lit de la rivière peut être érodé. Comme l'affouillement dû à des jets plongeants peut atteindre des profondeurs considérables même dans les rivières à lits rocheux, l'instabilité des pentes de la vallée est à craindre, ce qui peut mettre en danger dans certains cas la fondation et les culées du barrage lui-même. De tels problèmes d'affouillement surviennent surtout pour des ouvrages où les déversoirs sont combinés avec la structure du barrage et, par conséquent, la zone d'impact des jets plongeants à haute vitesse est relativement proche du barrage. C'est généralement le cas des barrages en béton, où des jets plongeants à haute vitesse peuvent être créés par des déversoirs sur le couronnement (barrages-voûtes seulement), des déversoirs en coursier suivis d'un saut de ski, des évacuateurs en orifice, ainsi que des vidanges de fond à grande capacité. Des conditions d'affouillement sévères surviennent surtout dans le cas des grands barrages-voûtes en béton situés dans des vallées étroites avec des débits de crue élevés.

Un tel dispositif typique d'évacuateur de crues est illustré dans les Figures 3.1 et 3.2 avec l'exemple du projet de barrage de Khersan III en Iran. La gestion des crues du barrage à double voûte de 175 m de hauteur est assurée par trois évacuateurs de crues distincts :

- un double coursier terminé en saut de ski en rive droite avec un déversoir standard à la cote 1 404.5 m et contrôlé par une vanne segment de 11.5 m de large et 13.5 m de haut (capacité de 4 240 m³/s au niveau 1 426.3 m de la crue maximale probable, CMP) ;

- un déversoir standard en crête non contrôlé divisé en 6 passes ; deux de 13.5 m de large, deux de 19.5 à la cote 1 418 m et deux de 12.5 à la cote 1 421 m (capacité total de 3 360 m3/s au niveau 1 426.3 m de la CMP) ; et

- deux vidanges de fond centrées à la cote 1 330 m et 1 345 m avec ouvertures de vanne de service de 3 m de large par 4 m de haut (capacité totale de 395 m³/s au niveau 1 426.3 m de la CMP)

(*) Ce chapitre correspond à une version mise à jour et améliorée de Schleiss (2002)

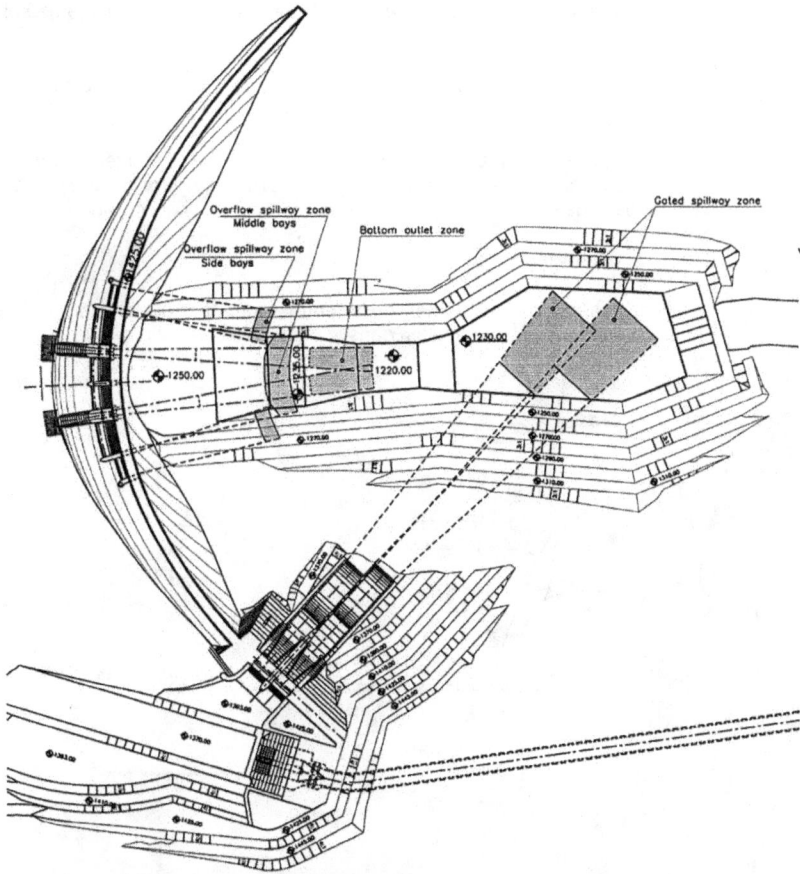

Fig. 3.1

Projet de barrage Khersan III en Iran (mis en service en 2004). Configuration du barrage-voûte et ses ouvrages d'évacuation de crues avec la zone d'impact des jets.

Dans la conception actuelle des évacuateurs de crues des barrages, la tendance est d'augmenter le débit unitaire du jet à haute vitesse sortant des structures annexes. Dans les déversoirs contrôlés en coursier et saut de ski, des débits spécifiques de l'ordre de 200 à 300 m³/s/m ne sont plus rares, car le risque de cavitation du coursier peut être atténué à l'aide d'aérateurs de fond. Les déversoirs libres en crête pour les barrages-voûtes sont aujourd'hui conçus pour des débits spécifiques jusqu'à 70 m³/s/m, et jusqu'à 120 m³/s/m en installant des vannes sur la crête. Grâce aux dernières technologies en vannes haute pression, les orifices de bas niveau peuvent évacuer des débits spécifiques dans la plage de 300 à 400 m³/s/m.

Cette tendance est également confirmée par de nombreux projets de grands barrages en Chine avec des débits importants et construits dans des vallées étroites. Des expériences spéciales sur les grands barrages-poids et voûte et les barrages en remblai rocheux élevé ont été rapportées par Gao et al. (2011).

Outre les questions de conception hydraulique des ouvrages d'évacuation des crues eux-mêmes, le défi de la dissipation d'énergie et du contrôle de l'affouillement doit répondre aux questions suivantes :

- Quelle sera l'évolution et l'étendue de l'affouillement à l'aval du barrage dans la zone d'impact du jet ?

- La stabilité des pentes de la vallée et la fondation du barrage lui-même sont-elles menacées ?

- Faut-il construire un bassin aval pour créer un coussin d'eau et comment cela affecte-t-il la profondeur d'affouillement ?

- Est-ce qu'une pré-excavation du lit rocheux est nécessaire et quelle devrait être sa forme ? Comment la pré-excavation affecte-t-elle l'écoulement et les turbulences dans la fosse de dissipation ?

- Le bassin doit-il être revêtu ?

- Le niveau aval et le fonctionnement de la centrale sont-ils influencés par l'affouillement et par les courants de recirculation dans le bassin de dissipation ?

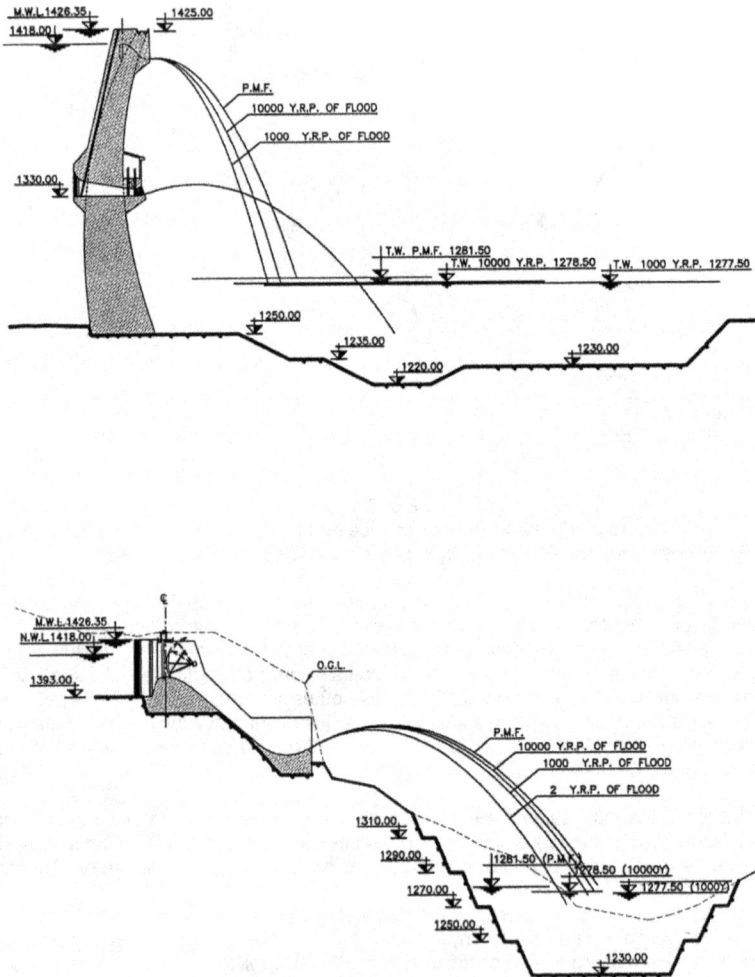

Fig. 3.2
Projet de barrage Khersan III en Iran. Profils des déversoirs en crête et saut de ski et des bassins de dissipations

3.2. LE PROCESSUS DE L'AFFOUILLEMENT

3.2.1. Le processus physique

L'affouillement est un problème complexe et étudié depuis longtemps. Comme l'illustre la Figure 3.3, il peut être décrit par une série de processus physiques comme suit (Bollaert, 2002; Mason, 2011) :

a. le comportement des jets en chute libre dans l'air et impact des jets aérés ;

b. le comportement des jets et de l'écoulement turbulent dans le bassin de dissipation ;

c. la fluctuation de pression à l'interface eau-rocher ;

d. la propagation des pressions dynamiques dans les joints rocheux ;

e. la fracturation hydrodynamique des joints rocheux fermés et la fragmentation de la roche en blocs ;

f. l'éjection des blocs de roche ainsi formés par soulèvement dynamique dans le bassin de dissipation et l'entraînement et la circulation de la roche excavée dans le bassin et verticalement sur sa hauteur ;

g. la fragmentation et la dégradation des blocs rocheux par l'effet de broyage à billes de l'écoulement turbulent dans le bassin de dissipation ;

h. le dépôt des roches à l'aval jusqu'à un point où elles ne peuvent pas retourner dans le bassin et donc la formation d'un monticule en aval ; et

i. le déplacement vers l'aval des matériaux affouillés par le transport en rivière.

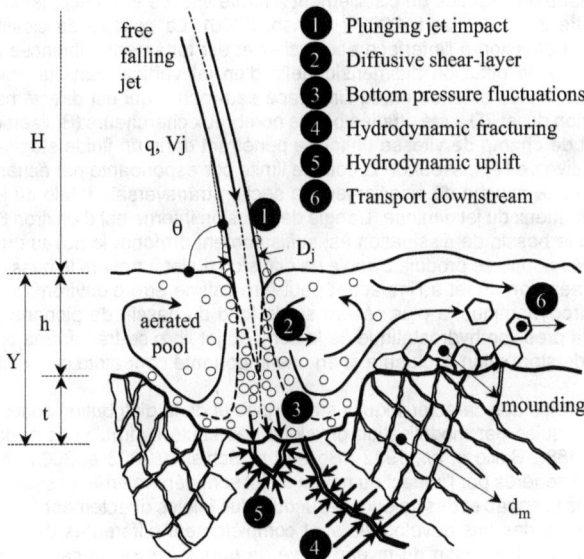

Fig. 3.3
Principaux paramètres et procédés physico-mécaniques impliqués dans le phénomène d'affouillement (Bollaert, 2002)

3.2.2. Comportement des jets dans l'air

Pour évaluer l'affouillement causé par les jets, il est d'abord nécessaire de prédire la trajectoire de ceux-ci afin de connaître l'emplacement de l'impact dans le bassin et la zone du trou d'affouillement (Whittaker et Schleiss, 1984). Le comportement d'un jet idéal peut être facilement évalué à l'aide d'équations balistiques. Néanmoins, pour les prototypes de jets, il faut tenir compte de l'effet de traînée de l'air, et donc de la désintégration du jet dans l'air et de l'aération initiale de l'écoulement le long des coursiers. Un certain nombre de chercheurs ont mis au point des équations pour prédire la trajectoire du jet, c'est-à-dire la longueur du trajet du jet qui tient compte de ces effets (Gun'ko et al., 1965 ; Kamenev, 1966 ; Kawakami, 1973 ; Zvorykin et al., 1975 ; Taraimovich, 1980 ; Martins, 1977).

La diffusion du jet pendant la chute est également un problème qui a été abordé de façon empirique par Taraimovich (1980) et U.S.B.R. (1978). Plus récemment, la diffusion latérale a été liée à son intensité initiale de turbulence (Ervine et Falvey, 1987 ; Ervine et al., 1997). Les angles typiques de diffusion des jets sont de 3 à 4 % pour les jets assez turbulents. L'intensité de turbulence est inférieure à 3 % pour les jets en chute libre, de 3 à 5 % pour les jets sortant de sauts de ski et de 3 à 8 % pour les jets sortant d'orifice (Bollaert, 2002). Sur la base des résultats expérimentaux et d'une revue de littérature approfondie, les paramètres d'émission des jets pertinents pour la pratique de l'ingénierie peuvent être trouvés dans Manso et al. (2008).

3.2.3. Comportement des jets dans le bassin de dissipation et fluctuations de pression

Plusieurs expressions définissant la teneur en air au point d'impact d'un jet dans un bassin de dissipation ont été développées à l'aide de tests sur modèles. Certaines d'entre elles peuvent être raisonnablement étendues à des vitesses prototypes (Van de Sande, 1973 ; Bin, 1984 ; Ervine et al., 1980 ; Ervine, 1998).

Les conditions d'écoulement dans le bassin peuvent être caractérisées par un écoulement turbulent biphasé dans une couche de cisaillement à haute vitesse et un écoulement macro-turbulent à l'extérieur de cette zone (Bollaert, 2002 ; Manso, 2006). La couche de cisaillement produit de fortes fluctuations de pression à l'interface eau-rocher et est fortement influencée par l'aération. Les théories existantes sur la diffusion bidimensionnelle d'un jet vertical dans un milieu semi-infini ou borné définissent les limites extérieures de l'interface eau-rocher qui est directement soumise à ces pressions. La diffusion du jet 2D a été étudiée par de nombreux chercheurs (Bollaert et Schleiss, 2003). Le concept d'un jet de champ de vitesse uniforme pénétrant dans un fluide stagnant est basé sur la croissance progressive de l'épaisseur de la couche limite correspondante par échange de quantité de mouvement. Dans cette couche de cisaillement, la section transversale totale du jet augmente alors que le noyau non visqueux du jet diminue. L'angle de diffusion interne est d'environ 8° pour les jets très turbulents. Lorsque le bassin de dissipation est suffisamment profond, le noyau au jet disparaît et un jet complètement développé se produit. L'angle de diffusion du jet à travers le bassin dépend du degré de turbulence et d'aération du jet à l'impact et peut être estimé être d'environ 15° (Ervine et Falvey, 1987). L'action hydrodynamique la plus sévère sur le fond du bassin de plongée se produit dans la zone d'impact, où la pression hydrostatique de la zone du jet libre se transforme progressivement en une forte pression de stagnation fluctuante et en une importante contrainte de cisaillement au fond.

La connaissance des caractéristiques statistiques et de la distribution spatiale des fluctuations de pression a été acquise par modélisation physique des sauts hydrauliques et des jets plongeants (Toso et Bowers, 1988 ; Ballio et al., 1992 ; Bollaert et Schleiss, 2003 et 2005 ; Manso, 2006). Les régimes de pression générés par l'impact du noyau du jet sont généralement assez constants avec des valeurs élevées dans le noyau et des valeurs beaucoup plus faibles directement à l'extérieur de celui-ci. Les pressions d'impact des jets développés sont complètement différentes des pressions générées par l'impact des noyaux. En raison du niveau élevé de turbulence et du caractère biphasique de la couche de cisaillement, les conditions d'impact de jets développées peuvent générer des pressions dynamiques beaucoup plus sévères au fond du bassin. Par conséquent, tous les coussins d'eau n'ont pas un effet retardateur sur la formation de l'affouillement. L'énergie spectrale d'un jet plongeant à haute vitesse s'étend sur une gamme de fréquences beaucoup plus large (> 100 Hz) que ce que l'on suppose généralement pour les écoulements macro-turbulents dans les bassins de dissipation (jusqu'à 25 Hz).

Des études expérimentales systématiques dans des bassins de dissipation avec différents confinements latéraux ont montré que la géométrie du bassin influence non seulement la diffusion des jets plongeants et l'entraînement de l'air dans le bassin, mais aussi les pressions d'impact à l'interface eau-rocher et à l'intérieur de la masse rocheuse fissurée (Manso *et al.*, 2009). La géométrie favorable des jets diffusés et des bassins peut limiter les fluctuations de pression à l'interface eau-rocher et donc réduire le potentiel d'affouillement (Manso, 2006 ; Bollaert *et al.*, 2012).

3.2.4. Propagation des pressions dynamiques dans les joints rocheux, fracturation hydrodynamique et soulèvement dynamique

Le transfert des pressions du fond du bassin dans les joints rocheux entraîne un écoulement transitoire qui est régi par la propagation des ondes de pression. Dans le cas de joints rocheux fermés, comme ceux que l'on rencontre dans une masse rocheuse partiellement fracturée, la réflexion et la superposition des ondes de pression peuvent accroître la charge hydrodynamique à l'intérieur de la fissure. Si les contraintes correspondantes à l'extrémité du joint dépassent la résistance à la rupture et les contraintes de compression initiales dans la roche, la roche se fissurera et le joint pourra se développer davantage. Dans le cas de joints rocheux ouverts dans une masse rocheuse entièrement fracturée, les ondes de pression à l'intérieur des joints créent une force de soulèvement dynamique importante sur les blocs rocheux. Cette force de soulèvement dynamique peut briser les ponts rocheux restants dans les joints par fatigue et, si elle est suffisamment élevée, éjecter les blocs rocheux ainsi formés dans l'écoulement macro-turbulent du bassin. La transmission de la charge hydrodynamique au joint rocheux dépend fortement du rapport entre la portée spatiale du champ de pression et la géométrie de l'ouverture du joint (Manso *et al.*, 2007).

La présence de telles ondes de pression et d'amplifications significatives dues à des phénomènes de résonance a pu être observée et mesurée par une étude expérimentale à une échelle proche du prototype dans des joints de roches simplifiés. (Bollaert, 2002 ; Bollaert et Schleiss, 2003 et 2005 ; Manso, 2006 ; Federspiel, 2011 ; Duarte, 2014).

3.2.5. Effet de broyage à billes de l'écoulement turbulent dans le bassin et formation d'un monticule en aval

Une fois que les blocs de roche sont formés et éjectés de la masse rocheuse par le soulèvement dynamique, ils peuvent être emportés par les tourbillons macro-turbulents. Si le bloc est trop gros pour être transporté par l'écoulement, il sera brisé après un certain temps par l'effet de broyage à billes des tourbillons dans le bassin de dissipation. Ayant atteint la taille minimale requise, les blocs de roche seront déplacés vers l'aval par l'écoulement et déposés sur le monticule ou transportés comme sédiments dans la rivière. Le monticule peut limiter la profondeur d'affouillement, mais aussi élever le niveau d'eau aval. Au-delà d'un niveau critique, l'augmentation peut nuire au fonctionnement des sorties de fond ou réduire la hauteur de chute nette de la centrale, si elle est située en amont du monticule. Si le monticule limite la profondeur d'affouillement, on considère que l'affouillement a atteint une limite dite dynamique. Toutefois, si la butte est enlevée et que l'affouillement se poursuit jusqu'au maximum possible, on considère qu'il a atteint la limite statique ultime (Eggenberger and Müller, 1944).

3.3. MÉTHODES D'ÉVALUATION DE L'AFFOUILLEMENT

3.3.1. Aperçu général

Les méthodes existantes d'évaluation de l'affouillement peuvent être regroupées comme suit (Bollaert et Schleiss, 2003) :

- des approches empiriques fondées sur des observations en laboratoire et sur le terrain ;

- des méthodes semi-empiriques combinant des observations en laboratoire et sur le terrain ;

- des approches basées sur des valeurs extrêmes de pressions fluctuantes au fond du bassin de dissipation ;

- des techniques fondées sur les différences de pression instantanées et moyennes dans le temps et tenant compte des caractéristiques de la roche ; et

- des modèles basés sur des pressions transitoires biphasées dans les joints rocheux considérant des effets de résonance et de fatigue.

Les méthodes les plus couramment utilisées pour l'évaluation de l'affouillement dû à la chute des jets à haute vitesse sont illustrées à la Figure 3.4 avec les différents paramètres physiques considérés impliqués dans le processus d'affouillement qui peuvent être reliés aux trois phases eau, roche et air. L'évolution temporelle est un paramètre supplémentaire.

3.3.2. Formules empiriques

Un grand nombre d'équations empiriques ont été développées pour prédire l'affouillement des jets plongeants. Ces formules empiriques sont pour la plupart dérivées d'essais sur modèles hydrauliques en laboratoire mais aussi de l'observation de prototypes et sont largement utilisées dans la pratique pour la conception d'ouvrage. Certaines expressions sont d'application générale, tandis que d'autres sont spécifiques aux jets à chute libre, aux sauts de ski ou aux déversoirs en orifices. Whittaker et Schleiss (1984), Mason et Arumugam (1985) et Bollaert (2002) ont donné un aperçu complet et comparé la plupart des formules connues.

Le processus complexe d'affouillement est réduit par les formules empiriques à quelques paramètres. La profondeur totale ultime de l'affouillement Y mesurée à partir du niveau aval est considérée comme étant en fonction des paramètres suivants :

- e débit spécifique q (débit par unité de largeur du jet) ;

- la chute H ;

- la profondeur aval h (mesurée du niveau du lit initial) ; et

- la taille caractéristique des sédiments ou le diamètre du bloc rocheux d.

La plupart des formules sont écrites sous la forme

$$Y = t + h = K \cdot \frac{H^y \cdot q^x \cdot h^w}{g^v \cdot d_m{}^z} \tag{1}$$

où t est la profondeur d'affouillement au-dessous du niveau initial du lit et K une constante.

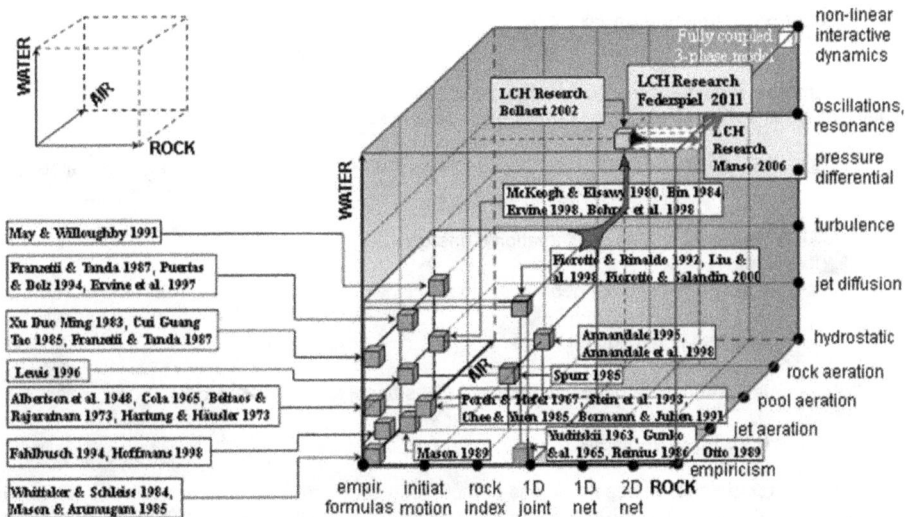

Fig. 3.4
Résumé des méthodes existantes d'évaluation de l'affouillement en tenant compte des paramètres physiques impliqués dans le processus d'affouillement qui peuvent être liés aux trois phases : eau, roche et air (selon Bollaert, 2002)

Mason et Arumugam (1985) ont appliqué cette forme de formule à un grand nombre de données d'affouillement, dont 26 sets provenant de prototypes et 47 d'essais sur modèles. Après d'autres travaux, Mason (1989) a suggéré les exposants et constantes suivants pour les conditions de modèle et de prototype :

$K = (6.42 - 3.10H^{0.10})$

$v = 0.30$

$w = 0.15$

$x = 0.60$

$y = 0.05$

$z = 0.10$

Il a été démontré que cela donnait la prédiction la plus cohérente et la plus précise pour l'ensemble des données d'essai des 46 modèles et une limite supérieure générale pour l'ensemble des données des prototypes. Il s'est également avéré que la formule était équilibrée sur le plan dimensionnel, ce qui constituait un bon indicateur de sa potentielle validité (Mason, 1989).

Selon les ensembles de données, la formule est applicable pour les modèles hydrauliques dont la hauteur de chute H est comprise entre 0.325 m et 2.15 m et entre 15.82 m et 109 m pour les prototypes dans le cas des déversoirs en jet libre, en saut de ski ou en orifice. L'utilisation de la taille moyenne des particules d_m a donné de meilleurs résultats que l'utilisation de la taille moyenne des particules d_{90}. Pour une roche prototype, une taille de particule équivalente $d_m = 0.25$ m est recommandée dans la formule ci-dessus.

3.3.3. *Équations semi-empiriques*

Comme il a déjà été mentionné, les méthodes semi-empiriques combinent les observations en laboratoire et sur le terrain avec certains principes physiques :

- le déclenchement du mouvement du matériau du lit par contrainte de cisaillement ;

- les équations de conservation d'énergie ;

- les caractéristiques géomécaniques ;

- l'angle d'impact du jet ;

- la théorie de la diffusion du jet bidimensionnel en régime stationnaire ; et

- les effets d'aérations.

Un aperçu détaillé de ces équations et méthodes semi-empiriques se trouve dans Whittaker et Schleiss (1984) et Bollaert, (2002). Le processus hydrodynamique de l'affouillement est souvent dérivé de la théorie de la diffusion du jet bidimensionnel. Le comportement géomécanique de la masse rocheuse est pris en compte par le concept d'initiation de mouvement par cisaillement pour les matériaux granulaires non cohésifs ou par un indice qui définit la résistance de la masse rocheuse à l'érosion. Les caractéristiques hydrodynamiques et géomécaniques sont par exemple combinées dans les méthodes de l'indice d'érodabilité des roches de Spurr (1985) et d'Annandale (1995 et 2006) et dans les équations de conservation de la quantité de mouvement établies par Fahlbush (1994) et Hoffmans (1998) pour les matériaux non cohésifs.

3.3.4. *Approches fondées sur des valeurs extrêmes de pressions fluctuantes au fond du bassin de dissipation*

Au fond du bassin de dissipation, des pressions dynamiques fluctuantes se produisent en raison de l'impact direct du noyau du jet dans le cas d'un coussin d'eau relativement petit. Pour des profondeurs de bassin élevées (supérieures à 4 à 6 fois l'épaisseur du jet entrant), un flux de cisaillement turbulent, ou impact de jet développé, selon la théorie de la diffusion bidimensionnelle du jet, est créé. Ces deux types d'impact de jet génèrent des modèles de pression complètement différents, comme il a déjà été mentionné.

Les pressions dynamiques au fond du bassin de plongée peuvent pénétrer dans les fissures de la masse rocheuse sous-jacente. Les approches basées sur les pressions extrêmes au fond supposent que les pressions maximales se produisant à l'interface eau-rocher sont transférées par les fissures sous les blocs rocheux. Néanmoins, cela n'est possible que pour les joints ouverts et lisses et pour les blocs petits par rapport à la structure turbulente typique qui génère l'impulsion de pression. Dans cette hypothèse, ces pressions maximales sous les blocs de roche combinées aux pressions minimales au fond du bassin de dissipation créent une pression nette de soulèvement Δp sur les blocs de roche (Figure 3.5). L'affouillement ultime est atteint lorsque cette différence de pression nette Δp sur le bloc n'est plus capable de l'éjecter. Néanmoins, cette approche néglige l'effet dissipatif de la non-corrélation spatiale du champ de pression. Comme les pressions maximale et minimale ne se produisent pas au même moment, la pression de soulèvement nette ainsi définie représente une limite supérieure physique des conditions de charge dynamique et est donc plutôt conservatrice.

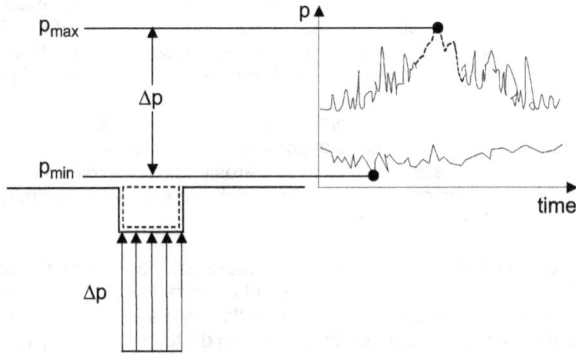

Fig. 3.5
Schéma de définition des pressions dynamiques extrêmes au fond de la piscine. Les pressions maximale et minimale sont définies au centre du bloc pendant un intervalle de temps suffisamment long (selon Bollaert, 2002)

Les études sur les fluctuations de pressions dans les bassins de dissipation ont principalement été réalisées par Ervine *et al.* (1997), Xu-Duo-Ming (1983) et Franzetti et Tanda (1984 et 1987) pour des jets circulaires et par Tao *et al.* (1985), Lopardo (1988), Armengou (1991), May et Willoughby (1991) et Puertas et Dolz (1994) pour des jets rectangulaires. Ces études fournissent des informations utiles sur les fluctuations de pression de fond mais ne décrivent pas leur propagation à l'intérieur des discontinuités de la masse rocheuse sous-jacente. L'application simultanée de pressions de fond minimales et maximales extrêmes au-dessus et au-dessous des blocs de roche pourrait hypothétiquement entraîner des différences de pression nette allant jusqu'à 7 fois la valeur moyenne quadratique ou jusqu'à 1.5 – 1.75 fois l'énergie cinétique à l'impact du jet. Même si cela semble être un critère de conception conservateur, il faut noter que les phénomènes violents de pression transitoire, qui pourraient se produire à l'intérieur des joints rocheux, ne sont pas considérés (Bollaert, 2002).

3.3.5. Techniques fondées sur la moyenne temporelle et les différences de pression instantanées et tenant compte des caractéristiques de la roche

Cette approche, contrairement à celle présentée dans la section précédente, prend en compte les différences de pression moyennes dans le temps ou instantanées se produisant à ou pendant un certain temps à la surface et sous les blocs rocheux. Cela signifie que les pressions fluctuantes doivent être mesurées non seulement au fond du bassin de plongée, mais aussi à l'intérieur des joints rocheux.

Yuditskii (1963) (rapporté par Gunko *et al.*, 1965) a probablement été le premier à affirmer que les pressions moyennes dans le temps et les pressions pulsées peuvent être responsables du soulèvement des blocs rocheux dans le processus d'affouillement. Il a mesuré les forces sur un seul bloc de roche sur le fond plat d'un bassin de dissipation, exercées par l'impact d'un jet produit par un déversoir en saut de ski sur une maquette à l'échelle. Les techniques de mesure de l'époque ne permettaient alors que d'obtenir des forces moyennes dans le temps. Otto (1989) a quantifié les pressions de soulèvement moyennes dans le temps agissant sur un bloc de roche pour les jets plans impactant obliquement. En fonction de la saillie relative du bloc et du point d'impact du jet, d'importants effets d'aspiration de surface se sont produits sur la surface supérieure du bloc, entraînant des pressions moyennes de soulèvement presque égale à la totalité de l'énergie cinétique à l'impact. Sans tenir compte de cet effet d'aspiration, les pressions de soulèvement maximales représentaient encore la moitié de cette énergie cinétique.

Hartung et Häusler (1973) ont mis pour la première fois en évidence les effets destructeurs des différences de pression instantanées entrant dans les minuscules joints rocheux. Kirschke, (1974) a d'abord effectué une analyse analytique et numérique de la propagation des coups de bélier dans les joints rocheux. Des différences de pression instantanées basées sur des hypothèses d'écoulements transitoires ont été mesurées et analysées pour la première fois pour le cas des revêtements de dalles en béton des bassins de dissipation (Fiorotto et Rinaldo, 1992 ; Bellin et Fiorotto, 1995 ; Fiorotto et Salandin, 2000 ; Fiorotto et Caroni, 2007). Cependant, l'échelle des essais sur modèle et le taux d'acquisition des données n'ont pas permis de mesurer les effets d'oscillation ou de résonance dans le joint sous les dalles.

Annandale et al. (1998) ont simulé l'érosion des roches fracturées à l'aide de blocs de béton légers, placés en une série de deux couches avec un angle de pendage de 45°. L'impact des jets a confirmé leur théorie selon laquelle le critère du seuil d'érosion pour les roches et les terrains meubles peut être défini au moyen d'un indice géomécanique, l'indice d'érodabilité. Le pouvoir érosif de l'eau peut être lié à la résistance à l'érosion du matériau.

Des études expérimentales et numériques axées sur les forces de soulèvement nettes fluctuantes agissant sur des blocs de roche simulés ont également été réalisées par Liu et al (1998). Un critère de conception pour le soulèvement des blocs de roche a été donné sur la base d'un modèle d'écoulement transitoire mais sans tenir compte des effets de résonance (Liu, 1999).

Toutes les études mentionnées sur les dalles de béton et les blocs de roche considèrent le champ de pression de surface en fonction de l'espace et du temps. Mais le champ de pression en dessous est supposé uniformément réparti le long de la surface de l'élément et égal à la pression à l'entrée des joints, c'est-à-dire à tout endroit de la surface. C'est pourquoi les conditions d'écoulement totalement transitoires dans les joints, telles que les réflexions et les amplifications des ondes de pression, sont négligées par ces techniques de pression extrême.

3.3.6. *Modèle d'affouillement basé sur des pressions d'eau transitoires biphasées dans les joints rocheux et tenant compte des effets de résonance et de fatigue*

Bollaert (2002) a mesuré pour la première fois les pressions transitoires dans les joints rocheux dues à l'impact d'un jet à haute vitesse avec des essais systématiques en laboratoire sur un dispositif expérimental reproduisant des conditions proches du prototype. Il a reproduit ces pressions également dans un modèle numérique. De nouveaux phénomènes ont pu être observés et expliqués comme la réflexion et la superposition des ondes de pression, les pressions de résonance et la libération quasi instantanée de l'air due aux chutes de pression dans le joint.

L'analyse a révélé que la vitesse de l'onde de pression est fortement influencée par la présence de bulles d'air libres dans les joints. Ces bulles peuvent être transportées par l'écoulement de la fosse dans le joint mais aussi être libérées de l'eau en cas de chute soudaine de pression en dessous de la pression atmosphérique.

Dans les joints ouverts, des pressions de soulèvement nettes instantanées de 0.8 à 1.6 fois l'énergie cinétique du jet à l'impact ont été mesurées. Ce chiffre est nettement plus élevé que toutes les hypothèses formulées précédemment dans la littérature (voir également la section 3.3.5) et souligne que les effets transitoires de pression dans les joints rocheux sont un processus physique clé pour la formation de l'affouillement.

Sur la base des résultats expérimentaux et de la simulation numérique, un nouveau modèle d'évaluation de la profondeur ultime de l'affouillement a été mis au point, la *Comprehensive Scour Method* (CSM), qui représente une évaluation complète des deux processus physiques : fracturation hydrodynamique des joints rocheux fermés et soulèvement dynamique des blocs (Bollaert, 2002 et 2004 ; Bollaert et Schleiss, 2005). Les processus pertinents comme les caractéristiques du jet (vitesse et diamètre à l'émission, intensité initiale de turbulence du jet), les fluctuations de pression au fond du bassin de dissipation et la charge hydrodynamique à l'intérieur des joints rocheux sont traités et comparés à la résistance de la roche à la propagation des fissures en considérant la fatigue.

Le modèle CSM a été amélioré en tenant compte de la géométrie du bassin de plongée et du confinement latéral du jet plongeant (Manso, 2006) ainsi que de la réponse dynamique d'un bloc rocheux due à une interaction fluide-structure (Federspiel, 2011 ; Asadollahi et al., 2011). Plus récemment, l'influence de l'aération du jet et, par conséquent, de l'aération de la fosse a été étudiée et mise en œuvre dans le modèle CSM (Duarte *et al.*, 2015 et 2016).

L'application du modèle CSM nécessite une connaissance de la distribution spatiale des pressions dynamiques de l'eau à l'interface rocheuse ainsi que des paramètres rocheux comme le nombre, l'espacement, la direction et la persistance des ensembles de fractures, les contraintes in situ dans la masse rocheuse, la résistance à la compression non confinée et la fracturation. Les pressions dynamiques de l'eau agissant au fond du bassin de plongée pour une certaine géométrie d'affouillement peuvent être évaluées à l'aide d'un modèle hydraulique d'essai et utilisées ensuite comme données d'entrée pour le modèle CSM. Les paramètres rocheux sont normalement évalués lors de la campagne de reconnaissance géologique et géotechnique pour tout projet de fondation de barrage et sont donc disponibles. Néanmoins, les incertitudes doivent être prises en compte par analyse de sensibilité et le jugement de l'ingénieur.

3.4. LES DIFFICULTÉS RENCONTRÉES LORS DE L'ESTIMATION DE LA PROFONDEUR D'AFFOUILLEMENT

3.4.1. *Quelle-est la formule ou la théorie appropriée*

Au stade de l'étude de faisabilité d'un projet de barrage, les formules empiriques et semi-empiriques facilement applicables sont habituellement utilisées pour obtenir une première idée de la profondeur d'affouillement ultime probable à l'aval des ouvrages d'évacuation pendant la durée de vie du projet. La plupart des formules ont été mises au point pour un cas particulier, comme un saut de ski, une chute libre de crête ou un orifice, tandis que d'autres sont d'application générale. Par conséquent, il convient de sélectionner soigneusement les équations appropriées pour chaque projet. Néanmoins, même après cette sélection, les résultats des formules restantes montrent très souvent une grande dispersion. Dans un tel cas, les décisions techniques peuvent être basées sur une analyse statistique des résultats en utilisant, par exemple, la moyenne de toutes les formules ou l'écart-type positif d'une manière plus conservatrice. Cette analyse est effectuée pour un certain débit d'évacuateur de crues avec le niveau d'eau aval correspondant et en faisant varier la taille caractéristique du bloc rocheux (si prise en compte) en fonction des conditions géologiques prévues.

Parfois, des mesures d'affouillement sur prototype d'un barrage existant, situé dans des conditions géologiques similaires à celles du nouveau projet prévu, sont disponibles. La formule prédisant le mieux la profondeur d'affouillement peut alors être identifiée et appliquée au nouveau projet.

3.4.2. *Comment les essais sur modèle devraient être effectués et interprétés ?*

Si l'on modélise en laboratoire des jets en chute libre impactant la roche au fond d'un bassin de dissipation, trois difficultés peuvent survenir (Whittaker et Schleiss, 1984):

- le choix du matériau approprié qui se comportera dynamiquement dans le modèle comme le fait la roche fissurée dans le prototype ;

- l'effet de la granulométrie ; et

- l'effet de l'aération.

Dans la plupart des modèles, le processus de désintégration, c'est-à-dire la fracturation hydrodynamique des joints rocheux fermés et la fracturation du massif en blocs de roche, est supposé avoir eu lieu. Ainsi, seule l'éjection des blocs de roche dans l'écoulement macro-turbulent du bassin de dissipation et le transport du matériau à l'extérieur de la fosse d'affouillement sont modélisés. Des résultats raisonnables peuvent être obtenus si la roche fissurée est modélisée par des éléments en béton de forme appropriée (Martins, 1973), mais leur configuration et leurs tailles régulières ne représentent pas véritablement une masse rocheuse avec plusieurs ensembles de fractures qui se croisent. Néanmoins, de la pierre concassée ayant au moins une distribution granulométrique similaire à celle des blocs de roche attendus devrait être utilisée à la place du gravier de rivière rond. Dans tous les cas, les essais sur modèle ne peuvent simuler la rupture des blocs de roche par l'effet de fraisage à billes (« marmites de géant ») de l'écoulement turbulent dans le bassin de dissipation. C'est pourquoi le modèle forme un monticule plus haut et plus stable que le prototype. Par conséquent, la profondeur d'affouillement du prototype est sous-estimée. Ceci peut être compensé dans une certaine mesure en choisissant soigneusement le matériau, y compris la réduction d'échelle. Normalement, il est possible d'obtenir de bons résultats prédictifs de la profondeur d'affouillement en utilisant un matériau non cohésif, mais l'étendue de l'affouillement peut ne pas être correcte car les pentes raides et quasi verticales ne peuvent être reproduites. C'est pourquoi il est proposé d'utiliser un matériau légèrement cohésif en ajoutant du ciment, de la pâte d'argile, de la poudre de craie, de la cire de paraffine, etc. au gravier concassé (Johnson, 1977 ; Gerodetti, 1982 ; Quintela et Da Cruz, 1982 ; Mason, 1984).

Il est connu que pour des granulométries inférieures à 2 à 5 mm, la profondeur d'affouillement ultime devient constante (Veronese, 1937, Mirtskhulava et al., 1967, Machado, 1982). Pour une échelle acceptable d'un modèle de barrage complet de 1:50 à 1:70, les plus petits blocs de roche prototypes reproductibles se situent entre 0.1 et 0.35 m.

Enfin, l'entraînement de l'air ne peut pas être correctement modélisé dans des modèles complets à moins d'utiliser une grande échelle peu pratique (de l'ordre du 1:10). L'entraînement de l'air a un caractère très aléatoire, ce qui influence considérablement le processus d'affouillement (voir section 3.2). Mason (1989) a étudié systématiquement l'effet de l'aération du jet sur l'affouillement dans un lit de gravier mobile à l'aide d'un appareil spécialement conçu en laboratoire pour les chutes de 2 m maximum. Duarte (2014) a étudié systématiquement l'effet de l'aération du jet sur la réponse dynamique d'un bloc rocheux impacté par un jet haute vitesse dans un dispositif expérimental reproduisant des conditions proches du prototype (Duarte et al., 2015 et 2016).

Grâce à la modélisation hybride, le comportement mécanique de l'affouillement des roches peut être correctement représenté. Cela a été fait avec succès pour définir les mesures de mitigation de l'affouillement du barrage de Kariba (Noret et al., 2013 ; Bollaert et al., 2012). Plusieurs géométries historiques d'affouillement ont été reproduites dans un modèle physique avec une surface de mortier fixe équipée de capteurs de pression. Quatre différentes pressions dynamiques d'exploitation de l'évacuateur de crues ont été mesurées à l'interface eau-rocher. Ces pressions dynamiques enregistrées ont ensuite été introduites dans la méthode d'affouillement globale (CSM) de Bollaert et Schleiss, (2005), qui permet d'estimer, à l'aide d'un étalonnage par observation sur prototype, jusqu'à quelle profondeur des blocs rocheux peuvent être formés par fatigue et éjectés dans le bassin de plongée par les pressions dynamiques.

3.4.3. Comment analyser les observations sur prototype correctement ?

Lors de l'analyse des observations sur prototype sur la profondeur d'affouillement afin d'obtenir une équation pour des conditions similaires, il convient de répondre aux questions suivantes :

- Quelle a été la durée de fonctionnement de l'évacuateur de crues pour différents débits spécifiques (courbe débit-durée) ? (La Figure 3.6 donne un exemple de courbe de débit-durée d'un évacuateur de crues).

- Quel était le débit spécifique prépondérant qui a formé la profondeur d'affouillement observée ?

- La durée de ce débit spécifique a-t-elle été assez longue pour créer une profondeur d'affouillement ultime ?

Dans la mesure où, dans la pratique, il est très souvent difficile de répondre précisément à ces questions, des incertitudes probablement importantes ont été introduites dans les formules existantes dérivées d'observations sur prototype. Ceci peut également expliquer la grande dispersion lors de la prédiction de l'affouillement pour d'autres prototypes.

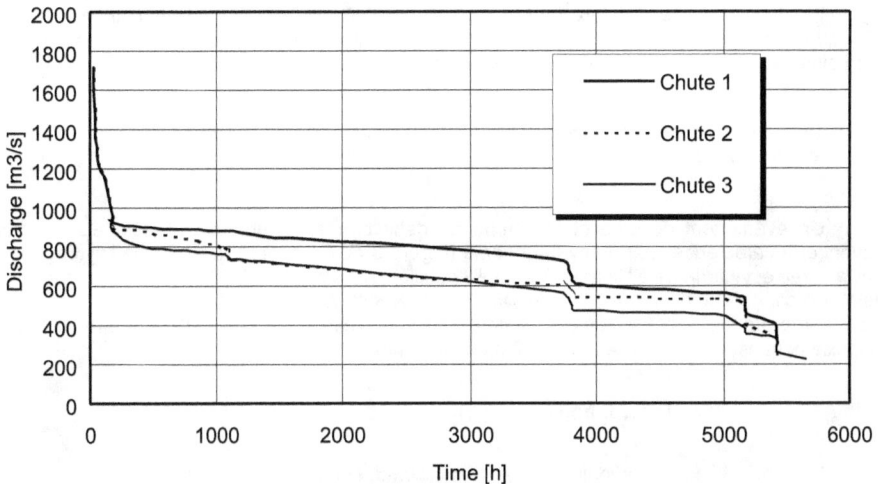

Fig. 3.6
Courbe débit-durée du déversoir sur coursier du barrage Karun I en Iran pendant la période Mars
1980 à Juillet 1988 (largeur par coursier 15 m) (Schleiss, 2002).

3.4.4. *La profondeur ultime d'affouillement peut-elle être atteinte pendant la période d'exploitation de l'évacuateur de crues et quel est la vitesse d'affouillement ?*

En principe, la profondeur d'affouillement estimée à l'aide d'une formule empirique et semi-empirique ne se produira que pour une longue durée d'exploitation de l'évacuateur de crues, après avoir atteint des conditions stables dans la fosse. Cela ne se produira qu'après une durée minimale de fonctionnement de l'évacuateur de crues, principalement en fonction de la qualité et du jointoiement de la masse rocheuse. Par conséquent, il convient d'évaluer le débit spécifique qui aura une durée suffisamment longue pour former l'affouillement ultime pendant la durée de vie technique du barrage. Des débits plus élevées et donc plus rares ne sont pas en mesure de créer un affouillement ultime.

Étant donné que l'affouillement de la fosse t + h se développe à un rythme exponentiel avec le temps T, le taux d'affouillement peut être estimé selon la relation suivante (Spurr, 1985) :

$$(t+h)(T) = (t+h)_{end}(1 - e^{-aT/T}e) \qquad (2)$$

où a est une constante propre au site. L'évaluation de T_e (moment auquel l'équilibre est atteint) dépend de la vitesse à laquelle l'hydrofracturation et le lavage du matériau du trou d'affouillement se produiront, compte tenu des caractéristiques primaires et secondaires de la roche. Les caractéristiques primaires de la roche comprennent le RQD, l'espacement des joints, la résistance à la compression uniaxiale et l'angle d'impact du jet par rapport aux failles principales ou aux stratifications En connaissant la profondeur d'un affouillement qui s'est produit pendant une certaine période de fonctionnement et en estimant la profondeur maximale par l'une des formules, on peut déterminer la constante spécifique a/T_e du site.

Les données de prototypes disponibles indiquent que l'affouillement ultime n'est généralement atteint qu'après Te = 100 à 300 heures de fonctionnement de l'évacuateur de crues pour un certain débit considéré.

Il est possible de conclure que la profondeur d'affouillement ultime pour un certain débit spécifique ne se produit que si la durée de ce débit est suffisamment longue. L'affouillement pour une durée plus courte peut être estimé à l'aide d'une relation de vitesse exponentielle.

3.4.5. Quel sera le débit prépondérant pour la formation d'affouillement lors d'une crue ?

Un événement de crue et la courbe de débit correspondante de l'évacuateur de crues peuvent être caractérisés par un hydrogramme (Figure 3.7). Pour tous les débits de l'hydrogramme dont la durée est inférieure à l'instant T_e auquel l'équilibre est atteint, l'affouillement ultime ne sera pas atteint. En connaissant la relation de vitesse d'affouillement [Équation (2)], le débit prépondérant qui produira la profondeur d'affouillement maximale pendant l'événement de crue peut être déterminé. L'affouillement est estimé successivement pour les débits :

$$q_u(T=T_e=T_u), q_1(T_1<T_u), q_2(T_2<T_1<T_u), ..., q_{peak}(T_{peak}<T_i<T_u)$$

Le débit, qui donne l'affouillement le plus profond, est le débit prépondérant.

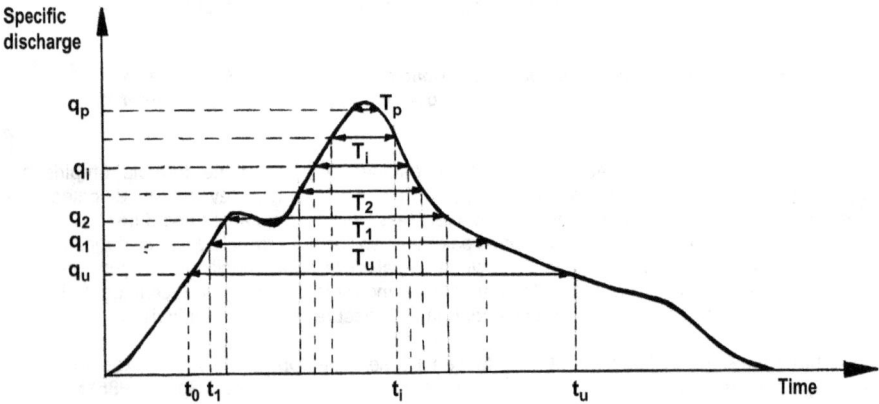

Fig. 3.7

Événement de crue et courbe de débit correspondante de l'évacuateur de crues, montrant des débits d'une durée inférieure à l'instant Te = Tu où l'équilibre de l'affouillement et sa profondeur ultime sont atteints, avec pour but la détermination du débit prépondérant (Schleiss, 2002).

Il faut noter que ces considérations ne s'appliquent qu'aux évacuateurs de crues en surface libre non vannés. Dans le cas des évacuateurs de crues vannés, le débit ne doit pas être directement lié à l'afflux du réservoir, mais être prescrit par les règles d'exploitation. Lorsque le niveau du réservoir est abaissé pendant les crues, les débits sortants sont supérieurs aux débits entrants du réservoir. Cela peut également être le cas pour les déversoirs à orifice sous pression.

3.5. DÉBIT DE DIMENSIONNEMENT DES ÉVACUATEURS DE CRUES ET L'ÉVALUATION DE L'AFFOUILLEMENT

Les évacuateurs de crues sont conçus pour une crue dite de dimensionnement ou de projet, généralement une crue de 1000 ans, et contrôlés pour une crue dite de sécurité, généralement entre une crue de 10 000 ans et une CMP. La question se pose de savoir pour quelle crue la profondeur de l'affouillement doit être évaluée et les mesures constructives de contrôle de l'affouillement basées.

Comme vu à la section 3.4.4, la profondeur ultime de l'affouillement ne se produira qu'après l'obtention de conditions stables dans la fosse, ce qui ne sera le cas que pour une certaine durée de fonctionnement du déversoir. Il est donc très conservateur de baser l'estimation de la profondeur de l'affouillement ou la conception des mesures de mitigation sur des crues de faible fréquence (CMP ou crue de 10 000 ans). Il est très peu probable que pendant la durée de vie du barrage, ces crues rares puissent produire une profondeur d'affouillement ultime.

Par conséquent, pour chaque crue avec une certaine période de retour, le débit prépondérant et la profondeur maximale de l'affouillement doivent d'abord être déterminés conformément au point 3.4.5. En outre, il convient de décider pour quelle période de retour la profondeur maximale d'affouillement atteinte pendant la durée de vie du barrage sera estimée. La probabilité d'occurrence d'une crue avec une période de retour donnée pendant la vie utile d'un barrage est la suivante :

$$r = 1-(1-1/n)^m \qquad (3)$$

où r est le risque ou la probabilité d'occurrence, n la période de retour de la crue (années) et m la durée de vie utile du barrage.

Dans la Figure 3.8, la probabilité d'occurrence des crues pour différentes durées de vie utile du barrage est illustrée. Il est possible de constater que la probabilité d'occurrence d'une crue de 200 ans pendant 200 ans d'exploitation est de 63 %, alors que pour une crue de 1000 ans elle n'est que de 20 %.

Pour l'évaluation de l'affouillement et les mesures de protection il semble raisonnable de choisir pour la serviabilité un débit de dimensionnement avec une probabilité d'occurrence d'environ 50 % pendant la durée de vie utile d'un barrage. Des débits de dimensionnement plus élevés avec une probabilité d'occurrence plus faible sont trop conservateurs. Néanmoins, la stabilité du barrage lui-même, y compris ses culées et ses fondations, ne doit pas être compromise jusqu'au niveau de crue de contrôle de sécurité du barrage. Dans des conditions de crue de contrôle de sécurité, des dommages sont acceptables, mais pas une rupture.

Il est à noter que dans le cas des évacuateurs vannés, des débits élevés peuvent être libérés à tout moment en ouvrant les vannes. De plus, pour les orifices de sortie à bas niveau, la vitesse d'impact du noyau du jet est presque indépendante du débit.

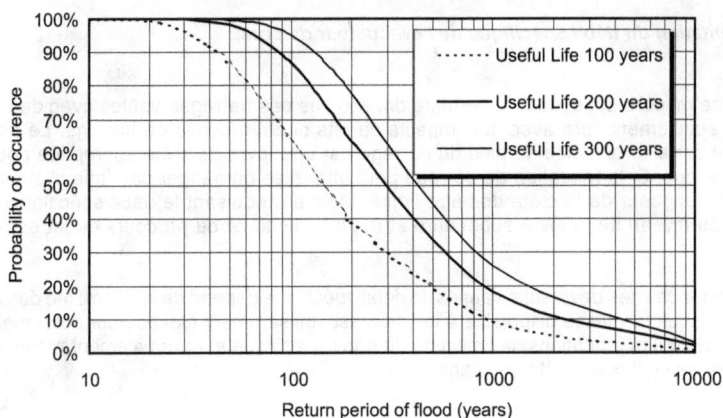

Fig. 3.8
Probabilité d'occurrence d'une crue avec une période de retour donnée pour différentes durées de vie utile du barrage (Schleiss, 2002).

3.6. MESURES DE CONTRÔLE DE L'AFFOUILLEMENT

3.6.1. Aperçu général

Afin d'éviter les dommages dus à l'affouillement, trois solutions actives peuvent être envisagées :

- éviter complètement la formation d'affouillement ;

- concevoir les évacuateurs de manière à ce que l'affouillement se produise loin des fondations et des culées du barrage ; et

- limiter l'étendue de l'affouillement.

Comme les dispositifs de lutte contre l'affouillement sont relativement coûteux, seules les deux dernières sont normalement économiquement réalisables (Ramos, 1982). En plus d'allonger autant que possible la zone d'impact des jets par une conception appropriée des évacuateurs, l'étendue de l'affouillement peut être influencée par les mesures suivantes :

- limitation du débit spécifique de l'évacuateur de crues ;

- l'aération forcée et le fractionnement des jets quittant la structure de l'évacuateur de crues ;

- l'augmentation de la profondeur d'eau aval par la création d'une digue qui peut être continue ou simplement réduire la section transversale de la rivière ; et

- pré-excavation du bassin de dissipation.

L'emplacement de l'affouillement dépend du type d'évacuateur choisi et de sa conception.

Pour complètement éviter l'affouillement, des mesures structurelles telles que des bassins de dissipation revêtus sont nécessaires. Outre les options actives, il est également possible de prévenir les dommages dus à l'affouillement par des mesures passives, par exemple en protégeant les culées de barrage avec des ancrages contre l'instabilité due à la formation d'affouillement.

3.6.2. Limitation du débit spécifique de l'évacuateur de crues

Cette mesure est surtout importante dans le cas des barrages-voûtes avec des déversoirs en crête à écoulement libre avec des impacts de jets plutôt proches du barrage. Le jet peut être guidé à une certaine distance du pied du barrage par une lèvre de crête appropriée pour un débit spécifique donné. Si la fondation du barrage peut être mise en danger par l'affouillement, le débit par unité de longueur de la crête doit être limité. Mais en réduisant le débit spécifique, la vitesse disponible au niveau de la lèvre supérieure et donc la distance de parcours du jet est également réduite.

Dans le cas des déversoirs vannés, le débit spécifique dépend de la géométrie des ouvertures de sortie. Puisque la vitesse disponible à la sortie est suffisamment élevée pour détourner le jet loin du barrage et de ses fondations, la limitation du débit spécifique est normalement moins importante que pour les déversoirs en crête non vannés.

3.6.3. Aération forcée et fractionnement des jets

Afin de fractionner et d'aérer les jets quittant un auget ou une lèvre de déversoir en crête, ceux-ci sont souvent équipés de blocs brise-charge, de séparateurs et de déflecteurs. De plus, les écoulements à grande vitesse dans le coursier des évacuateurs de crues sont normalement aérés par des rampes d'aération et des fentes le long du coursier. Toutes ces mesures augmenteront l'entraînement d'air mais aussi le développement des jets dans l'air et leur diffusion dans le bassin de dissipation, ce qui pourrait finalement réduire la capacité d'érosion des jets plongeants. Néanmoins, la quantité d'air entraîné est difficile à estimer. En raison des effets d'échelle, l'efficacité de ces mesures ne peut être vérifiée que qualitativement par des essais sur modèles hydrauliques.

Martins (1973) a suggéré une réduction de 25 % de la profondeur d'affouillement calculée dans le cas d'un entraînement d'air élevé et de 10 % pour un entraînement d'air intermédiaire. Mason (1989) a proposé une formule empirique tenant compte du rapport volumétrique air-eau β. L'équation empirique proposée basée sur les modèles de déversoirs ne dépend pas de la hauteur de chute H, car elle utilise une relation directe entre β et H telle que développée par Ervine (1976). La formule empirique est précise pour des données de modèle et semble donner une limite supérieure raisonnable de profondeur d'affouillement pour des conditions de prototype.

Néanmoins, dans le cas d'affouillement dans des roches fracturées de prototypes, l'air influence fortement la vitesse des coups de bélier et par conséquent les effets de résonance des ondes de pression dans les joints rocheux. Des recherches récentes révèlent que l'aération forcée peut réduire les fluctuations de pression à l'interface eau-rocher, mais peut augmenter les effets de résonance en raison de la célérité réduite des ondes de pression (Duarte *et al.*, 2015 et 2016). Ainsi, l'aération forcée ne réduit pas nécessairement le potentiel d'affouillement, mais peut même, dans certaines conditions, augmenter le risque de rupture des blocs rocheux.

3.6.4. Augmentation de la profondeur d'eau aval par la création d'une digue ou d'une contraction du cours d'eau

Une autre façon de contrôler l'affouillement par des jets est d'augmenter la profondeur d'eau aval en construisant un barrage en aval de la zone d'impact des jets ou en contractant la rivière avec un batardeau partiel ou bien une combinaison des deux. L'efficacité d'un coussin d'eau était souvent surestimée par le passé (Häusler, 1980). Pour une profondeur de bassin de dissipation inférieure à 4 à 6 fois le diamètre du jet, l'impact du noyau du jet (voir section 3.2.3) est normalement observé dans les bassins plats (Bollaert, 2002). Le noyau du jet est caractérisé par une vitesse constante et n'est pas influencé par l'état de la couche de cisaillement extérieure biphasée du jet d'impact. Les pressions sont également constantes avec de faibles fluctuations, qui ont une énergie spectrale importante aux très hautes fréquences (jusqu'à plusieurs 100 Hz). Pour des profondeurs d'eau supérieures à 4 à 6 fois le diamètre (ou l'épaisseur) du jet, l'impact d'un jet développé se produit sur des fonds plats avec une configuration de pression différente produite par une couche turbulente de cisaillement biphasée. Des fluctuations très importantes sont produites avec un contenu spectral élevé à des fréquences allant jusqu'à 100 Hz. Des fluctuations maximales ont été observées dans les bassins à fond plats pour des profondeurs d'eau de 5 à 8 fois le diamètre du jet (Bollaert et Schleiss, 2003). Des valeurs élevées subsistent jusqu'à une profondeur d'eau de 10 à 11 fois le diamètre du jet. De ces observations, il est possible de conclure que des coussins d'eau mesurant entre 5 et 11 fois le diamètre du jet peuvent générer des pressions dynamiques encore plus fortes sur des fonds plats que pour des profondeurs plus faibles. Par conséquent, de manière générale pour les bassins à fond plat, seuls les coussins d'eau plus profonds que 11 fois le diamètre du jet à l'impact ont un effet retardateur sur la formation de l'affouillement.

Cette tendance a également été confirmée par l'équation empirique d'affouillement de Martins (1973), qui donne une profondeur d'affouillement maximale pour une certaine profondeur d'eau aval.

Johnson (1967) a déjà constaté que des coussins d'eau trop petits sont encore pires que l'absence de coussin puisque le matériau peut être transporté plus facilement hors de la fosse.

Il faut noter que l'élévation du niveau aval à l'aide d'une digue peut interférer avec les vidanges de fond.

3.6.5. Pré-excavation de la fosse d'affouillement

En principe, la pré-excavation augmente la profondeur du bassin aval, et les mêmes remarques qu'à la section 3.6.4 s'appliquent par rapport au contrôle de l'affouillement.

En général, la pré-excavation de la zone d'affouillement peut également être appropriée pour le contrôle de l'érosion, en anticipant l'enlèvement des matériaux érodés et transportés par la rivière qui pourraient former des dépôts dangereux en aval, par exemple près de la sortie de la centrale. De tels dépôts pourraient augmenter le niveau à la sortie et réduire la production d'électricité. Ces problèmes ne devraient normalement pas être attendus, ou alors seraient gérables, si l'affouillement se produit à environ 200 m en amont de la sortie de la centrale.

La pré-excavation de la fosse d'affouillement est aussi souvent envisage lorsque la stabilité des berges est remise en question. Dans ce cas, l'excavation de la fosse doit être stabilisée par des ancrages ou autres mesures. Comme alternative, même dans de telles conditions, il est possible de ne pas procéder à une pré-excavation si les pentes de la vallée sont stabilisées par des mesures appropriées, de telle sorte que les pentes restent stables même après la formation de l'affouillement ultime.

Le choix du débit de dimensionnement pour une géométrie d'excavation de la fosse doit être fondé sur des considérations semblables à celles qui sont décrites à la section 3.3.5. En général, ce n'est pas économiquement réalisable de creuser plus loin que la profondeur de l'affouillement qui serait formée par des crues de 50 à 100 ans. Si l'instabilité des pentes de la vallée est considérée comme possible en raison de l'affouillement plus profond dû à l'augmentation des débits de l'évacuateur de crues, on peut utiliser des ancrages rocheux et des câbles précontraints, comme indiqué précédemment (Figure 3.9).

Au lieu de procéder à une excavation complète, la fissuration préalable de la roche de la fosse peut, dans certaines conditions, être suffisante. Cela rend la roche plus facilement érodable à certains endroits. La pré-fracturation peut donc être utilisée pour influencer la géométrie finale du bassin de plongée sans qu'il soit nécessaire de procéder à une excavation complète. Dans certains cas, par exemple lorsque l'affouillement affecte la production d'électricité, une excavation complète peut s'avérer plus appropriée.

La géométrie de la fosse pré-excavée doit être basée sur la géométrie d'affouillement naturelle attendue. Plusieurs auteurs ont proposé des formules empiriques pour l'estimation de la longueur (Martins, 1973 ; Kotoulas, 1967) et de la largeur (Martins, 1973) de la fosse d'affouillement. Amaniam et Urroz (1993) ont effectué un certain nombre d'essais sur un modèle d'évacuateur de crues terminé en auget avec un bassin de dissipation au lit de gravier afin d'élaborer des équations pour décrire la géométrie du trou créé par le jet entrant dans le bassin. Ils ont observé que la performance des fosses d'affouillement pré-excavées est meilleure lorsqu'elles sont proches de la fosse auto-excavée pour les mêmes paramètres d'écoulement.

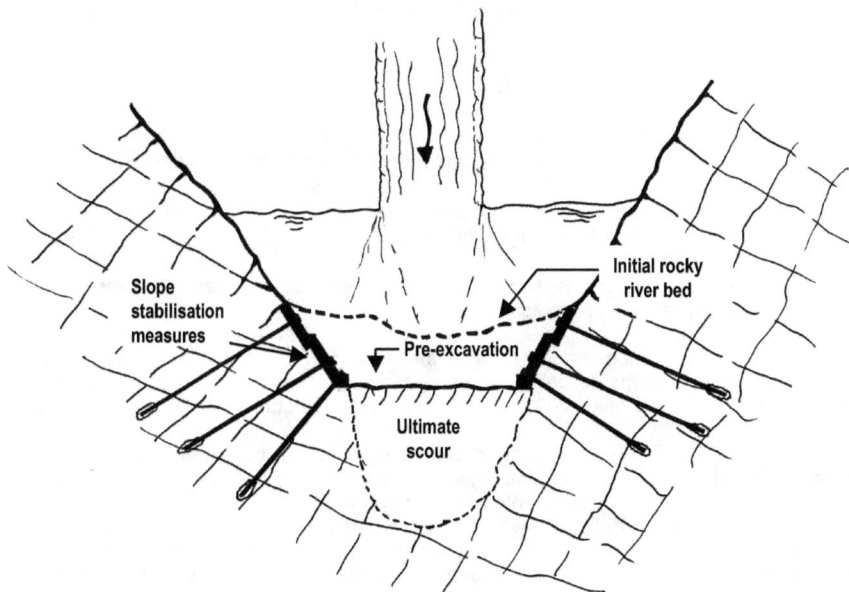

Fig. 3.9
Pré-excavation de la fosse pour une profondeur d'affouillement qui serait formée par une crue de 50 à 100 ans, et mesures de stabilisation des pentes en cas de formation d'affouillement ultime afin d'éviter des instabilités des berges (Schleiss, 2002).

Pour le cas bien connu de la fosse d'affouillement du barrage de Kariba, sur le Zambèze, qui a atteint une profondeur de plus de 80 m sous le niveau aval, une solution innovante a été mise au point en remodelant le trou existant avec une excavation principalement en direction aval (Noret et al., 2013). Ceci permet d'anticiper l'évolution future de la fosse de manière contrôlée et de réduire significativement les pressions dynamiques à l'interface eau-rocher dues à l'impact du jet. Avec des techniques de modélisation hybrides, il a pu être prouvé que le trou d'affouillement remodelé ne progressera plus à l'avenir (Bollaert et al., 2012).

3.6.6. Bassins de dissipation revêtus de béton

S'il est nécessaire qu'aucune formation d'affouillement ne puisse se produire dans le rocher à l'aval d'un barrage, le bassin de dissipation doit être protégé par un revêtement en béton. Étant donné que l'épaisseur du revêtement est limitée pour la mise en œuvre et pour des raisons économiques, des ancrages sont normalement nécessaire pour assurer la stabilité du revêtement face à la charge dynamique importante. Pour des charges dynamiques extrêmes, les ancrages peuvent également être précontraint (par exemple, le bassin de dissipation de la vidange de fond de Tarbela). De plus, la surface du revêtement doit être protégée avec une armature (grillages, fibres métalliques) et un béton à haute résistance à la traction et à l'abrasion. Les joints de construction doivent être fermés avec des waterstops efficaces. Enfin, la stabilité du revêtement contre la pression statique de soulèvement pendant l'assèchement du bassin doit être garantie par un système de drainage. Cela peut également limiter les pressions dynamiques de soulèvement dans le cas d'une fissuration limitée du revêtement ou d'une défaillance locale des waterstops.

Pour la conception des revêtements de bassins de dissipation, la séquence d'événements suivante doit être prise en compte (Fiorotto et Rinaldo, 1992 ; Fiorotto et Salandin, 2000) :

- des fluctuations de pressions peuvent endommager les joints d'étanchéité entre les dalles (joints de construction) ;

- à travers ces joints d'étanchéité, des valeurs de pression extrême peuvent se propager de la surface supérieure à la surface inférieure des dalles ;

- les différences de pression instantanées entre les surfaces supérieure et inférieure des dalles peuvent atteindre des valeurs élevées ; et

- la force résultante de la différence de pression peut dépasser le poids de la dalle et la résistance de l'ancrage.

De plus, la propagation des pressions dynamiques à travers les fissures du revêtement révèle la présence d'effets de coups de bélier, qui peuvent amplifier les pulsations de pression sous les dalles (Bollaert, 2002 ; Melo et al., 2006 ; Liu and Liu, 2007 ; Fiorotto et Caroni, 2007).

Étant donné qu'il n'est pas exclu que le revêtement du bassin soit fissuré ou que son étanchéité soit endommagée localement, l'hypothèse d'un revêtement absolument étanche et donc d'un soulèvement dynamique négligeable est une approche dangereuse. Si les hautes pressions dynamiques peuvent se propager à travers une petite fissure locale de la surface supérieure à la surface inférieure d'une dalle de béton, les pressions dynamiques sous cette dernière peuvent soulever la dalle localement et provoquer une défaillance progressive du revêtement du bassin de dissipation dans son ensemble. Ainsi, le concept d'un revêtement étanche de bassin en cas d'impact de jet à grande vitesse est aussi risqué qu'un concept de chaîne : la résistance du système est donnée par le maillon le plus faible, c'est-à-dire la perméabilité locale de la dalle en béton.

De plus, les pressions fluctuantes dues aux jets plongeants dans le bassin peuvent être élevées comparées au poids propre de la dalle, ce qui peut entraîner des vibrations de la dalle de béton et donc la formation de fissures dans le béton et une éventuelle rupture par fatigue des ancrages.

La fissuration d'une dalle de béton avant l'exploitation ne peut pas être totalement exclue, même en utilisant des joints de construction sophistiqués et des waterstops, en raison des effets de température lors du remplissage du bassin et de la déformation du sol. Par conséquent, les critères de conception du revêtement doivent tenir compte des éléments suivants :

- le cas de charge des pressions dynamiques de soulèvement du déversoir en fonctionnement ;

- l'armature de la dalle de béton doit être conçue pour limiter la largeur des fissures en fonction des modes de vibration dynamiques possibles (en fonction de la répartition et de la rigidité des ancrages) ; et

- les ancrages scellés et précontraints doivent être conçus pour résister à la fatigue.

Le système de drainage sous le revêtement est d'une très grande importance car il augmente considérablement la sécurité contre les pressions de soulèvement dynamique. Néanmoins, étant donné qu'une fissuration limitée du revêtement ne peut être exclue, comme déjà mentionné, les ondes de pression peuvent être transférées par les fissures dans le système de drainage. La réponse du système de drainage face à ces pressions dynamiques, avec une large gamme de fréquences à l'entrée d'une fissure, doit être contrôlée par une analyse transitoire, afin de s'assurer qu'aucune amplification des pressions dans le système ne se produit (Mahzari et al., 2002). Ces amplifications des pressions dans le système de drainage ou toute pression dynamique sous les dalles doivent être prises en compte dans la conception de la dalle et des ancrages (Mahzari et Schleiss, 2010).

En conclusion, il convient d'utiliser une approche "ceinture et brettelles" pour la conception d'un bassin de dissipation soumis l'impact d'un jet à grande vitesse, avec les recommandations suivantes :

- les dalles sont généreusement dimensionnées pour que les fluctuations instantanées localisées soient moyennées sur une grande surface et elles sont en général renforcées ;

- des waterstops sont prévues entre les dalles pour empêcher la pression dynamique de communiquer avec la face inférieure des dalles ;

- comme précaution supplémentaire, des barres d'ancrage sont utilisées et dimensionnées en fonction des différences de pression moyenne entre les différentes parties du tablier et de l'hypothèse que certains waterstops peuvent avoir cédé sur une certaine longueur ; et

- enfin, un drainage de dimensions généreuses est réalisé de manière à ce que les fluctuations de pression qui se produisent entre les dalles à travers de petits joints puissent être dissipées.

3.7. CONCLUSIONS

Bien que la compréhension physique du processus d'affouillement se soit considérablement améliorée au cours des 10 à 20 dernières années, son évaluation demeure un défi pour les concepteurs de barrages. Il existe maintenant des modèles d'affouillement qui tiennent compte des fluctuations de pression dans le bassin de dissipation et de la propagation des pressions transitoires dans les joints du massif rocheux, comme la *Comprehensive Scour Method*. Elle permet de modéliser le processus physique d'éjection des blocs rocheux dû au soulèvement dynamique et à l'influence des bulles d'air dans le processus. Néanmoins, une estimation raisonnable des fluctuations de pression à l'interface rocheuse sous différentes géométries d'affouillement est nécessaire, ce qui peut être obtenue par modélisation numérique dans une certaine mesure ou par modélisation physique, donc par une approche hybride.

Les paramètres de la roche, qui doivent être évalués pendant la campagne de reconnaissance de la fondation du barrage, ainsi que les conditions hydrologiques pendant la durée de vie considérée du barrage, influencent l'évolution de l'affouillement dans le temps. Il en résulte des incertitudes qui doivent être surmontées par le jugement de l'ingénieur.

Afin de vérifier et de calibrer des modèles d'affouillement complexes, des données de prototypes plus détaillées sur l'évolution de l'affouillement avec le suivi des débits sont encore nécessaires. Ces observations sont essentielles pour une évaluation continue de la sécurité d'un barrage et pour prédire l'évolution de l'affouillement.

3.8. REFERENCES

Amanian, N. and G. E. Urroz, (1993). 'Design of pre-excavated scour hole below flip bucket spillways'. *Proceedings of the ASCE International Symposium on Hydraulics*, San Francisco, USA

Annandale, G. W. (1995). 'Erodibility', *Journal of Hydraulic Research*, IAHR, vol. 33, N°4, pp. 471-494.

Annandale, G.W., R. J. Wittler, J. Ruff, and T. M. Lewis. (1998). 'Prototype validation of erodibility index for scour in fractured rock media', *Proceedings of the 1998 International Water Resources Engineering Conference*, Memphis, Tennessee, USA.

Annandale, G.W. (2006). *Scour Technology*, McGraw-Hill, New York, USA.

Armengou, J. (1991). *Disipacion de energia hidraulica a pie de presa en presas boveda*, PhD thesis, Universitat Politechnica de Catalunya, Barcelona, Spain.

Asadollahi, P., F. Tonon, M. Federspiel, M. and A. J. Schleiss, (2011). 'Prediction of rock block stability and scour depth in plunge pools' *Journal of Hydraulic Research*, vol. 49 (5), pp. 750-756.

Ballio, F., S. Franzetti, and M. G. Tanda, (1994). 'Pressure fluctuations induced by turbulent circular jets impinging on a flat plate', *Excerpta*, vol. 7.

Bellin, A. and V. Fiorotto, (1995). 'Direct dynamic force measurement on slabs in spillway stilling basins', *Journal of Hydraulic Engineering*, ASCE, vol. 121, N° HY 10, pp. 686-693.

Beltaos, S. and N. Rajaratnam, (1973). 'Plane turbulent impinging jets', *Journal of Hydraulic Research*, IAHR, vol. 11, N° 1, pp. 29-59.

Bin, A.K. (1984). 'Air entrainment by plunging liquid jets', Proceedings of IAHR Symposium on Scale Effects in Modelling Hydraulic Structures, Esslingen.

Bollaert, E. 2002. 'Transient water pressures in joints and formation of rock scour due to high-velocity jet impact'. *Communication N° 13 of the Laboratory of Hydraulic Constructions (LCH)*, EPFL, Lausanne. (Ed. By A. Schleiss).

Bollaert E. and A. J. Schleiss. (2003a). 'Scour of rock due to the impact of plunging high velocity jets - Part I: A state-of-the-art review'. *Journal of Hydraulic Research*, vol.41, No. 5, pp. 451-464.

Bollaert E. and A. J. Schleiss. (2003b). 'Scour of rock due to the impact of plunging high velocity jets. Part II: Experimental results of dynamic pressures at pool bottoms and in one- and two-dimensional closed end rock joints', *Journal of Hydraulic Research*, vol. 41, No. 5, pp. 465-480, 2003

Bollaert, E. (2004). 'A comprehensive model to evaluate scour formation in plunge pools'. *International Journal of Hydropower & Dams* 1, 94-101.

Bollaert E. and A. J. Schleiss. (2005). 'Physically based model for evaluation of rock scour due to high-velocity jet impact'. *Journal of Hydraulic Engineering* ASCE, March 2005, pp.153-165

Bollaert, E., R. Duarte, M. Pfister, A. J. Schleiss and D. Mazvidza. (2012). 'Physical and numerical model study investigating plunge pool scour at Kariba Dam'. *Proceedings of the 24th Congress of CIGB –ICOLD*, Kyoto, Japan, Q. 94 – R. 17, pp. 241-248

Duarte, R., Bollaert, E., Schleiss, A.J., Pinheiro, A. 2012. Dynamic pressures around a confined block impacted by plunging aerated high-velocity jets. Proceedings of the 2nd European IAHR Congress, 27-29 juin 2012, Munich, Allemagne, réf. B14.

Duarte R., Schleiss A.J., Pinheiro A. (2013) Dynamic pressure distribution around a fixed confined block impacted by plunging and aerated water jets. Proc. of the 35th IAHR World Congress, Chengdu, China, September 8-13, 2013.

Eggenberger, W. & Müller, R. 1944. Experimentelle und theoretische Untersuchungen über das Kokproblem. Mitteilungen Nr. 5 der VAW, ETH Zürich (in German)

Ervine, D.A. 1976. The entrainment of air in water, Water Power and Dam Construction, 28(12), pp. 27-30.

Ervine, D.A. 1998. Air entrainment in hydraulic structures: a review, Proceedings of the Institution of Civil Engineers Wat., Marit. & Energy, Vol. 130, pp. 142-153.

Ervine, D.A. & Falvey, H.R. 1987. Behavior of turbulent jets in the atmosphere and in plunge pools, Proceedings of the Institution of Civil Engineers, Part 2, Vol. 83, pp. 295-314.

Ervine, D.A. & Falvey, H.R. 1988. Aeration in jets and high velocity flows, Conference Proceedings, Model-Proto Correlation of Hydraulic Structures, P. Burgi, 1988, pp. 22-55.

Ervine, D.A.; Falvey, H.R.; Withers, W. 1997. Pressure fluctuations on plunge pool floors, Journal of Hydraulic Research, IAHR, Vol. 35, N°2.

Ervine, D.A.; McKeogh, E.; Elsawy, E.M. 1980. Effect of turbulence intensity on the rate of air entrainment by plunging water jets, Proceedings of the Inst. Civ. Eng., Part 2, pp. 425-445.

Fahlbusch, F. E. (1994). Scour in rock riverbeds downstream of large dams, Hydropower & Dams, pp. 30-32.

Federspiel, M.P.E.A., 2011. Response of an embedded block impacted by high-velocity jets. Communication 47 of Laboratory of Hydraulic Constructions, Lausanne, Switzerland, ISSN 161-1179 (Ed. Schleiss A.).

Federspiel, M., Bollaert, E., Schleiss, A. J. 2011. Dynamic response of a rock block in a plunge pool due to asymmetrical impact of a high-velocity jet. Proc. of 34th IAHR World Congress, 26 June - 1st July 2011, Brisbane, Australia, CD-Rom, ISBN 978-0-85825-868-6, pp. 2404-2411

Fiorotto, V. & Rinaldo, A. 1992. Fluctuating uplift and lining design in spillway stilling basins, Journal of Hydraulic Engineering, ASCE, Vol. 118, HY4.

Fiorotto, V. & Salandin, P. 2000. Design of anchored slabs in spillway stilling basins, Journal of Hydraulic Engineering, ASCE, Vol. 126, N° 7, pp. 502-512.

Fiorotto, V. & Caroni, E. 2007. Discussion of 'Forces on plunge pool slabs: influence of joints location and width' by J. F. Melo, A. N. Pinheiro, and C. M. Ramos, J. Hydraul. Eng., 133(10), 1182-1184.

Franzetti, S. & Tanda, M.G. 1984. Getti deviati a simmetria assiale, Report of Istituto di Idraulica e Costruzioni Idrauliche, Politecnico di Milano.

Franzetti, S. & Tanda, M.G. 1987. Analysis of turbulent pressure fluctuation caused by a circular impinging jet, International Symposium on New Technology in Model Testing in Hydraulic Research, India, pp. 85-91.

Gerodetti, M. 1982. Auskolkung eines felsigen Flussbettes (Modellversuche mit bindigen Materialen zur Simulation des Felsens). Arbeitsheft N° 5, VAW, ETHZ, Zürich (in German)

Gao, J.; Liu, Z. and Guo, J. 2011. Newly achievements on dam hydraulic research in China. In: Valentine, EM (Editor); Apelt, CJ (Editor); Ball, J (Editor); Chanson, H (Editor); Cox, R (Editor); Ettema, R (Editor); Kuczera, G (Editor); Lambert, M (Editor); Melville, BW (Editor); Sargison, JE (Editor). Proceedings of the 34th World Congress of the International Association for Hydro- Environment Research and Engineering (IAHR): 2436-2443.

Gunko, F.G.; Burkov, A.F.; Isachenko, N.B.; Rubinstein, G.L.; Soloviova, A.G. ; Yuditskii, G.A. 1965. Research on the Hydraulic Regime and Local Scour of River Bed Below Spillways of High-Head Dams, 11th Congress of the I.A.H.R., Leningrad.

Hartung, F. & Häusler, E. 1973. Scours, stilling basins and downstream protection under free overfall jets at dams, Proceedings of the 11th Congress on Large Dams, Madrid, pp.39-56.

Häusler, E. 1980. Zum Kolkproblem bei Hochwasser-Entlastungsanlagen an Talsperren mit freiem Überfall. Wasserwirtschaft 3.

Hoffmans, G.J.C.M. 1998. Jet scour in equilibrium phase, Journal of Hydraulic Engineering, ASCE, Vol. 124, N°4, pp. 430-437.

Johnson, G. 1967. The effect of entrained air in the scouring capacity of water jets, Proceedings of the 12th Congress of the I.A.H.R., Vol. 3, Fort Collins.

Johnson, G. 1977. Use of a weakly cohesive material for scale model scour tests in flood spillway design, Proceedings of the 17th Congress of the I.A.H.R., Vol. 4, Baden-Baden.

Kamenev, I.A., 1966. Alcance de jactos livres provenientes de descarregadores. (Trans. N° 487 L.N.E.C.) Gidrotekhnicheskoe Stroitel'stvo N° 3.

Kawakami, K. 1973. A study on the computation of horizontal distance of jet issued from ski-jump spillway. Trans. of the Japanese Society of Civil Engineers, Vol. 5.

Kirschke, D. 1974. Druckstossvorgänge in wassergefüllten Felsklüften, Veröffentlichungen des Inst. Für Boden und Felsmechanik, Univ. Karlsruhe, Heft 61 (in German)

Kotoulas, D. 1967. Das Kolkproblem unter besonderer Berücksichtigung der Faktoren "Zeit" und "Geschiebemischung" im Rahmen der Wildbachverbauung. Schweizerische Anstalt für das Forstliche Versuchswesen, Vol. 43, Heft 1 (in German).

Liu, P.Q. 1999. Mechanism of energy dissipation and hydraulic design for plunge pools downstream of large dams, China Institute of Water Resources and Hydropower Research, Beijing, China.

Liu, P.Q.; Dong, J.R.; Yu, C. 1998. Experimental investigation of fluctuating uplift on rock blocks at the bottom of the scour pool downstream of Three-Gorges spillway, Journal of Hydraulic Research, IAHR, Vol. 36, N°1, pp. 55-68.

Liu, P. Q. and Li, A. H. 2007. Model discussion on pressure fluctuation propagation within lining slab joints in stilling basins. J. Hydraul. Eng., 133(6), 618-624.

Lopardo, R.A. 1988. Stilling basin pressure fluctuations, Conference Proceedings, Model-Prototype Correlation of Hydraulic Structures, P. Burgi, pp. 56 – 73.

Machado, L.I. 1982. O sistema de dissipação de energia proposto para a Barragem de Xingo, Transactions of the International Symposium on the Layout of Dams in Narrow Gorges, ICOLD, Brazil.

Mahzari M; Arefi F.; Schleiss A. 2002. Dynamic response of the drainage system of a cracked plunge pool liner due to free falling jet impact. Proc. of Int. Workshop on Rock Scour due to falling high-velocity jets, Lausanne, Switzerland, 25-28 September, (Ed. Schleiss & Bollaert), pp 227 – 237.

Mahzari, M., Schleiss, A. J. 2010 Dynamic analysis of anchored concrete linings of plunge pools loaded by high velocity jet impacts issuing from dam spillways. Dam Engineering, Volume XX Issue 4, pp. 307-327

Martins, R. 1973. Contribution to the knowledge on the scour action of free jets on rocky river beds, Proceedings of the 11th Congress on Large Dams, Madrid, pp. 799-814.

Martins, R. 1977. Cinemática do jacto livre no âmbito das estruturas hidráulicas. Memória N° 486, L.N.E.C., Lisboa.

Manso, P. 2006. The influence of pool geometry and induced flow patterns in rock scour by high-velocity plunging jets. Communication 25, Laboratory of Hydraulic Constructions, Ecole Polytechnique Fédérale de Lausanne (EPFL), Lausanne, Switzerland (Ed. Schleiss A.).

Manso P., Bollaert E., Schleiss A.J. 2008. Evaluation of high-velocity plunging jet-issuing characteristics as a basis for plunge pool analysis. Journal of Hydraulic Research, Volume 46, No. 2, pp. 147-157, 2008

Manso, P. A., Bollaert, E. F. R., Schleiss, A. J. 2009. Influence of plunge pool geometry on high-velocity jet impact pressures and pressure propagation inside fissured rock media. Journal of Hydraulic Engineering, ASCE, Volume 135, Issue 10, pp. 783-792

Mason, P. J. 1984. Erosion of plunge pools downstream of dams due to the action of free-trajectory jets. Proc. Instn Civ. Engrs, Part 1, 76 May, 524-537.

Mason, P. J. & Arumugam, K. 1985. Free jet scour below dams and flip buckets, Journal of Hydraulic Engineering, ASCE, Vol. 111, N° 2, pp. 220-235.

Mason, P. J. 1989. Effects of air entrainment on plunge pool scour. Journal of Hydraulic Engineering, ASCE, Vol. 115, N° 3, pp. 385-399.

Mason P J. 1993. Practical guidelines for the design of flip buckets and plunge pools. International Water Power & Dam Construction, 45 (9/10), September/October 1993.

Mason P J. 2011. Plunge Pool Scour: An Update. Hydropower & Dams, Vol 18, (6), 2011.

May, R.W.P. & Willoughby, I.R. 1991. Impact pressures in plunge pool basins due to vertical falling jets, Report SR 242, HR Wallingford, UK.

Melo, J. F., Pinheiro, A. N. and Ramos, C. M. 2006. Forces on plunge pool slabs: influence of joints location and width." J. Hydraul. Eng., 132(1), 49-60.

Mirtskhulava, T.E.; Dolidze, I.V.; Magomeda, A.V. 1967. Mechanism and computation of local and general scour in non cohesive, cohesive soils and rock beds, Proceedings of the 12th IAHR Congress, Vol. 3, Fort Collins, pp. 169-176.

Noret, Ch., Girard J.-C., Munodawafa, M.C., Mazvidza, D.Z. 2013. Kariba dam on Zambezi river: stabilizing the natural plunge pool. La Houille Blanche, n°1, 2013, p. 34-41.

Otto, B. 1989. Scour potential of highly stressed sheet-jointed rocks under obliquely impinging plane jets, PhD thesis, James Cook University of North Queensland, Townsville.

Puertas, J. & Dolz, J. 1994. Criterios hidraulicos para el diseno de cuencos de disipacion de energia en presas boveda con vertido libre por coronacion, PhD thesis, Politechnical University of Catalunya, Barcelona, Spain.

Quintela, A.C. & Da Cruz, A.A. 1982. Cabora-Bassa dam spillway, conception, hydraulic model studies and prototype behaviour, Transactions of the International Symposium on the Layout of Dams in Narrow Gorges, ICOLD, Brazil.

Ramos, C.M. 1982. Energy dissipation on free jet spillways. Bases for its study in hydraulic models, Transactions of the International Symposium on the Layout of Dams in Narrow Gorges, ICOLD, Rio de Janeiro, Brazil, Vol. 1, pp. 263-268.

Schleiss, A.J. 2002. Scour evaluation in space and time – the challenge of dam designers, Proc. of Int. Workshop on Rock Scour due to falling high-velocity jets, Lausanne, Switzerland, 25-28 September, (Ed. Schleiss & Bollaert), pp 3-22.

Spurr, K. J. W. 1985. Energy approach to estimating scour downstream of a large dam, Water Power & Dam Construction, Vol. 37, N°11, pp. 81-89.

Tao, C.G.; JiYong, L.; Xingrong, L. 1985. Translation from Chinese by de Campos, J.A.P. Efeito do impacto, no leito do rio, da lâmina descarregada sobre uma barragem-abóbada, Laboratório Nacional de Engenharia Civil, Lisboa.

Toso, J. & Bowers, E.C. (1988). Extreme pressures in hydraulic jump stilling basin. Journal of Hydraulic Engineering, ASCE, Vol. 114, N° HY8, pp. 829-843.

Taraimovich, I.I. 1980. Calculation of local scour in rock foundations by high velocity flows, Hydrotechnical Construction N°8.

U.S.B.R. 1978. Hydraulic design of stilling basins and energy dissipators. Water Resources Technical Publication. Engineering Monograph N° 25, 4th Printing.

Van de Sande, E. & Smith, J.M. 1973. Surface entrainment of air by high velocity water jets, Chem. Engrg. Sci., 28, pp. 1161-1168.

Veronese, A. 1937. Erosion of a bed downstream from an outlet, Colorado A & M College, Fort Collins, United States.

Whittaker, J. & Schleiss, A. 1984. Scour related to energy dissipators for high head structures. Mitteilung Nr. 73, Versuchsanstalt für Wasserbau, Hydrologie und Glaziologie, ETH-Zurich: Zürich.

Xu-Duo-Ming 1983. Pressão no fundo de um canal devido ao choque de um jacto plano, e suas características de fluctuação, Translation from Chinese by J.A. Pinto de Campos, Lisboa.

Yuditski, G.A. 1963. Actual pressure on the channel bottom below ski-jump spillways, Izvestiya Vsesoyuznogo Nauchno – Issledovatel – Skogo Instituta Gidrotekhiki, Vol. 67, pp. 231-240.

Zvorykin, K.A., Kouznetsov, N.V., Akhmedov, T.K. 1975. Scour from rock bed by a jet spilling from a deflecting bucket of an overflow dam. 16th Congress of the IAHR, Vol.2, São Paulo.

4.   DÉVERSOIRS EN MARCHES D'ESCALIER

4.1.   INTRODUCTION

L'utilisation des déversoirs en marches d'escalier pour les barrages et les canaux n'est pas récente ; elle était fréquente dans les anciennes constructions hydrauliques et spécialement celles en maçonnerie (Chanson, 1994 and 2002). L'évolution de la technique du béton compacté au rouleau (BCR) pour les barrages a relancée son utilisation et la recherche dans le domaine. Tant par le passé qu'aujourd'hui, le déversoir en marches d'escalier est inhérent aux techniques de construction. Les recherches intensives qui ont été menées ces dernières années ont permis d'accroitre les connaissances sur les performances hydrauliques, les risques de cavitation, la durabilité et la dissipation d'énergie permettant une amélioration notable par rapport aux méthodes du passé.

Les déversoirs en marches d'escalier présentent deux avantages principaux ; premièrement une bonne adéquation avec les méthodes de construction des barrages en BCR dont les coffrages verticaux permettent la réalisation des marches et deuxièmement une dissipation d'énergie accrue de par l'effet des marches sur l'écoulement permettant ainsi de réduire les dimensions des ouvrages de dissipation aval.

Le taux de dissipation d'énergie élevé de ces ouvrages ouvre une nouvelle ère dans laquelle les déversoirs en marches d'escalier sont utilisés quelle que soit la technique de construction choisie.

Les déversoirs en marches d'escalier sont principalement utilisés pour les barrages poids en BCR (Figure 4.1), où le parement aval est utilisé comme évacuateur de crue et pour les barrages en remblai avec un risque hydrologique nécessitant la mise en place d'une protection anti-débordement sur la recharge aval (Figure 4.2).

Fig. 4.1
Barrage La Breña II (Espagne). Barrage poids en BCR. (ACUAES)

Fig. 4.2
Y Barrage Yellow River No.14 (USA). Evacuateur en BCR. (Golder Associates)

Récemment, les déversoirs en marches d'escaliers sont également combinés avec des barrages en remblai. Ils sont excavés et construits le long des appuis du barrage (Figure 4.3), ils peuvent être armé ou non et présenter une géométrie variable (hauteur des marches et pente) selon la topographie (Baumann *et al*. 2006).

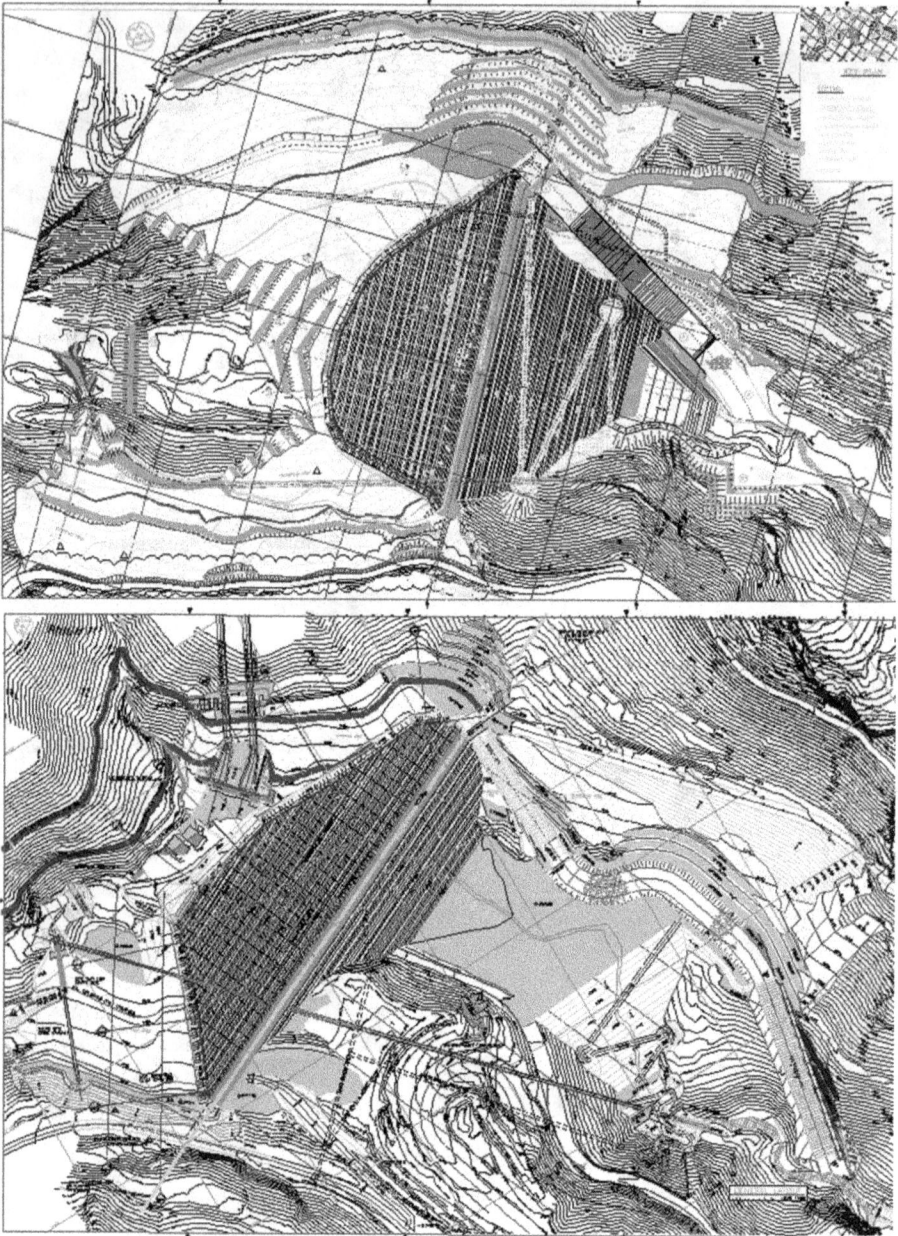

Fig. 4.3
Aménagement de pompage-turbinage de Siah Bishe (Iran). Barrage inférieur (en haut) et barrage supérieur (en bas). Déversoir en marches d'escalier le long des appuis du barrage en remblai. (Prof. A.J. Schleiss)

Il existe également de nombreux barrages plus anciens, toujours en service, présentant un déversoir en marches d'escalier en maçonnerie combiné à un barrage en remblai. Ces derniers nécessitent une maintenance et des contrôles réguliers pour s'assurer de leur bon fonctionnement.

Ce chapitre décrit les utilisations principales des déversoirs en marches d'escalier (section 4.2), leurs performances hydrauliques (section 4.3) ainsi que les aspects spécifiques qui requièrent une attention particulière lors de la conception (section 4.4).

4.2. PRINCIPALES UTILISATION DES DEVERSOIRS EN MARCHES D'ESCALIER

4.2.1. Déversoir en marches d'escalier pour les barrages en BCR

L'une des utilisations principales des déversoirs en marches d'escalier est pour les barrages poids en BCR. La technique de construction des barrages en BCR en elle-même offre une solution intéressante d'un point de vue économique et de délais de construction pour des aménagements hydrauliques importants. Pour les premiers ouvrages en BCR, le dimensionnement était conservatif avec l'utilisation de coursier lisse en béton vibré conventionnel. De nos jours et pour des raisons constructives, la plupart des barrages en BCR ont des parements aval en marches d'escalier. L'utilisation des déversoirs en marches d'escalier a augmentée mondialement de par la complémentarité aux méthodes de construction et la grande capacité de dissipation d'énergie.

Les recherches sur les performances hydrauliques ainsi que l'expérience acquise sur des prototypes existants ont permis de renforcer l'utilisation de ces évacuateurs. Sur la base des études et de l'expérience, le débit spécifique de dimensionnement a progressivement augmenté au fil des ans (Tableau 4.1). Les anciennes conceptions étaient limitées à des débits spécifiques relativement petits (jusqu'à 10 m³/s/m) (Sorensen, 1985) en raison des incertitudes sur les capacités hydrauliques du coursier et du risque de cavitation, alors que de nos jours des débits spécifiques plus importants sont utilisés. Selon Boes, (2012), des débits spécifiques de dimensionnement inférieurs à 30 m³/s/m sont usuels alors que ceux supérieurs requièrent une attention particulière.

Le schéma d'un déversoir en marches d'escalier est similaire à celui typique d'un barrage poids en béton conventionnel : structure de contrôle, coursier et ouvrage final. Cependant le dimensionnement de chaque partie doit être adaptée aux performances hydrauliques du déversoir en marches d'escalier : la transition crête-coursier doit être conçue avec soin afin d'éviter la formation de brume, la hauteur des parois latérales du coursier doit tenir compte du gonflement du flux, le risque de cavitation doit être analysé et de l'aération fournie si nécessaire, la structure finale doit tenir compte des caractéristiques de l'écoulement eau-air et de la dissipation d'énergie le long du coursier. Ces particularités sont discutées par la suite dans la section 4.4.

4.2.2. Déversoir de surverse en BCR sur les barrages en remblai

La protection contre les surverses des barrages en remblai par un évacuateur en BCR est une autre application importante des déversoirs en marches d'escaliers. Ils sont typiquement utilisés pour des barrages existants avec une sécurité vis-à-vis des crues limitée afin d'évacuer des crues importantes (Berga, 1995; PCA, 2002; FEMA, 2014; Toledo et al. 2015). Pour les petits barrages (hauteur < 15 m), cette alternative est en terme de coût intéressante en comparaison à d'autres mesures telle que l'élargissement de l'évacuateur existant, la construction d'un nouvel évacuateur séparé du corps du barrage ou la surélévation du barrage (Hansen, 2003; Bass et al. 2012). Les statistiques de 109 évacuateurs en BCR construits sur des barrages en remblai aux USA (FEMA, 2014), montrent que leur hauteur varie entre 5 et 20 m avec une hauteur maximale de 35 m. Le débit spécifique moyen est inférieur à 3 m³/s/m alors que celui maximal est d'environ 10 m³/s/m.

Table 4.1

Évolution du débit spécifique de dimensionnement des déversoirs en marches d'escalier des barrages poids

(Adapté et mis à jour à partir de Chanson (2002) and Matos (2003))

Barrage (Pays)	Année	Débit spécifique de dimensionnement (m³/s/m)	Observations
Monskvile (USA)	1986	9.3	
De Mist Kraal (Afrique du Sud)	1986	10.3	
Upper Stillwater (USA)	1987	11.4	
Les Olivettes (France)	1987	6.6	
Wolwedans (Afrique du Sud)	1990	12.4	
La Puebla de Cazalla (Espagne)	1991	9.0	
Shuidong (Chine)	1994	100	Déversoir en marches d'escalier non conventionnel. Utilisation de piles évasées pour dissiper l'énergie avant le coursier.
Rambla del Boquerón (Espagne)	1997	17.8	
Dona Francisca (Brésil)	2001	32	Débit spécifique maximal
Dachaosan (Chine)	2002	165	Déversori en marches d'escalier non conventionnel. Utilisation de piles évasées pour dissiper l'énergie avant le coursier.
Pedrogão (Portugal)	2005	39.9	
La Breña II (Espagne)	2008	22.3	Utilisation de coins arrondis dans la transition crête – coursier pour réduire la formation de brume.
Boguchany (Russie)	2012	44	Débit spécifique pour la capacité maximale. Utilisation d'une marche importante (3.6 m) à la fin des piles pour améliorer l'aération.
Enlarged Cotter (Australie)	2013	48	Débit spécifique du premier évacuateur. Utilisation de déflecteurs avec injection d'air.
De Hoop (Afrique du Sud)	2014	40	Débit spécifique maximal. Utilisation de saillies triangulaires sur le tiers supérieur pour améliorer l'aération.
El Zapotillo (Mexique)	En const.	38	Utilisation d'une grande marche pour assurer l'aération.

Ce type de déversoir est constitué de dalles de béton qui protègent la pente en aval et permettent le débordement. La technique de BCR est souvent utilisée pour construire ces dalles. Le BCR est typiquement mis en place sur le parement aval par couches horizontales superposées qui forment les marches du déversoir.

Bien que ces déversoirs soient également construits en BCR, les conditions de conception et de construction sont différentes de celles des barrages poids. Par conséquent, la taille et la finition des marches varient. La hauteur habituelle des marches est de 0.3 m et le parement peut être coffré ou non-coffré. Lorsque le béton n'est pas coffré, la face de la marche n'est pas compactée et doit être considérée comme du béton sacrificiel. De plus, les performances hydrauliques des marches non-coffrées vont différées de celles coffrées et il y aura moins de dissipation d'énergie (FEMA, 2014). La largeur des étapes de levage est déterminée par les exigences de construction et celles structurales. Une attention particulière doit être porté aux sous-pressions qui peuvent survenir sous les dalles, auquel cas un système de drainage doit être prévu. La largeur minimale de levage est d'environ 2.5 m en raison de la taille du rouleau vibrant. Des largeurs plus grandes sont nécessaires pour des débits spécifiques importants et pour des pentes plates, pour lesquelles une épaisseur de couche minimale de 0.6 m est recommandée (PCA, 2002).

Les particularités de conception de ces déversoirs, y compris la forme de la crête, la structure finale et les parois latérales sont présentées dans la section 4.4.

4.2.3. Déversoir en marches d'escalier le long des appuis des barrages en remblai

L'introduction de déversoirs en marches d'escalier sur des évacuateurs situés le long des appuis des barrages en remblai est relativement récente et pourrait être utilisée à l'avenir pour les projets dont le but est d'améliorer la dissipation d'énergie. Cette utilisation est issue des recherches menées sur les performances hydrauliques des déversoirs en marches d'escalier sur des pentes moyennes à plates (Ohtsu *et al.* 2001 and 2004; André, 2004; González and Chanson, 2007; Meireles and Matos, 2009).

Une particularité de ces évacuateurs est la nécessité d'adapter la pente des coursiers et la hauteur des marches à la topographie en place. Dans ces cas-là l'utilisation d'un canal de transition peut s'avérer utile pour maintenir des hauteurs de marches constantes sur des distances aussi longues que possible (Figure 4.3) (Baumann *et al.* 2006; Boes *et al.* 2015).

4.2.4. Déversoir en marches d'escalier en maçonnerie

Avant la période des bétons modernes, les barrages en terre étaient souvent munis d'évacuateur en marches d'escaliers en maçonnerie. La pente de ces évacuateurs était typiquement de l'ordre de 18.4°(3H/1V) à 5.7° (10H/1V). Des défaillances majeures de deux ouvrages en Angleterre (barrage de Boltby en 2005 et barrage de Ulley en 2007) ont conduit à des investigations sur l'origine de ces dernières (Winter *et al.* 2010). Les variations de pressions hydrodynamiques sur les parois latérales ont été considérées comme une vulnérabilité importante de ce type d'ouvrage de même que la nécessité de bien entretenir et maintenir la maçonnerie ainsi que le mortier. Cet aspect est abordé dans la section 4.3.7.

4.2.5. Structures en Gabion et d'autres types

Les déversoirs en marches d'escaliers peuvent également être vus sur des petits barrages en gabion. Ce type de barrage a quelques applications intéressantes parmi elles : barrages de rétention pour les matériaux solides et petits seuils pour la correction des pentes et la prévention des érosions. Les évacuateurs avec coursier en marches d'escalier peuvent également être utilisés pour les petits réservoirs agricoles et pour les barrages avec une faible retenue et un débit spécifique petit (jusqu'à 3 m³/s/m) (Peyras et al. 1991, 1992; Rice and Kadavy, 1997). Une application additionnelle est l'utilisation de gabions poreux pour la réduction du bruit dans les municipalités (Boes and Schmid, 2003).

Un autre type de déversoir en marches d'escalier est issu de l'utilisation de blocs en forme de coin. Les blocs préfabriqués sont assemblés en se chevauchant, formant une surface en gradins. Les blocs bénéficient de la succion produite sur la face verticale de la marche, pour améliorer son adhérence avec les fondations, ce qui permet d'obtenir des pièces plus fines. C'est une solution qui a été utilisée pour des petits barrages et des bassins agricoles avec des débits spécifiques allant jusqu'à 4 m³/s/m (Pravdivets and Bramley, 1989; Hewlett et al. 1997; FEMA, 2014; Morán and Toledo, 2014).

Les déversoirs en marches d'escalier directement excavés dans la roche doivent également être référencés. Ces ouvrages se retrouvent dans des anciennes constructions hydrauliques mais également dans des ouvrages récents pour des évacuateurs situés le long des appuis des barrages en remblai (Baumann *et al.* 2006; Lutz *et al.* 2015; Scarella and Pagliara, 2015). Le coursier est caractérisé par des grandes marches irrégulières de 2 à 3 m voir plus adaptées à la géologie du sol et à la morphologie (Chanson, 1994 and 2002; Felder and Chanson, 2011). En plus de l'étude des performances hydrauliques, il est important de vérifier la résistance du rocher à l'érosion et à l'affouillement car des pertes de matériaux pourraient modifier la géométrie et ainsi réduire l'énergie dissipée le long du coursier.

4.3.   COMPORTEMENT HYDRAULIQUE

La caractérisation hydraulique de l'écoulement est essentielle pour la conception de l'évacuateur. Le dimensionnement de n'importe quel évacuateur nécessite la détermination du type d'écoulement, des hauteurs d'eau, de la distribution des vitesses et de l'énergie résiduelle. Pour les coursiers en marches d'escalier il est également important de connaitre où et comment l'aération se développe et de définir le champ de pressions au-dessus des marches afin d'éviter la cavitation et les dommages.

4.3.1.   Types d'écoulement

L'hydraulique des coursiers en marches d'escalier est régi par plusieurs paramètres, les plus importants étant la hauteur critique (hc) (laquelle est fonction du débit spécifique (q) et de la section transversale), la hauteur de la marche (s) et la pente du coursier (Φ). On distingue trois types d'écoulement (Figure 4.4), qui dépendent principalement du rapport entre la profondeur d'eau et la hauteur de la marche. Lorsque ce rapport est petit (la hauteur de la marche est prédominante), l'écoulement est dit en nappe (*nappe flow*) et correspond à une succession de petites cascades plongeantes sur la partie horizontale de la marche aval (Essery and Horner, 1978). Lorsque le rapport est grand (la profondeur d'eau est prédominante), l'écoulement est dit en mousse (*skimming flow*). Il s'agit d'un écoulement cohérent le long du coursier, sans poches d'air dans les niches des marches, qui mousse au-dessus des coins des marches (Essery and Horner, 1978; Sorensen, 1985; Rajaratman, 1990). L'état intermédiaire, lorsque l'écoulement est instable partiellement en nappe et partiellement en mousse, est dit de transition (Ohtsu and Yasuda, 1997; Chanson and Toombes, 2004).

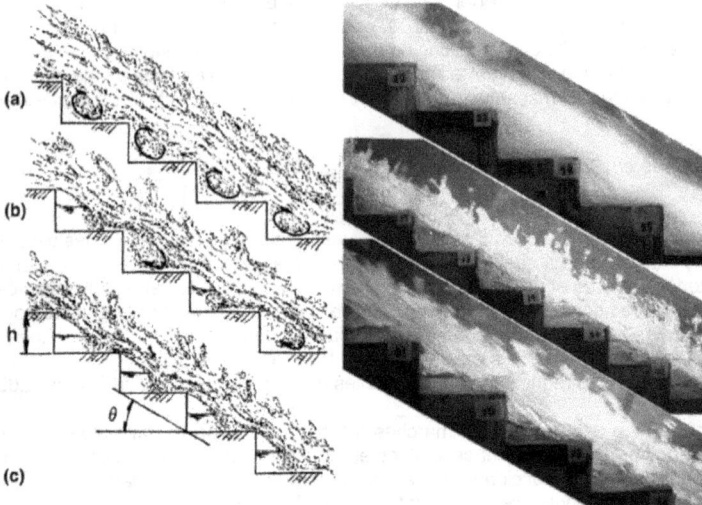

Fig. 4.4
Ecoulement en nappe (bas), écoulement en transition (centre) et écoulement en mousse (haut).
(Othsu *et al.*, 2001)

L'écoulement en nappe se produit lorsque la hauteur critique est petite en comparaison à celle de la marche pour une pente du coursier donné. Pour les déversoirs non vannés, c'est l'écoulement qui se développe au début et à la fin de l'évacuation des crues, et dans les périodes où le débit est relativement faible. Lorsque la profondeur d'eau est faible par rapport à la taille des marches, l'écoulement saute d'une marche à l'autre. Le jet frappe sur la face horizontale de la marche, provoquant la formation d'un petit ressaut hydraulique, avant de passer à la marche suivante, de sorte que l'énergie est progressivement dissipée à chaque saut. Le ressaut hydraulique peut être entièrement ou partiellement développé en fonction du débit et de la géométrie des marches

(Chanson, 1994 and 2002). La poche d'air délimitée par la face verticale de la marche et la nappe du jet tombant est une caractéristique de ce type d'écoulement.

L'écoulement en mousse se développe lorsque la hauteur critique est supérieure d'environ 80 % par rapport à celle des marches pour une pente de coursier donnée (Essery and Horner, 1978). Dans de telles conditions, un écoulement principal compact s'écoule sur les marches qui constituent un fond rugueux. La cavité définie par la lame d'eau inférieure de l'écoulement et les marches est complètement remplie d'eau. Les contraintes de cisaillement entre l'écoulement principal et l'eau qui remplit la marche conduisent à la formation d'un vortex de recirculation. Un transfert de moment continu alimente le mouvement circulaire du vortex; ce transfert est la cause principale de la dissipation d'énergie sur les déversoirs en marches d'escalier.

Ohtsu et al. (2004) ont étudié une grande diversité de pentes, de 5.7° (10H/1V) à 55° (0.7H/1V). Ils ont observé que l'écoulement en mousse peut être sous-classé en deux catégories. Pour les pentes comprises entre 19° (2,9H/1V) et 55° (0,7H/1V), la surface de l'eau est parallèle au fond virtuel défini par les bords des marches. Ce type d'écoulement est celui qui se développe dans la plupart des déversoirs en marches d'escalier, ceux situés sur les barrages poids et sur de nombreux barrages en remblai. Pour les pentes plus douces, de 5,7° (10H/1V) à 19° (2,9H/1V), la surface d'écoulement n'est pas complètement parallèle au fond virtuel, la face horizontale est tellement longue que l'influence de la gravité sur l'écoulement provoque une déflexion de l'écoulement et un impact avant le bord. Cet effet est reproduit par la surface de l'eau qui a un motif ondulé.

Une description détaillée des différents types et sous-types de flux se trouve dans Othsu et al (2004) et dans González et Chanson (2007).

4.3.2. Utilisation de l'écoulement en mousse pour le dimensionnement

Comme nous l'avons déjà mentionné, l'écoulement sur les marches d'escalier peut être de type nappe ou en mousse avec un régime de transition entre les deux. La détermination des limites et des conditions de chaque type de flux est un sujet de recherche éprouvé. Cette caractérisation reflète d'une grande importance car il est conseillé de s'assurer que le débit de dimensionnement soit distinctement dans un écoulement en nappe ou en mousse, mais pas en transition. Le régime de transition doit être évité lors de la conception, car il est caractérisé par d'importantes instabilités hydrodynamiques, ce qui peut mener à une brume élevée et à un mauvais rendement de l'évacuateur de crues (Chanson et Toombes, 2004).

Tant l'écoulement en nappe que celui en mousse ont été étudiés en profondeur, mais l'analyse a été davantage axée sur le débit en mousse puisqu'il est habituellement choisi pour le dimensionnement hydraulique. Selon ce critère, le débit de dimensionnement du déversoir et celui de la crue extrême sont considérés selon un écoulement en mousse. Pour les débits plus petits, qui se produisent pour des crues moins fréquentes et au début et à la fin de l'évacuation des crues, l'écoulement sera inévitablement de type nappe et éventuellement de type en transition. En conséquence, malgré l'ajustement des paramètres de dimensionnement pour les grands débits, il est également nécessaire de vérifier la bonne performance hydraulique pour les petits débits.

Généralement, le début de l'écoulement de transition devrait se produire pour une crue de période de retour de 100 ans. Ceci est raisonnable car pour des crues plus fréquentes, le régime d'écoulement en nappe avec une énergie de dissipation élevée se produira.

Le développement de l'un ou l'autre type d'écoulement dépend essentiellement du rapport entre la hauteur de la marche et la hauteur critique ainsi que de la pente du coursier. Le début de l'écoulement en mousse peut être estimé par :

(Boes and Hager, 2003b) Pour des pentes comprises entre 26°<Φ<55°.

$$\frac{h_c}{s} = 0.91 - 0.14 \tan\Phi$$

(Ohtsu et al. 2004) Pour des pentes comprises entre 5.7°<Φ<55°.

$$\frac{h_c}{s} = 0.857 \ (\tan \Phi)^{-0.1667}$$

(André, 2004) Pour des pentes comprises entre 18.6°<Φ<30°.

$$\frac{h_c}{s} = 0.939 \ (\tan \Phi)^{-0.364}$$

(Amador et al. 2006) Pour des barrages poids (Φ~51°).

$$\frac{h_c}{s} = 0.854 \ (\tan \Phi)^{-0.169}$$

Le tableau 4.2 montre le seuil du débit spécifique pour la formation de l'écoulement en mousse pour des pentes et des hauteurs de marches typiques, en utilisant les méthodes mentionnées.

Table 4.2
Débit spécifique (en m³/s/m) pour l'initiation de l'écoulement en mousse selon différentes configurations types de déversoirs en marches d'escalier

Méthode et domaine de validité*		Barrages poids			Barrages en remblai			
		(0.8H/1V)			(1.5H/1V)	(2H/1V)	(3H/1V)	(4H/1V)
		Φ =51.3°			Φ =33.7°	Φ =26.6°	Φ =18.4°	Φ =14.0°
		s=0.9 m	s=1.2 m	s=1.5 m	s=0.3 m	s=0.3 m	s=0.3 m	s=0.3 m
Boes and Hager (2003b)	26°< Φ <55°	1.7	2.6	3.6	0.4	0.4	---	---
Ohtsu et al. (2004)	5.7°< Φ <55°	2.0	3.1	4.3	0.4	0.5	0.5	0.6
André (2004)	18.6°< Φ <30°	---	---	---	---	0.7	0.9	---
Amador (2006)	Φ ~51°	2.0	3.1	4.3	---	---	---	---
* Selon les auteurs.								

Parmi les trois variables qui influencent le type d'écoulement, deux d'entre elles (c.-à-d. la hauteur des marches et la pente du coursier) sont habituellement difficiles à modifier vu qu'elles sont conditionnées par d'autres facteurs tels que la méthode de construction, le type de barrage ou la morphologie du sol. Par conséquent, pour ajuster le type d'écoulement, il est nécessaire de modifier la hauteur critique, qui à son tour peut être ajustée en changeant la longueur de la crête.

4.3.3. Emplacement et caractéristiques du point de début d'entrainement d'air

On distingue deux régions qui diffèrent dans l'aération pour le débit en mousse (figure 4.5). Dans la partie amont de l'évacuateur de crues, près de la crête, l'écoulement n'est pas aéré ; cette région est aussi connue sous le nom de région des eaux noires ou des eaux claires. En aval, généralement dans les parties moyenne et inférieure du coursier, l'écoulement est aéré ; cette région est aussi appelée région des eaux blanches. Dans la zone non aérée, le flux se caractérise par un aspect compact et transparent, sans bulles. L'écoulement d'approche de la crête de l'évacuateur de crues est laminaire. La couche limite se développe le long du coursier avec la crête comme origine. Lorsque la couche limite atteint la surface, la friction air-eau est suffisamment importante pour causer des irrégularités de surface, créant des poches d'air qui sont happées et rapidement distribuées dans l'écoulement (Amador et al. 2009). En aval de ce point de départ, le débit est aéré.

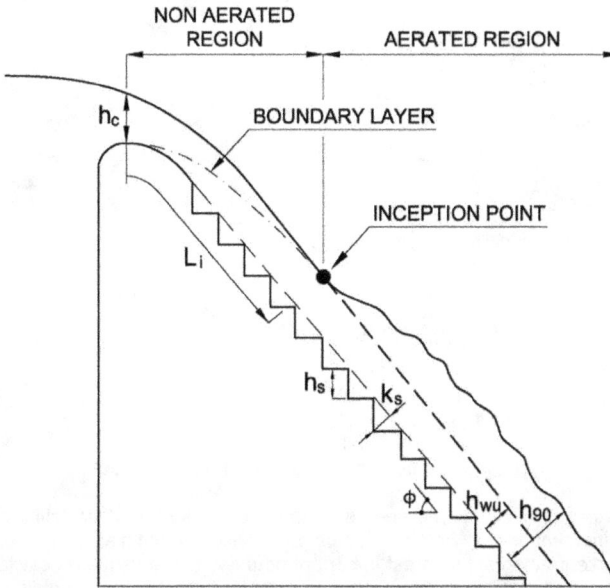

Fig. 4.5
Développement de l'écoulement en mousse. Régions non-aérée et aérée. Variables de dimensionnement.

Trois régions différentes peuvent être différenciées au sein de la zone aérée selon l'aération et le développement du débit (Ohtsu et al. 2001 ; Amador et al. 2006) (Figure 4.6). Dans la première, située immédiatement en aval du point de début d'entrainement d'air, l'air est rapidement distribué par l'effet de turbulence et le débit varie rapidement. Cette région dure jusqu'à ce que le processus d'aération se stabilise. Le débit dans la deuxième région varie graduellement jusqu'à devenir quasi uniforme, ce qui constitue la troisième région.

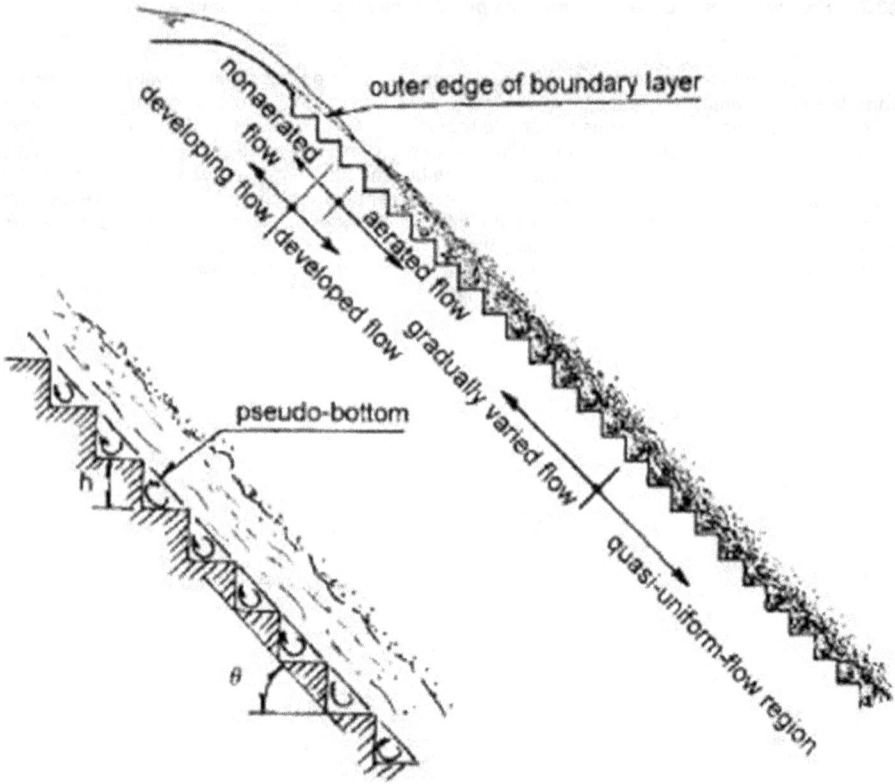

Fig. 4.6
Sous-régions de la zone aérée. (Ohtsu *et al.* 2001)

Comme expliqué ci-dessus, le point de début d'entrainement d'air définit où commence le processus d'entraînement de l'air. En amont de ce point, l'écoulement n'est pas aéré. L'emplacement du point de début d'entrainement d'air est important pour évaluer le risque de cavitation. La zone la plus sujette à la cavitation se situe autour du point de début d'entrainement d'air, car il s'agit de la zone non aérée avec les vitesses d'écoulement les plus élevées et donc les sous-pressions les plus élevées. En aval de cette zone, le risque de cavitation n'est pas significatif car le flux est très aéré.

Quelques équations utilisées pour déterminer l'emplacement du point de début d'entrainement d'air sont illustrées ci-dessous :

(Boes and Hager, 2003a) pour des pentes comprises entre 26°<Φ<55°:

$$L_i = \frac{5.90 \, h_c^{6/5}}{(\sin \Phi)^{7/5} \, s^{1/5}}$$

Avec L_i la distance depuis la crête jusqu'au point de début d'entrainement d'air.

L'analyse de cette formule montre l'influence des différents facteurs sur la région non aérée. Pour être précis, des hauteurs critiques plus grandes (signifiant des débits spécifiques plus élevés) conduisent à des distances plus importantes de l'écoulement non aéré, des pentes plus douces produisent des distances plus grandes (par exemple, les barrages en remblai ont des zones non aérées plus grandes que les barrages gravitaires), et des marches plus petites se traduisent également par des distances plus grandes. L'emplacement du point de début d'entrainement d'air est plus sensible à la hauteur critique et à la pente du coursier qu'à la hauteur des marches.

(Amador *et al.* 2009) pour les barrages poids ($\Phi\sim51°$):

$$\frac{L_i}{K_s} = 5.982\ F_*^{0.84}$$

avec K_s la rugosité du fond, c'est-à-dire la hauteur de la marche dans la direction de l'écoulement, et F_* le nombre de Froude de rugosité.

$$K_s = s\cos\Phi$$

$$F_* = \frac{q}{\sqrt{g\sin\Phi\ K_s^3}}$$

(Meireles *et al.* 2012) pour les barrages poids ($\Phi\sim53°$) :

$$\frac{L_i}{K_s} = 6.75\ F_*^{0.76}$$

(André, 2004) pour les déversoirs sur les barrages en remblai ($18.6°<\Phi<30°$) :

$$\frac{L_i}{K_s} = \frac{8}{\tan\Phi}\ F_{*\Phi}^{0.73}$$

Avec $F_{*\varphi}$ le nombre de Froude de rugosité corrigé pour tenir compte de l'effet de la pente :

$$F_{*\Phi} = \frac{q}{\sqrt{g\cos\Phi\ K_s^3}}$$

(Meireles and Matos, 2009) pour les déversoirs sur les barrages en remblai ($16°<\Phi<26.6°$) :

$$\frac{L_i}{K_s} = 5.25\ F_*^{0.95}$$

Hunt et Kadavy, (2013) ont analysé l'influence du nombre de Froude de rugosité, notant que pour les grands débits ($F^*>28$) le coursier a une performance différente, dans laquelle l'effet des marches diminue à mesure que la distance au point de début d'entrainement d'air à partir de la crête augmente. Ces auteurs ont proposé les équations suivantes pour déversoirs à crêtes épaisse en marches d'escalier au-dessus des barrages en remblai ($\Phi \leq 26.6°$):

$$\frac{L_i}{K_s} = 5.19\ F_*^{0.89} \quad 0.1<F_*<28$$

$$\frac{L_i}{K_s} = 7.48\ F_*^{0.78} \quad 28<F_*<10^5$$

Le Tableau 4.3 donne des valeurs L_i pour des configurations typiques de barrages (pente et hauteur des marches) ainsi que les débits spécifiques selon les méthodes référencées.

La hauteur d'eau au point de début d'entrainement d'air (h_i) at également été étudiée dans diverses recherches. Quelques formules sont proposées ici :

(Boes and Hager, 2003a) pour des pentes comprise entre $26°<\Phi<55°$:

$$\frac{h_i}{s} = 0.40\ F_{*1}^{0.6}$$

A mentionner que ces auteurs définisse le nombre de Froude de rugosité (F_{*1}) à partir de la hauteur de la marche (s) au lieu de la rugosité de fond (K_s).

$$F_{*1} = \frac{q}{\sqrt{g \sin \Phi s^3}}$$

Table 4.3

Distance du point de début d'entrainement d'air (en m) depuis la crête du déversoir, pour différentes configurations et débits spécifiques

Méthode et domaine de validité *		Barrages poids			Barrages en remblai					
		(0.8H/1V) Φ =51.3°			(2H/1V) Φ =26.6°			(3H/1V) Φ =18.4°		
		s=1.2 m			s=0.3 m			s=0.3 m		
		q=5 m³/s/m	q=10 m³/s/m	q=20 m³/s/m	q=1 m³/s/m	q=3 m³/s/m	q=5 m³/s/m	q=1 m³/s/m	q=3 m³/s/m	q=5 m³/s/m
Boes and Hager (2003b)	26°< Φ <55°	11.6	20.3	35.5	9.3	22.3	33.4	---	---	---
Amador et al. (2009)	Φ ~51°	10.6	18.9	33.9	---	---	---	---	---	---
Meireles et al. (2012)	Φ ~53°	11.0	18.6	31.5	---	---	---	---	---	---
André (2004)	18.6°< Φ <30°	---	---	---	8.3	18.4	26.8	12.1	26.9	39.0
Meireles and Matos (2009) (1)	16°< Φ <26.6° 1.9<F-<10	---	---	---	4.6	13.1	---	5.3	15.0	---

* Selon les auteurs.

(1) Cette formule a été développée avec un modèle de déversoir à seuil épais et l'emplacement du point de début d'entrainement d'air (Li) se réfère au bord aval du déversoir à seuil épais.

(Amador et al. 2009) pour les barrages poids (Φ~51°):

$$\frac{h_i}{K_s} = 0.385\ F_*^{0.58}$$

(Meireles and Matos, 2009) pour les déversoirs sur les barrages en remblai (16°<Φ<26.6°):

$$\frac{h_i}{K_s} = 0.28\ F_*^{0.68}$$

Selon Boes and Hager (2003a) la concentration moyenne d'air au point de début d'entrainement d'air $\left(\overline{C_i}\right)$ se calcule par

$$\overline{C_i} = 1.2 \cdot 10^{-3}\left(240° - \Phi\right)$$

4.3.4. Écoulement uniforme

Comme mentionné précédemment, la région aérée peut être divisée en trois zones en fonction de l'aération et du développement de l'écoulement (Figure 4.6). La première, située immédiatement en aval du point de début d'entrainement d'air, est l'endroit où le processus d'aération se développe et où la hauteur d'eau augmente considérablement en raison du processus d'aération très turbulent. En aval de cette région, l'écoulement est biphasique eau-air (Boes et Hager, 2003a) et varie graduellement vers un écoulement uniforme. La hauteur du mélange eau-air augmente progressivement jusqu'à atteindre une valeur constante dans des conditions d'écoulement uniforme. L'écoulement uniforme peut ne pas être atteint dans certains cas, lorsque le coursier n'est pas assez long.

(Boes and Hager, 2003b) définissent la hauteur relative du coursier (H_{chute}/h_c) pour atteindre un écoulement uniforme

$$\frac{H_{chute}}{h_c} \sim 24\,(\sin \Phi)^{2/3}$$

Selon cette équation, plus la pente est raide, plus le coursier doit être haut pour atteindre un écoulement uniforme. Par exemple, la hauteur relative du coursier nécessaire pour atteindre un écoulement uniforme sur un barrage poids typique (pente 0.8H/1V ; Φ= 51.34°) est H_{chute}/h_c = 20.35, alors que pour un barrage en remblai (pente 3H/1V ; Φ= 8.43°) elle est de 11.14. Par conséquent, les déversoirs de barrages poids nécessitent des hauteurs de coursier plus grandes pour atteindre un écoulement uniforme que les déversoirs au-dessus des barrages en remblai, pour un débit spécifique donné.

La hauteur uniforme de l'eau pure (h_{wu}) (équivalente à la hauteur d'eau sur un coursier lisse conventionnel) peut être approchée par :

(Boes and Hager, 2003b)

$$\frac{h_{wu}}{h_c} = 0.215\,(\sin \Phi)^{-1/3}$$

L'écoulement en mousse sur les déversoirs en marches d'escaliers est très aéré, ce qui donne des profondeurs plus grandes que celles correspondant à un coursier lisse. La turbulence importante dans les écoulements en mousse rend difficile la détermination précise de la profondeur du mélange air-eau. Pour faire face à cette limitation, la profondeur à laquelle la concentration locale d'air est de 90 % est considérée comme la profondeur d'écoulement caractéristique du mélange (h_{90}) (Chamani et Rajaratnam, 1999). Cette variable, qui tient compte du gonflement de l'écoulement, est souvent utilisée pour le dimensionnement des parois latérales moyennant une marge de sécurité pour éviter les déversements latéraux. Cette question est examinée plus en détail à la section 4.4.3. La hauteur du mélange eau-air h_{90} peut être déterminée de la façon suivante :

(Chanson, 1994 and 2002) hauteur caractéristique du mélange dans un écoulement uniforme (h_{90u}):

$$h_{90u} = h_c \sqrt[3]{\frac{f_e}{8\,(1-C_e)^3\,\sin \Phi}}$$

avec f_e le coefficient de frottement équivalent de Darcy pour le mélainge eau-air et C_e la concentration d'air à l'équilibre pour un écoulement uniforme:

$$\frac{f_e}{f} = 0.5\left(1 + \tanh\left(2.5\cdot\frac{0.5-C_e}{C_e\,(1-C_e)}\right)\right)$$

$$C_e = 0.9 \sin \Phi \text{ (for } \Phi<50°)$$

Avec f le coefficient de frottement de Darcy pour des écoulements non aérés qui s'obtient grâce à la formule suivante :

$$\frac{1}{f} = 1.42 \ln\left(\frac{D_h}{K_s}\right) - 1.25$$

avec D_h le diametre hydraulique $D_h = 4A/P$; A la section transversale et P le périmètre mouillé.

(Boes and Hager, 2003b) hauteur caractéristique du mélange dans un écoulement uniforme (h_{90u}):

$$\frac{h_{90u}}{s} = 0.5 \, F_{*1}^{(0.1\tan\Phi + 0.5)}$$

Une approximation de l'équation différentielle de la courbe de remous par Boes et Minor (2000) peut être utilisée pour calculer la hauteur d'eau caractéristique du mélange (h_{90}) dans la région des écoulements graduellement variés :

$$h_{90}(x) = 0.55\left(\frac{q^2 s}{g\sin\Phi}\right)^{1/4} \cdot \tanh\left(\frac{\sqrt{gs\sin\Phi}}{3q}(x - L_i)\right) + 0.42\left(\frac{q^{10}s^3}{(g\sin\Phi)^5}\right)^{1/18}$$

Avec x la distance le long du coursier depuis la crête.

Une analyse de l'hydraulique des déversoirs en marches d'escalier réalisée par UK Environment Agency (Winter et al. 2010) a mis en évidence que l'approche de Chanson est susceptible de donner les résultats les plus précis pour les petits coursiers associés aux barrages en remblai, tandis que les équations de Boes et Hager, ainsi que de Boes et Minor, sont applicables aux coursiers avec des pentes plus raides associés aux barrages en BCR et pour lesquels elles ont été développées à l'origine.

4.3.5. Dissipation d'énergie

Coefficient de frottement

La grande capacité de dissipation d'énergie des déversoirs en marches d'escalier est l'un de leur avantage principal. Une partie importante de l'énergie initiale est dissipée par le transfert de moments aux vortex qui se forment entre les marches et le fond virtuel. Ceci est fondamental pour le dimensionnement des structures de dissipation, c'est pourquoi plusieurs études et analyses ont été menées à ce sujet.

La dissipation d'énergie sur un coursier dépend du coefficient de frottement de la rugosité du fond (f_b). Dans le cas d'un coursier en marches d'escalier, celle-ci est liée à la rugosité du fond (K_s) et à l'espacement entre les bords des marches. Le coefficient de frottement peut être déterminé analytiquement à l'aide de la formule de Darcy-Weisbach en effectuant une correction de forme sur un canal rectangulaire ouvert et en déduisant l'effet latéral des parois lisses. En dehors de cette approche, il convient de noter qu'il est difficile d'ajuster une formule de coefficient de frottement pour les déversoirs en marches d'escalier en raison de la forte turbulence dans laquelle le flux biphasé se développe. Boes et Hager (2003b) ont proposé l'équation suivante pour déterminer le coefficient de frottement de la rugosité du fond pour les pentes comprises entre $19° \leq \Phi \leq 55°$.

$$f_b = \left(0.5 - 0.42\sin(2\Phi)\right)\left(\frac{K_s}{D_h}\right)^{0.2}$$

Pour $0.1 < K_s/D_h < 1.0$, cette équation devient :

$$\frac{1}{\sqrt{f_b}} = \frac{1}{\sqrt{0.5 - 0.42\sin(2\Phi)}}\left(1.0 - 0.25\log\left(\frac{K_s}{D_h}\right)\right)$$

Une autre caractéristique importante à prendre en compte lors de l'étude du coefficient de frottement est l'influence de l'aération. Les bulles d'air ont un effet lubrifiant, limitant les contraintes de cisaillement entre le courant principal et les vortex de recirculation, réduisant ainsi la dissipation d'énergie obtenue le long du coursier (Chanson, 1994). En tenant compte de cet effet, Boes et Hager (2003b) recommandent de calculer le diamètre hydraulique en utilisant la hauteur d'eau claire équivalente (h_w) au lieu de celle de l'écoulement caractéristique (h_{90}). Ce paramètre est appelé D_{hw}. Selon ces auteurs, l'utilisation de la hauteur d'eau de l'écoulement caractéristique peut conduire à des coefficients de frottement importants qui, à leur tour, peuvent produire une surestimation de la dissipation d'énergie et une conception non conservative de la structure finale.

Energie résiduelle

La dissipation d'énergie dépend du type d'écoulements. Mateos et Elviro (2000) ont constaté que le taux de dissipation d'énergie devient significatif lorsque l'écoulement est proche d'atteindre celui uniforme. Selon ces auteurs pour des hauteurs relatives du coursier $H_{chute}/h_c < 10$, la dissipation d'énergie est similaire à celle des coursiers lisses.

Boes and Hager, (2003b) proposent deux formules pour calculer l'énergie résiduelle (H_{res}) en fonction du type d'écoulement :

(Boes and Hager, 2003b) pour $H_{chute}/h_c < 15\text{-}20$ (écoulement uniforme non atteint):

$$\frac{H_{res}}{H_{max}} = \exp\left[\left(-0.045\left(\frac{K_s}{D_{hw}}\right)^{0.1}\left(\sin\Phi\right)^{-0.8}\right)\frac{H_{chute}}{h_c}\right]$$

Avec H_{max} la charge maximale – énergie potentielle dans la retenue– estimée par $H_{max} = H_{chute} + 1.5h_c$.

(Boes and Hager, 2003b) pour $H_{chute}/h_c \geq 15\text{-}20$ (écoulement uniforme atteint):

$$\frac{H_{res}}{H_{max}} = \frac{E}{\dfrac{H_{chute}}{h_c} + E}$$

Avec E égal à :

$$E = \left(\frac{f_b}{8\sin\Phi}\right)^{1/3}\cos\Phi + \frac{\alpha}{2}\left(\frac{f_b}{8\sin\Phi}\right)^{-2/3}$$

Avec α le coefficient de correction de l'énergie cinétique qui peut être approximé par $\alpha = 1.1$.

Selon André, (2004) l'énergie résiduelle à la fin d'un coursier de pente moyenne ($\Phi = 30°$) est donnée par:

$$\frac{H_{res}}{H_{max}} = \frac{1.5\tan\Phi}{10}\exp\left(25.26\frac{h_c}{N_s \cdot s}\right)$$

Avec N_s le nombre de marches.

Le Tableau 4.4 résume le potentiel de dissipation d'énergie pour différentes configurations de barrages (pente et hauteur de marches) pour des hauteurs de barrages et des débits spécifiques donnés.

Table 4.4

Energie résiduelle normalisée (H_{res}/H_{max}) et taux de dissipation (%), pour différentes configurations de déversoirs en marches d'escalier et pour différents débits spécifiques

Hauteur du coursier	Barrage poids (0.8H/1V) Φ =51.3°								
	s=0.9 m			s=1.2 m			s=1.5 m		
	q=5 m³/s/m	q=10 m³/s/m	q=20 m³/s/m	q=5 m³/s/m	q=10 m³/s/m	q=20 m³/s/m	q=5 m³/s/m	q=10 m³/s/m	q=20 m³/s/m
H_{chute}=30 m	0.32 / 68%	0.52 / 48%	0.68 / 32%	0.32 / 68%	0.51 / 49%	0.68 / 32%	0.31 / 69%	0.51 / 49%	0.67 / 33%
H_{chute}=50 m	0.22 / 78%	0.33 / 67%	0.52 / 48%	0.22 / 78%	0.32 / 68%	0.51 / 49%	0.21 / 79%	0.31 / 69%	0.50 / 50%
H_{chute}=100 m	0.13 / 87%	0.19 / 81%	0.28 / 72%	0.12 / 88%	0.19 / 81%	0.28 / 72%	0.12 / 88%	0.18 / 82%	0.28 / 72%
H_{chute}=150 m	0.09 / 91%	0.14 / 86%	0.21 / 79%	0.08 / 92%	0.13 / 87%	0.21 / 79%	0.08 / 92%	0.13 / 87%	0.20 / 80%

Méthode: Boes and Hager (2003b)

Les cellules grisées indiquent que l'écoulement uniforme a été atteint à la fin du coursier.

Hauteur du coursier	Barrages en remblai s=0.3 m							
	(3H/1V) Φ =18.4°			(2H/1V) Φ =26.6°			(2H/1V) Φ =26.6°	
	q=1 m³/s/m	q=3 m³/s/m	q=5 m³/s/m	q=1 m³/s/m	q=3 m³/s/m	q=5 m³/s/m	q=0.20 m³/s/m	q=0.30 m³/s/m
H_{chute}=15 m	0.09 / 91%	0.19 / 81%	0.35 / 65%	0.14 / 86%	0.26 / 74%	0.45 / 55%	0.10 / 90%	0.11 / 89%
H_{chute}=30 m	0.05 / 95%	0.10 / 90%	0.14 / 86%	0.07 / 93%	0.15 / 85%	0.21 / 79%	0.09 / 91%	0.09 / 91%
H_{chute}=50 m	0.03 / 97%	0.06 / 94%	0.09 / 81%	0.04 / 96%	0.10 / 90%	0.14 / 86%	0.08 / 92%	0.08 / 92%
Méthode	Boes and Hager (2003b)			Boes and Hager (2003b)			André (2004)	

Dans les calculs réalisés avec les équations de Boes and Hager (2003b), les cellules grisées indiquent que l'écoulement uniforme a été atteint à la fin du coursier.

La dissipation d'énergie peut être améliorée en introduisant des seuils d'extrémité, des blocs et des saillies. André et al (2003 et 2004) ont étudié l'effet des seuils et des blocs sur les déversoirs en marches d'escalier et ont conclu que ces derniers peuvent augmenter la dissipation d'énergie de 5 à 8 % et réduire le risque de cavitation au début du coursier. Wright, (2010) a étudié l'utilisation des saillies triangulaires, obtenant des conclusions similaires. Ce type de bloc, qui a été utilisé dans le barrage De Hoop en Afrique du Sud, s'est avéré être une mesure efficace pour augmenter le coefficient de frottement et pour améliorer l'aération et la dissipation d'énergie.

4.3.6. Champs de pression sur les marches

Une autre thématique importante concernant l'hydraulique des déversoirs en marches d'escalier est celle des champs de pressions hydrodynamiques. Il est important d'identifier les zones présentant un risque de cavitation. Les travaux d'André et al (2004) ; Amador et al (2006) ; Sánchez-Juny et al (2007) et Amador et al (2009) décrivent en détail le développement et les fluctuations du champ de pression sur les faces horizontales et verticales des marches, et fournissent des équations pour leur calcul dans les barrages poids, tant sur les faces horizontales que verticales. Deux régions doivent être distinguées pour décrire le champ de pression ; soit la zone externe située près du bord de la marche, qui est influencée par le courant principal, et la partie interne située autour du coin intérieur, qui est influencée le plus par les vortex de recirculation.

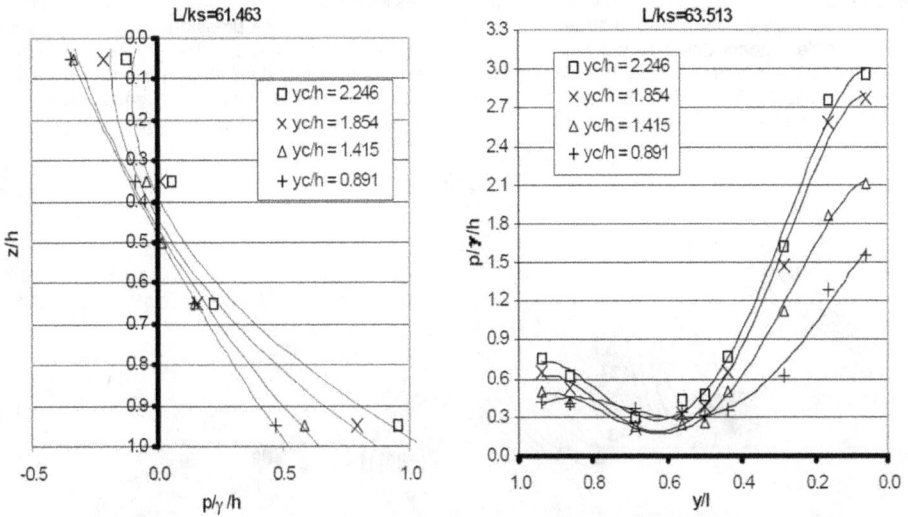

Fig. 4.7
Profils des pressions moyennes sur les faces horizontale et verticale de la marche. (Sánchez-Juny et al. 2007)

Sur la face horizontale, la pression maximale est située dans la zone externe en raison de l'impact de l'écoulement principal qui est influencé par l'action gravitationnelle, ce qui entraîne une diminution de la pression vers l'intérieur de la marche (Figure 4.7). Sur la face verticale, la pression minimale est située dans la zone externe, en raison de l'effet de traînée causé par le courant principal. La pression a un gradient positif jusqu'au coin intérieur, où sa valeur est similaire à celle de la face horizontale (Figure 4.7). La fluctuation sur les deux faces est très élevée et plus prononcée près des bords de la marche et pour des débits plus élevés.

En ce qui concerne l'évolution du champ de pression le long de l'évacuateur de crues, les chercheurs s'accordent à dire que la zone la plus délicate est celle située dans la région non aérée, alors que la région aérée n'est pas sujette aux dommages par cavitation (Zhang et al. 2012). Les pressions minimales et maximales ont été enregistrées autour du point de début d'entrainement d'air ; en aval de cette zone, la présence de bulles d'air dans le flux produit un effet de coussin qui réduit la pression et sa fluctuation (Amador et al. 2005). La recherche concernant la conception des aérateurs au début des déversoirs en marches d'escalier est en cours.

A noter que les mesures de pression effectuées sur les modèles réduits d'évacuateur de crues ne montrent pas les pressions à haute fréquence telles qu'elles se produisent sur les prototypes. Par conséquent, les résultats des modèles doivent être transposés avec soin aux prototypes et seulement si les pentes sont similaires en raison des effets d'échelle. Les effets d'échelle peuvent aussi avoir une incidence sur d'autres variables comme la concentration d'air et, à leur tour, influencer le coefficient de frottement et l'estimation de l'énergie résiduelle. Boes (2000) suggère d'utiliser une échelle de Froude minimale comprise entre 1:10 et 1:15 pour les modèles hydrauliques afin de minimiser ces effets d'échelle.

4.3.7. Champs de pression sur les parois latérales

Comme mentionné à la section 4.2.4, l'effondrement de deux anciens déversoirs en maçonnerie en marches d'escalier au Royaume-Uni, au barrage de Boltby en 2005 et au barrage d'Ulley en 2007, a donné lieu à un programme de recherche sur les effets hydrodynamiques sur les déversoirs en marches d'escalier (Winter et al. 2010 ; Winter, 2010). Ces études ont révélé qu'en présence d'écoulement en mousse, les centres des vortex horizontaux de chaque marche peuvent générer des pressions négatives élevées qui, à leur tour, agissent directement sur les zones locales associée de la paroi latérale. De plus, cela se propagera à proximité d'autres zones latérales soumises à des pressions positives élevées en raison de l'impact des marches (Figure 4.8). Un mauvais entretien et des fissures dans le mortier peuvent donc conduire à une contre-pression élevée de la maçonnerie dans les zones soumises à des pressions négatives externes. Le résultat étant des blocs "arrachés" du mur, une augmentation de la turbulence locale et, dans les cas extrêmes, un effondrement complet de la paroi latérale.

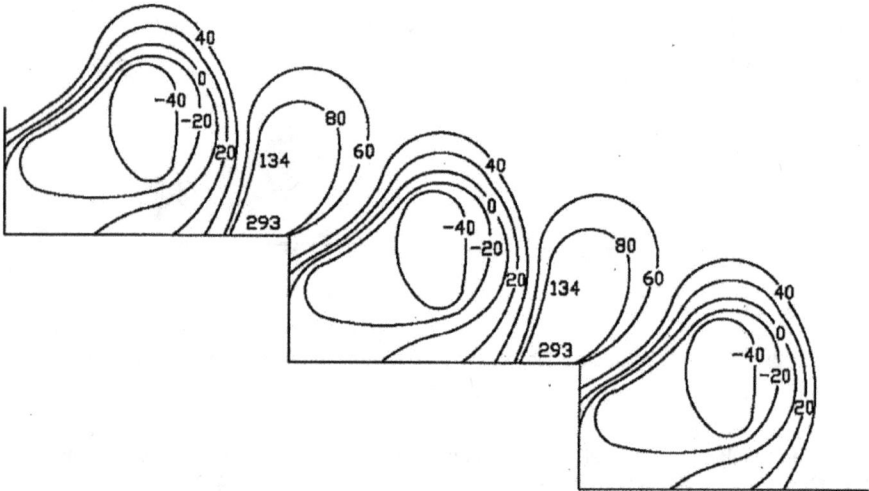

Fig. 4.8
Variation de pression hydrodynamique sur la paroi latérale d'un déversoir en marches d'escalier soumis à un écoulement en mousse (Winter et al., 2010)

4.4. CONCEPTS DE DIMENSIONNEMENT

4.4.1. Hauteur des marches

Les recherches existantes montrent que la relation optimale entre la hauteur de la marche et la dissipation d'énergie se produit lorsque $s/h_c \geq 0,3$ (Tozzi, 1992) ou $s/h_c \geq 0,25$ (Othsu et al. 2004). Cependant, il est important de noter que la hauteur des marches est principalement influencée par les techniques de construction actuelles. Pour les déversoirs en marches d'escalier en BCR, la hauteur des marches est un multiple de l'épaisseur de la couche de béton. L'épaisseur standard est de 0,3 m. Par le passé, des épaisseurs plus petites ont été utilisées sur les barrages pionniers en BCR, mais de nos jours, une épaisseur de 0,3 m prédomine. Par exemple : 23 des 27 barrages en BCR existants en Espagne ont été construits en utilisant des couches de 0,3 m d'épaisseur (de Cea et al. 2012). Dans les barrages poids, il est d'usage de déplacer le coffrage toutes les quatre couches, ce qui donne des hauteurs de marches de 1,2 m, alors que pour les déversoirs en BCR des barrages en remblai, la hauteur de marche la plus courante est de 0,3 m. Ceci est dû au fait que les marches plus petites s'adaptent mieux aux pentes plus douces (sinon la longueur des marches serait trop grande et du béton gaspillé) et parce que les marches ne sont pas toujours coffrées. Comme les hauteurs des marches sont largement influencées par la technologie de construction, l'épaisseur de la couche et la hauteur des marches pourront légèrement changer à l'avenir en fonction de l'évolution du BCR.

Dans les nouveaux projets, où la pente des marches n'est pas directement dictée par la méthode de construction, comme pour les barrages poids en béton conventionnel, tel que le barrage de Boguchany en Russie (Bellendir et al. 2012), ou dans les déversoirs latéraux des barrages en remblai comme les barrages Siah Bishe Lower et Upper Dams en Iran (Baumann et al. 2006), les hauteurs des marches ne sont pas déterminées par la technique de construction et peuvent être ajustées de manière plus flexible pour maximiser la dissipation d'énergie.

Certains auteurs indiquent que la hauteur de la marche peut être ajustée pendant la phase de conception afin de s'assurer que le débit de dimensionnement soit évacué selon un écoulement en mousse. Il convient de souligner que la hauteur de marche n'est pas le paramètre le plus adéquat pour le faire, car elle dépend fortement de la méthode de construction. Comme indiqué à la section 4.3.2, le paramètre le plus souple est la hauteur critique, qui peut être modifiée en ajustant la longueur de la crête (si possible).

4.4.2. Forme de la crête et transition vers le coursier

La plupart des déversoirs des barrages poids ont une crête standard (comme WES ou Creager). Cette forme de crête convient très bien à la section transversale triangulaire typique et aux pentes des barrages poids. La transition entre la crête et le coursier de l'évacuateur de crues doit être conçue de manière à prévenir la déviation de l'écoulement et l'impact sur les premières marches, ce qui entraîne par la suite la formation de brume et un mauvais rendement pour les petits débits, et un risque de cavitation pour les débits plus importants. Les travaux menés par CEDEX (Madrid) se sont largement concentrés sur cette question. Mateos et Elviro, (1995 et 2000) ont proposé une transition progressive qui suit le profil de la crête, au moyen de marches dont la hauteur et la longueur augmentent au fur et à mesure (figure 4.9).

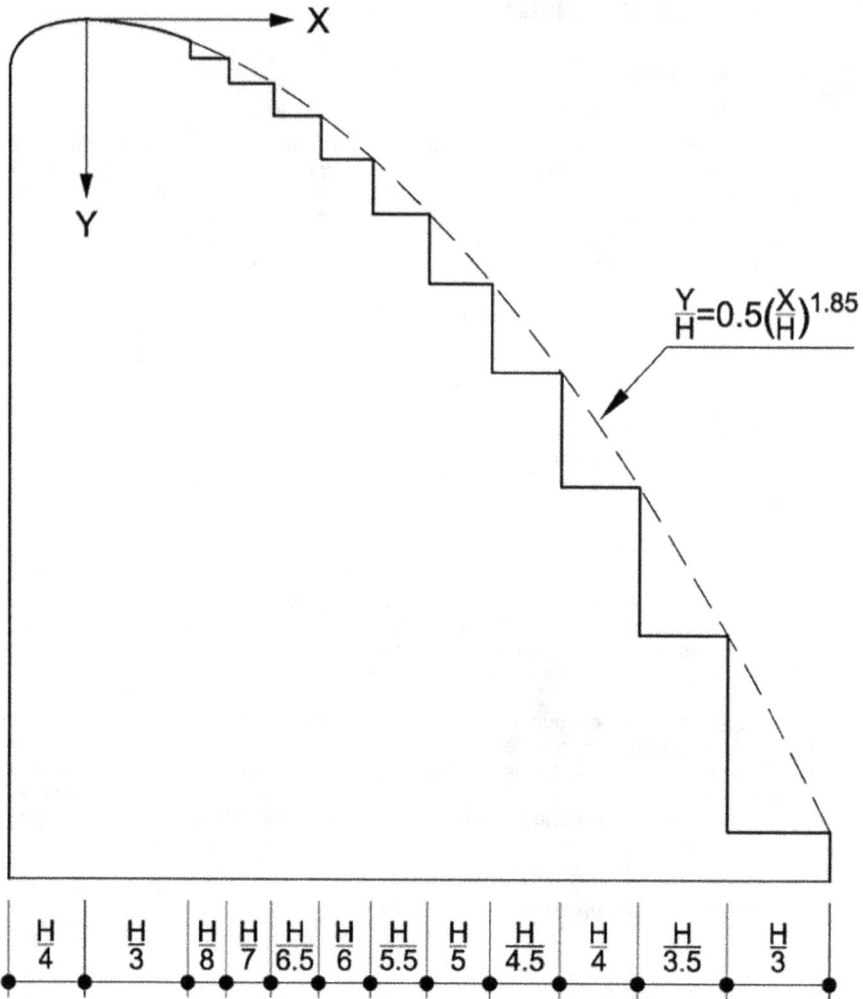

Fig. 4.9
Proposition de CEDEX pour la transition crête - coursier Avec H la charge de dimensionnement de l'évacuateur. (Mateos and Elviro, 1995)

Une autre alternative qui a été proposée pour réduire l'impact et la formation de brume est la modification de l'angle d'impact (Pfister et al. 2006a). Lorsque l'impact se produit, l'écoulement est dévié avec un angle à peu près symétrique par rapport à celui d'impact. Ainsi, le chanfreinage du bord des premières marches réduit la brume. Cette action peut également être reproduite au moyen d'un encart ou en arrondissant les bords des marches, comme cela a été fait sur les trois premières marches du barrage de La Breña II (figure 4.10) (Elviro et Balairón, 2008).

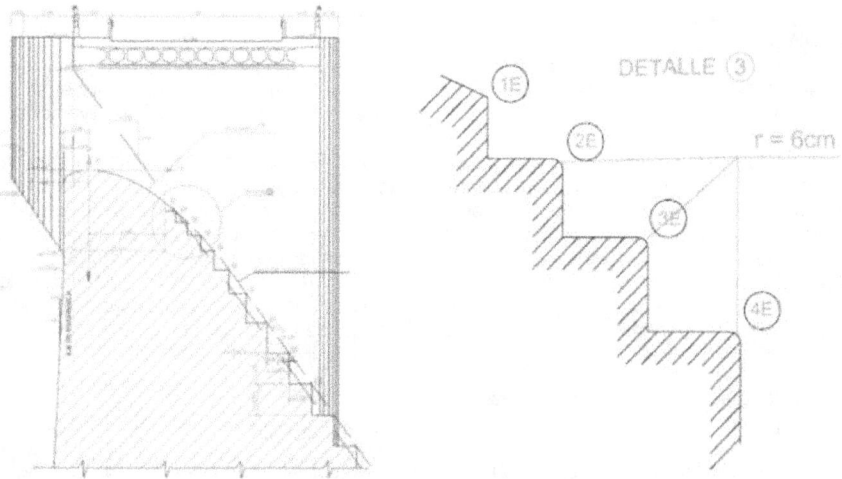

Fig. 4.10
Barrage La Breña II (Espagne). Utilisation d'arêtes arrondies pour réduire la brume. (Elviro and Balairón, 2008)

Les autres alternatives de crête pour les barrages poids en BCR sont les déversoirs à seuil épais, les déversoirs préfabriqués et les déversoirs en touches de piano. Ces solutions de rechange peu fréquentes peuvent être envisagées lorsque des conditions spéciales, telles que des restrictions économiques, temporelles ou hydrologiques, doivent être respectées. Dans ces cas, la performance de ces alternatives doit être vérifiée au moyen d'études spécifiques.

Dans le cas des déversoirs en marches d'escalier au-dessus des barrages en remblai, il est courant d'utiliser des déversoirs à seuil épais au-dessus du barrage (Hansen, 2003). Cette alternative s'adapte bien à la section transversale trapézoïdale des barrages en remblai (Figure 4.11). Bien que la capacité de ce type de crêtes soit plus petite, ceci n'est pas un problème car il est dans la plupart des cas rentable d'augmenter la largeur de l'évacuateur de crues. Généralement, ces déversoirs s'étendent sur l'ensemble du barrage. Si des capacités d'évacuation plus importantes sont nécessaires, on peut utiliser un déversoir à mince paroi, un déversoir de type labyrinthe ou un déversoir de type Ogee (Bass et al. 2012). La transition pour ces types de déversoirs n'a pas fait l'objet de recherches spécifiques. Dans les conceptions habituelles, le coursier en marches d'escalier commence directement après la crête (Figure 4.11). Les problèmes de saut et de brume qui surviennent sur les déversoirs des barrages poids ne font généralement pas l'objet d'une attention particulière, du fait que ces protections anti-débordement sont, dans de nombreux cas, des déversoirs d'urgence avec de faibles débits spécifiques.

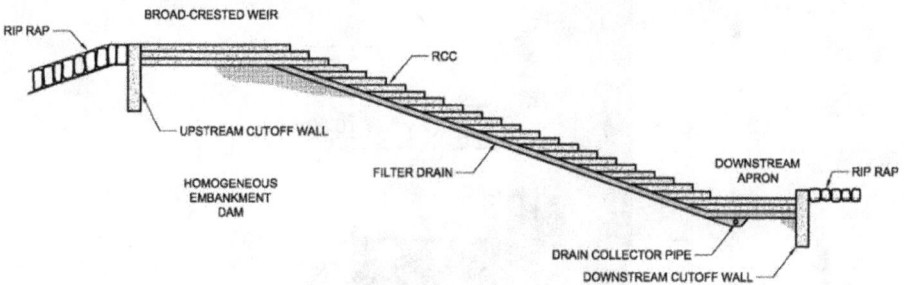

Fig.4.11
Section transversale typique d'un barrage en remblai avec un déversoir desurverse en BCR (Adapté de Bass et al. (2012))

Les déversoirs en marches d'escalier vanné ne sont pas courants car ils impliquent habituellement de plus grands débits spécifiques, de sorte que de nombreux déversoirs vannés sur les barrages en BCR ont des coursiers lisses. Des critères de conception spécifiques telles que l'agrandissement de la partie lisse pour couvrir entièrement la région de l'écoulement non aérée (Amador et al. 2006) ou l'aération artificielle (Guo et al. 2003 ; Boes, 2012) doivent être considérés pour les déversoirs vannés.

L'évacuateur de crues du barrage de Dachaosan (Chine) est un déversoir en marches d'escalier vanné qui a été dimensionné pour un très grand débit spécifique (165 m³/s/m). La partie amont de ce déversoir est lisse et l'entraînement d'air est assuré par des piliers évasés et un petit saut de ski. Les piliers évasés concentrent le flux, le basculant en position verticale, obtenant ainsi une aération latérale élevée due au frottement avec l'air. Le barrage de Dachaosan a une capacité de décharge de 6 173 m³/s et un débit spécifique de 93 m³/s/m attestant d'une bonne performance et d'aucun dommage significatif (Shen, 2003). Ce type de dissipateur d'énergie est une alternative intéressante pour les grands projets et a été utilisé avec succès pour d'autres déversoirs à grande hauteur de chute/vitesse en Chine.

4.4.3. Risque de cavitation et mesures d'aération

La zone sujette à la cavitation dans les déversoirs à marches d'escalier est celle située dans la zone d'écoulement non aérée. La zone la plus délicate est à proximité du point de début d'entrainement d'air, car le processus d'aération n'est pas complètement développé et la vitesse est plus grande qu'en amont. Boes et Hager (2003a) ont défini une limite supérieure pour la vitesse au point de début d'entrainement d'air de 20 m/s, basée sur les exigences de concentration d'air pour éviter la cavitation. Amador et al (2005) ont défini un seuil de vitesse inférieur au point de début d'entrainement d'air de 15 m/s, en se basant sur l'analyse du champ de pression. Frizell et al (2013) indiquent qu'aucun problème de cavitation ou dommage grave n'a été signalé jusqu'à présent sur les déversoirs en marches d'escalier ; par conséquent, ces limites peuvent être augmentées avec l'appui de recherches supplémentaires (utilisation de chambres à faible pression ambiante pour induire la cavitation) et aussi avec plus de données analysées à partir de prototypes. Il est toutefois conseillé d'inclure des mesures de prévention de la cavitation lorsque les limites mentionnées sont dépassées, ainsi que pour les débits spécifiques supérieures à 30 m³/s/m (Boes, 2012).

La meilleure mesure pour lutter contre la cavitation est l'aération. Les recherches menées au VAW de l'ETH Zurich se sont concentrées sur l'effet des aérateurs et déflecteurs de marche (Pfister et al. 2006b ; Schiess-Zamora et al. 2008). L'aérateur de marche est un conduit d'approvisionnement en air qui sort au droit de la première face verticale (première marche après la partie lisse en amont) (Figure 4.12). Les pressions négatives sur la partie supérieure de la face verticale contribuent à l'entrainement de l'air, qui est ensuite mélangé à l'écoulement. L'utilisation d'un aérateur de marche permet d'augmenter les débits spécifiques jusqu'à 40 m³/s/m (pour les barrages poids habituels et les hauteurs de marches de s=1,2 m). Pour les débits spécifiques plus importantes (q>40 m³/s/m), un second aérateur est nécessaire qui doit être situé en amont du point de début d'entrainement d'air.

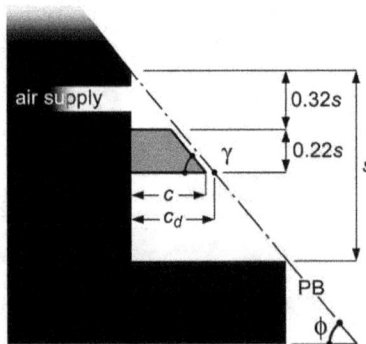

Fig. 4.12
Aérateur de marche pour l'approvisionnement en air. (Schiess-Zamora et al. 2008)

Une mesure plus efficace et plus rapide pour assurer l'aération est l'utilisation de déflecteurs. Le déflecteur est situé au début du coursier en marches d'escalier et produit un saut par-dessus les premières marches créant ainsi une grande cavité d'air sous le jet (Pfister et al. 2006b). L'air est entraîné dans l'écoulement lorsque le jet frappe le coursier en aval. Le déflecteur doit être complété par des conduits d'approvisionnement en air (Figure 4.13). La conception de ces dispositifs doit tenir compte du fait qu'une brume importante se produit autour du point d'impact. Une alternative à ce déflecteur consiste à dimensionner une marche plus haute dont le but est également de créer une cavité d'air sous l'écoulement.

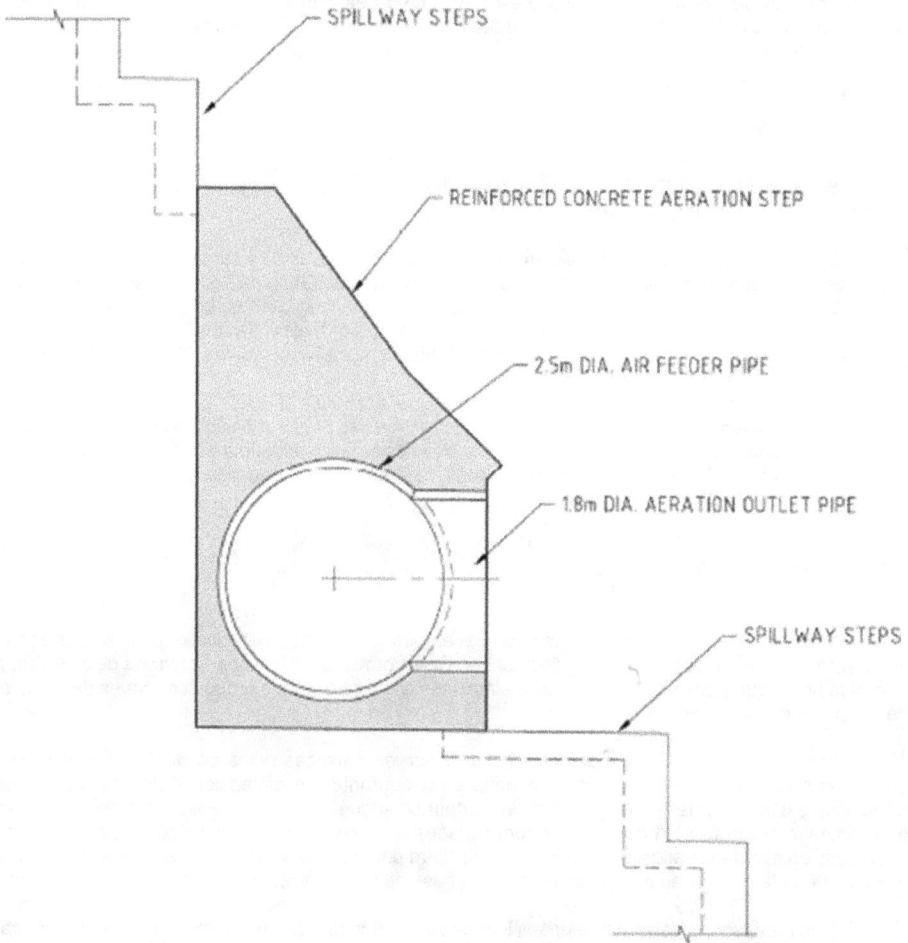

Fig. 4.13
Barrage de Enlarged Cotter (Australie). Déflecteur pour l'approvisionnement en air. (Willey *et al.*, 2010)

De plus, l'aération peut également être améliorée en utilisant des séparateurs et des saillies. Les travaux réalisés par Wright (2010) montrent que les saillies triangulaires (Robert's Splitters) situées sur la partie amont de l'évacuateur de crues contribuent à raccourcir la distance jusqu'au point de début d'entrainement d'air et à améliorer l'aération. Ainsi, pour une limite supérieure de vitesse de 20 m/s au point de début d'entrainement d'air, un déversoir équipé de saillies triangulaires sera efficace jusqu'à 40 m³/s/m.

4.4.4. Parois latérales

Le dimensionnement des parois latérales doit tenir compte de l'effet de gonflement de l'écoulement aéré. La hauteur d'eau de l'écoulement biphasé est supérieure à celle de l'écoulement équivalent sur un coursier lisse. Par conséquent, lors de l'évaluation de la hauteur de mur requise, un facteur de sécurité est généralement appliqué en fonction de la nature de la topographie environnante. Une attention particulière est nécessaire si le débordement de l'évacuateur de crues peut affecter un barrage en remblai.

Pour un déversoir en marches d'escaliers avec des parois latérales parallèle, la hauteur perpendiculaire au fond virtuel (h_t) peut être obtenue en appliquant un coefficient de bord libre (η) à la hauteur d'eau caractéristique.

$$h_t = \eta \cdot h_{90}$$

Boes and Hager, (2003b) suggère l'utilisation d'un coefficient de η =1.2 pour les barrages poids et de η=1.5 pour les barrages en remblai.

Le calcul de la hauteur de la paroi latérale pour les déversoirs convergents nécessite une approche spécifique (Hunt et al. 2012), puisqu'il y a une concentration de l'écoulement près du mur. Ces types de déversoirs convergents se trouvent dans les vallées étroites et, le plus souvent, dans les barrages en remblai où le dispositif de protection contre le débordement de l'évacuateur de crues peut aussi déborder par-dessus les culées du barrage constituant un canal en marches d'escalier convergent.

La performance sans parois latérales est une approche inédite. Elle se caractérise par l'expansion latérale de l'écoulement amenant une réduction des débits spécifiques au pied du barrage. Il s'agit d'une solution rentable dans les cas où le débit spécifique est faible et où les fondations sont bonnes. Les études réalisées à l'UPC (Barcelone) montrent que la réduction des débits spécifiques se situe entre 50% et 70% (Estrella et al. 2012).

4.4.5. Structure finales

L'un des principaux avantages des déversoirs en marches d'escalier est la dissipation d'énergie qui peut être obtenue le long du coursier. Cela permet la conception d'ouvrage de dissipation aval plus petits que ceux d'un coursier lisse similaire, ce qui se traduit par des économies de coûts et de temps de construction.

Dans les déversoirs en marches d'escalier sur les barrages poids en BCR, l'utilisation d'un bassin de dissipation comme structure terminale est courante. Le dimensionnement du bassin de dissipation est basé sur la détermination de la hauteur conjuguée. Pour les déversoirs lisses, la vitesse et la hauteur d'eau au pied du coursier sont utilisées pour calculer la hauteur conjuguée. Pour les déversoirs en marches d'escalier, la hauteur d'eau claire équivalente et la vitesse terminale équivalente doivent être déterminées au pied du coursier sur la base de l'énergie résiduelle (Boes et Hager, 2003b).

Des précautions doivent être prises lors de l'utilisation de bassins de dissipation standard, car ceux-ci ont été développés pour des coursiers lisses. Un soin particulier doit être pris pour ceux qui utilisent des blocs et des seuils pour réduire la longueur du bassin. Les travaux de Frizell et al (2009) concernant la performance des bassins de dissipation USBR-Type III (Peterka, 1978) montrent que la modification de la distribution de la vitesse dans un coursier en marches réduit l'effet des blocs, avec pour résultat qu'une longueur de bassin similaire voir plus grande est nécessaire.

Dans le cas des déversoirs de débordement pour les barrages en remblai construits en BCR, les débits spécifiques (moyenne de 3 m³/s/m) et les hauteurs des barrages (moyenne de 15 m) sont plus petits que ceux des barrages poids en BCR, de sorte que les structures terminales requises sont plus simples. Dans de nombreux cas, un tapis de réception en aval est suffisant pour prévenir l'érosion et les bassins de dissipation ne sont utilisés que pour les plus grandes protections.

4.5. NOTATION

Les symboles suivants sont utilisés dans ce chapitre:

A	Section mouillée
C_e	Concentration d'air à l'équilibre pour un écoulement uniforme
$\overline{C_i}$	Concentration moyenne d'air au point de début d'entrainement d'air
D_h	Diamètre hydraulique ($D_H = 4A / P$)
D_{hw}	Diamètre hydraulique calculé à partir de la hauteur d'eau claire équivalente
E	Paramètre utilisé pour déterminer l'énergie résiduelle lorsque l'écoulement uniforme est atteint
F_*	Nombre de Froude de rugosité ($F_* = q / (g \cdot \sin \Phi \cdot K_s{}^3)^{1/2}$)
F_{*1}	Nombre de Froude de rugosité ($F_{*\Phi} = q / (g \cdot \cos \Phi \cdot s^3)^{1/2}$)
$F_{*\Phi}$	Nombre de Froude de rugosité corrigé de l'effet des pentes ($F_{*\Phi} = q / (g \cdot \cos \Phi \cdot K_s{}^3)^{1/2}$)
f	Coefficient de frottement de Darcy pour des écoulements non-aérés
f_b	Coefficient de frottement de la rugosité du fond
f_e	Coefficient de frottement de Darcy équivalent pour le mélange eau-ai
g	Accélération terrestre
H_{chute}	Hauteur du coursier
H_{max}	Charge maximum – énergie potentielle de l'eau dans la retenue
H_{res}	Energie résiduelle
h_c	Hauteur critique
h_i	Hauteur d'eau au point de début d'entrainement d'air
h_t	Hauteur des parois perpendiculaire au fond virtuel
h_w	Hauteur d'eau claire équivalente
h_{wu}	Hauteur d'eau Claire équivalente dans un écoulement uniforme
h_{90}	Hauteur d'eau caractéristique du mélange eau-air
h_{90u}	Hauteur d'eau caractéristique du mélange eau-air dans un écoulement uniforme
K_s	Rugosité de fond ($K_s = s \cdot \cos \Phi$)
L_i	Longueur entre la crête du déversoir et le point de début d'entrainement d'air
N_s	Nombre de marches
P	Périmètre mouillé

q	Débit spécifique
s	Hauteur de marche
x	Distance de l'écoulement sur le coursier depuis la crête
α	Coefficient de correction de l'énergie cinétique
η	Coefficent de bord libre
Φ	Pente du coursier

4.6. RÉFÉRENCES

Amador A., Sánchez-Juny M. and Dolz J. (2005). "Discussion of 'Two phase flow characteristics of stepped spillways' by R.M. Boes and W.H. Hager" *Journal of Hydraulic Engineering*, 131(5):421-423.

Amador A., Sánchez-Juny M. and Dolz J. (2006). "Diseño hidráulico de aliviaderos escalonados en presas de HCR" *Ingeniería del Agua*, 13(4):289-302. (in Spanish)

Amador A., Sánchez-Juny M. and Dolz J. (2009). "Developing flow region and pressure fluctuations on steeply sloping stepped spillways" *Journal of Hydraulic Engineering*, 135(12):1092-1100.

André S. (2004). "High velocity aerated flows on stepped chutes with macro-roughness elements" *Communications du Laboratoire de Constructions Hydrauliques No. 20*, Ed. Schleiss A., EPFL, Lausanne (Switzerland).

André S., Manso P., Schleiss A.J. and Boillat J.L. (2003). "Hydraulic and stability criteria for the rehabilitation of appurtenant spillway structures by alternative macro-roughness concrete linnings" *Proc. 21st ICOLD Congress*, Montreal (Canada), Q.82, R.6.

André S., Matos J., Boillat J.L. and Schleiss A.J. (2004). "Energy dissipation and hydrodynamic forces of aerated flow over macro-roughness linings for overtopped embankment dams" *Proc. Intl. Conference on Hydraulics of dams & River structures*, Tehran (Iran), 189-196.

Bass R.P., Fitzgerald T. and Hansen K.D. (2012). "Lesson learned - More than 100 RCC overtopping spillways in the United States" *Proc. 6th Intl. Symposium on Roller Compacted Concrete Dams*, Zaragoza (Spain).

Baumann A., Arefi F. and Scheiss A. (2006). "Design of two stepped spillways for a pumped storage scheme in Iran" *Proc. HYDRO 2006 Conference*, Porto Carras (Greece), CD-ROM.

Bellendir E.N., Volynchikov A.N. and Sudolskiy G.N. (2012). "Boguchany HPP additional spillway: Necessity of construction and peculiar features of design" *Proc. 24th ICOLD Congress*, Kyoto (Japan), Q.94, R.12.

Berga L. (1995). "Hydrologic safety of existing embankment dams and RCC for overtopping protection" *Proc. 2nd Intl. Symposium on Roller Compacted Concrete Dams*, Santander (Spain), 639-652.

Boes R.M. (2000). "Scale effects in modelling two-phase stepped spillway flow", *Proc. Intl. Workshop Hydraulics of stepped spillways*, VAW-ETH Zurich, Eds. Minor, H.E., and Hager, W.H., Balkema, Rotterdam (The Netherlands), 53-60.

Boes R.M. (2012). "Guidelines on the design and hydraulic characteristics of stepped spillways" *Proc. 24th ICOLD Congress*, Kyoto (Japan), Q.94, R.15.

Boes R.M. and Hager W.H. (2003a). "Two-phase flow characteristics of stepped spillways" *Journal of Hydraulic Engineering*, 129(9):661-670.

Boes R.M. and Hager W.H. (2003b). "Hydraulic design of stepped spillways" *Journal of Hydraulic Engineering*, 129(9):671-679.

Boes R.M., Lutz N. and Lais A. (2015). "Upgrading spillway capacity at large, non-overtoppable embankment dams", *Proc. 25th ICOLD Congress*, Stavanger (Norway), Q.97, R.23.

Boes R.M. and Minor H.E. (2000). "Guidelines to the hydraulic design of stepped spillways" *Proc. Intl. Workshop Hydraulics of stepped spillways*, VAW-ETH Zurich, Eds. Minor, H.E., and Hager, W.H., Balkema, Rotterdam (The Netherlands), 163-170.

Boes R.M. and Schimd H. (2003). "Weir rehabilitation using gabions as a noise abatement option" *Proc. HYDRO 2003 Conference*, Cavtat (Croatia), CD-ROM.

Chanson H. (1994). Hydraulic design of stepped cascades, channels, weirs and spillways, Pergamon, Oxford (UK).

Chanson H. (2002), *The hydraulics of stepped chutes and spillways*, Balkema, Lisse (The Netherlands).

Chanson H. and Toombes L. (2004). "Hydraulics of stepped chutes: the transition flow" *Journal of Hydraulic Research*, 42(1):43-54.

Chamani M.R. and Rajaratmam N. (1999). "Characteristics of skimming flow over stepped spillways" *Journal of Hydraulic Engineering*, 125(4):361-368.

de Cea J.C., Ibáñez de Aldecoa R., Polimón J., Berga L. and Yagüe J. (2012). "30 years constructing RCC dams in Spain" *Proc. 6th Intl. Symposium on Roller Compacted Concrete Dams*, Zaragoza (Spain).

Elviro V. and Balairón L. (2008). "Recrecimiento de la presa de La Breña. Estudio en modelo reducido del aliviadero escalonado" *Proc. VIII Jornadas Españolas de presas*, SPANCOLD, Córdoba (Spain). (in Spanish)

Essery I.T.S. and Horner M.W. (1978). "The hydraulic design of stepped spillways" *CIRIA-Report 33*, London (UK).

Estrella S., Sánchez-Juny M., Pomares J., Dolz J., Ibáñez de Aldecoa R., Domínguez M., Rodríguez J. and Balairón L. (2012). "Recent trends in stepped spillways design: behaviour without sidewalls" *Proc. 24th ICOLD Congress*, Kyoto (Japan), Q.94, R.28.

Felder S. and Chanson H. (2011). "Energy dissipation down a stepped spillway with nonuniform step heights" *Journal of Hydraulic Engineering*, 137(11):1543-1548.

FEMA (2014). Technical Manual: Overtopping protection for dams FEMA P-1015, Federal Emergency Management Agency (USA).

Frizell K.W., Kubitschek J.P. and Matos J. (2009). "Stilling basin performance for stepped spillways of mild to steep slopes - Type III basins" *33rd IAHR Congress*, Vancouver (Canada).

Frizell K.W., Renna F.M. and Matos J. (2013). "Cavitation potential of flow on stepped spillways" *Journal of Hydraulic Engineering*, 139(6):630-636.

González C.A. and Chanson H. (2007). "Diseño hidráulico de vertedores escalonados con pendientes moderadas: Metodología basada en un estudio experimental" *Ingeniería Hidráulica en México*, 22(2):5-20. (in Spanish)

Guo J., Liu Z., Liu J. and Lu Y. (2003). "Field observation on the RCC stepped spillways with the flaring pier gate on the Dachaoshan project." *Proc. 30th IAHR Biennial Congress*, Eds. Ganoulis J. and Prinos P., Thessaloniki (Greece), 473-478.

Hansen K.D. (2003). "RCC use in dam rehabilitation projects" *Proc. 4th Intl. Symposium on Roller Compacted Concrete Dams*, Madrid, Eds. Berga *et al.*, Balkema, Rotterdam (The Netherlands), 79-89.

Hewlett H.W.M., Baker R., May R.W.P. and Pravdivets Y. (1997). "Design of stepped-block spillways" *CIRIA-Special publication 142*, London (UK).

Hunt S.L, Temple D.M., Abt S.R., Kadavy K.C. and Hanson G. (2012). "Converging stepped spillways: simplified momentum analysis approach" *Journal of Hydraulic Engineering*, 138(9):796-902.

Hunt S.L and Kadavy K.C. (2013). "Inception point for embankment dams stepped spillways" *Journal of Hydraulic Engineering*, 139(1):60-64.

Lutz N., Lucas J., Lais A. and Boes R.M. (2015). "Stepped chute of Tränsglet Dam: Physical model study" *Journal of Applied Water Engineering and Research*, 3(2):166-176.

Mateos C. and Elviro V. (1995). "Stepped spillways. Design for the transition between the spillway crest and the steps" *Proc. XXVI IAHR Congress. HYDRA 2000*, Thomas Telford, London (UK), 260-265.

Mateos C. and Elviro V. (2000). "Stepped spillway studies at CEDEX" *Proc. Intl. Workshop Hydraulics of stepped spillways*, VAW-ETH Zurich, Eds. Minor, H.E., and Hager, W.H., Balkema, Rotterdam (The Netherlands), 87-94.

Matos J. (2003). "Roller compacted concrete and stepped spillways: from new dams to dam rehabiltation" *Proc. Intl. Congress on Dam Rehabilitation and Maintenance*, Madrid, Eds. Llanos J.A. *et al.*, Balkema, Lisse (The Netherlands), 553-559.

Meireles I. and Matos J. (2009). "Skimming flow in the nonaerated region of stepped spillways over embankment dams" *Journal of Hydraulic Engineering*, 135(8):685-689.

Meireles I., Renna F., Matos J. and Bombardelli F. (2012). "Skimming, nonaerated flow on stepped spillways over roller compacted concrete dams" *Journal of Hydraulic Engineering*, 138(10):870-877.

Morán R. and Toledo M.A. (2014) "Design and construction of the Barriga Dam spillway through an improved wedge-shaped block technology" *Canadian Journal of Civil Engineering*, 41(10):924-927.

Ohtsu I. and Yasuda Y. (1997). "Characteristics of flow conditions on stepped channels" *Proc. 27th IARH Congress*, San Francisco (USA), 583-588.

Ohtsu I., Yasuda Y. and Takahashi M. (2001). "Discussion of 'Onset of the skimming flow on stepped spillways' by M.R. Chamani and N. Rajaratman" *Journal of Hydraulic Engineering*, 127(6):522-524.

Ohtsu I., Yasuda Y. and Takahashi M. (2004). "Flow characteristics of skimming flow in stepped channels" *Journal of Hydraulic Engineering*, 130(9):860-869.

PCA (2002). *Design manual for RCC spillways and overtopping protection*, prepared by URS Greiner Woodward Clyde, Portland Cement Association, Illinois (USA).

Peterka A.J. (1978). "Hydraulic design of stilling basins and energy dissipators" *Engineering monograph No. 25*, Bureau of Reclamation, Colorado (USA).

Peyras L., Royet P. and Dégoutte G. (1991). "Ecoulements et dissipation sur les déservoirs en gradins de gabions" *La Hoille Blanche*, 46(1) :37-47. (in French)

Peyras L., Royet P. and Dégoutte G. (1992). "Flow and energy dissipation over stepped gabion weirs" *Journal of Hydraulic Engineering*, 118(5):707-717.

Pfister M., Hager W.H. and Minor H.E. (2006a). "Stepped chutes: pre-aeration and spray reduction" *Intl. Journal of Multiphase Flow*, 32(2):269-284.

Pfister M., Hager W.H. and Minor H.E. (2006b). "Bottom aeration of stepped spillways" *Journal of Hydraulic Engineering*, 132(8):850-853.

Pravdivets Y.P. and Bramley M.E. (1989). "Stepped protection blocks for dam spillways" *Water Power and Dam Construction*, 41(7):60-66.

Rajaratman N. (1990). "Skimming flow in stepped spillways" *Journal of Hydraulic Engineering*, 116(4):587-591.

Rice C.E. and Kadavy K.C. (1997). "Physical model study of the proposed spillway for Cedar Run Site 6, Fauquier County, Virginia" *Applied Engineering in Agriculture*, 13(6):723-729.

Sánchez-Juny M., Blade E. and Dolz J. (2007). "Pressures on stepped spillways" *Journal of Hydraulic Research*, 45(4):505-511.

Scarella M. and Pagliara S. (2015). "A challenging solution for Zarema May Day Dam spillway design and model tests", *Proc. 25th ICOLD Congress*, Stavanger (Norway), Q.97, R.37.

Schiess-Zamora A., Pfister M., Hager W.H. and Minor H.E. (2008). "Hydraulic performance of step aerator" *Journal of Hydraulic Engineering*, 134(2):127-134.

Shen C. (2003). "RCC dams in China" *Proc. 4th Intl. Symposium on Roller Compacted Concrete Dams*, Madrid, Eds. Berga *et al.*, Balkema, Rotterdam (The Netherlands), 15-25.

Sorensen R.M. (1985). "Stepped spillway hydraulic model investigation" *Journal of Hydraulic Engineering*, 111(12):1461-1472.

Toledo M.A., Morán R. and Oñate E. (Eds.) (2015), *Dam protections against overtopping and accidental leakage*, CRC press, Leiden (The Netherlands).

Tozzi M.J. (1992). *Caracterização/comportamento de escoramentos em vertedouros com paramento em degraus*, PhD thesis, Universidade de São Paulo, São Paulo (Brazil). (in Portuguese)

Willey J., Ewing T., Lesleighter E. and Dymke J. (2010). "Refinement of hydraulic design of a complex stepped spillway arrangement through numerical and physical modelling" *Proc. ASIA 2010 Conference*, Sarawak (Malaysia).

Winter C., Mason P.J., Baker R. and Ferguson A. (2010). *Guidance for the design and maintenance of stepped masonry spillways,* UK Environment Agency Project SC080015, Bristol (UK).

Winter C. (2010). "Research into the hydrodynamic forces and pressures acting within stepped masonry spillways" *Dams and Reservoirs*, 20(1):16-26.

Wright H.J. (2010). "Improved energy dissipation on stepped spillways with the addition of triangular protrusions" *Proc. 78th ICOLD Annual Meeting*. Hanoi (Vietnam).

Zhang. J., Chen J. and Wang Y. (2012). "Experimental study on time-averaged pressures in stepped spillways" *Journal of Hydraulic Research*, 50(1):236-240.

5. ÉVACUATEURS LABYRINTHE

5.1. INTRODUCTION

Pour une cote de retenue donnée et une largeur d'emprise fixe, la débitance des évacuateurs labyrinthe est supérieure à celle d'un seuil classique. Ce type d'évacuateur peut s'adapter à plusieurs problématiques de site : augmentation significative de la débitance, amélioration du passage des crues, augmentation de la capacité de stockage. Ce type d'évacuateur « 3D » permet également d'avoir un fonctionnement passif, fiable, à dissipation d'énergie améliorée, à nappe naturellement aérée et d'aspect esthétique sympathique. L'optimisation hydraulique de ces évacuateurs est communément un compromis économique / complexité technique de réalisation. Les évacuateurs labyrinthe sont plus particulièrement adaptés aux projets nécessitant une diminution de la largeur d'un évacuateur ou une optimisation de l'emprise des ouvrages d'évacuation.

La Figure 5.1 présente un exemple d'évacuateur labyrinthe. Équipé d'évents d'aération et de piles de pont, sa débitance est de 15 000 m³/s.

Fig. 5.1
Évacuateur labyrinthe du barrage de Maguga (Royaume de Swaziland) (Photo Aurecon).

Gentilini (1941) est semble-t-il crédité d'une des premières études sur les déversoirs labyrinthe. La première publication sur la méthode de conception des évacuateurs labyrinthe, réalisée par Hay et Taylor en 1970 fut précédée par celle de Kozák et Sváb (1961). Depuis, de nombreux modèles physiques ont été réalisés et de nombreuses publications techniques ont été conduites. Ce chapitre se base sur les publications de l'U.S Bureau of Reclamation (Denver, Colorado, USA), du Laboratório Nacional de Engenharia Civil (Lisbon, Portugal), et de Utah Water Research

Laboratory (Logan, Utah, USA). Falvey (2003) a résumé une grande partie des contributions du XXième siècle concernant l'hydraulique des évacuateurs labyrinthe. Crookston (2010) a enrichi la base de connaissances techniques en développant une méthode de conception plus complète et en étudiant les influences de forme de crête, le comportement de la nappe d'écoulement, la configuration d'alimentation (alimentation canal ou retenue), le comportement des géométries arquées et enfin les effets d'échelle sur l'hydraulique des déversoirs labyrinthe.

5.2. DESCRIPTION GÉNÉRALE

Les déversoirs labyrinthe ont été installés aussi bien dans des chenaux qu'en connexion directe avec la retenue et ont été conçus avec une grande variété de géométries et de forme de crête différentes. Comme présenté dans la Figure 5.2, ils sont souvent caractérisés par une répétition de cycle identique placés de façon linéaire et calé à une seule cote de crête. En général, les évacuateurs labyrinthe sont des structures symétriques, suivis d'un coursier, et finalisés par un bassin de dissipation aval. Cependant, des configurations de cycles placés de manière arquée ont également été réalisés comme aux barrages Avon (Australie, Darvas 1971), Kizilcapinar (Turquie, Yildiz et Uzecek 1996), Maguga (Swaziland, Van Wyk *et al* 2006), María Cristina (Espagne, Page *et al.* 2007) et Weatherford (USA, Tullis 1992). A noter également qu'un grand évacuateur labyrinthe arqué est également en projet par l'U.S. Army Corp au barrage d'Isabella en Californie (États-Unis). Certains déversoirs labyrinthe ont également incorporé des calages altimétriques de crête variables, ainsi que diverses structures annexes comme des piles de pont ou passerelle, des échancrures en crête permettant le déversement contrôlé d'un débit minimum, des orifices d'alimentation aval en pied de bajoyer en cas de niveau de réservoir inférieur à la cote de déversement pouvant également servir de restitution aval d'eau à température profonde plus fraîche que l'eau de surface.

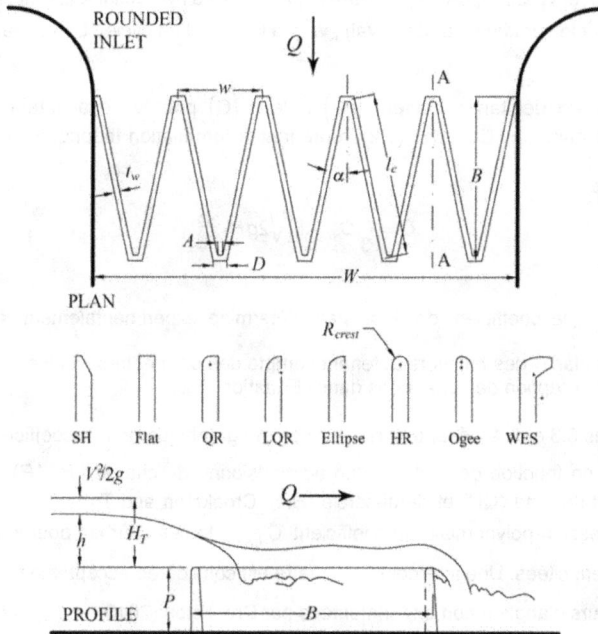

Fig. 5.2
Définition de la géométrie d'un évacuateur labyrinthe suivant Crookston, 2010

Crookston (2010) offre un aperçu complet de la conception hydraulique des évacuateurs labyrinthe. Il présente un historique détaillé concernant la recherche et le design des évacuateurs labyrinthe, y compris les premières méthodes de conception, les nombreuses études de cas et une bibliographie complète des contributions scientifiques sur ce type d'évacuateur. Les résultats de Crookston (2010) ont été résumés et publiés dans Crookston et Tullis (2013a, b) et Crookston et Tullis (2012a,b,c). Ces articles techniques comprennent les principales références pour la discussion suivante concernant la conception hydraulique des évacuateurs labyrinthe et sont généralement appelés : Méthode de Conception « Crookston et Tullis ».

5.3. DÉBITANCE

La capacité déversante d'un évacuateur labyrinthe est fonction des conditions d'écoulements, de la géométrie de l'ouvrage et peut être exprimée sous la forme :

$$Q = f(g, v, H_T, H_d, L_c, \alpha, A_c, l_c, P, B, w, t_w, crest, approach, nappe) \tag{5.1}$$

Avec Q = la débitance totale ; g = la constante de gravité ; ν = la viscosité cinématique de l'eau, H_T = la charge totale mesurée par rapport à la crête de l'ouvrage avec $H_T = \dfrac{V^2}{2g} + h$; V = la vitesse moyenne dans la section d'approche ; h = la hauteur piézométrique amont mesurée par rapport à la cote de crête de l'ouvrage ; H_d = la charge totale aval ; L_c = la longueur médiane totale développée de la crête avec $L_c = 2N(A_c + l_c)$; N = le nombre de cycle de labyrinthe ; A_c = longueur médiane de l'apex (petite base du trapèze) ; l_c = la longueur médiane des bajoyers ; α = l'angle des bajoyers par rapport à l'écoulement ; P = la pelle hydraulique amont ; B = la profondeur d'un cycle (longueur longitudinale amont/aval) ; w = la largeur d'un cycle ; t_w = l'épaisseur de la crête au sommet.

La relation de débitance hauteur (H_T) / débit (Q) pour un évacuateur labyrinthe a été caractérisée sur la base de l'Équation (5.2), reprenant la formulation théorique d'un seuil classique (Henderson 1966) :

$$Q = \frac{2}{3} C_{d(\alpha°)} L_c \sqrt{2g} H_T^{3/2} \tag{5.2}$$

Avec $C_{d(\alpha°)}$ le coefficient de débitance, déterminé expérimentalement, adimensionné en fonction de l'angle plan α des bajoyers et tenant compte des paramètres d'influence présentés dans l'Équation (5.1) à l'exception de ceux repris dans l'Équation (5.2).

Les Figures 5.3 et 5.4 présentent respectivement graphiquement le coefficient $C_{d(\alpha°)}$ suivant Crookston (2010) en fonction de α et du ratio adimensionné de charge (H_T / P) pour des formes de crête dite quart-de-rond (QR) et demi-cercle (HR). Crookston and Tullis (2013a) fournissent les équations de régression polynomiale du coefficient $C_{d(\alpha°)}$ de ces courbes pour les deux formes de crête précédemment citées. Une interpolation linéaire est considérée acceptable pour estimer $C_{d(\alpha°)}$ pour d'autres valeurs d'angle α non expérimentées par Crookston (2010).

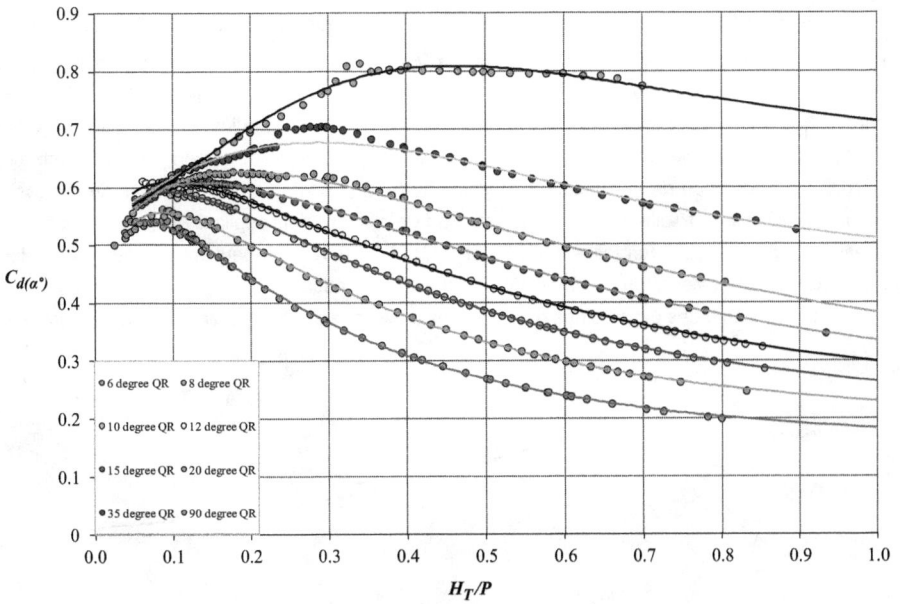

Fig. 5.3
$C_{d(\alpha°)}$ versus H_T/P pour une forme de crête dite quart-de-rond (Quater-Round QR)

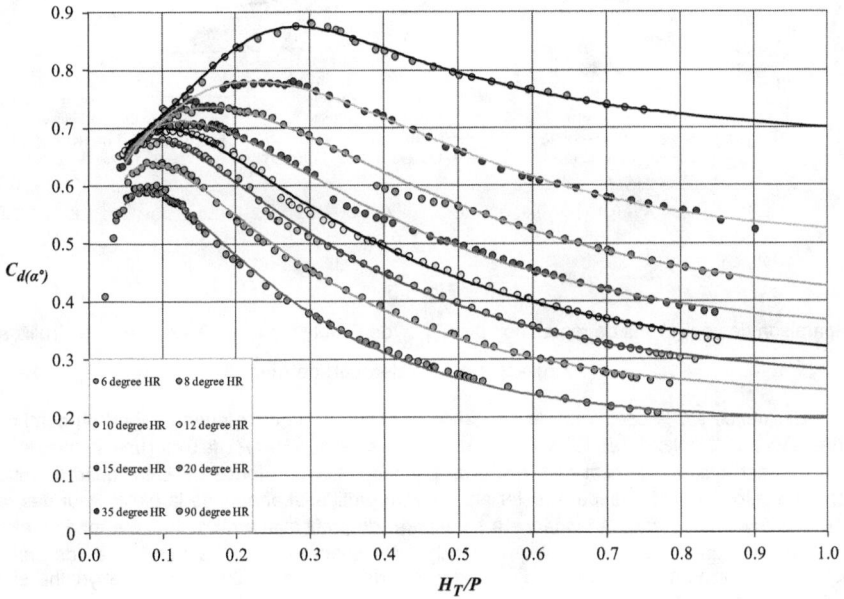

Fig. 5.4
$C_{d(\alpha°)} = f(H_T/P)$ pour une forme de crête dite demi-cercle (Halft-Round HR)

Les équations d'ajustement du coefficient $C_{d(\alpha°)}$ présentées par Crookston et Tullis (2013a) sont basées expérimentalement sur une plage de valeur $(H_T / P) \leq 1$. La forme spécifique des polynômes d'ajustement a cependant été choisie pour faciliter une extrapolation du coefficient $C_{d(\alpha°)}$ au-delà du rapport $(H_T / P) = 1$ (voir Figure 5.5). L'équation pour $\alpha=15°$ a été validée jusqu'à une plage de valeur $(H_T / P) \leq 2,1$ par Crookston et al. (2012) par modélisation physique et numérique confirmant ainsi la solidité de la Méthode de Conception Crookston et Tullis. A contrario, les équations de régression polynomiale utilisées par Tullis *et al* (1995) (crête quart-de-rond) ne sont applicables que dans la gamme des essais expérimentaux dont elles sont issues $((H_T / P) \leq 0,9)$ et ne peuvent pas être extrapolées au-delà (c.-à-d., les valeurs du coefficient $C_{d(\alpha°)}$ deviennent négatives lorsque H_T / P augmente du fait de la nature même des expressions des fonctions polynomiales choisies en 1995, voir Figure 5.5).

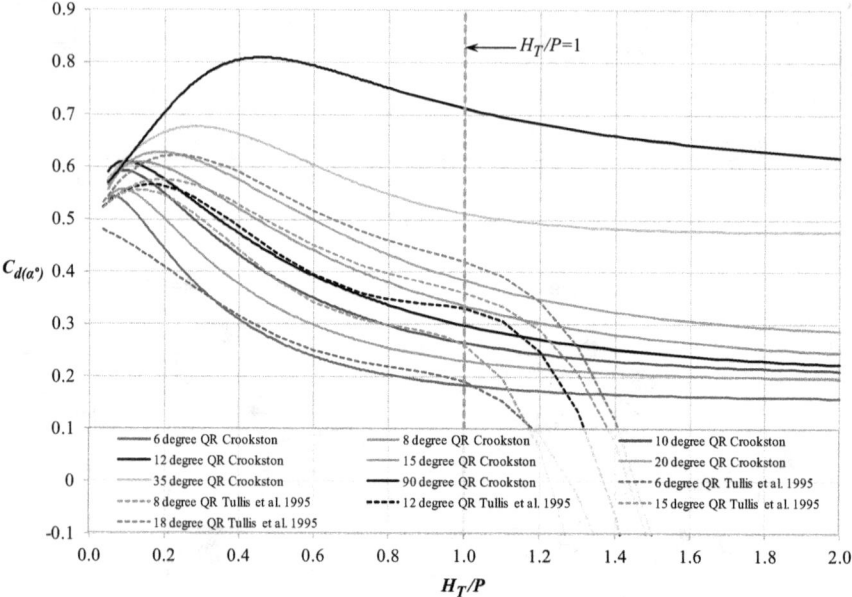

Fig. 5.5

Comparaison des équations de régression de $C_{d(\alpha°)}$ de Crookston et Tullis (2013a) avec Tullis *et al.* (1995) – forme de crête quart-de-rond

L'influence sur le coefficient de débitance de la forme de crête quart-de-rond ou demi-cercle est discutée par Crookston et Tullis (2013a). Pour des ratios (H_T / P) faibles, une forme de crête demi-cercle est hydrauliquement plus efficiente par rapport à une forme de crête quart-de-rond en raison de la capacité de la nappe à rester attachée au profil aval arrondi de la crête. Pour des ratios (H_T / P) supérieurs, la nappe a tendance à se séparer du profil aval arrondi de crête rendant alors le comportement hydraulique similaire à un profil de crête quart-de-rond. D'autres formes de crête ont également été mises en œuvre dans la pratique sur des bajoyers d'évacuateurs labyrinthe et sont présentées en Figure 5.2.

Les formes de crête arrondies ont été construites efficacement par le biais de coffrages et une finition manuelle ou via la mise en place de pièces préfabriquées. La rentabilité économique du choix de construction augmente généralement avec le nombre de cycles. Le barrage Brazos (Texas, USA) est un excellent exemple d'évacuateur labyrinthe ayant fait l'objet d'une optimisation économique utilisant des éléments préfabriqués de crête afin de respecter une conception basée sur des apexs arrondis et une forme de crête profilée de type Ogee.

5.4. COMPORTEMENT DE LA NAPPE ET AÉRATION ARTIFICIELLE

En plus de la débitance hydraulique, le comportement de la nappe des évacuateurs labyrinthe doit également être pris en compte dans la conception. La Figure 5.6 illustre les diverses conditions d'aération et d'instabilité des nappes pour des formes de crête quart-de-rond ; des informations similaires pour des crêtes demi-cercle sont présentées dans Crookston and Tullis (2013b). La Figure 5.6 ne présente cependant pas le comportement des nappes vibrantes qui peuvent également apparaître (Crookston *et al.* 2014, Anderson 2014). Quatre conditions d'aération de nappe existent pour les déversoirs : adhérente, aérée, partiellement aérée et partiellement noyée. L'influence de l'aération artificielle sur la capacité débitante et le comportement de la nappe, via des conduites d'aération et des briseurs de nappe, est également discutée par les deux auteurs incluant également des informations sur le placement et la forme de ces artifices.

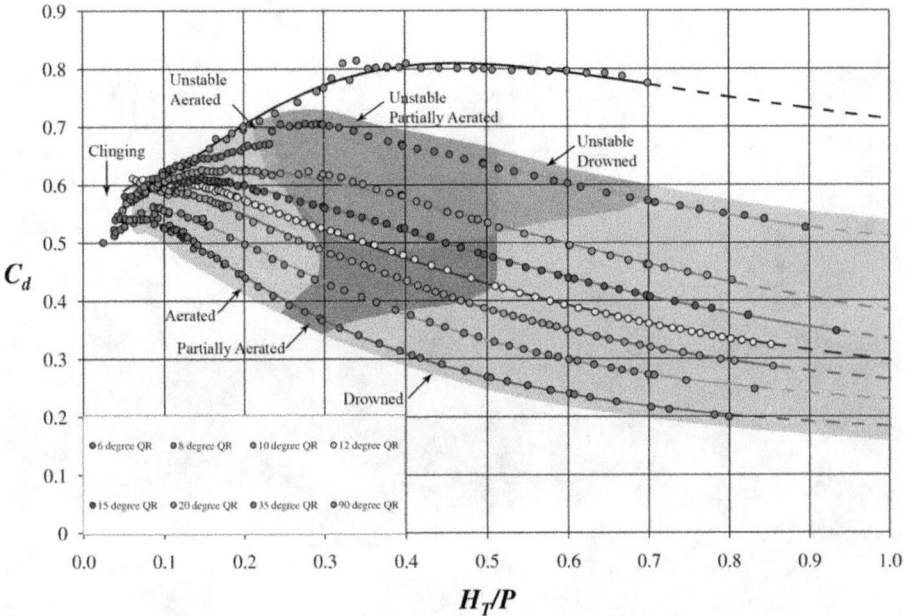

Fig. 5.6
Comportement de la nappe pour des évacuateurs labyrinthe à forme de crête quart-de-rond

Pour des conditions d'écoulement particulières, la taille de la cavité aérée située derrière la nappe peut fluctuer temporairement amenant à un comportement de nappe instable. Il s'agit d'un phénomène de fluctuation de la trajectoire de la nappe à faible fréquence qui se produit à des ratios (H_T / P) modérés, dépendant de l'angle α et de la forme de crête et produisant alors du bruit. Des séparateurs de nappe permettent de diminuer l'amplitude de l'instabilité mais les essais en laboratoire ont mis en évidence qu'ils ne peuvent entièrement empêcher ce phénomène (Crookston 2010). Des informations supplémentaires concernant le phénomène d'instabilité de nappe sont disponibles dans Crookston and Tullis (2013b).

Indépendamment des instabilités de nappe, les nappes vibrantes (voir Figure 5.7) peuvent apparaître à faible charge (observées sur prototypes avec des lames d'eau sur crête de 15 cm environ voire moins dans certains cas) et peuvent se développer sur des géométries aussi bien linéaires que non-linéaires (Crookston et al. 2014). La vibration des nappes peut entraîner d'intenses ondes de pression acoustiques et du bruit (Casperson, 1995), bruit pouvant être à juste titre assimilé au son d'un hélicoptère. En général, le phénomène de vibration de nappe est visuellement observé via des bandes horizontales étroitement espacées qui se développent là où la nappe quitte la crête. Les nappes très minces peuvent être ondulées ou battantes. Enfin, les vibrations peuvent être amplifiées par le vent.

Indépendamment des instabilités de nappe, les nappes vibrantes (voir Figure 5.7) peuvent apparaître à faible charge (observées sur prototypes avec des lames d'eau sur crête de 15 cm environ voir moins dans certains cas) et peuvent se développer sur des géométries aussi bien linéaires que non-linéaires (Crookston et al. 2014). La vibration des nappes peut entraîner d'intenses ondes de pression acoustiques et du bruit (Casperson, 1995), bruit pouvant être à juste titre assimilé au son d'un hélicoptère. En général, le phénomène de vibration de nappe est visuellement observé via des bandes horizontales étroitement espacées qui se développent là où la nappe quitte la crête. Les nappes très minces peuvent être ondulées ou battantes. Enfin, les vibrations peuvent être amplifiées par le vent.

Fig. 5.7
Nappe vibrante (Schnabel Engineering,USA).

L'influence de la forme de crête, de la longueur du labyrinthe, de la hauteur de chute et des conditions d'écoulements sur le phénomène de vibrations de nappe, reste encore aujourd'hui un sujet de recherche. L'ajout de rugosité sur la surface de crête pour atténuer les vibrations de nappe est un exemple de méthode efficace reportée dans la littérature (Metropolitan Water, Sewerage and Drainage Board 1980). Les vibrations de nappe ont déjà été également atténuées par l'utilisation de multiples déflecteurs avec un espacement définit en fonction d'essais.

5.5. ÉFFICACITÉ D'UN CYCLE

Conformément à l'Equation 5.2, Q est proportionnel à $C_{d(\alpha°)}$ [$C_{d(\alpha°)}$ diminuant avec la diminution de α (voir Figures 5.3 et 5.4)] et à L_c [L_c augmentant avec la diminution de α pour une largeur de cycle fixe]. Pour une valeur H_T / P donnée, l'efficacité d'un cycle ($\varepsilon' = C_{d(\alpha°)}L_{c-cycle} / w$) est représentative de la débitance d'un cycle pour une géométrie de labyrinthe et peut être utilisée pour illustrer l'effet réel sur la débitance des deux paramètres à évolution en opposition de phase en fonction de α. La Figure 5.8 (Crookston et Tullis 2013a) montre l'efficacité d'un cycle ε' équipé de forme de crête quart-de-cercle pour différentes valeurs de l' α et (H_T / P). Les données de ε' montrent que l'augmentation de la longueur déversante d'un labyrinthe à mesure que l'angle α diminue compense largement la diminution des valeurs de $C_{d(\alpha°)}$ pour une même diminution de l'angle α. Cela signifie que la plus grande débitance par cycle se produit aux valeurs d'angle α les plus petites. Ce bénéfice se réduit avec l'augmentation de (H_T / P). ε' évalue seulement l'efficacité de la débitance par cycle et devrait être jumelé avec une analyse des coûts de construction pour choisir la géométrie optimale du déversement puisqu'une géométrie moins hydrauliquement efficace mais répondant à des exigences de conception et construction peut s'avérer être une option plus rentable. Les comparaisons d'efficience ε' entre les structures sont particulièrement utiles lorsque les structures partagent le même ratio (H_T / P). La débitance effective par largeur de cycle exige de multiplier ε' par $H_T^{3/2}$.

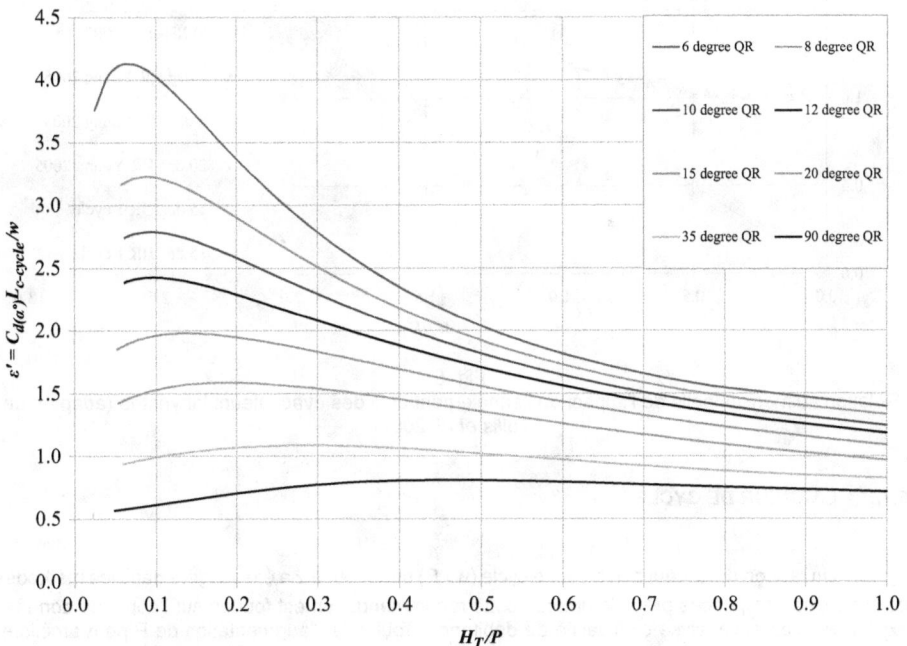

Fig. 5.8
Efficacité de cycle pour des évacuateurs labyrinthe à crête quart-de-rond

5.6. ENNOIEMENT AVAL

L'ennoiement aval des évacuateurs labyrinthe se produit lorsque le niveau aval atteint la cote de crête des modules. On appelle un ennoiement modulaire une condition de submergence aval sans impact sur le niveau amont. Des niveaux d'ennoiement supérieurs à la limite d'ennoiement modulaire conduiront à un niveau amont supérieur pour un même débit donné. À noter que l'ennoiement local notamment au niveau de l'apex amont diffère de l'ennoiement aval dans le sens où il est indépendant du niveau aval. La Figure 5.9 (Tullis *et al.*, 2007) présente une relation adimensionnée charge/ débit en fonction du niveau d'ennoiement aval. Les variables H^* et H_d utilisées dans la Figure 5.9 correspondent respectivement à la charge totale amont et aval dans des conditions d'écoulement submergées. H_T correspond à la charge totale amont pour le même débit mais pour une condition d'écoulement dénoyée (Q représente une variable indépendante dans cette analyse). La limite ennoiement modulaire correspond pour un évacuateur labyrinthe à $H^*/H_T = 1$. Les données de la Figure 5.9 peuvent être utilisées pour déterminer H^* en connaissant H_T et H_d. Dabling et Tullis (2012) présente pour des PKW une relation similaire charge/débit en fonction du niveau d'ennoiement aval.

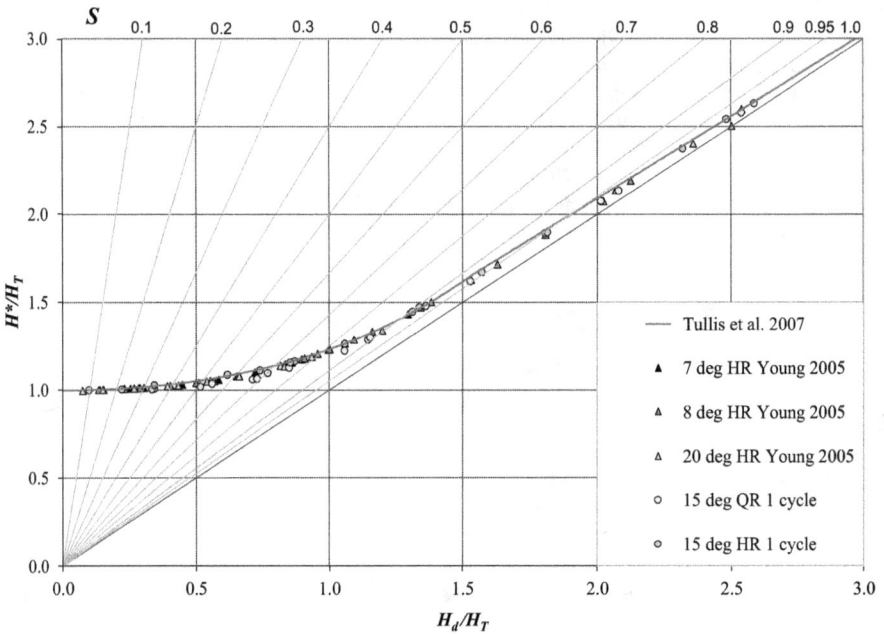

Fig. 5.9
Relation adimensionnelle de l'impact de l'ennoiement aval des évacuateurs labyrinthe (adaptée de Tullis *et al.* 2007)

5.7. LARGEUR DE CYCLE

Un rapport de largeur minimale de cycle (w / P) supérieur à 2 a été suggéré dans les méthodes de conception proposées précédemment ; cette recommandation est fondée sur l'appui de données expérimentales et l'analyse d'efficacité de débitance. Toutefois, l'augmentation de P peut améliorer les conditions d'écoulement et donc améliorer la débitance. Cependant, la réduction de w et par conséquent de w / P, entraîne généralement une réduction des coûts de construction pour un débit de conception donné et, par conséquent, augmente l'intérêt économique de l'évacuateur, malgré une réduction de l'efficacité hydraulique (Crookston *et al.* 2013, Paxson *et al.* 2013). À noter que la méthode de conception Tullis *et al.* (1995) incluait des données $C_{d(\alpha°)}$ pour des ratios (w / P) inférieurs à 2 malgré des ratios suggérés de 3 à 4. Compte tenu de cela, la recherche effectuée par Crookston (2010) a exploré l'influence de w et P sur la débitance, étant donné que des ratios (w / P) identiques peuvent exister pour différentes combinaisons de valeurs de w et P. Des directives générales concernant l'influence de w et P sur la débitance des évacuateurs labyrinthe sont présentées dans Crookston *et al.* (2013) et Seamons (2014).

5.8. POSITIONNEMENT ET EFFETS DES ENTONNEMENTS SUR LA DÉBITANCE DES CYCLES

Des recherches effectuées par l'USBR pour des prototypes d'évacuateurs labyrinthe ont permis de comprendre l'effet de l'emplacement des ouvrages au sein des réservoirs (Houston, 1983). Des recherches appliquées supplémentaires sur les effets des culées de rives pour les évacuateurs labyrinthe avec une crête de forme demi-cercle sont présentées par Crookston & Tullis (2012b). La Figure 5.10 présente les configurations géométriques d'implantation testées aussi bien en chenal qu'en connexion directe avec la retenue.

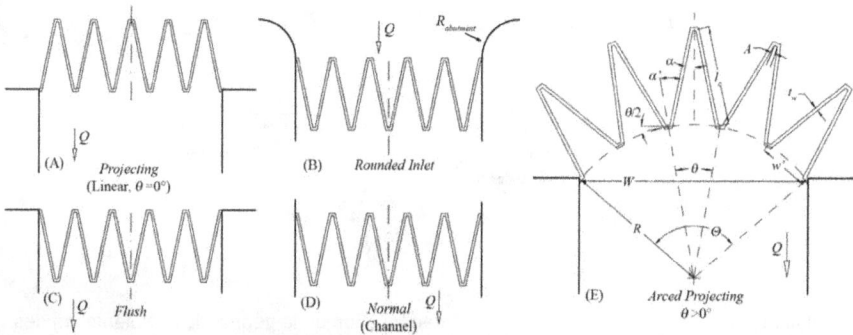

Fig. 5.10

Crookston et Tullis (2012b) configurations géométriques (A : apex axial déporté vers l'amont, B : entonnement de culée circulaire, C : apex axial déporté vers l'aval, D : situation de référence en chenal, E : implantation arquée.

Une comparaison de $C_{d(\alpha°=12°)}$ pour quatre géométries d'implantation différente au sein de réservoirs [Figure 5.10 (A), (B), (C) et (E)] par rapport à une implantation de référence en chenal [Figure 5.10 (D)] est présentée à la Figure 5.11 (Crookston et Tullis, 2012b). Sur la base des géométries spécifiques testées, la configuration d'implantation avec l'apex axiale déporté vers l'aval (implantation dite « Flush », Figure 5.10 (C)] conduit au moins bon coefficient de débitance par rapport à l'implantation en chenal de référence, en raison de la séparation des flux au niveau des culées. L'ajout de culées profilées de forme arrondie [Figure 5.10 (B)] permet une amélioration du coefficient de débitance. Néanmoins, les configurations [Figure 5.10 (B) et (A)] conduisent toujours à une valeur de coefficient de débitance 3% à 7% moins efficace que la configuration de référence. Des informations supplémentaires sur la performance hydraulique de ces configurations sont présentées par Crookston et Tullis (2012b).

Une implantation arquée d'un déversoir de labyrinthe avec un angle de cycle de θ = 10 ° [voir la Figure 5.10 (E)] conduit quant-à-elle à une efficacité d'environ 5% à11 % supérieure par rapport à la configuration de référence en canal et a constitué la géométrie la plus efficace testée. Une géométrie arquée augmente ainsi la débitance des évacuateurs labyrinthes installés au sein de réservoir en orientant les cycles vers le flux d'approche. La configuration arquée permet également d'optimiser la largeur des ouvrages en aval (canal ou chute) réduisant potentiellement les coûts de construction. Crookston et Tullis (2012a) présente une nomenclature normalisée pour les évacuateurs labyrinthes arqués et discutent des schémas de configuration comparables. Plusieurs configurations arquées ont été testées et ont été jugées environ 5 à 30% plus efficaces que des configurations linéaires en canaux (H_T/P dépendant). L'avantage hydraulique des configurations arquées diminue avec l'augmentation de H_T/P (influence accrue de l'augmentation de la submergence locale).

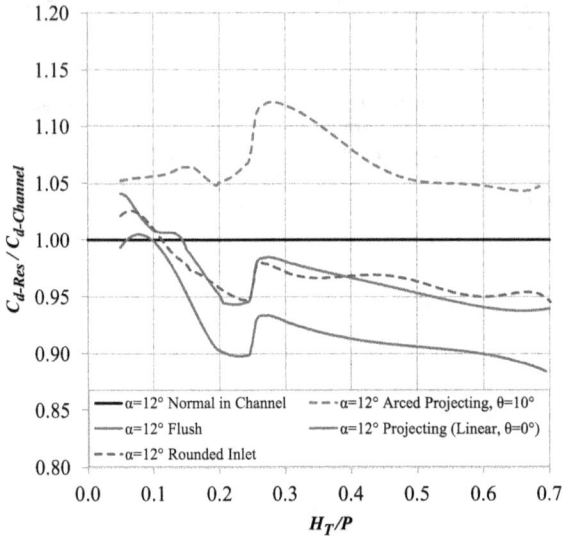

Fig. 5.11
Évolution relative du coefficient de débitance en fonction de la géométrie d'implantation des évacuateurs labyrinthe ($\alpha=12°$; implantation de référence en chenal)

5.9. DISSIPATION D'ÉNERGIE ET ÉNERGIE RÉSIDUELLE

Les évacuateurs labyrinthe sont très efficaces concernant la dissipation d'énergie. Lopes et al. (2006, 2008) présente une analyse concernant l'énergie résiduelle relative (H_1) en aval du déversoir. L'énergie résiduelle augmente à mesure que H_T / P augmente (c.-à-d. moins de dissipation d'énergie relative) ; les évacuateurs labyrinthe produisent une énergie résiduelle plus faible qu'une chute verticale (peu importe l'angle α). L'estimation de l'énergie résiduelle en aval des évacuateurs labyrinthe peut être réalisée sur la base des publications de Lopes et al (2006, 2008).

5.10. CORPS FLOTTANTS

Les corps flottants sont souvent transportés vers les structures hydrauliques durant les crues (Pfister et al., 2013). Par exemple, les tempêtes dans les bassins versants forestiers peuvent entraîner des débits chargés en bois flottants. Selon les conditions d'approche topographique et hydraulique, ce type de débris peut passer au-dessus des labyrinthes pendant les crues et nécessite in fine peu d'entretien. Une enquête sur la gestion des embâcles au niveau de 75 évacuateurs labyrinthe aux États-Unis et au Portugal a récemment été réalisée (Crookston et al., 2015). Les résultats indiquent qu'en général la gestion des embâcles pour les évacuateurs labyrinthe est acceptable avec peu de rapports d'exploitation y faisant référence. Cependant, sous certaines conditions, une grande accumulation de bois flottants peut se produire et soulever alors des problèmes de sécurité. Par conséquent, l'accumulation de débris sur les déversoirs même labyrinthe ne doit pas être négligée pour des situations où il existe un potentiel réel d'une réduction de débitance de l'évacuateur pour cause d'embâcles. La recherche sur les bois flottants pour les évacuateurs d'évacuateurs PKW fournit des aperçus uniques qui s'appliquent généralement aux déversoirs labyrinthe (Pfister et al. 2013).

5.11. CONCEPTION HYDRAULIQUE

La procédure recommandée pour la conception d'un évacuateur labyrinthe est celle présentée dans le Tableau 5.1 (Crookston et Tullis 2013a). Ce format a été introduit pour la première fois par Tullis et al. (1995), poursuivi par Falvey (2003) et a été raffiné et élargi par Crookston et Tullis. La partie supérieure du tableau présente les conditions hydrauliques définies par l'utilisateur ou les exigences structurales pour le labyrinthe. Le débit de conception ($Q_{conception}$) représente généralement un débit de crue (par exemple, une crue 100 ans, une crue de sûreté, etc.) estimé à partir d'une étude hydrologique amont. Dans la pratique, $Q_{conception}$ est généralement estimé à l'aide de programmes informatiques tels que HEC-HMS, son prédécesseur HEC-1 ou d'autres méthodes appropriées. H représente l'élévation maximale autorisée du réservoir. H_T est la différence entre l'élévation du réservoir et la cote de la crête, $H_{crête}$ (terme cinétique négligeable). Toutefois, il peut être approprié d'examiner les conditions de d'approche et les pertes mineures lors de l'estimation de H_T. H_D est la charge totale aval mesurée par rapport à la crête de l'évacuateur et peut être déterminée par un calcul de courbe de remous. Les paramètres géométriques définis par l'utilisateur sont présentés dans la deuxième section ; l'utilisation d'un dispositif d'aération de nappe peut être spécifiée si souhaité. La troisième section calcule les données de performance hydraulique du labyrinthe et détermine les paramètres géométriques supplémentaires. Bien que la Méthode de Conception Crookston et Tullis ne fournisse que $C_{d(\alpha°)}$ pour des formes de crête de quart-de-cercle (QR) et demi-cercle (HR), la méthode s'accommode facilement de coefficients $C_{d(\alpha°)}$ déterminés expérimentalement pour des conditions spécifiques au site ou d'autres formes de crête. Le coefficient $C_{d(90°)}$ et la longueur de crête linéaire correspondante (même forme de crête) requise pour correspondre au débit de conception et aux exigences H sont également reportés pour comparaison avec un seuil linéaire. La dernière section traite l'ennoiement aval conformément aux relations charge/débit développées par Tullis et al. (2007). Le modèle de calcul présenté au Tableau 5.1 peut être élargi afin d'aboutir à une relation complète Charge/Débit pour une géométrie du labyrinthe spécifiée. Cette méthode de conception peut également être utilisée pour estimer la relation Charge/Débit pour les évacuateurs labyrinthes existants. Cette procédure, qui s'adapte facilement sous format tableur, est décrite par Crookston et Tullis (2013a) ; les paramètres géométriques des labyrinthes sont spécifiés plutôt que calculés. Les effets de l'ennoiement aval sont évalués, connaissant H_T et H_d, en déterminant H^* comme discuté par Tullis et al. (2007).

La présente méthode de conception et les données de base servant de support sont limitées aux géométries (Crookston 2010) et aux conditions hydrauliques testées (p.ex. $0,05 \leq H_T/P \leq 0,9$). Ces résultats peuvent être appliqués de manière conservative (avec jugement technique appliqué) à d'autres géométries d'évacuateurs labyrinthe et d'autres conditions de débit (une vérification du design via une étude sur modèle hydraulique est recommandée). L'interpolation linéaire est recommandée pour déterminer $C_{d(\alpha°)}$ pour d'autres valeurs d'angle α non expérimentées par Crookston (2010). D'après les données de base servant de support disponible à l'étude actuelle, la méthode de conception (Tableau 5.1) peut être également utilisée pour $H_T/P \leq 2.0$, puisque les données expérimentales, pour $\alpha=15°$, ont validé l'équation d'interpolation jusqu'à une plage de valeur $(H_T/P) \leq 2,1$; les données $C_{d(\alpha°)}$ résultantes des essais concordent bien avec l'équation de la courbe d'interpolation polynomiale proposée par Crookston et Tullis (2013a).

Table 5.1
Modèle de calcul d'un évacuateur labyrinthe

Parameter	Symbol	Value	Units	Notes
Hydraulic Conditions – Input Data				
Design Flow	Q_{design}	1,500.00	(m³/s)	g = 9.81 m/s²
Design Flow Water Surface Elevation	H	1,680.00	(m)	
Approach Channel Elevation	H_{apron}	1,674.00	(m)	
Crest Elevation	H_{crest}	1,678.00	(m)	
Unsubmerged Total Upstream Head	H_T	2.00	(m)	Piezometric Head + Velocity Head - Losses
Downstream Total Head	H_d	0.50	(m)	
Labyrinth Weir Geometry – Input Data				
Angle of Side Legs	α	12	(°)	α ~ 6° - 35°
Number of Cycles	N	9	-	whole or half cycles
Crest Height	P	4.00	(m)	P ~1.0H_T
Thickness of Weir Wall at the Crest	t_w	0.50	(m)	t_w ~ P/8
Inside Apex Width	A	0.50	(m)	A ~ t_w
Crest Shape	Crest Shape	Quarter	-	Quarter- or Half-Round
Aeration Device (Nappe Breakers, Vents)	-	Breakers	-	Breakers, Vents, or None
Calculated Data				
Headwater Ratio	H_T/P	0.50	-	Data for H_T/P<1.0, extrapolation for H_T/P≤2.0
Labyrinth Weir Discharge Crest Coefficient	$C_{d(\alpha°)}$	0.429	-	$C_{d(\alpha°)} = f(H_T/P, \alpha, \text{Crest Shape})$
Total Centerline Length of Weir	L_c	418.28	(m)	$L_c = 3/2 Q_{design}/[(C_{d(\alpha°)} H_T^{3/2})(2g)^{1/2}]$
Centerline Length of Sidewall	l_c	2.33	(m)	$l_c = (B-t_w)/\cos(\alpha)$
Outside Apex Width	D	1.30	(m)	$D = A+2t_w\tan(45-\alpha/2)$
Cycle Width	w	11.10	(m)	$w = 2l_c\sin(\alpha)+A+D$
Width of Labyrinth (Normal to Flow)	W	99.87	(m)	$W = Nw$
Length of Apron (Parallel to Flow)	B	22.35	(m)	$B = [L_c/(2N)-(A+D)/2]\cos(\alpha)+t_w$
Magnification Ratio	M	4.19	-	$M = L_c/(wN)$
Cycle Width Ratio	w/P	2.77	-	Normally 2 ≤ w/P ≤ 4
Relative Thickness Ratio	P/t_w	8.0	-	~8
Apex Ratio	A/w	0.05	-	<0.08
Cycle Efficiency	ε'	1.80	-	$\varepsilon' = C_{d(\alpha°)}M$
Efficacy	ε	2.23	-	$\varepsilon = C_{d(\alpha°)}M/C_{d(90°)}$
# of Nappe Breakers or Vents	-	9	-	Breaker on ds Apex, 1 Vent per Sidewall
Linear Weir Discharge Coefficient	$C_{d(90°)}$	0.808	-	$C_{d(90°)} = f(H_T/P, \alpha, \text{Crest Shape})$
Length of Linear Weir for same Flow	$L_{c(90°)}$	222.33	(m)	$L_{c(90°)} = 3/2 Q_{design}/[(C_{d(90°)} H_T^{3/2})(2g)^{1/2}]$
Submergence (Tullis et al. 2007)				
Downstream/Upstream Ratio of Unsubmerged Head	H_d/H_T	0.25	(m)	
Submerged Head Discharge Ratio	H^*/H_T	1.013	-	Piecewise function Tullis et al. (2007)
Submerged Upstream Total Head	H^*	2.025	(m)	
Submergence Level	S	0.247	-	$S = H_d/H^*$
Submerged Weir Discharge Coefficient	C_{d-sub}	0.421	-	$C_{d(\alpha°)}(H_T/H^*)^{3/2}$

†Design limited to extent of experimental data; designs that exceed these limits may warrant a physical model study

La méthode de conception et les données de soutien sont limitées aux géométries (Crookston, 2010) et aux conditions hydrauliques testées (par exemple, $0{,}05 \leq HT/P \leq 0{,}9$). Ces résultats peuvent être appliqués de manière conservatrice (avec un jugement technique solide) à d'autres géométries de déversoirs labyrinthes et conditions d'écoulement (la vérification de la conception par une étude de modèle hydraulique est toutefois recommandée). Une interpolation linéaire est recommandée pour déterminer $C_{d(\alpha°)}$ pour des valeurs de l'angle α autres que celles présentées. Sur la base des données de soutien disponibles dans le cadre de la présente étude, la méthode de conception (tableau 5.1) peut être utilisée pour $HT/P \leq 2{,}9$, car les résultats expérimentaux sont cohérents et le $\alpha = 15°$ a été testé jusqu'à $HT/P = 2{,}1$; les données $C_{d(\alpha°)}$ résultantes étaient bien en accord avec l'équation d'ajustement de courbe proposée par Crookston et Tullis (3013a).

5.12. MODÉLISATION PHYSIQUE ET NUMÉRIQUE (CFD)

D'autres discussions sur l'utilisation de la modélisation physique et de la CFD dans la conception des évacuateurs de crues sont présentées au chapitre 9.

5.13. SÉLECTION D'EXEMPLE DE PROTOTYPE D'ÉVACUATEUR LABYRINTHE

La bibliographie incluse dans le présent document contient des références pour de nombreuses études de cas de déversoirs labyrinthes ; un résumé de ces structures est présenté dans le tableau 5.2.

Table 5.2
Prototype d'évacuateur labyrinth

Name	Location	Q_{design}	H_3	N	Source
		(m³/s)	(m)	( )	
Accord Pond Dam	USA	-	0.46	2	Crookston et al. (2015)
Agua Branca	Portugal	124	1.65	2	Quintela et al. (2000)
Alfaiates	Portugal	99	1.60	1	Quintela et al. (2000)
Alijó	Portugal	52	1.23	1	Magalhães & Lorena (1989)
Alloway Lake Dam	USA	-	-	5	Crookston et al. (2015)
Antelope Creek Channel	USA	-	-	3	Crookston et al. (2015)
Arcossó	Portugal	85	1.25	1	Quintela et al. (2000)
Avon	Australia	1790	2.80	10	Darvas (1971)
Bartletts Ferry	USA	5920	2.19	20.5	Mayer (1980)
Belia	Zaire	400	2.00	2	Magalhães & Lorena (1989)
Beni Bahdel	Algeria	1000	0.50	20	Afshar (1988)
Berg	South Africa	270	2.50	2	Fourie (1999)
Boardman	USA	387	1.77	2	Babb (1976), Lux (1985)
Bospoort 1	South Africa	2465	4.20	7.5	ARQ (2008)
Bospoort 2	South Africa	1613	4.20	2.5	ARQ (2008)
Boyde Lake	USA	1209	0.59	59	Brinker (2005)
Calde	Portugal	21	0.60	1	Quintela et al. (2000)
Capital City Country Club	USA	-	2.44	1	Crookston et al. (2015)
Carty	USA	387	1.80	2	Afshar (1988)
Cloe Dam	USA	-	-	7	Crookston et al. (2015)
Castelletto-Nerv. Canal	Italy	25	0.12	24	Magalhães & Lorena (1989)
Cimia	Italy	1100	1.5	4	Lux & Hinchliff (1985)
Concourse Lake Dam	USA	-	1.07	2	Crookston et al. (2015)
Crystal Bridges	USA	-	-	2	Crookston et al. (2015)
Dog River	USA	1572	2.74	8	Savage et al. (2004)
DRA Detention Structure	USA	-	1.22	4	Schnabel (2013)
Dungo	Angola	576	2.4	4	Magalhães & Lorena (1989)
Eikenhof	South Africa	190	3.5	2	ARQ (1998)
Elmendorf Lake	USA	-	-	6	Crookston et al. (2015)
Estancia	Venezuela	661	3.01	1	Magalhães & Lorena (1989)
Forestport	USA	76	1.02	2	Lux (1989)
Fort Miller	USA	-	-	45	Crookston et al. (2015)
Garland Canal	USA	25.5	0.37	3	Lux & Hinchliff (1985)
Gema	Portugal	115	1.12	2	Magalhães & Lorena (1989)
Glen Park	USA	-	-	25	Crookston et al. (2015)
Grahamstown	Australia	628	1.4	8	Barker et al (2001)
Greystone	USA	-	1.45	1	Schnabel (2013)
Hard Labor Creek	USA	-	3.66	4	Schnabel (2013)
Harrezza	Algeria	350	1.9	3	Lux (1989)
Hollis G Lathem Reservoir	USA	-	2.44	3	Crookston et al. (2015)
Huntington Hills	USA	-	0.58	3	Crookston et al. (2015)

Indian Run	USA	-	2.44	2.5	Crookston et al. (2015)
Infulene Canal	Mozambique	60	1.00	3	Magalhães & Lorena (1989)
Juturnaiba	Brazil	862	0.70	4	Afshar (1988)
Kauffman	USA	128	5.0	5	Crookston et al. (2015)
Keddara	Algeria	250	2.46	2	Magalhães & Lorena (1989)
King Falls	USA	-	-	6	Crookston et al. (2015)
Kizilcapinar	Turkey	2270	4.6	5	Yildiz (1996)
Lake Brazos	USA	24609	-	24	Tullis and Young (2005)
Lake Natalie	USA	-	2.44	4	Crookston et al. (2015)
Lake Paupacken	USA	-	-	2	Crookston et al. (2015)
Lake Sovereign	USA	-	1.22	2	Crookston et al. (2015)
Lake Townsend	USA	3483	4.57	7	Tullis & Crookston (2008)
Lake Upchurch	USA	-	1.52	6	Crookston et al. (2015)
Leaser Lake	USA	289	3.26	2	Crookston et al. (2015)
Linville Land Harbor	USA	693	2.61	4	Schnabel (2013)
Little Blue Run	USA	-	-	10	Crookston et al. (2015)
Lyman Run	USA	-	-	8	Crookston et al. (2015)
Maguga	Swaziland	15000	8.25	9	Van Wyk et al (2006)
Midmar	South Africa	3052	4.15	10	ARQ (2002)
María Cristina Dam	Spain	5444	-	7	Page et al. (2007)
Meacham Grove	USA	-	-	4	Crookston et al. (2015)
Mercer	USA	239	1.83	4	CH2M-Hill (1976)
Navet Pumped Storage	Trinidad	481	1.68	10	Phelps (1974)
New London	USA	-	-	4	Crookston et al. (2015)
Ohau C Canal	New Zealand	540	1.08	12	Walsh (1980)
Opossum	USA	-	-	4	Crookston et al. (2015)
Pacoti	Brazil	3400	2.72	15	Magalhães & Lorena (1989)
Pine Run	USA	-	1.40	4	Crookston et al. (2015)
Pisão	Portugal	50	1.00	1	Quintela et al. (2000)
Pye Lake	USA	-	0.34	3	Schnabel (2013)
Quincy	USA	552	2.13	4	Magalhães & Lorena (1989)
Rapp Run Flood Control	USA	-	-	6	Crookston et al. (2015)
Rocklands-Berg	South Africa	-	-	2	Fourie (1999)
Roy F. Varner Reservoir	USA	-	2.13	8	Crookston et al. (2015)
São Domingos	Portugal	160	1.84	2	Magalhães & Lorena (1989)
Sam Rayburn Lake	USA	-	-	16	USACE (1991)
Santa Justa	Portugal	285	1.35	2	Magalhães & Lorena (1989)
Sarioglan	Turkey	490.7	1.06	7	Yildiz (1996)
Sarno	Algeria	360	1.5	8	Afshar (1988)
South River No 29	USA	-	3.05	3	Schnabel (2013)
Standley Lake	USA	1539	1.98	13	Tullis (1993)
Teja	Portugal	61	1.05	1	Quintela et al. (2000)
Upper Dam - Rangeley	USA	-	1.19	4	Schnabel (2013)
Ute	USA	15570	5.79	14	Houston (1982)
Weatherford	USA	-	-	4	Tullis (1992)
Woronora	Australia	1020	1.36	11	Darvas (1971)

Les extraits de vues en plan et profils de quelques-unes de ces installations sont présentés dans les pages suivantes pour illustrer les différentes configurations expérimentées.

• **Barrage de MAGUGA, Royaume de Swaziland** (Informations recueillies par Aurecon, Afrique du Sud)

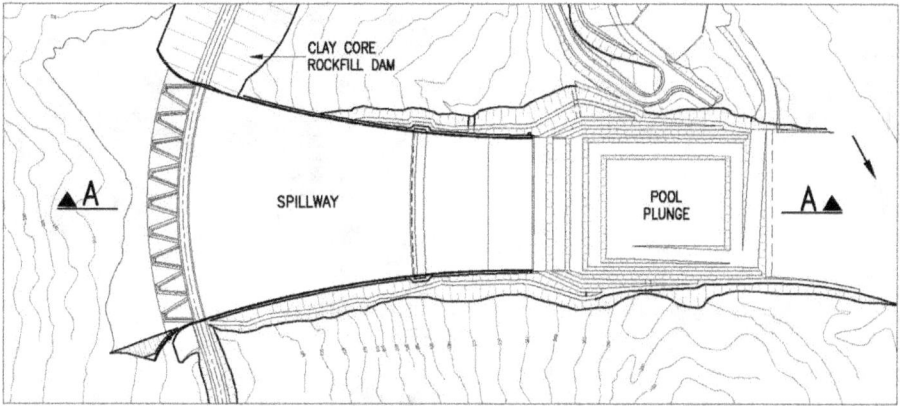

Fig. 5.12
Barrage de Maguga, Royaume de Swaziland (Informations recueillies par Aurecon, Afrique du Sud)

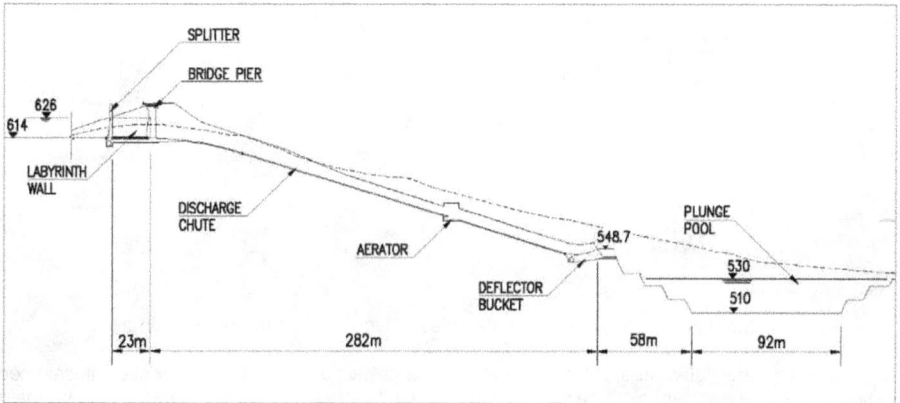

Fig. 5.13
Barrage de Maguga, Royaume de Swaziland (Informations recueillies par Aurecon, Afrique du Sud)

Un pont devait être construit au-dessus du déversoir. Afin d'adapter l'alignement routier sur l'évacuateur et le remblai, l'implantation géométrique du labyrinthe s'est faite suivant une courbe de rayon 300 m. Le linéaire développé hydraulique passe de 181 m de large au droit du labyrinthe à 100 m de large au niveau du saut à ski. Afin de rendre la transition aussi graduelle que possible, les parois latérales du coursier suivent une courbe de rayon 876 m ajustées tangentiellement à la fois au labyrinthe et à la cuillère de dissipation au niveau du saut à ski. Cela a permis de diriger les flux individuels de chaque cycle de labyrinthe vers la cuillère de dissipation de telle manière qu'aucune onde stationnaire ne se développe.

BARRAGE DE MIDMAR, AFRIQUE DU SUD (Informations recueillies par Aurecon, Afrique du Sud)

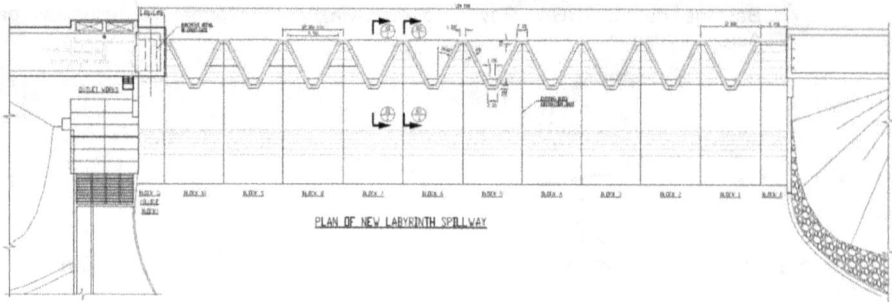

Fig. 5.14
Configuration générale de l'évacuateur de crues labyrinthe du barrage de Midmar –
Vue en plan

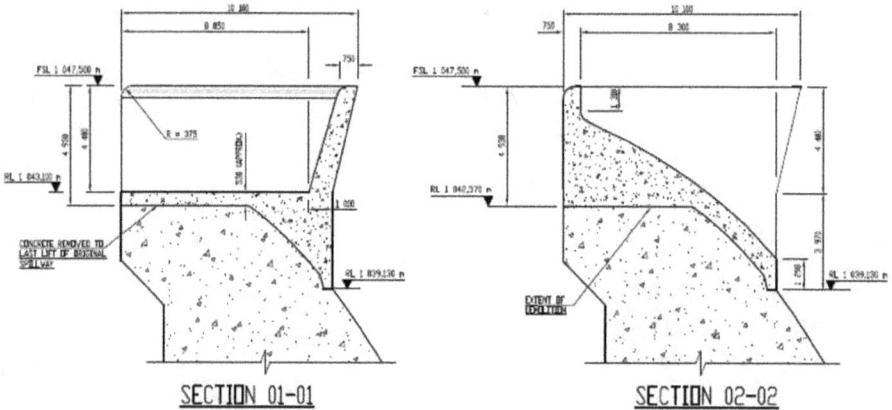

Fig. 5.15
Coupe au droit de l'évacuateur de crues labyrinthe du barrage de Midmar

L'évacuateur labyrinthe a été incorporé sur la crête du barrage poids existant initialement destiné à être surélevé par l'ajout de vannes radiales. Cette surélévation est réputée être la première mise en œuvre dans le monde d'un labyrinthe dans cette configuration. Le pont, les piles avec les massifs supports des axes des vannes, ainsi qu'une partie de la crête profilée, ont été démolis pour laisser place au nouveau labyrinthe. Une masse supplémentaire de béton a été intégrée au sein de la géométrie de l'alvéole aval (apex amont) entre les bajoyers du labyrinthe pour aider à la stabilité globale au renversement.

La principale motivation d'une modification de solution de surélévation par rapport à la conception originale vannée fut la fiabilité d'un évacuateur labyrinthe fixe et sa capacité à atténuer en toute sécurité la crue de projet et la crue de sûreté.

La conception du labyrinthe fut basée sur la hauteur d'eau amont de surélévation disponible. La structure a été testée sur modèle réduit à l'aide de trois cycles complets à l'échelle 1:27 pour confirmer les capacités théoriques.

- **BARRAGE DU LAC TOWNSEND, ÉTATS UNIS** ((Informations recueillies par Schnabel Engineering, USA

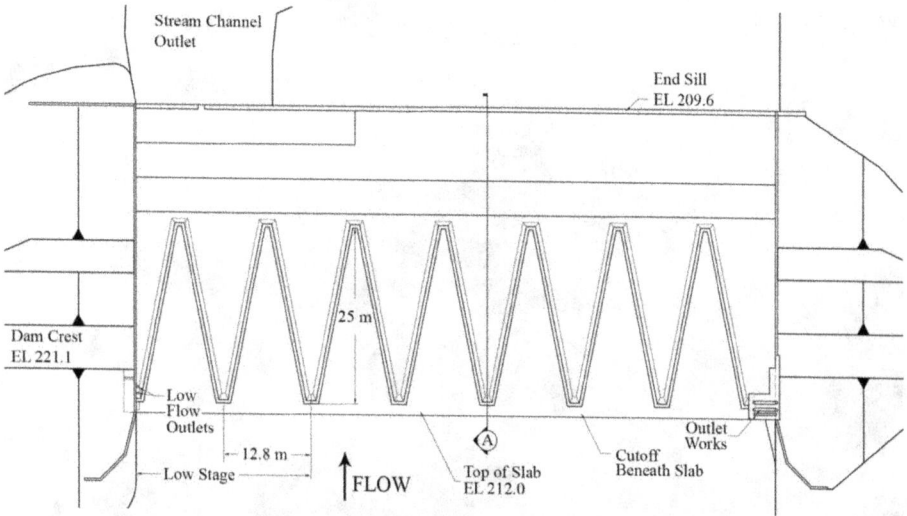

Fig. 5.16
Configuration générale de l'évacuateur de crues labyrinthe du barrage du lac Townsend– Vue en plan

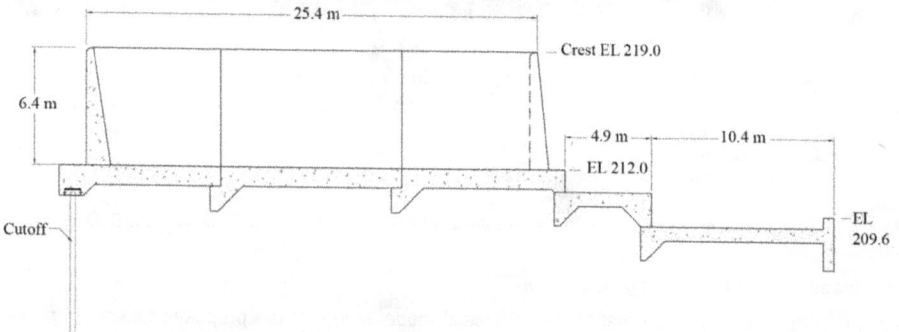

Fig. 5.17
Coupe au droit de l'évacuateur de crues labyrinthe du barrage du lac Townsend

L'évacuateur de crues labyrinthe (6,4 m de haut) au barrage du lac Townsend a remplacé un déversoir vanné de 84 m de large. Le labyrinthe de 7 cycles dispose d'une crête échelonnée avec 2 cycles à la cote normale et de 5 cycles à une cote supérieure. Pour des fortes crues, il est sujet à un ennoiement aval. La conception a été testée avec une modélisation physique réalisée au Laboratoire Utah Water Research Laboratory et une modélisation numérique CFD réalisée à l'Université Idaho State University (Paxson et al., 2007). Le débit de dimensionnement est d'environ 3 483 m³/s ; le barrage en remblai associé comprend une protection aval contre les déversements au-dessus de sa crête pour permettre le passage combiné de la crue de projet associée aux débits résultant des hydrogrammes de rupture de trois barrages situés en amont.

Fig. 5.18
Évacuateur de crues labyrinthe du barrage du lac Townsend

5.14. RÉFÉRENCES

Pour de plus amples références sur les évacuateurs labyrinthe voir aussi Crookston (2010) et Falvey (2003).

Liste identique à la version anglaise.

ARQ, (2002) - Raising of Midmar Dam – Physical model study for the proposed labyrinth spillway. Final Report and Addendum, ARQ, South Africa.

ARQ, (2008) - Dam safety rehabilitation programme. Second draft discharge capacity report for Bospoort Dam. ARQ, South Africa.

Barker, M.B., R.M. Holroyde and T. Qui (2001) - Grahamstown Dam Stage 2. Augmentation selection and design of a labyrinth spillway and baffle chute. ANCOLD Bulletin, Issue No. 118, August 2001.

Crookston, B. and B. Tullis (2008) - "Labyrinth weirs." In: S. Pagliara (ed.) *Hydraulic Structures*, 2nd IJREWHS on Hydraulic Structures, Edizioni Plus, University of Pisa, Italy.

Crookston B. and B. Tullis (2010) - "Hydraulic performance of labyrinth weirs." In: *3rd International Junior Researcher and Engineer Workshop on Hydraulic Structures*, Edinburgh, Scotland (currently unpublished).

Falvey, Henry T. (2003) - *Hydraulic design of labyrinth weirs*. ASCE Press, 1801 Alexander Bell Drive, Reston, Virginia, USA.

Fourie, S. (1999) - Revised design report for Rocklands-Berg Dam enlargement and upgrading of spillway. Report No 3690/5972, October 1999.

Lopes, R., J. Matos and J.F. Melo (2006) - "Discharge capacity and residual energy of labyrinth weirs." *International Junior Researcher and Engineer Workshop on Hydraulic Structures*, J. Matos and H. Chanson (Eds), Report CH61/06, Div. of Civil Eng., The University of Queensland, Brisbane, Australia.

Lopes, R., J. Matos and J.F. Melo (2008) - "Characteristic depths and energy dissipation downstream of a labyrinth weir." *Hydraulic Structures*, 2nd IJREW on Hydraulic Structures, S. Pagliara (ed.), Edizioni Plus, University of Pisa, Italy.

Lux III, F.L. and D. Hinchcliff (1985) - "Design and Construction of Labyrinth Spillways", *Transactions of the Fifteenth International Congress on Large Dams*, Vol. 4, Q. 59, R. 15, pp. 249-274, International Commission on Large Dams, Paris, France.

Paxson, G., D. Campbell and J. Monroe (2011) - "Evolving Approaches and Considerations for Labyrinth Spillways". *21st Century Dam Design – Advances and Adaptations*, 31st Annual USSD Conference, San Diego, California, April 11 – 15, 2011.

Savage, B., K. Frizell and J. Crowder (2001) - *Brains versus Brawn: The Changing World of Hydraulic Model Studies*. United States Bureau of Reclamation, www.usbr.gov/pmts/hydraulics_lab/pubs/PAP/PAP-0933.pdf.

Tullis, J.P., N. Amanian and D. Waldron (1995) - "Design of Labyrinth Spillways", *Journal of Hydraulic Engineering*, ASCE Volume 121, No 3, March 1995.

Van Wyk, D., A. Officer, W. Schwartz, G. Goodey and A. Rooseboom (2006) - "Design and Model Testing of a Labyrinth Spillway for Maguga Dam", *Transactions of the Twenty Second Congress on Large Dams*, Q.84 – R.54, ICOLD, Barcelona.

Amanian, N. (1987) - *Performance and design of labyrinth spillways*. M.S. Thesis, Utah State University, Logan, Utah, USA.

Anderson, A.A. (2014) - *Causes and Countermeasures for nappe oscillation: An experimental approach*. M.S. Thesis. Utah State University, Logan, Utah, USA.

Cassidy, J.J., C.A. Gardner and R.T. Peacock (1983) - "Labyrinth-crest spillway – planning, design, and construction." *Proceedings of the International Conference of the Hydraulic Aspects of Flood and Flood Control*, London, England.

Cordero-Page, D., V. García and C. Nonot (2007) - "Aliviaderos en laberinto. Presa de María Cristina." *Ingeneiería Civil*, 146, 5-20 (in Spanish)

Crookston, B.M. (2010) - *Labyrinth weirs*. Ph.D. Dissertation. Utah State University, Logan, Utah, USA.

Crookston, B.M., G.S. Paxson and B.M. Savage (2012) - *It can be done! Labyrinth weir design guidance for high headwater and low cycle width ratios*. Proc. of the 2012 ASDSO Annual Conference, Denver, Colorado. CD-ROM.

Crookston, B.M., A.A. Anderson, L. Shearin-Feimster and B.P. Tullis (2014) - "Mitigation investigation of flow-induced vibrations at a rehabilitated spillway." *5th International Symposium on Hydraulic Structures*, Brisbane, Australia.

Crookston, B.M., G.S. Paxson, B.M. Savage and B.P. Tullis (April 2013) - "Increasing hydraulic design flexibility of labyrinth spillways." *ICOLD 2013 International Symposium*. Seattle, Wash.Crookston, B.M and B.P. Tullis (2012a) - "Arced labyrinth weirs." *Journal of Hydraulic Engineering, ASCE*, 138(6). 555-562.

Crookston, B.M and B.P. Tullis (2012b) - Discharge efficiency of reservoir-application-specific labyrinth weirs. *J. Irrig. Drain. Engr., ASCE*, 138(6). 773-776.

Crookston, B.M and B.P. Tullis (2012c) - "Labyrinth weirs: Nappe interference and local submergence." *J. Irrig. Drain. Engr., ASCE*, 138(8), 757-765.

Crookston, B.M and B.P. Tullis (2013a) - "Hydraulic design and analysis of labyrinth weirs. Part 1: Discharge relationships." *J. Irrig. Drain. Engr., ASCE*, 139(5), 363-370.

Crookston, B.M and B.P. Tullis (2013b) - "Hydraulic design and analysis of labyrinth weirs. Part 2: Nappe aeration, instability, and vibration." *J. Irrig. Drain. Engr., ASCE*, 139(5), 371-377.

Dabling, M.R. and B.M. Crookston (2012) - "Staged and Notched Labyrinth Weir Hydraulics." *4th International Junior Researcher and Engineer Workshop on Hydr. Structures*. B. Tullis and H. Chanson (eds.).

Dabling, M.R., B.P. Tullis and B.M. Crookston (2013) - "Staged labyrinth weir hydraulics." *J. Irrig. Drain. Eng.*, posted ahead of print May 22, 2013.

Darvas, L. (1971) - "Discussion of performance and design of labyrinth weirs, by Hay and Taylor." *J. of Hydr. Engrg.*, ASCE, 97(80), 1246-1251.

Easterling, D., K. Kunkei and X. Yin (2013) - "Observed increases in probable maximum precipitation over global land areas." *Proceedings of 25th Conference on Climate Variability and Change*, Austin, TX, USA.

Falvey, H. and Trielle (1995) - "Hydraulics and design of fusegates." *J. of Hydr. Engrg.*, ASCE, 121(7), 512-518.

Frizell, K. (2003) - *Dog River Dam Hydraulic Model Study Results*. USBR Water Resources Research Laboratory Report, Denver, Colo.

Gentillini, B. (1940) - *Stramazzi con cresta a planta obliqua e a zig-zag*. Memorie e Studi del Instituto di Idraulica e Construzioni Idrauliche del Regil Politecnico di Milano, No. 48 (in Italian).

Hay, N. and G. Taylor (1970) - "Performance and design of labyrinth weirs." *J. of Hydr. Engrg.*, ASCE, 96(11), 2337-2357.

Henderson, F.M. (1966) - *Open Channel Flow*. MacMillan Company, New York, USA.

Hinchliff, D. and K. Houston (1984) - *Hydraulic design and application of labyrinth spillways*. Proc. of 4th Annual USCOLD Lecture. Dam safety and Rehabilitation, Bureau of Reclamation U.S. Dept. of the Interior, Washington, DC, USA.

Houston, K. (1982) - *Hydraulic model study of Ute Dam labyrinth spillway*. Report No. GR-82-7, US Bureau of Reclamation, Denver, Colo., USA

Houston, K. (1983) - *Hydraulic model study of Hyrum Dam auxiliary labyrinth spillway*. Report No. GR-82-13, US Bureau of Reclamation, Denver, Colo., USA.

Laugier, F. (2007) - "Design and construction of the first piano-key weir (PKW) spillway at the Goulours Dam." *Hydropower and Dams*, Vol 12, Issue 5.

Lempérière, F. and A. Ouamane (2003) - "The PK Weir: a new cost-effective solution for spillways." *Hydropower & Dams*, Vol. 8, Issue 5.

Magalhães, A. and M. Lorena (1989) - *Hydraulic design of labyrinth weirs*. Report No. 736, National Laboratory of Civil Engineering, Lisbon, Portugal.

Melo, J., C. Ramos and A. Magalhães (2002) - "Descarregadores com soleira em labirinto de um ciclo em canais convergentes. Determinação da capacidade de vazão." *Proc. 6° Congresso da Água*, Porto, Portugal. (in Portuguese).

Paxson, G., D. Campbell and J. Monroe (2011) - "Evolving design approaches and considerations for labyrinth spillways." *Proc. 31st Annual USSD Conf.*, San Diego, CA, USA. CD-ROM.

Paxson, G., J. Monroe, B.M. Crookston and D. Campbell, D. (2013) - "Balancing site considerations with hydraulic efficiency for labyrinth spillways." *ICOLD 2013 International Symposium*. Seattle, Washington, USA.

Pfister, M., A. Schleiss and B. Tullis (2013) - "Effect of driftwood on hydraulic head of Piano Key weirs." *Labyrinth and Piano Key Weirs II*.

Quintela, A., A. Pinheiro, J. Afonso and M. Cordeiro (2000) - "Gated spillways and free-flow spillways with long crests. Portugese dams experience." *20th ICOLD Q79-R12*, Bejing, China. 171-89.

Savage, B., K. Frizell and J. Crowder (2004) - "Brian versus brawn: The changing world of hydraulic model studies." *Proc. of the ASDSO Annual Conference*, Denver, CO, USA, CD-ROM.

Seamons, T. R. (2014) - Labyrinth weirs: A look into geometric variation and its effect on efficiency and design method predictions. M.S. Thesis. Utah State University, Logan, Utah. USA

Taylor, G. (1969) - *The performance of labyrinth weirs*. PhD Thesis, University of Nottingham, Nottingham, England.

Tullis, B. and B.M. Crookston (2008) - *Lake Townsend Dam spillway hydraulic model study report*. Utah Water Research Laboratory, Logan, Utah, USA.

Tullis, B., J. Young and M. Chandler (2007) - "Head-discharge relationships for submerged labyrinth weirs." *J. of Hydr. Engrg.*, ASCE, 133(3), 248-254.

Tullis, J.P. (1992) - *Weatherford spillway model study.* Hydraulic Report No. 311, Utah Water Research Laboratory, Logan, Utah, USA.

Tullis, J.P. (1993) - *Standley Lake service spillway model study.* Hydraulic Report No. 341, Utah Water Research Laboratory, Logan, Utah, USA.

Tullis, J.P., N. Amanian and D. Waldron (1995) - "Design of labyrinth weir spillways." *J.of Hydr. Engrg., ASCE,* 121(3), 247-255.

Waldron, D. (1994) - *Design of labyrinth spillway.* M.S. Thesis, Utah State University, Logan, Utah, USA.

6.   ÉVACUATEUR DE CRUE PKW

6.1.   DESCRIPTION GÉNÉRALE

Les évacuateurs de crue de type PKW ont été récemment développés. Ces structures innovantes, qui peuvent faire transiter des débits spécifiques très importants, constituent une variante des seuils labyrinthes traditionnels, afin de répondre aux inconvénients de ces derniers (Barcouda *et al.* 2006). Conçu selon une vue en plan rectangulaire et des radiers inclinés qui créent des porte-à-faux amont et aval, le seuil PKW est structurellement simple et efficace. Il peut être installé sur la crête de barrages béton existant ou neufs. Comparé à un seuil libre standard occupant la même emprise en crête, le seuil PKW augmente de manière importante la débitance (Ouamane and Lempérière, 2003).

Fig. 6.1
Vue du PKW du barrage des Gloriettes en France durant sa construction (Photo EDF)

Vue en plan, la succession des radiers inclinés vers l'amont et vers l'aval, pourrait ressembler à une succession de touches de piano, ce qui a amené à la dénomination de ces évacuateurs de crue. Comparé à un seuil labyrinthe traditionnel rectangulaire qui aurait la même vue développée en plan, l'évacuateur de crue PKW possède l'avantage de pouvoir être plus facilement installé sur des sites avec des emprises de fondation limitées (par exemple la crête d'un barrage poids). Par ailleurs, les radiers inclinés réduisent la hauteur des murs verticaux et donc la quantité de ferraillage requise dans le béton. Il s'agit des principales raisons pour lesquelles les évacuateurs de crue PKW sont efficaces et une solution économique pour augmenter la capacité d'évacuation des crues des barrages existants.

Fig. 6.2
Vue de l'évacuateur PKW du barrage de Malarce en France Durant un déversement (Photo EDF)

Les évacuateurs de crue PKW ont initialement été proposés par Hydrocoop en collaboration avec laboratoire hydraulique d'Électricité de France (LNHE, France), l'université de Roorkee (Inde) et l'université de Biskra (Algérie) (Ouamane and Lempérière, 2003). Depuis son invention, de nombreux travaux ont été menés dans le monde afin de comprendre son comportement hydraulique, optimiser sa conception et objectiver ses avantages et inconvénients (cf par exemple Blanc and Lempérière, 2001; Barcouda *et al.*, 2006; Ouamane and Lempérière, 2006; Truong Chi *et al.*, 2006; Machiels *et al.*, 2011a & 2014; Leite Ribeiro *et al.*, 2012a & b; Machiels, 2012; Anderson and Tullis, 2012 & 2013). Des études récentes ont été par exemple menées avec des vues en plan trapézoïdales (Cicéro *et al.*, 2013a).

Le premier évacuateur de crue PKW a été construit en 2006 par EDF en France au barrage de Goulours (Laugier, 2007). Depuis, les seuils PKW ont été utilisés pour augmenter la capacité d'évacuation de 5 autres barrages EDF, à savoir St. Marc (2008), Etroit (2009), Gloriettes (2010), Malarce (2012) and Charmine (2014), ou comme nouvelle structure de décharge, (Escouloubre (2011). Les enseignements tirés de la conception de ces premiers seuils PKW sont données dans Laugier *et al.* (2013 and 2017a). D'autres PKW sont présentement opérationnels au Vietnam (Ho Ta Khanh *et al.* 2017); Sri Lanka (Jayatillake and Perera, 2013 and 2017); Suisse (Eichenberger, 2013) et Ecosse (Ackers *et al.* 2013). De nouveaux PKW sont en cours d'étude ou de construction au Vietnam (Ho Ta Khan *et al.* 2017); France (Laugier *et al.* 2017b; Erpicum *et al.* 2011b; Bail *et al.* 2013); Algérie (Erpicum *et al.* 2012); Afrique du Sud (Botha *et al.*, 2013) et Inde (Das Singhal and Sharma, 2011). Ces travaux sont associés soit à des projets de réhabilitation de barrages existants (afin d'augmenter leur débitance), soit à des projets de barrages neufs. Les PKW peuvent être positionnés dans la rivière (seuil de dérivation), sur la crête d'un barrage en béton (en général poids), ou en rive du réservoir.

Jusqu'à présent, une trentaine de PKW prototypes ont été construit dans 8 pays durant ces dix dernières années. Les projets incluent une large variété de débits spécifiques et de charges hydrauliques allant de 30 cm à plus de 9 m et des hauteurs structurelles de 1 à 10 m. De plus, les débitances totales associées varient de quelques dizaines de m³/s à presque 10 000 m³/s.

La plupart des informations disponibles à ce jour sur les évacuateurs PKW, ont été publiés dans trois livres (Erpicum *et al*. 2011a, 2013a and 2017a), à la suite de trois workshop spécialisés tenus en Belgique (2011), France (2013) et Vietnam (2017). Un registre mondial des PKW avec une carte mondiale des projets a également été établie et est disponible sur un site internet dédié (http://www.pk-weirs.ulg.ac.be/?q=content/world-register-pkw).

6.2.   TYPES ET GÉOMÉTRIE

La géométrie des PKW peut apparaitre complexe. Elle inclut un large jeu de paramètre. Afin d'unifier les notations, une nomenclature spécifique à cette structure a été développée (Pralong *et al*. 2011a).

Selon cette nomenclature, « l'unité de PKW » peut être définie comme la structure de base du seuil. Elle est composée de deux murs latéraux, une alvéole d'entrée et deux demi-alvéoles de sortie (Figures 6.3a et 6.3b). Les principaux paramètres géométriques de la structure sont :

- La hauteur des alvéoles d'entrée et de sortie P_i aetnd P_o, (très souvent Pi = Po nommée "P");

- Leurs largeurs W_i et W_o, la largeur d'une unite PKW W_u;

- Le nombre d'unités PKW N_u;

- La longueur longitudinale en crête de barrage B_h, les longueurs B_o et B_i des porte-à-faux amont et aval, la longueur de la base du PKW B_b; et

- L'épaisseur des murs latéraux T_s.

Les indices *i, o* et *s* se referent respectivement aux alvéoles d'entrée (inlet), les alvéoles de sortie (oulet) et les murs latéraux (sidewall). W_u est égal à $W_i + W_o + 2T_s$ et la largeur totale W du seuil est égale à N_u fois W_u. La longueur développée en crête L_u d'une unite PKW vaut $W_u + 2B_h$ et la longueur développée totale L du seuil vaut N_u fois L_u. Les murs parapets (extension verticale à partir de la crête) peuvent être ajoutés au seuil. Leur hauteur est dénommée P_p.

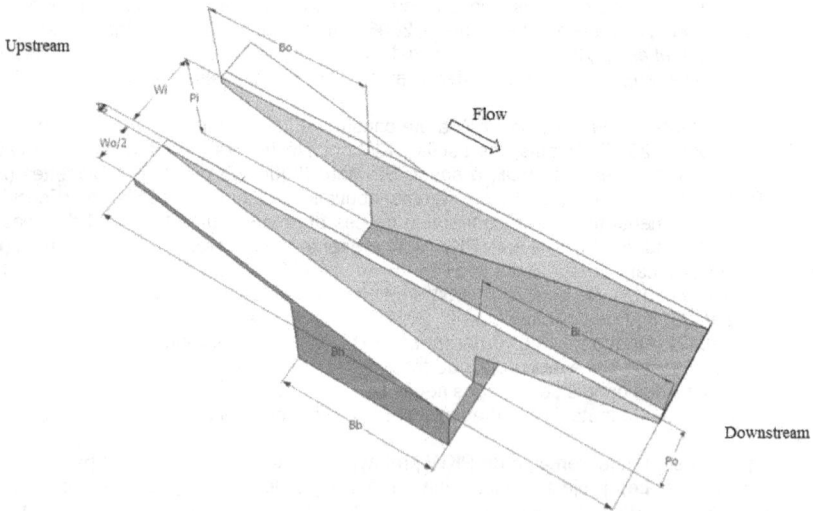

Fig. 6.3
(a) Schéma d'une unite PKW et principals notations (Erpicum *et al*. 2013b)

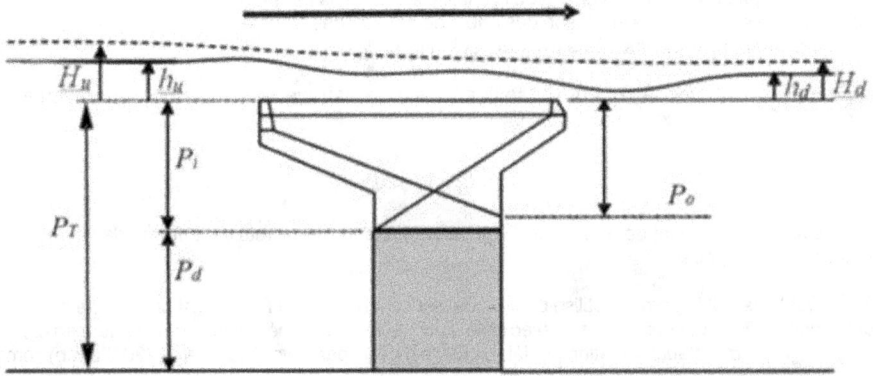

Fig. 6.3
(b) Coupe type d'un PKW (adapted from Pralong et al. 2011a)

Les seuils PKW ont été classifies en 4 types en fonction de la disposition des alvéoles amont ou/et aval (Truong *et al.*, 2006): type-A avec des porte-à-faux amont et aval; type-B avec un simple porte-à-faux amont; type-C avec un simple porte-à-faux aval; et type D sans porte-à-faux (Figures 6.4).

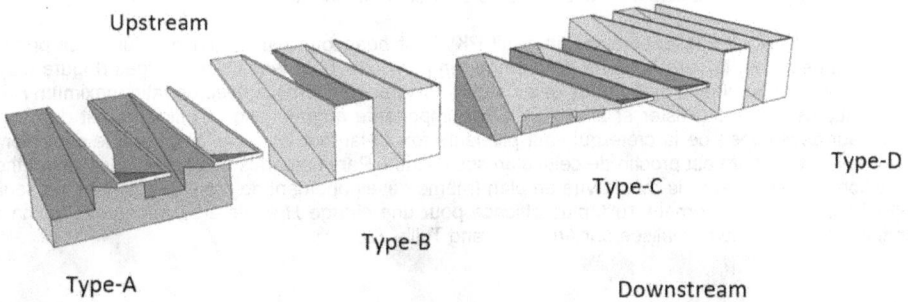

Fig. 6.4
Types of PK weir (adapted from Erpicum *et al.* 2013b)

6.3.   DÉBITANCE

L'évacuateur de crue PKW est un seuil à surface libre et sa débitance Q_P est ainsi dépendante de la charge amont H selon la relation

$$Q_p = \propto \sqrt{2gH^3}$$ (6.1)

Comme résumé par par Leite Ribeiro *et al.* (2012a), deux approches peuvent être choisies pour décrire le facteur de proportion, qui représente les effets de la forme et de la longueur de crête.

En relation avec la longueur développée L, le coefficient de débitance $C_{P,L}$ est en lien étroit avec la forme de la crête. Eq. 6.1 donne (Leite Ribeiro *et al.* 2012b)

$$Q_p = C_{p,L}L\sqrt{2gH^3}$$ (6.2)

Dans cette approche, L varie avec la charge alors que la longueur effective de la crête diminue avec les charges croissantes en raison des effets de noyage des alvéoles amont. $C_{P,L}$ varie ainsi avec la charge incluant à la fois les effets frontaux et latéraux du seuil.

En relation avec la largeur du seuil en crête de barrage W, Eq. 6.1 donne (Ouamane & Lempérière, 2006; Machiels et al. 2011a):

$$Q_p = C_{p,W} W \sqrt{2gH^3} \tag{6.3}$$

avec un coefficient de débitance $C_{P,W}$ incluant à la fois les effets de la forme de la crête et de la longueur développée.

Quel que soit l'approche utilise pour modéliser la débitance du PKW, il est habituel de considerer la débitance du PKW par comparaison avec celle d'un seuil standart linéaire de même largeur en crête de barrage (Q_s). Le ratio d'augmentation de débitance r est défini par Leite Ribeiro et al. (2012a & b) comme

$$r = \frac{Q_P}{Q_S} \tag{6.4}$$

Considérant Eq. 6.3, Eq. 6.4 s'écrit aussi

$$r = \frac{C_P, W}{C_S} \tag{6.5}$$

où C_S est le coefficient de débitance du seuil linéaire standard.

Quel que soit la géométrie, un seuil PKW est beaucoup plus efficace qu'un seuil profilé standard de même largeur en crête de barrage, en particulier pour les faibles charges (Figure 6.5). Cela explique pourquoi la plupart des évacuateurs PKW ont été conçus avec un ratio maximum H/P ratio plus petit que 1 (Pfister et al. 2012). Cette importante augmentation d'efficacité est due à la longueur développée de la crête qui vaut plusieurs fois la largeur en crête, alors que le coefficient de débitance unitaire est proche de celui d'un seuil mince. Par comparaison avec un seuil labyrinthe traditionnel vertical avec la même vue en plan (même développement de crête et emprise), un seuil PKW est approximativement 10% plus efficace pour une charge H égale à sa hauteur P_i, comme indiqué dans une étude réalisée par Anderson and Tullis, (2012).

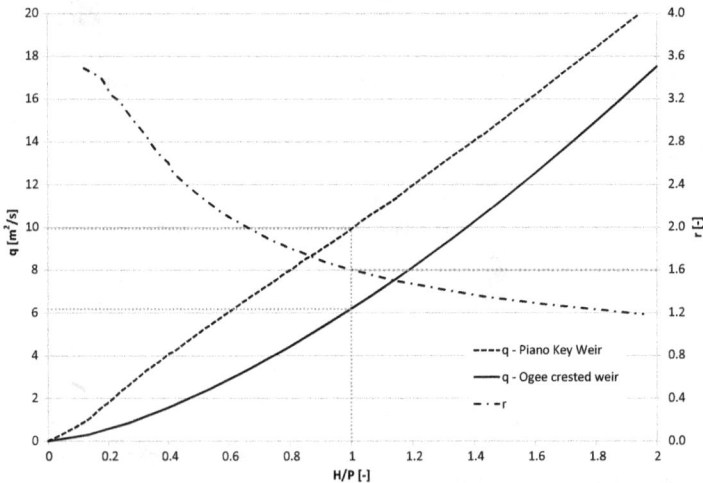

Fig. 6.5
Débitance spécifique (par unité de largeur de crête de barrage) d'un PKW comparé à un seuil profile standard PKW: $P=P_i=P_o=2$ m et $L/W=5$; seuil profile standard: hauteur de dimensionnement=2 m, $C_S=0.494$ (Erpicum et al. 2013b)

6.4. PRINCIPAUX PARAMÈTRES GÉOMÉTRIQUES

6.4.1. Principaux paramètres géométriques

Les principaux paramètres géométriques gouvernant la débitance des PKW sont :

- Ratio "L/W". Ce ratio "labyrinthe" représente la longueur développée. Il s'agit du facteur majeur pour les faibles charges.

- Hauteur du PKW "P". Les hautes valeurs de P améliorent les conditions d'approche et de distribution du débit dans les alvéoles d'entrée. Cela réduit également les effets de submergence dans les alvéoles de sortie.

- Le ratio de largeur des alvéoles "Wi/Wo". De forts ratios Wi/Wo améliorent les conditions d'approche et de distribution dans les alvéoles d'entrée. En revanche, ces forts ratios augmentent les effets de submergence dans les alvéoles de sortie.

- Largeur des murs latéraux Ts. Des murs latéraux épais réduisent significativement l'espace disponible en crête de barrage. Ce facteur structurel peut impacter de 10 à 15% la débitance spécifique du PKW.

6.4.2. Discussion

L'efficacité de la débitance du PKW provident d'effets cumulatifs : (i) trois différents types de surverse (écoulement de seuil linéaire par-dessus la crête de l'alvéole d'entrée, écoulement de seuil linéaire par-dessus la crête de l'alvéole de sortie, et écoulement de seuil latéral par-dessus la crête des murs latéraux, (ii) la longueur développée de ma crête et (iii) la charge hydraulique amont. Des nappes déversantes accrochantes, bondissantes ou élastiques peuvent de produire, en fonction des paramètres géométriques de la structure et de la charge (Machiels et al. 2011a). Plus la charge est importante, plus l'efficacité du seuil PKW, comparée à un seuil linéaire standard, se réduit, par le biais de la réduction progressive de la longueur développée efficace et de la saturation progressive des alvéoles de sortie.

Ouamane et Lempérière, (2006) et Leite Ribeiro et al. (2012a) ont montré le ratio d'amplification de la longueur de crête L/W est le paramètre principal contrôlant la débitance. Une valeur de 5 semble être un compromis raisonnable entre l'efficacité du seuil et la complexité de la structure à construire (Lempérière, 2009; Lempérière et al. 2011), alors que ce ratio L/W ratio des PKW existants est compris entre 4 et 8 (Pfister et al. 2012). Dans une approche complémentaire, Machiels et al. (2011a and 2012b) ont identifié la hauteur du PKW P, le ratio de largeur des alvéoles W_i/W_o et le ratio relatif des porte-à-faux amont et aval B_o/B_i comme les principaux paramètres géométriques influençant l'efficacité hydraulique des PKW pour un ratio donné L/W.

Ces études ont montré l'importance majeur d'augmenter tous les ratios géométriques qui augmentent la section hydraulique de l'alvéole d'entrée, dans la mesure où cette dernière peut être vue comme le « moteur » du PKW. Augmenter la section hydraulique de l'alvéole d'entrée diminue la vitesse des écoulements le long de la crête latérale, et ainsi améliore son efficacité. La valeur maximum des paramètres est atteinte lorsque la capacité de débitance des alvéoles de sortie est affectée. En fait, l'alvéole de sortie peut être vue comme le "frein" du PKW. Ainsi, des sections hydrauliques ou des pentes trop faibles pour les alvéoles de sortie, augmentent le niveau de la surface libre dans les alvéoles de sortie et au niveau des crêtes déversantes des murs latéraux (submergence locale) en limitant significativement l'efficacité du seuil PKW.

Machiels (2012) a trouvé qu'une conception de PKW avec un ratio P/W_u égal à 1.3, un ratio de largeur d'alvéoles W_i/W_o égal à 1.25 et un ratio de longueur de porte-à-faux B_o/B_i égal à 3 fournit la meilleure débitance lorsque le ratio L/W ratio vaut 5. Ce résultat est cohérent avec ceux de Leite Ribeiro et al. (2012a), Anderson and Tullis (2013) et Lempérière et al. (2011).

Cependant, les travaux de Machiel montrent également l'importance des critères techniques et économiques dans la définition d'une conception optimale de PKW. Un PKW haut (P/W_u = 1.3) est plus efficace d'un point de vue hydraulique et pourrait être considérée pour les projets de nouveau barrage. Des PKW plus petits ($P/W_u \approx 0.5$), bien qu'hydrauliquement moins efficaces, pourraient se révéler plus pratiques dans le cadre de projets de réhabilitation de barrages existants. Pour ces derniers, les ratios W_i/W_o et B_i/B_o égaux à 1 semblent pertinents (Erpicum *et al.* 2014).

Chaque conception de PKW est un compromis des paramètres précités afin de trouver une réponse appropriée en fonction des caractéristiques spécifiques de chaque projet. Dans tous les cas, pour une largeur disponible donnée, le PKW a (i) une grande efficacité pour les faibles charges (H/P_i), et (ii) peut fournir une débitance de 2 à 5 fois supérieure à celle d'un seuil profilé standard utilisant la même largeur en crête.

6.5. SUBMERGENCE AVAL

La submergence aval du seuil se produits lorsque le niveau d'eau aval dépasse le niveau de la crête et augmente le niveau amont pour une débitance donnée. Pour la plupart des applications, la submergence de PKW sera à considérer pour les structures en canal ou en rivière et non pour les applications en crête de barrage.

De manière similaire aux seuils labyrinthes traditionnels, la submergence aval des PKW a été étudié avec des modèles physiques en configuration de canal (Belaabed et Ouamane, 2011 ; Cicéro et Delisle, 2013b; Dabling et Tullis, 2012). Il apparait que le comportement des PKW vis-à-vis de la submergence aval est sensible à la géométrie et au type (Figure 6.6), et mérite une attention particulière pour être correctement prédite.

Dabling and Tullis, (2012) confluent que pour les niveaux de submergence relativement faibles, le PKW, le PKW demande moins de charge amont qu'un seuil labyrinthe pour passer un débit donné. Cette augmentation d'efficacité est inférieure à 6%, et cette tendance s'inverse pour des niveaux de submergence plus importants.

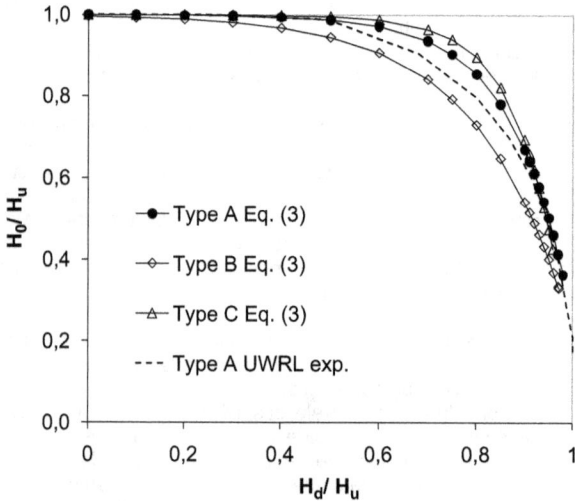

Fig. 6.6
Comparison of the sensitivity to submergence of three PKW different types (Cicéro et Delisle, 2013b)

6.6.   CORPS FLOTTANTS

Les corps flottants sont souvent charriés jusqu'aux structures hydrauliques durant les crues (Pfister *et al.* 2013b). L'accumulation de corps flottants devant les PKW devrait être considérée dans les situations où il y a un potentiel que des corps flottants réduisent la débitance de l'évacuateur. Par exemple, les épisodes orageux dans des bassins versants boisés peuvent induire des écoulements chargés en débris boisés. Selon les conditions de site et les caractéristiques hydrauliques des évacuateurs de crue, ce type de corps flottant peut effectivement transiter à travers le seuil durant les crues et ne nécessiter que peu de maintenance. A l'inverse, certaines conditions peuvent induire de larges accumulations de bois flottant réduisant la débitance de l'évacuateur (-25% peut être observé) et soulever des préoccupations de sécurité des ouvrages.

Une étude systématique conduit par Pfister *et al.* (2013b) afin d'évaluer les interactions entre différentes géométries de PKW et différents types et tailles de bois flottants, indique que la probabilité de blocage par les bois flottant est fortement influencée par le diamètre des troncs et la charge amont. Les effets d'accumulation des bois flottants sur la charge amont varient avec la valeur des conditions de référence de charge amont sans bois flottant. Pour les faibles charges amont de référence, les tests avec accumulation de bois flottant, indiquent une augmentation relative de la charge amont associée avec bois flottants, d'environ 70%. Pour des charges amont de référence plus importantes, l'augmentation relative de charge amont est limitée autour de 20%. Par ailleurs, comme pour la plupart des évacuateurs à seuil libre, la plupart des corps flottants sont naturellement évacués lorsque la charge amont augmente.

6.7.   AÉRATION ET DISSIPATION D'ÉNERGIE

Les observations réalisées des nappes déversantes aval, sur les PKW prototypes existants, pour la plupart équipés de tuyau d'aération sous les porte-à-faux aval des alvéoles d'entrée, montrent généralement une bonne aération et qu'ils ne sont pas sujets à des phénomènes d'adhérence ou de vibration, même pour les très faibles charges (Figures 6.7 et 6.8).

Il n'est pas clair si les organes d'aération sont nécessaires ou non pour les seuils PKW. Certains prototypes n'en sont pas équipés et semblent avoir déversé plusieurs fois sans difficulté de ce genre. Ce point reste à confirmer.

Par défaut et pour être conservative, beaucoup de prototypes PKW ont été équipés avec des organes d'aération dans la mesure où les coûts associés restaient mineurs. C'est normalement le cas pour les PKW fonctionnant sous faible charge. Pour les fortes charges, des méthodes simples de dimensionnement peuvent conduite à des tuyaux d'aération de grand diamètre avec un coût associé non négligeable. Les critères de dimensionnement utilisés étaient basés sur une hypothèse standard de ratio débit d'air / débit d'eau et une vitesse limite de vitesse de l'air dans les tuyaux.

Très peu de mesures d'entrainement d'air dans les tuyaux d'aération ont été réalisées sur les PKW prototypes ou des modèles physiques hydrauliques avec une échelle adéquate. Sur le PKW de Malarce (France) en 2014, EDF a équipé les tuyaux d'aération prototypes avec des appareils de mesure de l'air entraîné (Pinchard *et al.* 2013). Cet évacuateur PKW a déversé une douzaine de fois entre 2014 et 2017 et des résultats préliminaires ont été récemment publiés par by Vermeulen, (2017) incluant une méthode plus raffinée pour dimensionner les tuyaux d'aération. Cette méthode nouvellement développée permet de réduire significativement les caractéristiques du système d'aération.

Fig. 6.7
Écoulement sur le PKW de Malarce (France) avec une charge amont de quelques cm (Photo EDF)

Fig. 6.8
Écoulement sur le PKW d'Escouloubre (France) avec une charge amont de quelques cm (Photo ULg-HECE)

Des tests en laboratoire ont été réalisés avec des PKW installés en amont de coursier enn marche d'escalier (Ho Ta Khanh *et al.* 2011b; Silvestri *et al.* 2013a 2013b). Ils montrent que l'écoulement en aval du PKW est oujours bien aéré, i.e., que le point d'initiation de l'aération est situé immédiatement au pied du seuil PKW (Figure 6.9).

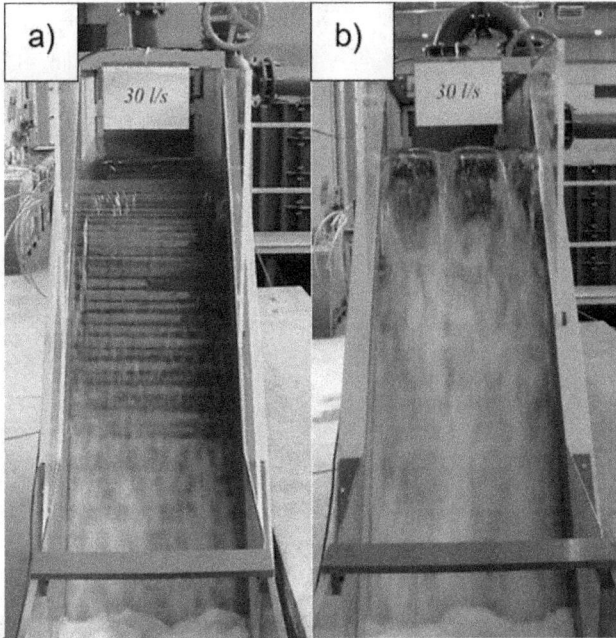

Fig. 6.9
Modification de la localisation du point d'initiation de l'aération point le long d'un coursier
en marche d'escalier avec (a) un seuil profile standard, et (b) un PKW (q = 0,06 m²/s) (adapté de
Silvestri *et al.* 2013a)

Les forts débits spécifiques à l'aval d'un évacuateur PKW impliquent qu'une attention particulière soit portée à la conception des structures aval de dissipation d'énergie, en particulier lorsque le seuil est situé sur la crête d'un grand barrage.

Les solutions de dissipation d'énergie déjà conçues et construites pour les évacuateurs PKW prototypes ne peuvent pas être généralisées et ont toutes été trouvées de manière innovante en utilisant des modèles physiques (Figures 6.10 et 6.11) (Bieri *et al.* 2011 ; Erpicum *et al.* 2011b ; Leite Ribeiro *et al.* 2011).

Fig. 6.10
coursier aval à faible pente en marche d'escalier du PKW du barrage des Gloriettes (Photo EDF)

Fig. 6.11
Coursier en béton lisse et système de saut de ski aval incliné à l'aval du PKW du barrage de Saint-Marc (Photo EDF)

Dans le cadre du projet du barrage des Gloriettes, Bieri *et al.* (2011), ont montré qu'un PKW combine avec un coursier en marche d'escalier conduit à une dissipation d'énergie importante.

6.8. ÉQUATIONS HYDRAULIQUE DE DIMENSIONNEMENT POUR UN PKW DE TYPE A ET MÉTHODE DE DIMENSIONNEMENT

Hydrocoop a proposé une méthode préliminaire très simple pour déterminer la débitance spécifique par unité de largeur de barrage q pour une charge amont H, pour un PKW de type A avec une hauteur maximum des murs latéraux égale à Pm (Lempérière, 2015). Cette formule simplifiée (Eqn. 6.6) s'applique seulement dans la gamme des paramètres définie par l'auteur.

$$q = 4.3\, HP_m^{\,0.5} \tag{6.6}$$

Avec q en m²/s, H et P_m en m (attention à ne pas confondre Pm et P).

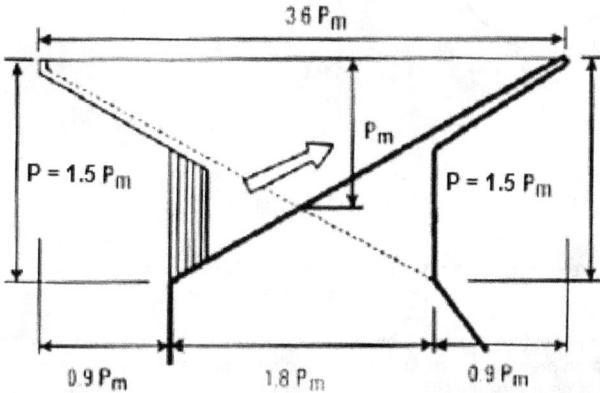

Fig. 6.12
Section transversale type d'un PKW applicable à la formule simplifiée de Lempérière. D'aprés Lempérière (2015).

En parallèle, des séries complètes et systématiques de tests sur des modèles physiques, ont été réalisées par plusieurs laboratoires (Schleiss, 2011). Sur la base de ces séries de test, trois équations hydrauliques générales de débitance des PKW de type A ont été proposées et validées (Kabiri-Samani et Javaheri, 2012 ; Leite Ribeiro *et al.* 2012a ; Machiel *et al.* 2014). Il est important de noter que ces équations obtenues sur base expérimentales ne doivent être utilisées que dans les limites des gammes de paramètres spécifiées par les auteurs, comme cela a été clairement démontré par Pfister *et al.* (2012). Il convient d'insister sur le fait que ces limitations doivent être strictement respectées dans la mesure où de faibles écarts à ces limites, peuvent conduire à des modifications importantes du coefficient de débitance du PKW.

Pfister & Schleiss (2013a) ont comparé ces trois formules hydrauliques de dimensionnement pour un PKW de type A hypothétique, place sur un barrage poids en béton compacté au Rouleau (BCR) avec une largeur de coursier aval W = 100 m, une hauteur de barrage P_d = 30 m sous la fondation du PKW. Un débit de dimensionnement Q_D = 2 500 m³/s a été considéré, amenant à un débit spécifique à évacuer de 25 m/s/m tel que couramment utilise sur un barrage à marche d'escalier.

La géométrie du PKW a une longueur totale amont/aval B = 8.00 m, une hauteur vertical $P = P_i = P_o$ = 5.00 m, une épaisseur des murs latéraux T_S = 0.35 m, R = 0 m (pas de mur parapet), W_i = 1.80 m largeur d'alvéole d'entrée, W_o = 1.50 m largeur d'alvéole de sortie, et $B_i = B_o$ = 2.00 m longueur des porte-à-faux amont et aval. Les caractéristiques suivantes découlent de cette géométrie de PKW :

- largeur d'une unite PKW $W_u = W_i + W_o + 2T_S$ = 4.00 m;

- nombre d'unités $N = W/W_u$ = 25;

- longueur développée en crête $L = W + (2NB) = 500$ m;

- et $L/W = 5.00$, $B/P = 1.60$, $W_i/W_o = 1.20$, $B_i/B = B_o/B = 0.25$, $S_i = S_o = 0.83$.

Dans la mesure où les trois formules de dimensionnement hydrauliques mentionnées précédemment ont été déterminées avec différentes formes de crête, les résultats ont été normalisés pour une forme de crête horizontale. Les courbes résultantes de débitance sont indiquées Figure 6.13 dans leur limite d'application du ratio H/P. Comme on peut le voir, ces trois courbes sont très similaires. De manière générale, l'équation de Kabiri-Samani and Javaheri (2012) prévoit les débitances les plus hautes. Par ailleurs, la courbe de débitance du seuil standard profilé (ogee) est également indiquée Figure 6.14, tire de Vischer and Hager (1999) considérant une charge de dimensionnement $H_D = 5.00$ m pour Q_D.

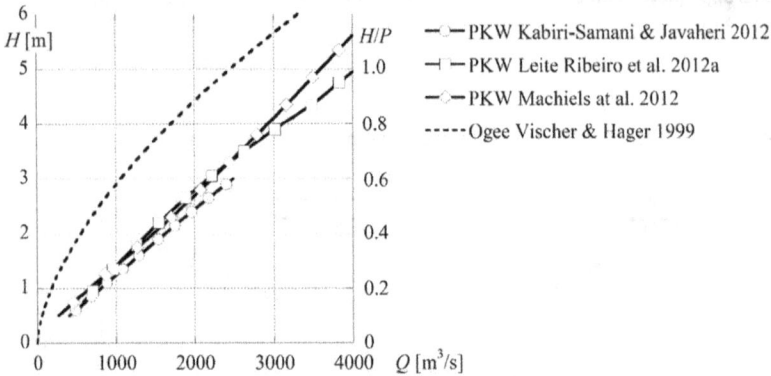

Fig. 6.13
Comparaison de trois formules de débitance pour un PKW de type A et d'un seuil profile standard
(Pfister et Schleiss, 2013a)

Pour une application pratique, on peut conclure que la formule hydraulique PKW la plus appropriée dépend en premier lieu de son domaine d'application et des caractéristiques du prototype. La forme de la crête peut être utilisée comme second critère de sélection de la formule. Si plusieurs formules sont dans le même domaine d'application du prototype, l'approche la plus conservative pourra être utilisée en première instance, afin de rester du côté sécuritaire.

En complément, Machiels et al. (2011b & 2012a) a proposé une méthode de dimensionnement préliminaire très simple afin de sélectionner la géométrie d'un PKW à partir des courbes de débitance existantes et des contraintes spécifiques du projet (débitance maximum, niveau du réservoir, largeur disponible etc.).

6.9. MODÉLISATION PHYSIQUE ET NUMÉRIQUE (CFD)

Le développement récent des évaluateurs PKW a conduit à une comparaison intensive entre la modélisation numérique et les essais en laboratoire hydraulique sur de nombreuses géométries. Ces travaux montrent clairement que l'utilisation de logiciels 3D (commerciaux) est très prometteur pour prédire avec une bonne précision (+/- 10%) la débitance d'une géométrie de PKW (Pralong et al. 2011b; Lefebvre et al. 2013).

Par ailleurs, le logiciel libre WOLF1D PKW (disponible sur http://www.pk-weirs.ulg.ac.be) a été développé par le groupe de recherche de l'HECE de l'Université de Liège, basé sur des essais de modèles physiques à grande échelle de Machiels (2012). A partir d'une géométrie de PKW, ce logiciel libre est capable de calculer très rapidement les écoulements à travers une demi-unité PKW pour une gamme donnée de débitance amont de la structure. Les résultats de calcul sont la charge amont / courbe de débitance (+/-10% de précision), le niveau d'eau à l'amont du seuil dépendant de la débitance et de la distribution de la profondeur d'eau le long des alvéoles d'entrée et de sortie.

Néanmoins, la modélisation numérique CFD ne pas reproduire toutes les caractéristiques des écoulements tels que les instabilités par exemple, les conditions complexes d'approche des écoulements ou des géométries spécifiques des piles, ainsi que les conséquences du blocage par les corps flottants. De plus, les écoulements aval et la dissipation d'énergie sont des questions critiques pour l'efficacité globale de l'évacuateur, qui ne peuvent pas être évalués précisément avec les modèles numériques.

Ainsi, la géométrie finale du PKW avec le système de dissipation d'énergie devrait être testé avec un modèle physique, qui reste le seul moyen de reproduire correctement les caractéristiques complètes des écoulements avec le niveau requis de précision. Il est utile de mentionner que la plupart des grands prototypes de PKW ont été validés et optimisés avec la modélisation physique.

Cependant, quelques PKW récents ont été dimensionnés sans modèle physique : Campauleil and Record (France, Dabertrand *et al.* 2017), Black Esk (UK, Acker *et al.* 2013). La débitance est effectivement bien connue pour les configurations classiques de PKW. Pour ces projets, la configuration aval était simple sans question sérieuse de dissipation ou d'effet complexe 3D. Ainsi, pour ces cas, il a été considéré qu'un modèle physique n'était pas nécessaire.

6.10. PROBLÈMATIQUES STRUCTURELLES ET CONSTRUCTIVES

Les questions structurelles dépendent fortement de la géométrie du PKW (taille et nombre des porte-à-faux), cas de charge (influence de la glace ou sismicité de site par exemple) et taux unitaires de chargement, caractéristiques du projet (tels que localisation sur une crête de barrage de poids ou isolé en rive). Les références existantes illustrent quelques ratios spécifiques, tels que :

- 50 à 100 kg d'acier de ferraillage par m^3 de béton ; et

- 60 à 130 m^2 de coffrage par m de largeur

C'est une pratique courant d'ajouter du béton ou des ancrages afin d'atteindre les facteurs de stabilité requis. Les cas de charge thermique (une face est au soleil, l'autre protégée par l'eau) ou les charges de poussée de la glace, peuvent facilement devenir les paramètres de contrôle du dimensionnement. Une attention particulière doit être donnée aux modes de construction, tels que la préfabrication (Figure 6.14) ou l'utilisation d'acier à la place du béton, comme pour des prototypes construit jusqu'à présent.

Fig. 6.14
Évacuateur de crue du réservoir de Back Esk– Évacuateur PKW constitué d'éléments béton préfabriqué (Photo Black et Veatch Ltd)

6.11. INGÉNIÉRIE

Jusqu'à présent, les évacuateurs PKW ont été conçus et construits pour la maîtrise des crues dans les configurations suivantes :

1. Vallée étroite avec seulement un espace disponible limité pour installer une nouvelle structure ;

2. Sur la crête d'un barrage poids, où le PKW est facilement implantable grâce à sa faible emprise au sol, permettant une maximisation de la capacité active de stockage du réservoir pour un niveau maximum donné du réservoir ;

3. Sur la crête de longs barrages (barrages vannés de type barrage mobile en rivière) dans des régions de plaine. Dans ce cas, l'utilisation conjointe de PKW et de vannes (dont le nombre est minimisé par rapport aux alternatives traditionnelles entièrement vannées), permet de minimiser les zones inondées durant les crues en limitant le niveau maximum du réservoir, en comparaison des solutions avec un seuil linéaire.

Les pratiques futures de dimensionnement devraient être orientées vers la maîtrise des coûts sur la base des connaissances hydrauliques disponibles. Par exemple, les recherches en cours explorent : (i) les profiles hydrauliques standards (permettant des économies sur la conception et la construction), et (ii) un meilleur compromis entre l'utilisation des PKW, des vannes et des hausses fusibles (afin de retarder le basculement du premier élément fusible).

À ce titre, le barrage de Van Phong au Vietnam, 475-m de long and 7- m de haut sur le terrain naturel, avec une crue de dimensionnement de 15 400 m³/s, constitue un bon exemple de la troisième configuration (Figure 6.15) mentionné ci-dessus. La conception initiale incluait 28 vannes radiales de 15 m de large et 7,5 m de haut. La conception finale, pour le même niveau maximum de réservoir, inclut 10 vannes radiales (et 8 auraient probablement suffi) en partie centrale et 60 unité PKW avec une longueur totale de 302 m de chaque côté. Comparé à une solution labyrinthe traditionnelle, le PKW offre une emprise de fondation réduite sur le toit du rocher, relativement profond, ce qui requiert moins d'excavation et de volume de béton. Les principaux avantages de l'alternative finale furent : (i) des économies sur les investissements et la maintenance, (ii) un meilleur niveau de sécurité en exploitation, et (iii) une meilleure intégration dans l'environnement. Considérant les récents dysfonctionnements de plusieurs vannes durant des crues, beaucoup d'ingénieurs Vietnamiens pensent qu'une solution sûre consiste maintenant à combiner, autant que possible, des vannes et des évacuateurs de crue à seuils libres. Dans ce contexte, les PKW constituent souvent une bonne alternative, particulièrement lorsque la longueur disponible en crête est limité.

Fig. 6.15
Le barrage de Van Phong barrage au Vietnam. (Photo M. Ho Ta Khanh)

Un nouveau projet avec une configuration similaire est actuellement en cours d'études pour le barrage de Xuân Minh.

6.12. RECHERCHES EN COURS

Les recherches en cours et les démarches de conception s'orientent vers des nouveaux développements dé géométrie de PKW (mur latéral réduisant l'alvéole d'entrée et élargissant celle de sortie ; nouvelle configuration hybride associant labyrinthe et PKW) et l'analyse des caractéristiques des écoulements aval (aération et dissipation d'énergie). Les recherches concernent aussi les aspects structurels et constructifs tels que l'utilisation de l'acier ou de structures sandwich acier/béton, ainsi que la préfabrication.

Des recherches plus récentes concernent l'application des PKW et des labyrinthes en rivière pour remplacer les vannes de régulation. Pfister (2017) a étudié la formation d'érosion aval en pied. Belzner (2017) a conduit une étude générale sur les PKW et labyrinthes concernant l'influence du niveau aval, le passage des corps flottants et le passage des sédiments à travers les seuils. Les premiers résultats sont prometteurs et pourraient permettre de remplacement en partie les vannes en rivières par des seuils labyrinthes ou PKW.

Il est à noter que le passage des sédiments à travers les PKW a été initialement étudiée par Das Singhal *et al.* (2011) pour le projet Sawra Kuddu qui a été construit en 2013. Le passage des sédiments a été étudié avec un modèle physique. L'article montre que les PKW type C (pas de porte-à-faux amont) sont légèrement plus efficaces et permettent de faire passer la plupart des sédiments par-dessus le seuil. Les résultats importants étaient nécessaires pour valider la faisabilité des PKW pour les aménagements au fil de l'eau avec un taux important de sédiments à faire transiter.

6.13. EXEMPLES DE PROTOTYPE PIANO KEY WEIR

La bibliographie incluse à la fin du chapitre content des références pour plusieurs prototypes PKW. Un résumé de ces structures est présenté dans le tableau suivant. Presque 30 projets de PKW ont été finalisés durant les 10 dernières années dans 8 pays différents. Il s'agit d'un développement très rapide considérant les constantes de temps des projets hydrauliques.

Table 6.1
Évacuateurs de crue Piano Key Weir Prototypes

Nom	Pays	Configuration	Q_{projet} m³/s	P m	H_{projet} m	Date d'achèvement	Source
Beaufort	France	Crête de barrage	70	1.30	0,70	2013	http://www.pk-weirs.ulg.ac.be/
Black Esk	Ecosse	Crête de tulipe	183	2.10	0.97	2013	Ackers et al. 2013
Campauleil	France	Crête de barrage	120	5.35	0.90	2014	Laugier et al. 2017a
Charmines	France	Crête de barrage	300	4.38	1.00	2015	Valley et al. 2017a
Da Gang3	Vietnam	Crête de barrage	7300	5.50	5.00	2016	Ho Ta Khanh et al. 2017
Dak Mi 4B	Vietnam	Crête de barrage	500	3.75	2.00	2013	Ho Ta Khanh et al. 2017
Dak Mi 3	Vietnam	Crête de barrage	6550	5.00	3.50	2013	Ho Ta Khanh et al. 2017
Dak Rong 3	Vietnam	Crête de barrage	6550	5.00	3.50	2013	Ho Ta Khanh et al. 2017
Emmenau	Suisse	Rivière	4.35	1.20	0.45	2013	Eichenberger, 2013
Escouloubre	France	Rive	13	1.80	0.65	2011	Erpicum et al. 2013c
Etroit	France	Crête de barrage	82	5.30	0.95	2009	Laugier et al. 2011
Gage	France	Right bank	455	6.00	1.75	2017	Laugier et al. 2017b
Giritale	Sri Lanka	Rivière		2.40	0.45	2013	Jayatillake and Perera. 2013
Gloriettes	France	Right bank	90	3.00	0.80	2010	Bieri et al. 2011
Gouillet	France	Right bank	18	1.20	0.35	2011	http://www.pk-weirs.ulg.ac.be/
Goulours	France	Right bank	68	3.10	0.95	2006	Laugier, 2007
Hazelmere	Afrique du Sud	Crête de barrage	4300	9.00	3.23	2017	Booyse et al. 2017
Loombah	Australie	Rive	416	2.50	2,5	2013	http://www.pk-weirs.ulg.ac.be/
Malarce	France	Crête de barrage	570	4.40	1.50	2012	Pinchard et al. 2013
Oule	France	Rive	72	1.00	1.00	2017	http://www.pk-weirs.ulg.ac.be/
Ouljet Mellegue*	France	Rive	3800	6.80	5.00	*	http://www.pk-weirs.ulg.ac.be/
Rambawa Tank	Sri Lanka	Crête de barrage	28	1.00	0,35	2015	Jayatillake H.M. & al 2017
Rattling Lake	Canada	Crête de barrage				2011	www.cda.ca
Rassisse	France	Crête de barrage	306+ +101	3.8/2. 7	1.30	2015	Bail et al. 2013
Raviège	France	Crête de barrage	300	4.67	1.40	2015	Cubaynes et al. 2017
Record	France	Crête de barrage	1350	4.67	3.00	2016	Dabetrand et al. 2017
Saint Marc	France	Crête de barrage	138	4.20	1.35	2008	Laugier et al. 2009
Sawra Kuddu	Inde	Crête de barrage	2500	10.45		2013	Das Singhal et al. 2011
Van Phong	Vietnam	Barrage (Rivière)	8750	5.00	5.20	2015	Ho Ta Khanh et al. 2014
Xuan Minh*	Vietnam	Barrage (Rivière)	9700	7.50	9.00	*	Ho Ta Khanh et al. 2017

* under construction

Les commentaires suivants peuvent être faits :

- Les prototypes PKW concernent une large gamme de débit spécifique avec des charges hydrauliques variant de 30 cm à plus de 9 m. En parallèle, la débitance totale va de quelques douzaines de m³/s à plus de 10 000 m³/s;

- Les PKW ont été utilisés à la fois pour des projets de réhabilitations et de barrages neufs; et

- Les PKW ont été installés sur la crête de plusieurs sortes de barrages : poids béton, voûte, remblais, barrage en rivière.

Ces points soulignent la grande flexibilité des PKW qui peuvent être utilisés à la fois pour des petits et des grands projets, pour des projets neufs ou de réhabilitation.

6.14. PKW OU SEUIL LABYRINTHE VERTICAL TRADITIONNEL ?

Comparer les PKW et les seuils labyrinthes verticaux traditionnels est un exercice difficile. Des tentatives pour comparer les deux options sont proposées par Blancher, (2011) ; Anderson, (2013) et Paxson (2014). Cela amène aux considérations suivantes :

- Blancher (2011) et Anderson (2013) ont proposé une comparaison hydraulique de structures comparables basée sur des modèle numériques (Blancher) et des modèles physiques (Anderson). La comparaison est effectuée pour la même hauteur structurelle P et la même vue en plan (longueur développée égale)

- Les résultats hydrauliques montrent que les PKW ont une débitance 10 à 15% plus forte que les labyrinthes traditionnels en fonction de la charge, dû au fait que les murs verticaux créent un obstacle induisant des pertes de charge. Cela est clairement visible avec les schémas d'écoulement de recirculation devant les murs verticaux.

- Paxson (2014) a proposé une approche globale essayant d'inclure à la fois les aspects hydrauliques et structurels afin d'obtenir une solution économique optimisée.

- Les murs verticaux pourraient être plus simples à construire, mais ils demandent plus de ferraillage interne. En fait, la hauteur moyenne des murs de labyrinthe est P alors que c'est P/2 pour les PKW. Ainsi, les efforts de moments moyens dans les structures labyrinthes sont beaucoup plus importants que pour les PKW, et donc le ferraillage.

- Un évacuateur labyrinthe requiert un radier de fondation alors que le PKW ne requiert pas la même épaisseur grâce à sa géométrie.

- A cause de leur large emprise en fondation, les labyrinthes conventionnels ne sont pas compatibles avec une construction sur la crête d'un barrage en béton (poids ou voûte).

- Les labyrinthes conventionnels peuvent être une option si les unités de labyrinthe sont petites (1 ou 2 m) ou si le seuil est construit sur une rive relativement plate. Le choix dépendra des accès de site, des méthodes et matériels des entreprises de construction.

6.15. RÉFÉRENCES

Ackers, J.C., F.C.J. Bennett, T.A. Scott and G. Karunaratne, (2013) - "Raising of the bellmouth spillway at Black Esk reservoir using Piano Key weirs". *Labyrinth and piano key weirs II - PKW 2013*, CRC Press, London, 235-242

Anderson, R.M. and B.P Tullis (2012) - "Comparison of Piano Key and Rectangular Labyrinth Weir Hydraulics", *J. Hydraulic Eng.* 138, 358-361

Anderson, R.M. and B.P Tullis (2013) - "Piano Key Weir Hydraulics and Labyrinth Weir comparison". *Journal of Irrigation and Drainage Engineering*, 139(3), 246-253.

Bail, A., L. Deroo and J.P Sixdenier (2013) - "Designing a new spillway at Rassisse Dam". *Labyrinth and piano key weirs II - PKW 2013*, CRC Press, London, 169-176.

Barcouda, M., F. Laugier, O. Cazaillet, C. Odeyer, P. Cochet, B.A. Jones, S. Lacroix and J.P. Vigny, J.P. (2006) - "Cost effective increase in storage and safety of most existing dams using fusegates or P.K.Weirs", *Proceedings of the 22nd ICOLD Congress*. (Q84, R78). Barcelona, Spain.

Belaabed, F. and A. Ouamane (2011) - "Contribution to the study of the Piano Key weir submerged by the downstream level". *Labyrinth and Piano Key Weirs – PKW 2011*, CRC Press, Leiden, 89-96.

Belzner F., J. Merkel, M. Gebhardt and C. Thorenz (2017) - "Piano Key and Labyrinth Weirs at German Waterways: recent and future research of the BAW". *Labyrinth and piano key weirs III - PKW 2017*, CRC Press, London

Bieri M., M. Federspiel, J.L. Boillat, B. Houdant, L. Faramond and F. Delorme (2011) - "Energy dissipation downstream of Piano Key Weirs - Case study of Gloriettes Dam (France)", *Labyrinth and piano key weirs-PKW 2011*, CRC press, London, 123-130.

Blanc, P. and F. Lempérière (2001) - "Labyrinth spillways have a promising future", *Hydropower & Dams*, 8(4):129-131.

Blancher B., F. Montarros, and F. Laugier F (2011) "Hydraulic comparison between piano-keys weir and labyrinth spillways". *Labyrinth and Piano Key Weirs – PKW 2011*, CRC Press, Leiden, 89-96.

Booyse D. (2017) - "The raising of Hazelmere Dam by means of Piano Key Weir". *Labyrinth and piano key weirs III - PKW 2017*, CRC Press, London.

Botha A.J., J.P. Fitz, A.J. Moore, F.E. Mulder and N.J. Van Deventer (2013) - "Application of the Piano Key weir spillway in the Republic of South Africa". *Labyrinth and piano key weirs II - PKW 2013*, CRC Press, London, 185-194.

Cicéro, G-M., J-R. Delisle, V. Lefebvre and J. Vermeulen (2013a) - "Experimental and numerical study of the hydraulic performance of a trapezoidal Piano Key weir", *Labyrinth and Piano Key Weirs II – PKW 2013*, CRC Press, Leiden, pp. 265-272.

Cicéro, G-M. and J-R. Delisle (2013b) - "Discharge characteristics of Piano Key weirs under submerged flows". *Labyrinth and Piano Key Weirs II – PKW 2013*, CRC Press, Leiden, pp. 101-109.

Cubaynes M., F. Laugier and V. Nagel (2017) - "Construction of a Piano Key Weir spillway at La Raviège dam". *Labyrinth and Piano Key Weirs III – PKW 2017*, CRC Press

Dabling, M. and B. Tullis (2012) - "Piano Key Weir Submergence in Channel Applications." *J. Hydraul. Eng.*, 138(7), 661–666.

Dabertrand F., J. Vermeulen and B. Blancher (2017) - "Construction of a Piano Key Weirs spillway at Record dam". (2017). "*Labyrinth and Piano Key Weirs III – PKW 2017*, CRC Press

Das Singhal G. and N. Sharma (2011) - "Rehabilitation of Sawara Kuddu Hydroelectric Project - Model studies of Piano Key Weir in India". *Labyrinth and piano key weirs-PKW 2011*, CRC Press, London, 241-250.

Dugué V., F. Hachem, J.L. Boillat, V. Nagel, J.P. Roca and Laugier F. (2011) - "PK Weir and flap gate spillway for the Gage II Dam". *Labyrinth and piano key weirs-PKW 2011*, CRC Press, London, 35-42.

Eichenberger, P. (2013) - "The first commercial Piano Key weir in Switzerland". *Labyrinth and piano key weirs II - PKW 2013*, CRC Press, London, 227-234

Erpicum, S., F. Laugier, J-L. Boillat, M. Pirotton, and B. Reverchon (2011a) - *Labyrinth and piano key weirs – PKW 2011*. Schleiss, A.J, eds. CRC Press, Boca Raton Fl, USA.

Erpicum S., V. Nagel and F. Laugier (2011b) - "Piano Key Weir design study at Raviege dam". *Labyrinth and piano key weirs-PKW 2011*, CRC Press, London, 43-50.

Erpicum, S., O. Machiels, B.J. Dewals, M. Pirotton and P. Archambeau (2012) - "Numerical and physical modelling of Piano Key Weirs", *Proceedings of Asia 2012 Conference*, Chiang Mai, Thailand.

Erpicum, S., F. Laugier, M. Pfister, M. Pirotton and G-M. Cicéro (2013a) - *Labyrinth and piano key weirs II – PKW 2013*. Schleiss, A.J, eds. CRC Press, Boca Raton, Fl, USA

Erpicum, S., O. Machiels, B. Dewals, P. Archambeau and M. Pirotton (2013b) - "Considerations about the optimum design of PKW", *Proc. Intl. Conf. Water Storage and Hydropower Development for Africa (Africa 2013)*, Addis Ababa (Ethiopia), CD 13.04.

Erpicum, S., A. Silvestri, B.J. Dewals, P. Archambeau, M. Pirotton, M. Colombié and L. Faramond (2013c) - "Escouloubre Piano Key weir: prototype versus scale models". *Labyrinth and piano key weirs II - PKW 2013*, CRC press, London

Erpicum, S., P. Archambeau, M. Pirotton and B.J. Dewals (2014) - "Geometric parameters influence on Piano Key Weir hydraulic performances". *5th IAHR International Symposium on Hydraulic Structures*, Brisbane, Australia, (1-8). 25-27 June 2014. doi:10.14264/uql.2014.31

Erpicum, S., F. Laugier, M. Ho Ta Khan and Pfister (2017) - *Labyrinth and piano key weirs III – PKW 2017*. CRC Press, Boca Raton, Fl, USA

Ho Ta Khanh M., D. Sy Quat and D. Xuan Thuy (2011a) - "P.K weirs under design and construction in Vietnam (2010)". *Labyrinth and piano key weirs-PKW 2011*, CRC Press, London, 225-232.

Ho Ta Khanh M., T.C. Hien and T.N. Hai (2011b) - "Main results of the P.K weir model tests in Vietnam (2004 to 2010)", *Labyrinth and piano key weirs-PKW 2011*, CRC press, London, 191-198.

Ho Ta Khanh, M., T. Chi Hien and D. Sy Quat (2012) - "Study and construction of PK Weirs in Vietnam (2004 to 2011)", *Proceedings of Asia 2012 Conference*, Chiang Mai, Thailand.

Ho Ta Khanh, M. (2017) - "History and development of Piano Key weirs in Vietnam from 2004 to 2016". *Labyrinth and piano key weirs III - PKW 2017*, CRC press, London.

Jayatillake H.M. and K.T.N. Perera (2013) - "Design of a Piano-Key Weir for Giritale dam spillway in Sri Lanka". *Labyrinth and piano key weirs II - PKW 2013*, CRC Press, London, 151-158.

Jayatillake H.M. and K.T.N. Perera (2017) - "Adoption of a type-D Piano Key Weir spillway with tapered noses at Rambawa Tank, Sri Lanka". *Labyrinth and piano key weirs III - PKW 2017*, CRC Press.

Kabiri-Samani, A. and A. Javaheri (2012) - "Discharge coefficient for free and submerged flow over Piano Key weirs". *Journal of Hydraulic Research*, 50(1), pp. 114-120.

Laugier F. (2007) - "Design and construction of the first Piano Key Weir (PKW) spillway at the Goulours dam", *International Journal of Hydropower and Dams* 14 (5), 94-101.

Laugier F., A. Lochu C. Gille M. Leite Ribeiro and J-L Boillat (2009) - "Design and construction of a labyrinth PKW spillway at Saint-Marc dam, France", *International Journal of Hydropower and Dams* 16 (5), 100-107.

Laugier F., C. Gille and O. Cazaillet (2011) - "Adaptation of Piano Key Weir (PKW) spillway solution to upgrade l'Etroit dam affected by concrete swelling pathology", *Proceedings of Hydro 2011 conference*.

Laugier F., J. Vermeulen and V. Lefebvre (2013) - "Overview of Piano Key Weirs experience developed at EDF during the past few years". *Labyrinth and piano key weirs II - PKW 2013*, CRC press, London, 213-226.

Laugier F. and J. Vermeulen (2017a) - "Overview of design and construction of 11 Piano Key Weirs spillways developed in France by EDF from 2003 to 2016". *Labyrinth and piano key weirs III - PKW 2017*, CRC press, London.

Laugier F., B. Blancher, S. Bouassida and V. Nagel (2017b) - "Hydrothermal - season based design of a new flood spillway at Gage II dam". In French. *Proceedings of CFBR / SHF conference, Chambery, France.*

Lefebvre, V., J. Vermeulen and B. Blancher (2013) - "Influence of geometrical parameters on PK-weirs discharge with 3D numerical analysis". *Labyrinth and piano key weirs II - PKW 2013*, CRC press, London, 49-56

Leite Ribeiro M., M. Bieri, J-L. Boillat, A.J. Schleiss, F. Delorme and F. Laugier (2009) - "Hydraulic capacity improvement of existing spillways - Design of Piano Key Weirs", in *Proceedings of 23rd ICOLD Congress*, Brasilia, Brazil, Q.90, R.43.

Leite Ribeiro M., J-L. Boillat, A. Schleiss and F. Laugier (2011) - "Coupled spillway devices and energy dissipation system at St-Marc Dam (France)", *Labyrinth and piano key weirs-PKW 2011*, CRC press, London, 113-121.

Leite Ribeiro, M., M. Pfister, A.J. Schleiss and J-L.Boillat (2012a) - "Hydraulic design of A-type Piano Key Weirs", *Journal of Hydraulic Research*, 50(4):400-408.

Leite Ribeiro, M., M. Bieri, J-L. Boillat, A.J. Schleiss, G. Singhal and N. Sharma (2012b) - "Discharge capacity of Piano Key Weirs", *Journal of Hydraulic Engineering*, 138:199-

Lempérière, F., J-P. Vigny and A. Ouamane, A. (2011) - "General comments on Labyrinths and Piano Key Weirs: the past and present", *Labyrinth and Piano Key Weirs – PKW 2011*, CRC Press, Leiden, 17-24.

Lempérière F. and A. Ouamane (2015) - "Increasing the discharge capacity of free-flow spillway fivefold", *Hydropower & Dams*, issue 65):80-82.

Loisel, P.E., P. Valley and F. Laugier, F. (2013) - "Hydraulic physical model of Piano Key weirs as additional flood spillways on the Charmine dam". *Labyrinth and piano key weirs II - PKW 2013*, CRC Press, London, 195-202.

Machiels, O., S. Erpicum, B. Dewals, P. Archambeau and M. Pirotton (2011a) - « Experimental observation of flow characteristics over a Piano Key Weir". *J. Hydraulic Res.* 49(3), 359-366.

Machiels O., S. Erpicum, P. Archambeau, B.J. Dewals and M. Pirotton M. (2011b) - "Piano Key Weir preliminary design method - Application to a new dam project", *Labyrinth and piano key weirs-PKW 2011*, CRC Press, London, 199-206.

Machiels, O. (2012) - *Experimental study of the hydraulic behaviour of Piano Key Weirs*, PhD thesis. HECE research unit, University of Liège, Belgium. http://hdl.handle.net/2268/128006.

Machiels O., S. Erpicum, P. Archambeau, B.J. Dewals and M. Pirotton (2012a) - "Method for the preliminary design of Piano Key weirs". *La Houille Blanche*, 4-5, 14-18

Machiels, O., S. Erpicum, P. Archambeau, B.J. Dewals and M. Pirotton (2012b) - "Parapet wall effect on Piano Key Weirs efficiency". *Journal of Irrigation and Drainage Engineering*, 139(6), 506-511.

Machiels, O., M. Pirotton, P. Archambeau, B.J. Dewals and S. Erpicum (2014) - "Experimental parametric study and design of Piano Key Weirs". *Journal of Hydraulic Research*, 52(3), 326-335

Ouamane, A. and F. Lempérière (2003) - "The piano keys weir: a new cost-effective solution for spillways", *Hydropower & Dams*, 10(5):144-149.

Ouamane, A. and F. Lempérière (2006) - "Design of a new economic shape of weir". *Dams and Reservoirs, Societies and Environment in the 21st Century*, Berga et al. (eds), Taylor & Francis, London: 463-470.

Paxson G.S., B.P. Tullis and D.J. Hertel (2013) - "Comparison of Piano Key Weirs with labyrinth and gated spillways: hydraulics, cost, constructability and operations". *Labyrinth and piano key weirs II - PKW 2013*, CRC Press, London

Paxson, G. and F. Laugier (2014) - Labyrinth and Piano Key Weirs, Perspectives and Case Histories from the USA and France). *Proceedings of ASDSO national Conference*

Pfister, M., S. Erpicum, O. Machiels, A. Schleiss and M. Pirotton (2012) - "Discharge coefficient for free and submerged flow over Piano Key weirs - Discussion", *Journal of Hydraulic Research*, 50(6):642-645.

Pfister, M. and A.J. Schleiss (2013a) - "Comparison of hydraulic design equations for A-type Piano Key weirs". Proc. Intl. Conf. Water Storage and Hydropower Development for Africa (Africa 2013), Addis Ababa (Ethiopia), CD 13.05.

Pfister, M., D. Capobianco, B. Tullis and A. Schleiss (2013b) - "Debris-Blocking Sensitivity of Piano Key Weirs under Reservoir-Type Approach Flow." *J. Hydraul. Eng.*, 139(11), 1134–1141.

Pinchard, T., J-L.Farges, J-M. Boutet, A. Lochu and F. Laugier (2013) - "Spillway capacity upgrade at Malarce dam: construction of an additional piano key weir spillway". *Labyrinth and piano key weirs II - PKW 2013*, CRC Press, London, 243-252

Pralong, J., J. Vermeulen, B. Blancher, F. Laugier, S. Erpicum, O. Machiels, M. Pirotton, J-L. Boillat, M. Leite Ribeiro and A. Schleiss (2011a) - "A naming convention for the Piano Key Weirs geometrical parameters". *Labyrinth and piano key weirs - PKW 2011*, CRC press, London, 271-278.

Pralong J., F. Montarros, B. Blancher and F. Laugier (2011b) - "A sensitivity analysis of Piano Key Weirs geometrical parameters based on 3D numerical modelling", *Labyrinth and piano key weirs-PKW 2011*, CRC Press, London, 133-139.

Schleiss, A.J. (2011) - "From labyrinth to piano key weirs: A historical review". *Labyrinth and piano key weirs-PKW 2011*, CRC Press, London, 3-15.

Silvestri, A., S. Erpicum, P. Archambeau, B. Dewals and M.Pirotton (2013a) - "Stepped spillway downstream of a Piano Key weir – Critical length for uniform flow". *International Worskhop on hydraulic structures*. Bundesanstalt für Wasserbau, Karlsruhe, Germany, 99-107.

Silvestri, A., P. Archambeau, M. Pirotton, B. Dewals and S. Erpicum (2013b) - "Comparative analysis of the energy dissipation on a stepped spillway downstream of a Piano Key weir". *Labyrinth and piano key weirs II - PKW 2013*, CRC Press, London, 111-120.

Truong Chi, H., S. Huynh Thanh and M. Ho Ta Khanh (2006) - "Results of some piano keys weir hydraulic model tests in Vietnam", *Proceedings of the 22nd ICOLD Congress*. (Q87, R39). Barcelona, Spain.

Valley P. and B. Blancher (2017) - "Construction and testing of two Piano Key Weirs at Charmines dam". *Labyrinth and piano key weirs III - PKW 2017*, CRC Press, London

Vermeulen, J., F. Laugier, L. Faramondand and C. Gille (2011) - "Lessons learnt from design and construction of EDF first Piano Key Weirs". *Labyrinth and piano key weirs - PKW 2011*, CRC press, London, 215-224

Vermeulen J., C. Lassus and T.C. Pinchard (2017) - "Design of a Piano Key Weir aeration network". *Labyrinth and piano key weirs III - PKW 2017*, CRC press, London

Vischer, D. and W.H. Hager (1999) - *Dam Hydraulics*. Wiley, Chichester UK.

7.   ÉVACUATEURS DE CRUE EN TUNNEL, EN PUITS ET À VORTEX

7.1.   INTRODUCTION

Ce chapitre considère différents types d'évacuateurs de crue en tunnel, tels que les tunnels à prise « haute », ou à prise « basse » (ou sortie basse), en vortex, en puits et en charge. Les stratégies d'aménagement général, la structure de contrôle du débit, l'ouvrage de transfert du débit et l'ouvrage aval de sortie, ainsi que les problèmes d'exploitation, l'analyse des risques et des considérations seront discutées. Des nouveaux développements, innovations et applications sur la conception des évacuateurs en tunnel pour des grands barrages seront donnés, avec attention particulière sur ce qui a été accompli au cours des dernières 20 ou 30 années. L'analyse des risques concernant soit le contrôle du risque de cavitation, soit les opérations d'urgence sera abordé. Des études de cas appropriées et valides sont données comme référence.

7.2.   STRATÉGIES D'AMENAGEMENT GENERAL POUR LES EVACUATEURS A HAUTER D'EAU ELEVEE

7.2.1.   Emplacement, type et dimension

Un évacuateur en tunnel peut être aménagé sur un ou sur les deux côtés des appuis du barrage. Le dernier cas se passe lorsque le barrage est construit dans une vallée encaissée avec grandes crues et où les installations de décharge superficielle ne sont pas suffisantes pour évacuer les crues. La sélection et la conception des évacuateurs en tunnel dépendent du type de barrage, de la capacité de débit totale, du système de dérivation de la rivière et des besoins en matière de gestion et exploitation. Le tableau 7-1, à la fin de ce chapitre, donne les caractéristiques de certains grands évacuateurs en tunnel typiques dans le monde qui sont en cours de conception, construction et/ou exploitation.

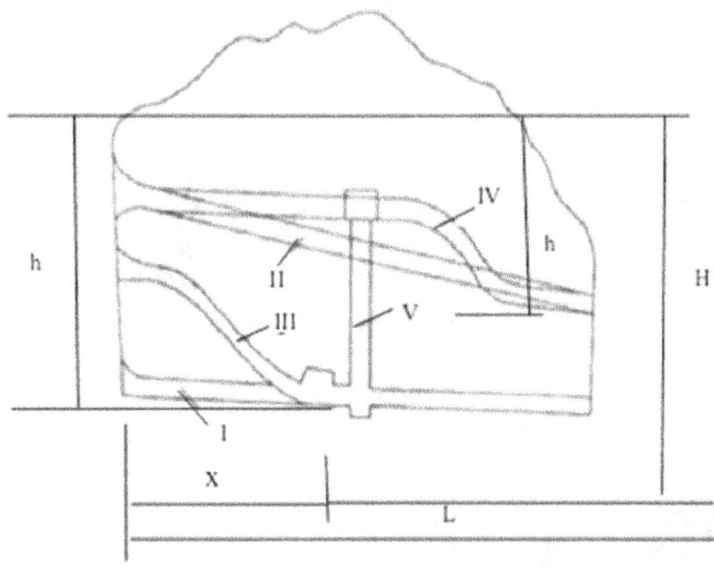

Fig. 7.1
Classification des dispositions d'un évacuateur en tunnel dans une coupe longitudinale

La Figure 7.1 représente une classification de cinq dispositions typiques des tunnels, basée sur un agencement selon une coupe longitudinale. Les tunnels peuvent aussi être classifiés par type de dissipation d'énergie, comme illustré dans la Figure 7.2.

La dissipation du flux d'énergie dans les tunnels de Type I jusqu'à IV à lieu à l'extérieur du tunnel. L'écoulement à forte vitesse conserve une grande quantité d'énergie dans le tronçon aval à surface libre. Par ailleurs, l'énergie de l'écoulement dans les tunnels de Type V est dissipée à l'intérieur du tunnel et le débit atteint la rivière avec une vitesse faible.

Les dispositions des évacuateurs en tunnel de Type I et II sont relativement simples. Le tunnel est presque droit et la différence d'élévation entre l'entrée et la sortie, dans la plupart des cas, n'est pas trop grande, permettant son application à la fois pour les grands barrages ou ceux de plus faible hauteur, en fonction des objectifs d'exploitation dans les actions de décharge ; voir Figure 7.3.

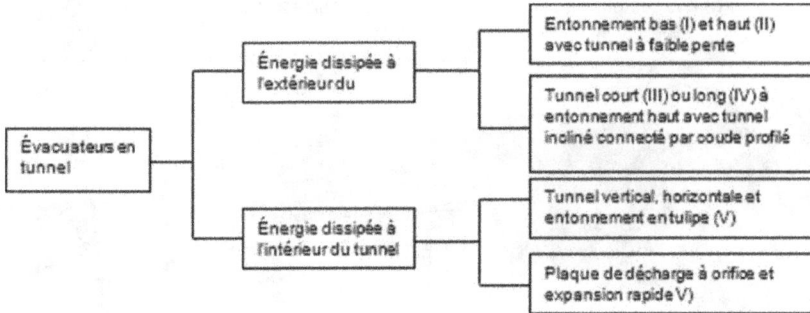

Fig. 7.2
Classification des dispositions d'un évacuateur en tunnel par type de dissipation d'énergie

Les Figures 7.3 (a), (b) et (c) montrent des exemples de dispositions des tunnels de type I et II. Le tunnel avec entonnement bas (de fond) dans la Figure 7.3 (a) (Moore, 1989) et avec entonnement de de demi-fond de niveau intermédiaire dans la Figure 7.3 (b) (Shakirov, 2013) sont construits de manière à répondre à l'exigence de démarrer le remplissage du réservoir et créer un débit qui garantit la sécurité du barrage et le débit environnemental. Pour le barrage brésilien d'Irapé, illustré dans la Figure 7.3 (c) (CBDB, 2009) il y deux évacuateurs de surface en tunnel et un évacuateur de demi-fond. L'évacuateur de demi-fond a permis le démarrage du remplissage du réservoir avant l'achèvement du remblai du barrage, ce qui était fondamental pour respecter le programme de construction très serré. Dans ce projet avec un barrage de plus de 200 m de haut, une grande partie de l'énergie est dissipée par un jet plongeant issu d'un saut de ski, atterrissant dans une fosse de réception au niveau de la rivière.

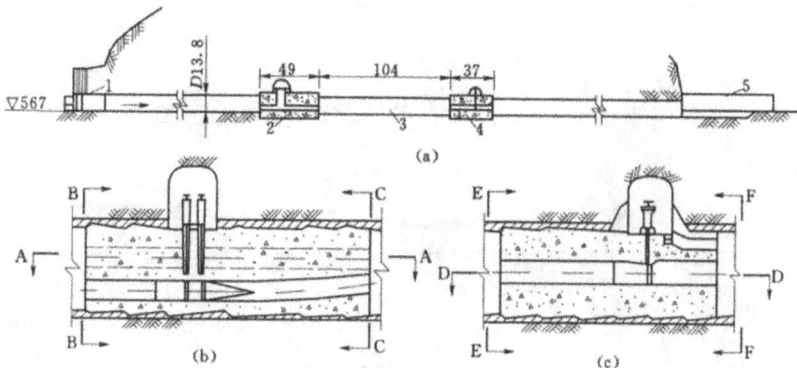

Fig. 7.3
(a) Évacuateur de fond en tunnel de la Mica Dam, Canada, avec expansion rapide

357

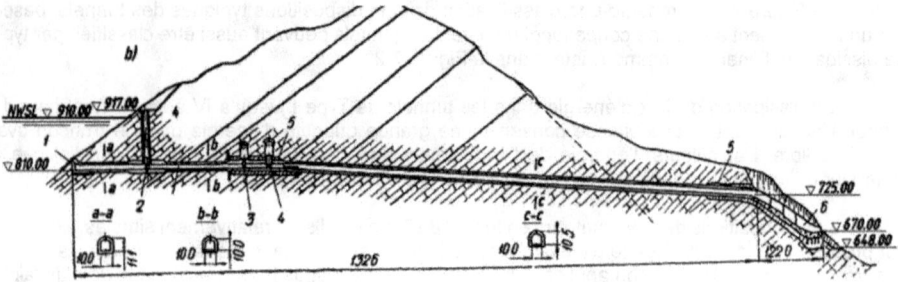

Fig. 7.3
(b) Évacuateur de demi-fond en tunnel du barrage de Nurek, Tajikstan

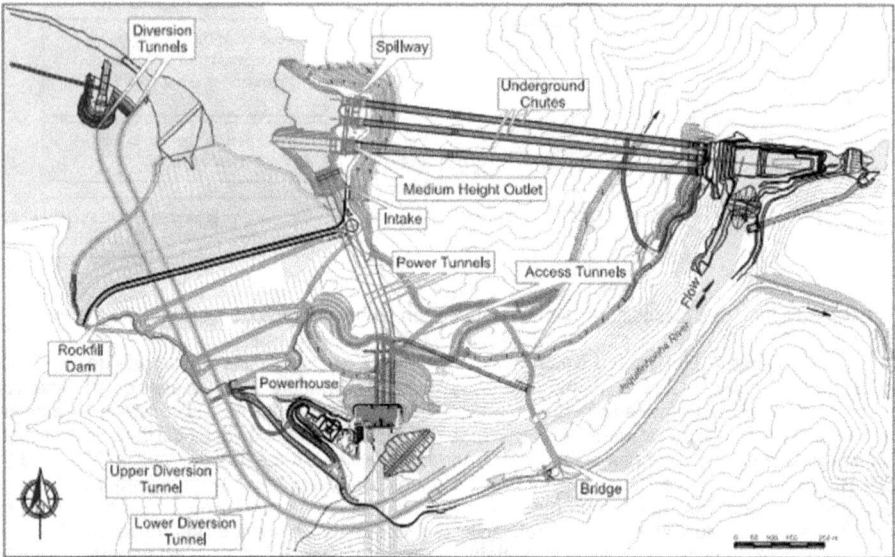

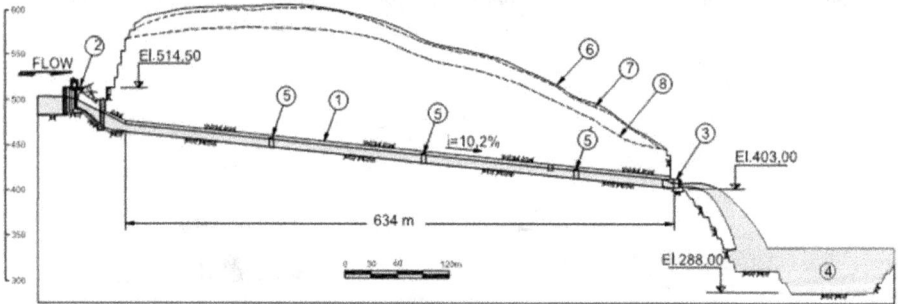

(1) Underground chute
(2) Spillway control structure
(3) Flip bucket
(4) Plunge pool
(5) Aeration structures
(6) Ground surface
(7) Weathered rock surface
(8) Slightly weathered to sound rock

Fig. 7.3
(c) Évacuateur en tunnel du barrage Irapé, Brésil

La configuration des évacuateurs en tunnel de Type III est largement appliquée dans les grands barrages, depuis qu'elle a été adoptée aux USA pour le barrage Hoover en 1930's ; voir

358

Figure 7.4. Un tunnel supérieur, plus court de 100 m, est construit et connecté à une prise d'eau constituée d'un seuil profilé, suivi par un tunnel incliné, puis une section courbe de raccordement vers un tunnel rectiligne subhorizontal à faible pente. Comme la hauteur entre les tunnels supérieurs et inférieurs peut atteindre entre 80 et 100 m avec une section transversale de l'ordre de 150 m², ce type d'évacuateur en tunnel peut être appliqué dans des barrages d'une hauteur supérieure à 200 m et pour des capacités de débit supérieures à 3000 m³/s. Les applications typiques sont les évacuateurs en tunnel du barrage Hoover (Falvey, 1990) et du barrage de Glen Canyon, (Burgi et Eckley, 1988) aux USA et le barrage Ertan en Chine (Gao 2014).

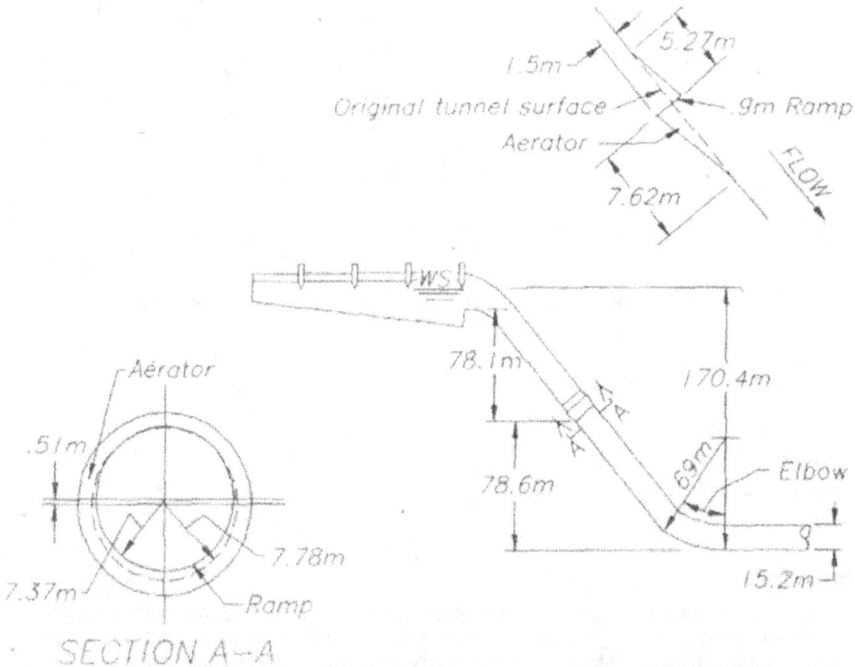

Fig. 7.4
Disposition d'un évacuateur de type III dans la Hoover Dam, États-Unis

Des grandes galeries de dérivation sont nécessaires lorsqu'un barrage est construit dans une vallée encaissée où une grande capacité d'évacuation est nécessaire. Les considérations de coût peuvent conduire à la construction d'un tunnel incliné à prise supérieure à connecter avec la galerie de dérivation originale pour constituer un évacuateur en tunnel permanent. Cet agencement est le même que celui d'une disposition de type III. Il a été utilisé pour l'évacuateur en tunnel à prise supérieure du barrage Mica, l'évacuateur de fond en tunnel du barrage Liujiaxia et dans beaucoup d'autres projets.

Comme le risque de cavitation augmente dans les tunnels de Type III lorsque la hauteur du barrage est supérieure à 250 ou 300 m, la Type IV a été conçu. Le concept de cette disposition est de réduire la longueur du tunnel avec écoulement à forte vitesse en construisant un tunnel à prise supérieure en charge et en déplaçant la structure de contrôle de la chambre des vannes autant à l'aval que possible dans la limite de coûts raisonnables.

Ce type d'évacuateurs en tunnel a été appliqué récemment dans plusieurs grands projets Chinois, comme par exemple, dans les quatre grands évacuateurs en tunnel conçus pour le grand barrage-voûte de Xiluodu. La hauteur du barrage est de 285 m et la capacité de débit de chaque tunnel est 4 000 m³/s, (Figure 7.5). Le tunnel supérieur en charge est long 550 m avec un diamètre de 17 m, et il y a deux coudes verticaux à 90° avant la chambre des vannes. La longueur de la dernière partie du tunnel en charge doit être assez longue pour réajuster la pression jusqu'à ce que la distribution de pression soit suffisamment uniforme avant d'entrer dans la section de transition. La longueur avec débit à haute vitesse dans la dernière partie du tunnel est seulement d'environ 20-30% de la longueur totale. La section transversale dans la partie du tunnel en écoulement libre est de 14x12.5m².

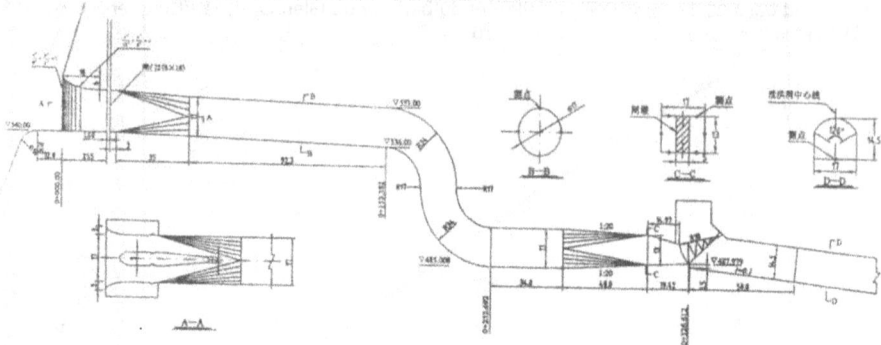

Fig. 7.5
Disposition de l'évacuateur en tunnel dans la Xiluodu Arch Dam (Chine)

1 – Tunnel de dérivation.

2 – Prise de surface.

3 – Puits vertical.

4 – Unité de mise en rotation de l'écoulement.

5 – Section circulaire de la vidange du tunnel;

6 – Section de transition;

7 – Section en fer à cheval du tunnel;

8 – Désaérateur.

L'autre façon pour transformer un tunnel de dérivation en un évacuateur de crue en tunnel est la configuration de Type V. Il peut être conçu comme évacuateur en puits à vortex ou évacuateur en charge. La caractéristique distinctive de ce type d'évacuateur en tunnel est de dissiper l'énergie à l'intérieur du tunnel. L'évacuateur en puits à vortex peut facilement changer la direction de l'écoulement par action vortex et réduire la longueur du tunnel de raccordement.

L'étude hydraulique d'évacuateur en puits à vortex à grande échelle a démarré dans l'ancienne Union Soviétique, ensuite en Russie, Chine, Suisse et quelques autres pays. L'évacuateur en puits à vortex du barrage de Shapai Dam a été la première application de ce type d'évacuateur en Chine, construit en 1990s, avec une charge hydraulique de 110 m et une capacité de débit de 250 m³/s.

La Figure 7.6 montre des exemples d'évacuateur en tunnel de Type V avec évacuateur en puits à vortex.

La Figure7.6(a) (Chen et al. 2007) illustre l'évacuateur en puits à vortex du barrage Gongboxia, qui a été la deuxième application en Chine d'un évacuateur à vortex. Il a une charge hydraulique d'environ 110 m et une capacité d'évacuation de 1 000 m³/s. Il a été utilisé en 2007 dans les conditions de conception.

Le barrage de Tehri, haut de 260 m en Inde, a adopté quatre évacuateurs en puits à vortex, deux dans chaque berge : Figure 7.6(b) (Sharma et al. 2006). Chacun utilise une galerie de dérivation précédemment construites. La capacité de débit pour chaque tunnel est comprise entre 1 800 m³/s et 1 900 m³/s avec une charge hydraulique totale de 200 m (Fink, 2012 and Shakirov, 2013).

Les Figures 7.6(c) et (d) montrent les évacuateurs en puits à vortex vertical et incliné conçus pour le barrage de Rogun au Tajikistan, avec des prises en surface et de fond (Shakirov, 2013). Ces évacuateurs en puits à vortex ont une capacité de débit totale de 3 800 m³/s (2 000+1 800 m³/s). L'évacuateur en puits à vortex incliné de ce projet avec prise de niveau haut, a une capacité de débit totale de 4 040 m³/s. Ce type de conception a l'avantage d'opérer avec plusieurs niveaux de charge hydraulique, spécialement pour les contraintes d'évacuation liés à la phase de mise en eau du réservoir.

Dans les évacuateurs en vortex, environ le 80% de l'énergie de l'écoulement, peut être dissipé à l'intérieur du tunnel de vidange sans avoir des effets dynamiques significatifs sur le revêtement du tunnel (Galant *et al.* 1995 et Riquois *et al.* 1967). Le débit d'eau sort du tunnel dans la rivière avec une vitesse proche de la vitesse de l'eau de la rivière.

L'évacuateur en charge du barrage Xiaolangdi, en Chine, a aussi été bâti par reconstruction d'un tunnel de dérivation et l'énergie était dissipée par trois plaques de décharge à orifice, comme illustré dans la Figure 7.7. Pour le barrage Mica, au Canada, l'évacuateur de fond en tunnel utilise ce type de dissipation d'énergie en fournissant une chambre à expansion rapide, comme montré dans la Figure 7.3 (a).

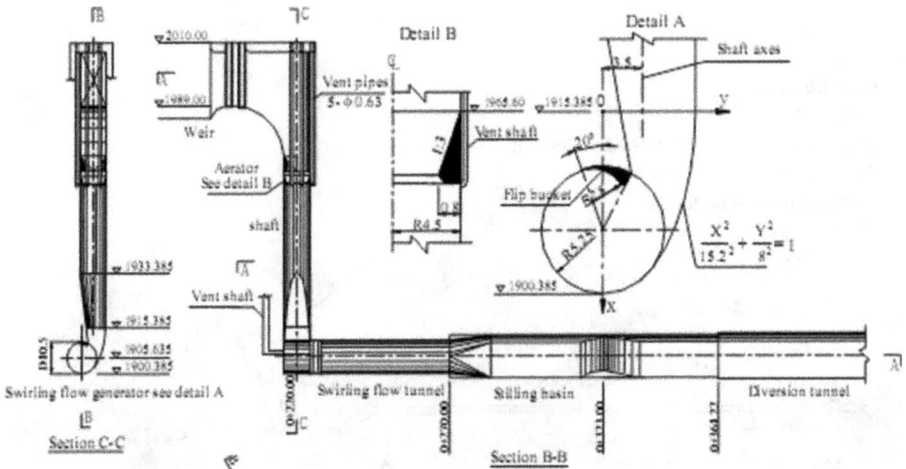

Fig. 7.6
(a) Évacuateur horizontal en tunnel à vortex dans la Gongboxia Dam en Chine

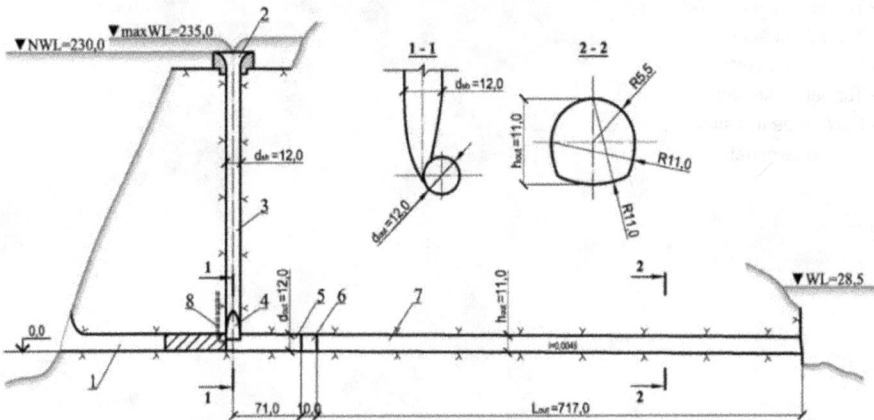

Fig. 7.6
(b) Évacuateur vertical en puits au barrage de Tehri, Inde

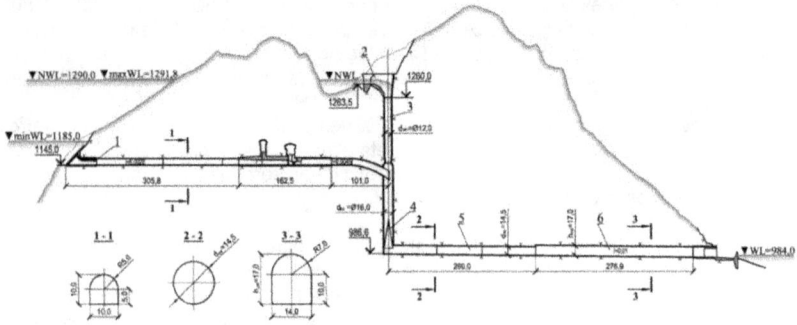

Fig. 7.6
(c) Évacuateur en puits à vortex avec prise de surface et de fond, Rogun Dam, Tajikistan

1 – Prise de fond;
2 – Prise de surface;
3 – Puits vertical;
4 – Unité de mise en rotation de l'écoulement;
5 – Section circulaire du tunnel;
6 – Section en canal du tunnel.

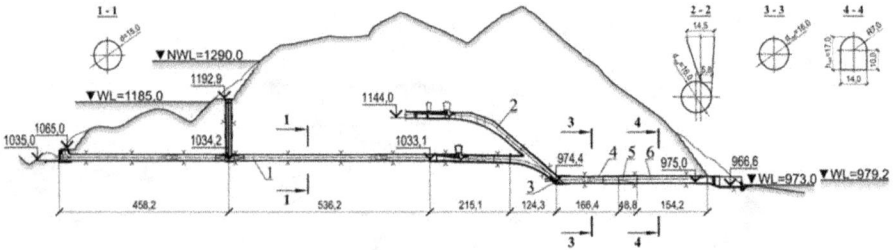

Fig. 7.6
(d) Évacuateur en puits à vortex avec prise de surface et de fond, Rogun Dam, Tajikistan

1 – Tunnel inférieur DT-3;
2 – Tunnel supérieur;
3 – Dispositif à vortex;
4 – Tunnel de section circulaire;
5 – Section de transition;
6 – Tunnel terminal.

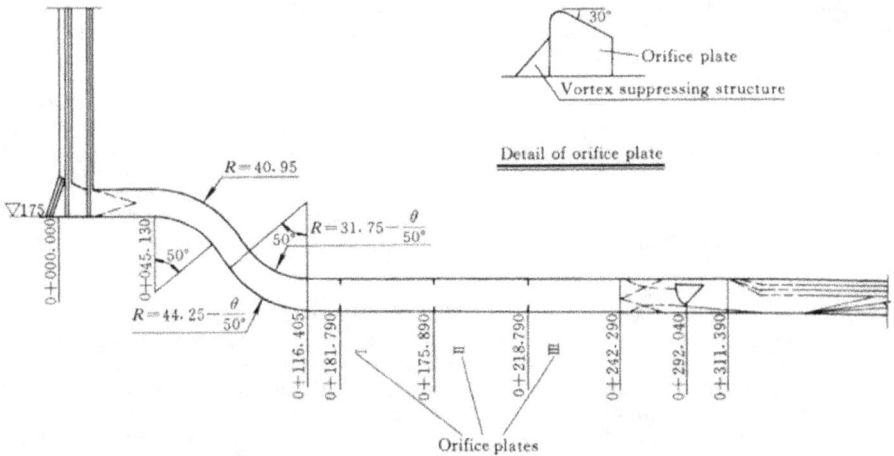

Fig. 7.7
Évacuateur en tunnel, en charge du barrage de Xiaolangdi, Chine

L'évacuateur en tulipe a rarement été utilisé pour des projets de grands barrages. Il possède un déversoir à crête circulaire connecté à un puits et un tunnel inférieur. Le déversoir circulaire peut être divisé par piles et contrôlé par des vannes. La charge hydraulique et la capacité de débit sont normalement faibles et ce type d'évacuateur n'est pas adéquat lorsqu' il faut évacuer de grandes crues.

7.2.2. Structure unique vs. différents composants fonctionnels

La conception et la disposition de différents types d'évacuateur en tunnel permettent de répondre à différents objectifs et exigences, depuis la construction jusqu'aux étapes d'exploitation. Les tunnels à prise haute (en partie supérieure) (Type II à Type IV) pourront se charger de la plupart des actions d'exploitations liées à la maitrise des crues puisqu'ils ont des grandes capacités de débit. Les évacuateurs de fond en tunnel (Type I) fonctionnent soit pour la maitrise des crues soit pour l'approvisionnement en eau pendant le remplissage du réservoir, ou pour l'abaissement du niveau d'eau du réservoir pour la maintenance du barrage et aussi, quelque fois, pour des opérations d'urgence.

Un déversoir intermédiaire ou temporaire en tunnel est parfois nécessaire pour la construction d'un barrage de grande hauteur, en particulier lorsque le barrage est construit par étapes. C'est le cas du barrage de Rogun au Tadjikistan. Dans ce cas, l'évacuateur de crues à mi-hauteur est relié à l'évacuateur de crues permanent, comme le montre la figure 7.6 (c).

7.3. STRUCTURE DE CONTROLE

7.3.1. Évacuateurs de surface

Pour les évacuateurs en tunnel à prise haute (Type II à Type IV), la chambre des vannes est située juste en aval du pertuis d'entrée. Deux vannes sont utilisées normalement : une vanne plane pour la maintenance et une vanne segment de service pour l'exploitation courante. Cependant, la disposition de trois vannes est une solution plus sûre et recommandée : la première - une vanne plane ou un batardeau (pour une faible charge) pour la maintenance ; la deuxième la vanne d'urgence, et la troisième – une vanne segment de service pour l'exploitation courante. La charge hydraulique pour cette conception devrait être d'environ 40-50 m. L'écoulement est à surface libre après la vanne d'exploitation courante et dans l'ensemble du tunnel. La vanne de service peut aussi être de type plane quand la charge hydraulique est petite.

La chambre des vannes est divisée en deux sections pour réduire la charge agissante sur chaque vanne lorsque la charge hydraulique est haute ou la capacité de débit est grande. Cette disposition a été appliquée dans les évacuateurs en tunnel du barrage Hoover et du barrage Xiluodu comme la capacité de débit pour ces deux tunnels sont respectivement 5 000 m³/s and 4 000 m³/s.

7.3.2. Évacuateurs de fond

La vanne de service de l'évacuateur de fond en tunnel (ou de prise inférieure) est généralement de type segment comme la charge hydraulique peut être supérieure à 60 m ou 80 m, en atteignant même 100 m ou 120 m, avec une débitance autour de plusieurs centaines ou milliers de mètres cubes par seconde. Une vanne à glissières peut être utilisée quand la charge hydraulique est plus grande et la débitance plus petite. Par exemple, des vannes à glissières ont été construites dans le bouchon en béton de l'évacuateur de fond en tunnel du barrage Mica, comme montré dans la Figure 7.3 (a).

7.4. STRUCTURE DE TRANSFERT DU DÉBIT

7.4.1. Coursier en tunnel

La disposition longitudinale du coursier en tunnel peut être conçue pour chaque type d'évacuateur en tunnel de la Figure 7.1. La pente de la section inférieure est en général inférieure à 10%, facteur en général déterminé en fonction de la capacité des moyens de transport et manutention durnat la construction. Cette pente a été augmentée jusqu'à environ 12-13% avec l'introduction de grands camions puissants. La pente du coursier en tunnel du barrage l'Irape (Figure 7.3 (c)) est 10,2% et la longueur moyenne est 634 m. Une forte pente du tunnel permet de réduire la longueur totale du coursier en tunnel.

La section transversale du coursier en tunnel peut être soit un cercle, soit en forme de D ou en forme de fer à cheval. La forme circulaire a été utilisée pour l'évacuateur en tunnel du barrage Hoover, du barrage de Glen Canyon et quelques autres barrages. Mais le type en forme de D est plus pratique pour l'excavation et pour le transport pendant la construction. Le revêtement en béton doit être conçu pour supporter un débit à haute vitesse. Les travaux de réalisation de revêtement dans un tunnel en forme de D sont beaucoup plus simples que pour un tunnel de forme circulaire.

Un écoulement à grande vitesse se produira nécessairement dans les goulottes des tunnels de type I à III. Par conséquent, la bonne gestion de ces écoulements à grande vitesse nécessite un contrôle efficace de la cavitation, qui est un aspect crucial de la conception hydraulique du revêtement du tunnel.

7.4.2. Systèmes d'aération

Les dommages par cavitation des évacuateurs en tunnel ainsi que dans évacuateurs de surface sont illustrés par de plusieurs études de cas importantes. L'atténuation de la cavitation par aération a été reconnue après les dommages par cavitation dans le tunnel du barrage Hoover, et plusieurs études ont été effectués depuis lors. La Figure 7.8 montre les dommages dans l'évacuateur en tunnel gauche du barrage de Glen Canyon en 1983 (Burgi et Eckley, 1988 et Falvey, 1990). Le "grand trou" de 11 m de profondeur a été découvert durant une inspection sur site.

Un système standard d'aération a été élaboré dans les années 1980et par la suite, par différents pays sur la base des conceptions particulières de leurs tunnels et exigences opérationnelles (ICOLD, 1987), (voir Figure 7.9). Les systèmes d'aération ont été largement appliqués dans le monde entier et les dommages par cavitation dans les tunnels avec débit à haute vitesse ont été considérablement réduits, bien que le mécanisme de mitigation de la cavitation soit encore un objet d'étude.

Dans le dispositif d'aération standard, un évidement marqué est créé dans le radier du tunnel alimenté par un évent d'aération, quelque fois combiné avec une petite rampe où une pression négative se développe. L'air est forcé à entrer dans la zone à basse pression afin d'aérer le débit d'eau. Il est très important de maintenir les conduites d'air propres et d'assurer que l'aération est continue et suffisante pendant l'utilisation du tunnel. La distance d'intervalle entre deux aérateurs doit être d'environ 120 m à 150 m. Plusieurs aérateurs peuvent être nécessaires le long du radier du tunnel dans un évacuateur en tunnel de grande dimension.

Fig. 7.8
Dommages de cavitation dans l'évacuateur en tunnel de la Glen Canyon Dam

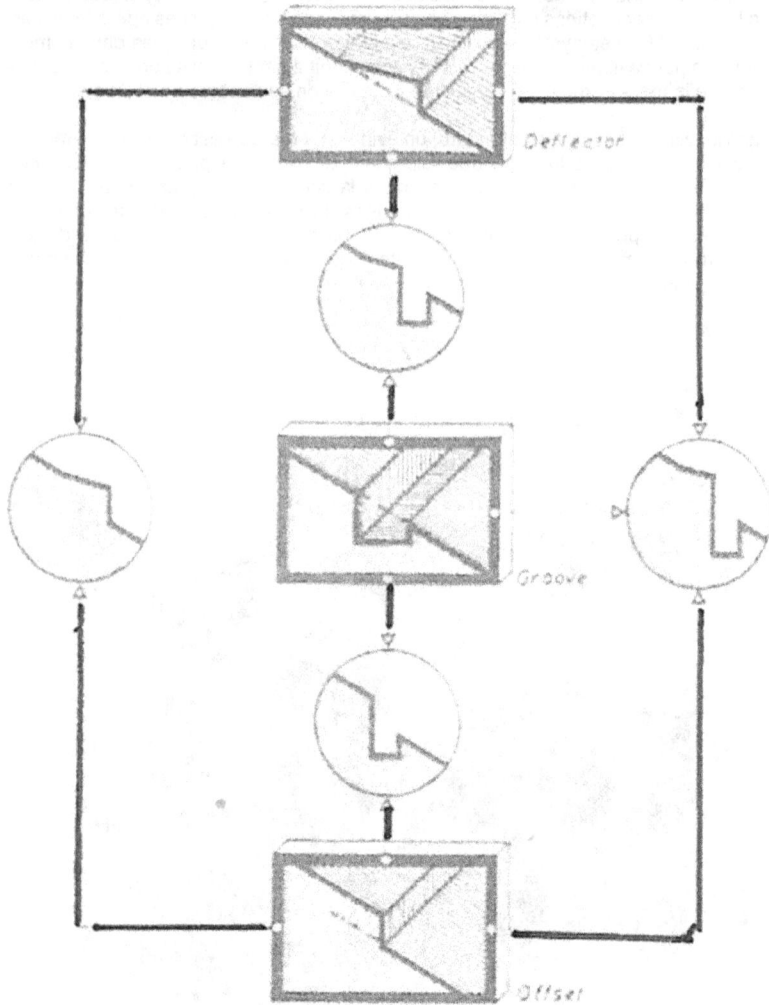

Fig. 7.9
Comparaison de la dissipation d'énergie entre deux types de cuillère de dissipation
Profil longitudinal de la chute étagée

Un type spécial d'aérateur a été développé pour les cas où il est difficile de forcer l'air à entrer dans la zone de basse pression dans un tunnel avec une faible pente ou lorsque la section transversale du tunnel est en forme de cercle. L'aérateur construit le long du périmètre mouillé provoque l'entraînement d'air pour ce cas particulier. Un aérateur en forme de demi-cercle a été appliqué dans l'évacuateur en tunnel du barrage Hoover, du barrage de Yellowtail et du barrage de Glen Canyon, aux USA. Un aérateur circulaire peut aussi être appliqué dans les évacuateurs en puits, comme par exemple dans le puits du barrage de Gongboxia (Chine) où l'étude du modèle physique a montré que la pression dans la partie basse du puits est assez basse. La longueur de la zone à basse pression mesurée sur site, en aval du dispositif d'aération circulaire était plus grande de celle mesurée en laboratoire.

Un tunnel ou puits d'aération intermédiaire peut être nécessaire si l'évacuateur en tunnel est trop long.

7.5. STRUCTURE TERMINALE (CARACTÉRISTIQUES DE LA DISSIPATION D'ENERGIE)

7.5.1. Bassin de dissipation

Un bassin de dissipation d'énergie peut être utilisé dans un schéma d'évacuateur en tunnel a basse pression si les conditions géologiques ne sont pas assez bonnes – sol ou rocher de mauvaise qualité – pour permettre l'utilisation d'une cuillère de dissipation. La présence de fortes vitesses et charges pulsatoires dans les bassins de dissipation exige une attention particulière pour leur résistance structurelle. Une bonne conception d'un bassin de dissipation est généralement basée sur les résultats d'une étude par modèle physique.

L'expérience d'exploitation des installations hydrauliques donne plusieurs exemples de dommages importants dans les bassins de dissipation (Malpaso en Mexique, Tarbella en Pakistan). Le plus grave dommage est caractérisé par le déplacement vers le haut de l'intégralité du radier, qui peut être une conséquence de l'apparition de la pression hydrodynamique totale sous le radier ou sous-pression liées aux fuites.

7.5.2. Cuillère de dissipation et fosse d'amortissement

Une cuillère de dissipation est plus souvent utilisée dans les schémas d'évacuateurs en tunnel à haute charge si les conditions géologiques sont bonnes, ce qui implique un rocher de bonne qualité. Le jet avec un grand pouvoir d'érosion peut être dirigé loin de la structure, plus en aval dans le lit de la rivière.

Comme discuté dans le chapitre 3 (section 3.5.6) de ce bulletin, une fosse d'amortissement peut être nécessaire et dans certaines situations peut être construite par une pré-excavation. Les charges hydrauliques dans le tunnel de la Figure 7.3 (b) et 7.3 (c) sont supérieurs à 200 m et la charge hydraulique entre la cuillère de dissipation et la surface de l'eau en aval est aussi de plus de 100 m ; par conséquent, la grande énergie de l'écoulement doit être dissipée dans une grande masse d'eau.

Normalement, en vue en plan, angle aigu existe entre l'évacuateur en tunnel et le lit de la rivière. La conception de la cuillère demande une grande attention. La cuillère peut être conçue avec formes différentes et complexes en 3D, incluant des déflecteurs obliques ou à fente. Le but de la conception des cuillères est de diriger correctement l'écoulement en jet dans le lit de la rivière, en obtenant une grande efficacité de dissipation d'énergie, en réduisant la profondeur d'affouillement du lit de la rivière et les effets sur les berges. Une bonne conception de cuillère est généralement combinée avec des études sur modèles physiques. Une cuillère avec déflecteurs à fente est très pratique à appliquer dans une vallée encaissée pour distribuer longitudinalement l'écoulement en jet le long du lit de la rivière (Guo 2014).

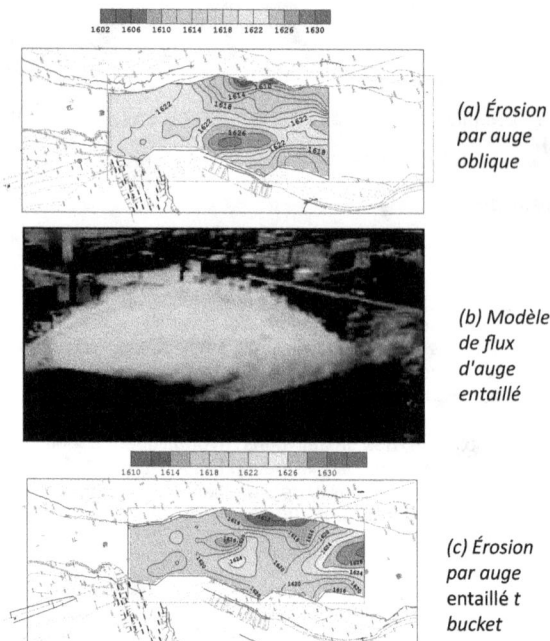

Fig. 7.9
Dispositifs d'aération communs et eux combinaison (Volkart et Rutschmann, 1984)

Comme discuté au chapitre 3 de ce bulletin, la configuration et la profondeur de la fosse d'amortissement peuvent risquer d'atteindre le pied aval du barrage et dans des configurations encaissées engendrer la perte de stabilité des rives de la vallée.

7.5.3. Énergie dissipée à l'intérieur du tunnel

L'énergie de l'écoulement peut aussi être dissipée à l'intérieur du tunnel dans les évacuateurs en puits et en charge, tel que dans le barrage de Xiaolangdi illustrée dans la Figure 7.6 (e). Dans les évacuateurs en charge à vortex plus de 70-80% de l'énergie de flux peut être dissipée à l'intérieur du tunnel et la vitesse dans le dernier tronçon horizontal du tunnel peut être réduite à moins de 20 m/s.

Le dissipateur à orifice est construit à l'intérieur du tunnel en créant une contraction soudaine, puis une expansion rapide à travers une chambre ou un orifice. Il s'agit de la solution retenue pour la vidange du tunnel à bas niveau du barrage de Mica pour dissiper l'énergie et elle a montré des bons résultats. Au barrage de Xiaolangdi, l'évacuateur en tunnel a été bâti en réalésant les tunnels de dérivation avec un diamètre de 14.5 m. Trois orifices ont été installés dans chaque tunnel en charge avec un espacement de trois diamètres et un ratio des orifices de 0.690, 0.724 et 0.724 respectivement. Plus de 40% de l'énergie est dissipée par les trois orifices.

Un autre exemple d'agencement de dissipation d'énergie à l'intérieur d'un évacuateur en tunnel est donné par le le cas du projet hydroélectrique de Sayano – Shushenskaya en Russie comme illustré dans la Figure 7.10. Cette configuration plutôt inhabituelle dissipe l'énergie de l'évacuateur de surface auxiliaire avec une charge hydraulique de 240 m en une séquence de 5 bassins de dissipation souterrains dans une chute étagée à la suite de l'évacuateur en tunnel subhorizontal.

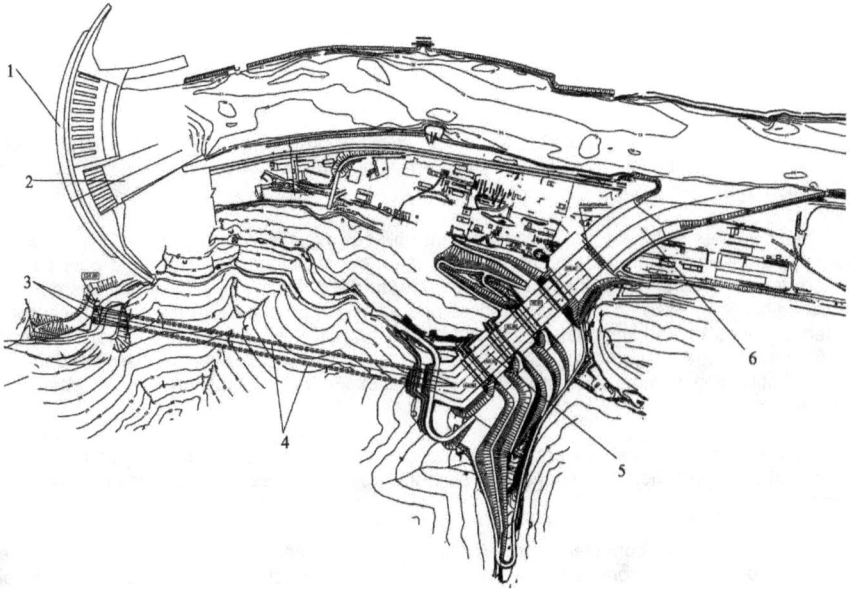

1 – barrage poids-voûte;
2 – évacuateur de service
3 – structure de prise de l'évacuateur auxiliaire
4 – tunnels
5 – chute étagée;
6 – canal de fuite

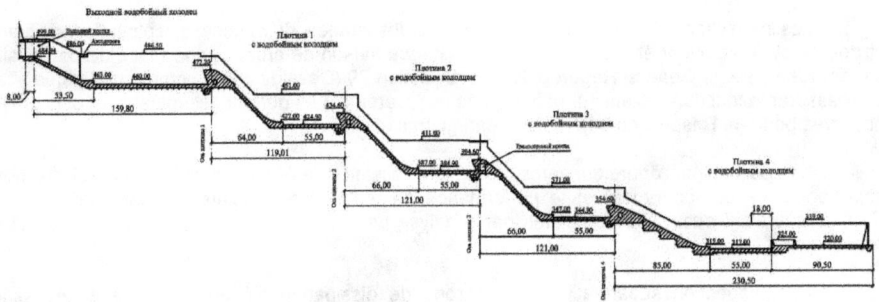

Profil longitudinal de la chute étagée

Fig. 7.10
L'évacuateur auxiliaire du Sayano – Shushenskaya HPP (Russie)

7.6. PROBLÈMATIQUES OPERATIONNELLES

7.6.1. Fonctionnement des vannes

La sécurité des évacuateurs en tunnel est principalement affectée par anomalies de fonctionnement des vannes, des systèmes de ventilation, aération et dissipation d'énergie pendant l'évacuation des crues.

Les vannes segment sont généralement utilisées pour le contrôle du débit à la fois pour les évacuateurs en tunnel à haute charge, ainsi que pour les évacuateurs de fond pour permettre la flexibilité par ouverture partielle dans le contrôle du débit. L'exploitation des vannes doit suivre une consigne rigoureuse de régulation. Toute opération inappropriée ou erronée peut causer des accidents graves et endommager la vanne et/ou le tunnel. Les petites ouvertures, par exemple 5-10%, doivent être évitées dans la mesure où des vibrations peuvent apparaître dans cette gamme d'utilisation comme l'ont montré le retour d'expérience opérationnel et les observations sur le terrain.

La vanne plate est communément conçue pour les évacuateurs en tunnel de fond. Les ouvertures partielles ne sont généralement pas autorisées car la charge hydraulique est importante.

Le système de contrôle commande électrique et mécanique doit avoir le même degré de fiabilité que la vanne comme cela a été montré lors des opérations d'urgence de l'évacuateur de fond du barrage de Zipingpi Dam (Chine) à la suite du tremblement de terre du Wenchuan en 2008. Cela est nécessaire pour garantir que le système des vannes reste opérationnel en cas d'urgence.

La ventilation derrière la vanne est aussi importante. La vitesse d'écoulement de l'air doit être contrôlée en sélectionnant des reniflards de section appropriée car des vibrations et bruits peuvent être induits par une forte vitesse d'écoulement de l'air.

7.6.2. Inspection et maintenance

Les inspections et la maintenance régulières des vannes, du matériel d'exploitation, du tunnel et des aérateurs doivent être effectuées après chaque saison de crue. L'importance de ce travail a été abordée dans le General Report of ICOLD Question 79 (Cassidy 2000) comme une partie vitale pour assurer le fonctionnement sûr et fiable des évacuateurs. Les dommages mineurs et plus graves observés pendant l'inspection doivent être enregistrés et réparés.

L'inspection par l'opérateur après la première utilisation de l'évacuateur du tunnel est d'extrême importance. Le comportement et la performance de toutes les structure doivent être contrôlés. Ceux-ci devraient aussi être complétés par l'analyse des résultats tracés de la surveillance et de l'auscultation.

Une inspection subaquatique de la zone de dissipation d'énergie peut être nécessaire dans le cas d'un fort débit de décharge ou d'une longue durée de déversement. Cette inspection cherchera à déterminer la position et la forme des cavités d'érosion. Elle sera la base d'une analyse ayant l'intention de définir le besoin de modification du système de dissipation d'énergie existant.

7.7. ANALYSE DES RISQUES ET CONSIDERATIONS

7.7.1. Maîtrise du risque de cavitation

La figure 7.10 donne les caractéristiques géométriques adimensionnelles des évacuateurs en tunnel à grande échelle de Type III et IV (Guo 2006). Les hauteurs des barrages représentés sont toutes supérieures à 200 m. Les trois paramètres, que sont la hauteur relative (h/H), la longueur relative (x/L) et la vitesse d'écoulement, sont prises en compte dans l'analyse des risques des différents types d'évacuateurs en tunnel.

Les dommages de cavitation dans les tunnels de Type III visés à la Figure 7.10 se produisent dans un tronçon horizontal du tunnel juste en aval du coude, zone où l'on trouve 40 à 80% de la charge hydraulique. Dans environ 70 à 80% de la longueur du tunnel, la vitesse d'écoulement est supérieure à 40 m/s. Cette disposition présente un risque important de dommage de cavitation. En revanche, dans la disposition des tunnels de Type IV, la longueur avec vitesse d'écoulement élevée se limite à environ 20 à 30% de la longueur totale du tunnel ; par conséquent, le risque de dommages de cavitation dans ces tunnels est fortement réduit. Ce type d'agencement a été appliqué dans plusieurs grands projets de barrages en Chine avec des barrages d'environ 300 m de hauteur et des capacités de débit de tunnels supérieures à 3000 m³/s par tunnel.

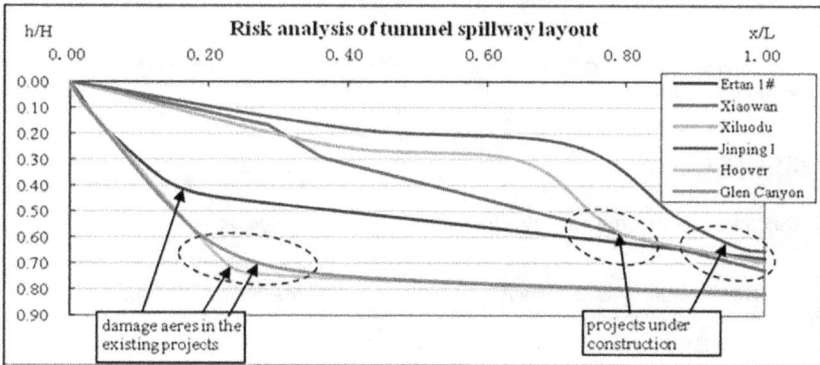

Fig. 7.10
Analyse des risques des dispositions des évacuateurs en tunnel

7.7.2. Action abrasive des sédiments sur les revêtements du tunnel

L'action abrasive des sédiments sur la surface du revêtement en béton est observée lorsque les prises des tunnels sont positionnées près du fond du réservoir. Elle se produit généralement dans les tunnels de dérivation des rivières, au début du période de construction lorsque le réservoir est trop petit pour diminuer la vitesse de la rivière et la capacité de transport solide de l'écoulement. Ainsi de grandes quantités de sédiments solides dans la rivière, avec dimensions jusqu'à 100 mm, sont déchargées par le tunnel pendant la dérivation de la rivière et les périodes de construction.

Les dommages de la cuillère de dissipation de l'évacuateur du barrage de Grand Coulee (USA) sont un exemple de dommages significatifs du béton dû à l'action abrasive (Keener 1994). La protection contre l'action abrasive peut aussi être garantie par l'utilisation de types spéciaux de béton avec durcissement de la partie du radier ou par un blindage en acier.

7.7.3. Charges dynamiques

Les charges dynamiques dans les évacuateurs en tunnel pourraient être le résultat, parmi d'autres causes, des changements constants des modes d'exploitation des tunnels, des phénomènes transitoires, ou des formations de ressauts hydraulique en absence d'aération derrière les piles de la chambre des vannes provoquant des zones de vide, cavitation et décollement du revêtement métallique (Ilyushin 2002).

Des charges dynamiques significatives ont été la cause d'un accident grave dans l'évacuateur en tunnel auxiliaire du barrage de San Esteban (Del Campo A., 1967). Le tunnel était conçu pour opérer en conditions d'écoulement libre, mais pendant une période avec hauts niveaux d'eau en aval, il été inondé. Il en est résulté la formation d'un ressaut hydraulique dans le tunnel, qui a conduit à des dommages du revêtement dans la zone d'une défaillance du massif rocheux, et la chute des roches a causé l'obstruction de l'évacuateur en tunnel.

7.7.4. Opération d'urgence

Une opération d'urgence peut exiger qu'un ou plusieurs tunnels de vidange de fond soient utilisés pour abaisser le niveau du réservoir, en particulier pour des barrages en remblai, en cas de séisme ou de glissement de terrain. Cela a été le cas du barrage de Zipingpu en Chine, qui a été exposé à un intense tremblement de terre (Cas 2, dessous). Cet exemple est très instructif et illustre l'importance de pouvoir décharger un débit important, afin de contrôler le niveau du réservoir dans des situations d'urgence.

Ceci met en outre en relief le rôle spécial des tunnels de dérivation et des vidanges de fond pour la sécurité du barrage.

7.8. ÉTUDES DE CAS

7.8.1. Déversement en crue par l'évacuateur en puits du barrage de Tehri (Inde)

Le niveau d'eau dans le réservoir du barrage de Tehri a augmenté jusqu'à El. 882.0 m en septembre 2009 ; les évacuateurs en puits de la berge rive droite ont commencé automatiquement à décharger un débit de 480 m³/s chacun, pour une capacité de débit maximale de 1 850 m³/s (avec MWL à El. 835.0 m). La Photographie 7.1 montre les évacuateurs en puits en fonction libérant le mélange d'eau et air par la chambre de désaération avec un jet dont la hauteur excède 50 m. Ceci démontre l'efficacité de l'exploitation des installations de désaération de l'évacuateur en puits (Sharma et al. 2006).

Photo 7.1
- Jets de désaération des évacuateurs de la Tehri Dam en 2009

7.8.2. **Opérations d'urgence de l'évacuateur de fond du barrage de Zipingpu après le tremblement de terre du "Wenchuan " (Chine)**

Le barrage de Zipingpu est situé dans le cours supérieur de la rivière Minjiang en Chine, 9 km en amont de la ville de Dujiangyan et 60 km nord-ouest de la ville de Chengdu. Le barrage est de type CFRD avec une hauteur de 156 m, un évacuateur de surface avec coursier, une usine hydroélectrique, un tunnel de délimonage et deux évacuateurs de fond en tunnel. Les deux évacuateurs de fond en tunnel ont été construits en reconfigurant les tunnels de dérivation. Les objectifs principaux de ce projet sont l'irrigation et l'approvisionnement en eau, ainsi que la production d'énergie, le contrôle des crues et la protection environnementale. Le projet a été terminé en 2006.

Un fort tremblement de terre avec magnitude 8.0 s'est produit le 12 mai 2008 (appelé "Wenchuan Earthquake"). Le barrage se trouve à 17 km seulement de l'épicentre du tremblement de terre (ICOLD Experts, 2009).

Le barrage a subi des dommages mineurs mais la production d'énergie et l'approvisionnement en eau ont été arrêtés. Reprendre l'approvisionnement en eau et en énergie et contrôler le niveau d'eau du réservoir étaient les tâches les plus importantes dans la gestion de la situation d'urgence. La première unité de production d'énergie a été mise en service seulement 7 minutes après le séisme et deux unités ont pu fonctionner peu. Le débit total déchargé a alors atteint 100 m³/s m, desservant les villes de Dujiangyan et Chengdu en aval, avec une population de 20 millions de personnes. Le tunnel de délimonage été remis en service 24 heures après le tremblement de terre et le débit déchargé a été porté à 280 m³/s, comme illustré dans la Photographie 7.2 (a). Un évacuateur de fond en tunnel a été ouvert 27 heures après le tremblement de terre. Le débit total déchargé était alors de 850 m³/s, valeur proche du débit entrant. La production d'énergie totale a repris 128 heures après le tremblement de terre. Toutes les vannes des tunnels ont été ouvertes avec succès 190 heures après le tremblement de terre, et le niveau d'eau du réservoir a été efficacement contrôlé du 14 au 20 mai, en réduisant fortement le risque de rupture du barrage. L'objectif de cette opération était d'assurer la sécurité du barrage en permettant les inspections des dommages dans la structure du barrage et les réparations des équipements mécaniques et électriques. L'autre objectif était de fournir une voie navigable pour les secours dans la zone du réservoir, comme indiqué dans la Photographie 7.2 (b).

Un groupe international d'experts de barrages de l'ICOLD a visité le barrage CFRD de Zipingpu en avril 2009. Leurs observations ont conclu que des mesures d'urgence furent prises immédiatement après le tremblement de terre et les règles existantes de gestion des situations d'urgence pour la sécurité du barrage, ont été correctement appliquées dans le cas du tremblement de terre du "Wenchuan " (ICOLD Experts, 2009).

Photo 2
Action d'urgence en Zipingpu CFRD après le "Wenchuan Earthquake".
(a) tunnel de délimonage fonctionnant 24 heures après le séisme ; (b) quai provisoire sur le rivage du réservoir pour fournir une voie navigable pour les sauvetages dans la zone du réservoir

Table 7.1
A Caractéristiques des évacuateurs en tunnel à grande échelle dans le monde (ordonnés par hauteur du barrage)

Nom	Pays	Type de barrage	Hauteur du barrage H (m)	Type de tunnel	Dimension du tunnel (largeur × hauteur)	Q_{max} (m³/s)	Type d'aération	Achèvement *	Expériences d'utilisation
Rogun	Tadjikistan	ER	335	V(a)	D14×17	4040	aérateur en vortex	u.d.	
Rogun	Tadjikistan	ER	335	V	D14×17	1800	aérateur en vortex	u.d.	
Rogun	Tadjikistan	ER	335	V	D14×17	2000	aérateur en vortex	u.d.	
Jinping I	Chine	AV	305	IV	14×12	1×3651	encoche en radier, rampe	u.c.	
Nurek	Tadjikistan	ER	300	III	11×10	1×2000	aérateur en aval vanne segment	1980	
Nurek	Tadjikistan	ER	300	II	10×10.5	1×2020	aérateur en aval vanne segment	1980	
Xiaowan	Chine	AV	292	IV	13×13.5	1×3811	Double encoche en radier	2011	
Xiluodu	Chine	AV	278	IV	14×12.5	4×3860	encoche en radier, rampe	2014	
Tehri	Inde	ER	261.5	IV	D=8.5	1×1100	-	2008	2010
Tehri	Inde	ER	260.5	V	4 horse-shoe, D=11	2×1850, 2×1800	désaérateur à la jonction entre puits vertical et tunnel horizontal	2008	2010
Mica	Canada	ER	244	V	-	1×850	obturateur en béton	1977	bon fonctionnement
Mica	Canada	ER	244	III	-	-	encoche en radier	1977	bon fonctionnement
Ertan	Chine	AV	240	III	13×13.5	2×3700	encoches, rampe espacées de 3 diamètres	1999	endommagée en 2001
Sayano-Shushenshkaya	Russie	AV	240	II	2 - 10×12	2×1900	5 bassins de dissipation étagés	2011	bon fonctionnement
Chirkeyskaya	Russie	AV	232	III	11.2×12.6	2900	-	1978	bon fonctionnement

Table 7.1
B Caractéristiques des évacuateurs en tunnel à grande échelle dans le monde (ordonnés par hauteur de barrage)

Nom	Pays	Type de barrage	Hauteur du barrage H (m)	Type de tunnel	Dimension du tunnel (largeur × hauteur)	Q_{max} (m³/s)	Type d'aération	Achèvement *	Expériences d'utilisation
Hoover	États-Unis	PG	221	III	D15.56	2×5500	aérateur circulaire	1936	endommagé en 1941
Glen Canyon	États-Unis	AV	216	III	D12.5	2×3900	aérateur circulaire	1966	endommagé en 1983
Irape	Brésil	TE	208	II	10×11.4	2×2000	encoche en radier	2005	bon fonctionnement
Irape	Brésil	TE	208	II	12×11.4	1×2000	encoche	2005	bon fonctionnement
Longyangxia	Chine	PG	178	III	5×7	1×1340	décalage à la vanne	1989	endommagé en 1989
Charvakskaya	Ouzbékistan	ER	168	IV, V	D=9.0 D=11.0	1100 1200	Blindage acier en aval	1976	
Xiaolangdi	Chine	ER	160	II	10×12, 8×9, 8×9.5	1×2680, 1×1973 1×1796	Encoche en radier	2000	bon fonctionnement
Xiaolangdi	Chine	ER	160	V	D14.5	1×1727, 2×1549	décalage à la vanne	2000	bon fonctionnement
Yellowtail	États-Unis	AV	160	III	2-D15.56	1×2600	aérateur circulaire	1968	endommagé en 1967
Zipingpu	Chine	CFRD	156	III	horse-shoe, D=10.7	1×1672	Encoche et rampe sur paroi latérale en forme de U	2006	bonne aération, observée
Liujiaxia	Chine	PG	147	III	8×12.9	1×2105	aérateur	1974	endommagé en 1972
Cousar	Iran	PG	140	II	2 - 10×10	2×1900	-	2006	
Aldeadavila	Espagne		139.5	III	D=10.4	2800	-	1962	
Gongboxia	Chine	CFRD	132.2	V	11×14	1×1090	aérateur circulaire dans la partie basse du puits	2006	bon fonctionnement
Shapai	Chine	RCC VA	132	V	4×5.5	1×242	Reniflard en aval du puits	2002	bon fonctionnement en 2008

*: u.c. = en construction; u.d. = en conception;

**:n.c. = pas clair

Table 7.2
A Classification des évacuateurs en Tunnel, en Puits et à Vortex

Type	Niveau de prise	Alignement	Critère entrée/sortie	Condition hydraulique	Structure de contrôle	Vanne de contrôle	Apport d'air/Aération/Reniflard	Dissipation d'énergie
I(a)	bas	Rectiligne	Tunnel rectiligne à faible pente, avec prise de fond. Sortie du tunnel libre sans submergence aval	Écoulement libre par gravité à forte vitesse	Contrôle en amont à la prise d'eau	vanne segment en amont pour fort débit. Vanne plate à glissières possible pour petit débit	Doit avoir un apport d'air suffisant pour l'écoulement à surface libre. A cause des très grandes vitesses, des rainures/dispositifs d'aération spécifiques doivent être réalisées en radier du tunnel.	Dissipation d'énergie à l'extérieur du tunnel
I(b)	bas	Après le tronçon bas en charge, le reste est rectiligne	Prise de fond avec tunnel à faible pente. Prise pour écoulement en charge. Sortie du tunnel libre sans submergence aval	Écoulement amont en charge par chambre d'expansion rapide	Par expansion rapide	En amont	Doit avoir un apport d'air suffisant sur la surface libre.	Par expansion rapide
II	haut	Rectiligne	Tunnel rectiligne avec prise en surface et tunnel à faible pente. Sortie du tunnel libre sans submergence aval	Écoulement à surface libre par gravité à forte vitesse	Contrôle en amont à la prise d'eau	vanne segment en amont pour fort débit. Vanne plate à glissières possible pour petit débit	Doit avoir un apport d'air suffisant pour l'écoulement à surface libre. A cause des très grandes vitesses, des rainures/dispositifs d'aération spécifiques doivent être réalisées en radier du tunnel.	Dissipation d'énergie à l'extérieur du tunnel
III	haut et bas	Rectiligne	Court tunnel supérieur connecté par tunnel incliné et coudé amont profilé, coude aval reliant le tunnel aval à faible pente.	Écoulement à surface libre par gravité à forte vitesse.	Contrôle en amont à la prise d'eau	vanne segment en amont pour fort débit. Vanne plate à glissières possible pour petit débit	Doit avoir un apport d'air suffisant pour l'écoulement à surface libre. A cause des très grandes vitesses, des rainures/dispositifs d'aération spécifiques doivent être réalisées en radier du tunnel.	Dissipation d'énergie à l'extérieur du tunnel
IV(a)	haut	L'alignement du tunnel en charge peut être changé soit horizontalement soit verticalement	Long tunnel en charge en partie supérieure, court tunnel à surface libre en aval.	Grande longueur du tunnel en charge. Petite longueur du tunnel en écoulement à surface libre.	Contrôle à la fin de la section charge par une vanne	Vanne segment à la fin du tunnel en charge, déchargeant l'écoulement dans le tunnel à surface libre	A cause des courtes longueurs de tunnel à surface libre, il peut ne pas y avoir de problèmes d'apport d'air et cavitation.	Dissipation d'énergie à l'extérieur du tunnel
V(a)	haut et bas	Puits en vortex avec changement de direction d'écoulement à la connexion avec le tunnel	Prise supérieure pour tunnel en charge à vortex et tunnel à surface libre en aval.	Écoulement par vortex en puits vertical, puis écoulement à surface libre dans le tunnel horizontal avec système de dissipation d'énergie dans la zone à surface libre.	Contrôle en amont du puits à vortex	Vanne radiale ou verticale en amont du puits à vortex	Système de ventilation adéquat pour le tuyau d'aération dans le puits vertical à vortex et le tunnel horizontal pour extraire l'air/ ou mélange air-eau	Dissipation d'énergie à l'intérieur du tunnel et bassin de dissipation
V(b)	haut et bas	Puits en vortex avec changement de direction d'écoulement à la connexion avec le tunnel	Long tunnel supérieur en charge et tunnel à surface libre en aval.	Écoulement en charge jusqu'à la vanne de contrôle. Puits incliné rejoignant tangentiellement le tunnel aval à surface libre	Contrôle en amont du puits incliné à vortex	Vanne radiale ou verticale en amont du puits à vortex	Système de ventilation adéquat pour le tuyau d'aération dans le puits vertical à vortex et le tunnel horizontal pour extraire l'air/ ou mélange air-eau	Dissipation d'énergie à l'intérieur du tunnel et bassin de dissipation
V(c)	haut et bas	Prise supérieure Écoulement amont en charge avec coude horizontal et vertical. Tunnel droit à surface libre en aval.	Prise supérieure avec écoulement en charge. Écoulement à surface libre en aval	Écoulement en charge en amont de la vanne de contrôle. Tunnel en aval à surface libre.	Contrôle en aval du tunnel en charge.	Vanne segment à la fin du tunnel en charge, déchargeant l'écoulement dans le tunnel à surface libre	L'aération et l'apport d'air libre ne devraient pas poser de problèmes critiques, mais doivent être étudiées	Dissipation d'énergie par une ou plusieurs plaques de décharge à orifice, à l'intérieur du tunnel horizontal en charge.

7.9. RÉFÉRENCES

CBDB (2009) - *Main Brazilian Dams III - Design, Construction and Performance*, Brazilian Committee on Dams, Rio de Janeiro, Brazil

ICOLD (1987) - *Spillways for Dams*, Bulletin 58, CIGB-ICOLD, Paris, France

Cassidy J. (2000) - "Gated spillways and other controlled release facilities and dam safety", General Report of Q.79, *Transactions of the 20 Congress of the ICOLD*, Vol. IV, Beijing, China, 735-758

Burgi, P.H. and M.S. Eckley (1988) - "Tunnel spillway performance at Glen Canyon Dam", *Proceedings of International Symposium on Hydraulics for High Dams*, Beijing, China, 810-818

Chen, W.X., G.F. Li, S.Z. Xie and K.L. Yang (2007) - "Study on aerators of high head spillway tunnels", Proceedings of the IAHR XXXII Congress. Theme D, Venice, Italy, 748-755

ICOLD Experts (2009) - "China shares experience and lessons from the Wenchuan earthquake, Part III: Comments from Symposium Participants", International Journal on Hydropower & Dams, Vol. 16, Issue 3, pp 111-112

Fink (2012) - Experience in design, construction and initial operation of shaft spillways at the Tehri HEP, Personal communication with Dr. Fink

Shakirov, R. (2013) - *Vortex spillways with energy dissipation inside the tunnel*. Contribution presented to the ICOLD Committee on Hydraulic for Dams.

ASCE (1989) - Civil Engineering Guidelines for Planning and Designing Hydroelectric Developments, edited by Edgar T. Moore Vol.1, New York, USA

Falvey, H. T. (1990) - *Cavitation in Chute and Spillways*, Engineering Monograph No. 42, US Bureau of Reclamation, Denver, USA

Gao, J., Z.P. Liu and J. Guo (2014) - "Energy dissipation and high velocity flow" *Dam Construction in China - A Sixty-Year Review*, Editor in Chief: Jia Jinsheng. China WaterPower Press, Beijing, China

Guo, J. (2013) - "Recent achievements in hydraulic research in China Hydro Power", Vol. 6 of *Comprehensive Renewable Energy*, Editor-in-Chief: Ali Sayigh, Elsevier Ltd. UK: Vol. 6, pp 485-505

Guo, J., D. Zhang, Z.P. Liu and L.Fan (2006) - "Achievements in hydraulic problems in large spillway tunnel with a high head and large discharge flow and its risk analysis" (in Chinese) *Journal of Hydraulic Engineering*, 37(10), pp1193-1198

Pugh, C. A. and T.J. Rhone (1988) - "Cavitation in the Bureau of Reclamation tunnel spillways", *Proceedings of International Symposium on Hydraulics for High Dams*, pp 645-652, Beijing, China.

Sharma, R.K., P.P.S. Mann and R.K. Vishnoi (2006) - Tehri Project Shaft Spillways - Example of an effective solution based on analytical and observational design approaches, *http://www.istt.com/doks/pdf/b09_bristane_06.pdf*

Sen, Siba P. (2013) - *Classification of tunnels*. Contribution presented to the ICOLD Committee on Hydraulic for Dams.

Galant M.A., B.A. Zhivotovskiy, I.S. Novikova, V.B. Rodionov and N.N. Rozanova (1995) - "Special features of swirl-type tunnel spillways and hydraulic conditions of their operation". *Gidrotekhnicheskoye stroitelstvo*, № 9, pp. 16-22.

Novikova I.S., V.B. Rodionov, B.A. Zhivotovskiy and N.N. Rozanova (2002) - *Shaft spillways with swirl-type water diversion*. Proceedings abstract. Volume paper on CD-ROM. St.-Petersburg, Russia

Rodionov V., V. Kupriyanov, S. Paremud, V. Vedosov, V. Vladimirov and A.Tolochinov (2006) - "Construction of auxiliary spillways at existing dams". *Transactions of the ICOLD Twenty-Second Congress on Large Dams,* vol. 3, pp.1257-1268 Barcelona, Spain.

Ilyushin V.F. (2002) - "Lessons of the accident at the diversion tunnel during construction of Rogun HPP". *Hydraulic Structures*, No 4, pp.51-56.

Keener K.B. (1944) - "Spillway erosion of Grand Coulee dam". *Engineering News-Record*, vol. 133, № 2, pp.95-101.

Del Campo A., I. Trincado and I.G. Rosello (1967) - "Some problems in operation of San Esteban dam spillways". *Transactions of the ICOLD Ninth International Congress on Large Dams*, vol. II, R. 33, Istanbul, Turkey.

Volkart, P. And P. Rutschmann (1984) – *Air entrainment devices (Air Slots)*. Report Nr. 72. Mitteilungen der Versucsanstalt für Wasserbau, Hydrologie und Glaziologie, Technischen Hochshule Zurich.

8. PROBLÈMES SPÉCIFIQUES DES ÉVACUATEURS DANS DES CLIMATS FROIDS

8.1. INTRODUCTION

Il est connu que la neige et la glace peuvent bloquer partiellement ou complètement les évacuateurs de crues et ainsi réduire la capacité de l'évacuateur et du système d'évacuation d'eau dans les régions froides. Ces problèmes, mais aussi d'autres défis en lien avec la régulation des vannes, les efforts spéciaux et les impacts de la glace sur les structures, sont examinés dans ce chapitre.

8.2. CAPACITÉ RÉDUITE DU DÉVERSOIR ET DU SYSTÈME D'ÉVACUATION

L'exploitation fiable des ouvrages hydrauliques, en particulier des déversoirs, des prises d'eau et des structures d'adduction d'eau, doit être assuré non seulement en hiver lorsque la couverture de glace est stable, mais aussi pendant la formation et la destruction de la couche de glace.

Il est usuel de séparer l'origine d'une réduction de la capacité des évacuateurs de crues en raison d'un climat froid selon les thèmes suivants :

- le frasil ;

- le « aufeis » qui provoque l'obstruction de la section transversale ;

- le bouchon de neige dans l'évacuateur de crues ; et

- la dérive des glaces.

8.2.1. Frasil

La période de formation de la couche de glace est caractérisée par la baisse de la température de l'air en dessous de 0°C. Pendant cette période, la couche superficielle de l'eau est refroidie à une température d'environ 0°C, ce qui provoque la formation d'un phénomène appelé "frasil" ou "boue", en particulier en présence de précipitations sous forme de neige humide. Le frasil est de l'eau en transition vers un autre état solide d'agrégation, c'est-à-dire de la glace qui n'a toutefois pas encore les propriétés physiques de la glace. Il faut garder à l'esprit que la densité de la glace est de 917 kg/m³ et la densité de l'eau est de 1 000 kg/m³, et la densité du frasil est donc une moyenne entre ces valeurs. Ce fait explique certaines propriétés du frasil : une masse visqueuse à la surface de l'eau ressemblant à de la colle. Lorsque l'eau est surfondue, le frasil liquide adhère aux structures telles que les barrages flottants, les grilles, les murs déflecteurs, les structures avec des fentes, les bords de vannes et les batardeaux. Ce qui va réduire la capacité de décharge. La masse des structures va également augmenter et rendre difficile le fonctionnement des vannes et des mécanismes de levage, voire même les endommager.

Le frasil est particulièrement problématique pour les structures de prise d'eau des centrales hydroélectriques, les stations de pompage et les équipements des centrales hydroélectriques. Des mesures préventives pour contrôler les effets néfastes du frazil peuvent être prises en mettant en œuvre les dispositifs décrits ci-dessous.

- Les prises d'eau et les passes à glace (figure 8.1) qui détournent l'écoulement chargé de glace en aval de la chute. La technologie la plus efficace pour le passage du frasil est choisie en fonction de la vitesse de dérive du frasil, des débits d'eau et d'autres conditions spécifiques au site. Pour réduire au minimum les débits spécifiques d'eau au travers des passes à glace tout en assurant l'entraînement maximal du frasil, le niveau optimal du seuil de déversement peut être obtenu en modifiant l'inclinaison du clapet ou la position du canal pivotant comme indiqué sur les figures ci-dessous.

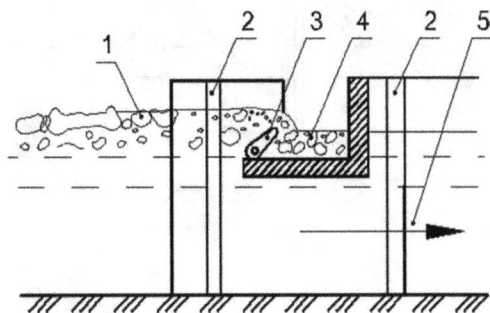

a)   Passe à glace avec clapet de régulation

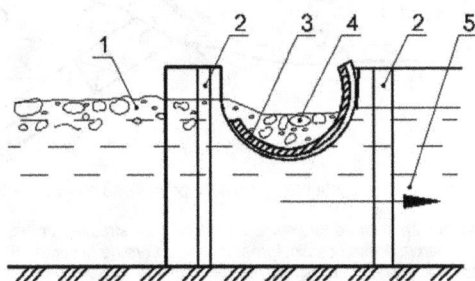

b)   Passe à glace avec canal pivotant de A. Gostunsky

1 — frasil flottant; 2— rainuret; 3— seuil de déversement avec niveau variable au travers duquel est entraîné le frasil; 4 — écoulement avec frasil dans le canal; 5— écoulement sans frasil.

Fig. 8.1
Disposition type d'une passe à glace

- Les déflecteurs à débris sont des structures qui dévient la surface de l'écoulement et les objets qui flottent à la surface, à l'écart du site à protéger, comme une prise d'eau par exemple (figure 8.2). L'élément principal d'un déflecteur à débris est une paroi plongeante ou un écran déflecteur. La profondeur d'immersion doit empêcher que le frasil passe sous son bord inférieur. La surface du déflecteur est placée à un angle précis par rapport au sens d'écoulement afin que le déflecteur ne limite pas le flux, mais le dévie seulement. La profondeur d'immersion du déflecteur sous l'eau varie entre environ 0,8 m à et 1,5 m, en fonction de son utilisation.

- Si la prise d'eau est située près de la zone de transition de débit de la rivière, il peut être protégé du frasil à l'aide d'un mur « racleur ». Le principe de fonctionnement est similaire à celui de la paroi déflectrice. Pour faire passer le frasil ou d'autres formations de glace non massive, 1 ou 2 ouvertures des vannes d'évacuation de crues sont réalisées en mode déversement.

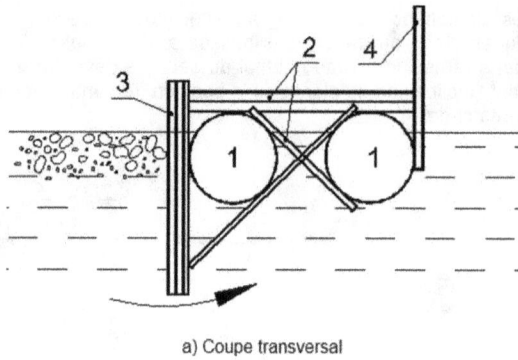

a) Coupe transversal

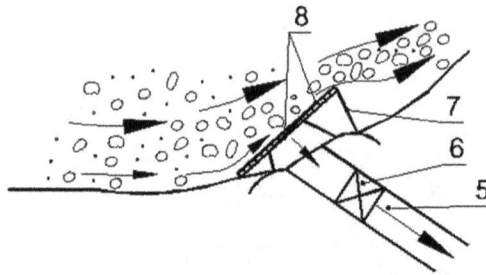

b) Dispositif pour la protection de la prise d'eau en rivière – Situation

1 —éléments flottants (tuyaux en acier de 600 mm de diamètre); 2—structure porteuse; 3— écran déflecteur (par exemple profilés en I avec parois déflectrices intégrées); 4 —barrière de sécurité; ,5 — canal; 6 –vanne d'entrée de la prise d'eau; 7 — Structure stabilisatrice de l'écran; 8 — éléments de liaison du déflecteur à débris.

Fig. 8.2
Déflecteur à débris

- La disposition de la paroi plongeante ou des poutres tranquilisatrices et des murs (Figures 8.3a) sont conçus pour résister aux impacts du frasil

Il est également connu que les systèmes de production de bulles à haut débit sont utilisés pour fabriquer un rideau qui empêche la formation de glace sur la partie amont des vannes ou des écluses (Tuthill, 2002).

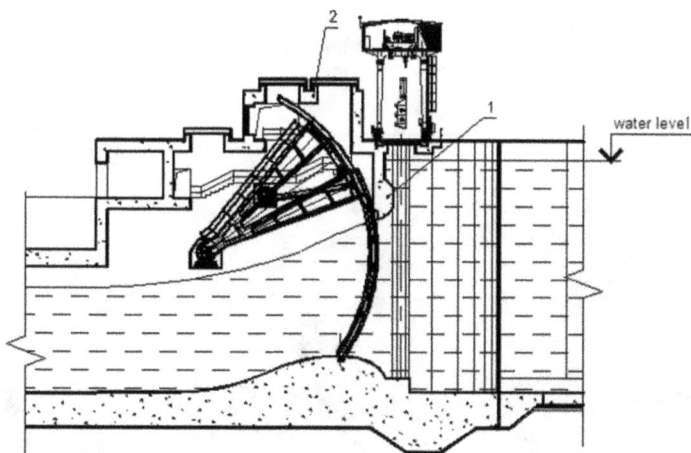

a) Infrastructure au-dessus de la vanne et paroi plongeante (poutres tranquilisatrices);

1 – paroi plongeante installée de manière correcte;
2 – Infrastructure générant une température supérieure à 0° C dans l'ouvrage

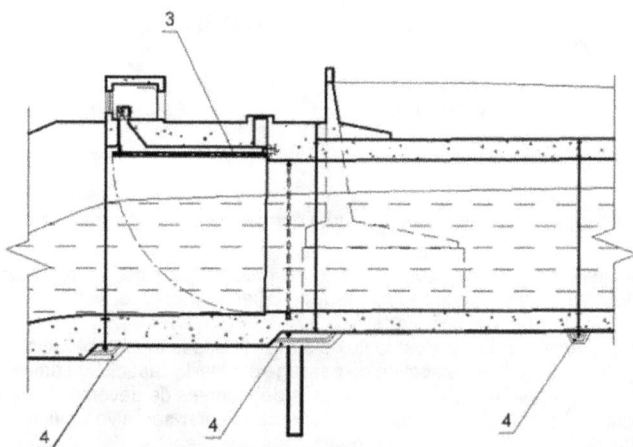

b) Concept d'écran thermique dans la section de transition Design of thermal shield in the transit section

3 – Ecran thermique qui protège l'ouvrage contre l'air chaud en provenance de l'aval
4 – Drainage derrière l'écran thermique (l'épaisseur de la plaque est supérieur à l'épaisseur de pénétration du gel, assurant ainsi la stabilité de la structure et le fonctionnement ininterrompu du drainage).

Fig. 8.3
Disposition de la paroi plongeante ou des poutres tranquilisatrices

8.2.2. Obstruction par l'Aufeis"

L'« Aufeis » est un terme utilisé pour désigner de la glace qui se forme par un courant d'eau de faible profondeur qui coule sur une couverture de glace. D'autres termes comme naled et icing sont également utilisés pour ce même phénomène. Le phénomène de « Aufeis » se produit dans des conditions hivernales normales.

La situation la plus fréquente provoquant le phénomène de « aufeis » dans un déversoir de crues est une fuite au travers des vannes (Figure 8.4). Cette formation de glace peut bloquer la vanne et réduire de manière significative la capacité d'évacuation.

Fig. 8.4
Pont Arnaud : Fuite des vannes en hiver engendrant aufeis (accumulation de glace) et réduisant de manière significative la capacité de l'évacuateur de crues (Hydro-Québec)

Une autre situation qui provoque l'aufeis est le flux d'air froid dans le tunnel d'évacuation du déversoir. Les fuites d'eau gèlent et réduisent la section de passage et ainsi la capacité du tunnel. Une accumulation d'Aufeis de 3 m d'épaisseur a été observée dans des tunnels de déversoirs. Pendant la période de fonte, l'eau qui arrive dans le réservoir peut faire augmenter son niveau jusqu'à une altitude inacceptable à cause de la capacité réduite de l'évacuateur de crues.

La description du phénomène d'accumulation et d'élimination de l'aufeis est basée sur les recherches de Schohl et Ettema (1986). La quantité d'accumulation d'aufeis est fonction du temps, de la surface, du flux de chaleur, du débit d'eau, de la pente et des propriétés de l'eau et de la glace.

L'élimination de l'aufeis peut être réalisée par deux procédés : la fusion et la rupture mécanique. Il a été constaté que si l'eau s'écoule au-dessus de l'aufeis, la vitesse de fusion dépend de la température de l'eau. S'il n'y a pas d'eau qui coule, la fonte dépendra du rayonnement, de la vitesse du vent et de la température de l'air. Une rupture soudaine de l'aufeis est possible uniquement pour des dépôts locaux et entraînera une dérive de glace. Sokolov (1973) a constaté qu'une rupture mécanique n'a jamais enlevé plus de 30 à 35 % de la glace. Selon Ashton (1986), la fonte de l'aufeis durerait plusieurs semaines.

Lia (1998) a présenté plusieurs solutions pour réduire les problèmes d'Aufeis dans les tunnels d'évacuateurs :

- Isolation - Un mur non porteur situé à l'exutoire (et dans certains cas aussi à l'entrée) empêche l'air froid de pénétrer dans le tunnel ou au moins de limite le flux traversant. En complément, il est possible d'isoler les parois du tunnel.

- Étanchéité - Afin d'empêcher l'eau de s'infiltrer dans le tunnel pendant les périodes froides, une étanchéité peut être appliquée. Pour la même raison, il est recommandé d'utiliser un dynamitage contrôlé près de l'exutoire, du pertuis d'entrée et dans la zone où le déversoir de l'évacuateur est à construire.

- Profil en travers amélioré – L'aufeis dans un évacuateur de crues augmente si l'eau s'écoule sur l'ensemble de la surface du tunnel et gèle. Si une rigole étroite et profonde est construite depuis la zone hors gel du tunnel jusqu'à la sortie du tunnel, une couche de glace va se former à la surface de l'eau et protégera l'écoulement situé en dessous, comme dans une rivière naturelle, voir figure 8.5.

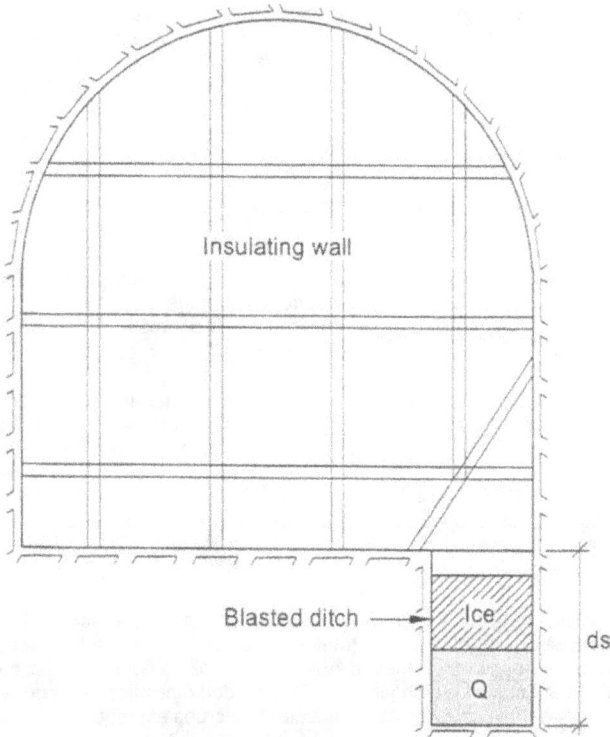

Fig. 8.5
Profil en travers avec une rigole profonde avec formation d'une couche de glace protectrice
(Lia, 1998)

8.2.3. Neige

La neige change de comportement et de propriétés avec le temps. La neige peut bloquer l'entrée, le tunnel/canal et la sortie de l'évacuateur de crues et peut empêcher l'inspection et l'entretien de l'évacuateur de crues pendant l'hiver.

L'épaisseur moyenne de la couche de neige dépend des précipitations, de la température et de l'accumulation de la neige. Localement, la topographie joue un rôle important dans l'accumulation de neige. Le vent enlève la neige d'un endroit et l'accumule dans un autre. Si le vent vient d'une direction principale (p. ex. l'ouest) pendant l'hiver, la neige se déposera sur les versants dont la face est dans la direction opposée (p. ex. l'est). C'est typique dans zones montagneuses soumises à des forts vents. Bien que la direction du vent change au cours des années, la direction principale est très souvent la même. Dans cette optique, deux ou trois ans d'observation sont nécessaires pour décrire les tendances locales.

La neige transportée par le vent change de propriétés et devient plus compacte lorsqu'elle s'accumule dans des congères. Après quelques heures, un matériau cohésif se développe. La neige accumulée dans des congères, et qui peut rester plusieurs étés avant de fondre, est appelée amas de neige pluriannuel. Une telle accumulation de neige a bloqué un tunnel d'évacuation de crues d'un aménagement hydroélectrique en Norvège pendant plusieurs années.

La neige va former un barrage de neige dans l'évacuateur de crues, ce qui réduira la capacité de l'évacuateur de crues ou, dans le pire des cas, bloquera complètement celui-ci.

Si de l'eau de fonte arrive à pénétrer la neige froide, l'eau va s'infiltrer au travers du barrage de neige. Des recherches ont été menées pour trouver la perméabilité de la neige. La pluie et l'eau de fonte s'écoulent vers le bas dans la neige, sous l'effet de la gravité, de la même manière que dans le sol, et peuvent être exprimées par la loi de Darcy, où la perméabilité de la neige dépend de la taille des cristaux de neige et de la densité, (Sommerfeld, 1989). Le barrage de neige va finir par se briser par des procédés thermiques, m^3 d'infiltration, de dérive, de submersion, de glissements ou une combinaison de ces phénomènes.

Lia (1998) présente différentes solutions pour réduire les problèmes dus à la neige dans les évacuateurs de crues :

- Barrières pare-neige - Il s'agit d'un système pratique pour modifier l'accumulation de la neige. Les barrières n'arrêtent pas la neige poudreuse, mais elles changent la direction du vent de telle sorte que la neige en suspension s'accumule derrière les barrières.

- Toits - En Norvège, l'évacuateur de crues est généralement un canal latéral alimenté par un déversoir de trop-plein. Une solution efficace pour limiter la quantité de neige dans le canal est de le couvrir avec un toit. Le toit va également réduire le compactage de la neige. De tels toits ont été construits sur certains canaux latéraux, voir la figure 8.6. Les contrôles ont montré que cette solution est efficace, bien que d'autres études soient recommandées. Il n'est pas nécessaire de recouvrir tout le canal, car lorsque le canal commence à se remplir d'eau, la neige devient saturée et s'érode facilement.

- Câbles chauffants - Cette méthode a été utilisée avec succès dans les buses (ponceaux) et elle a été transférée à différentes parties des déversoirs (pas observée dans les tunnels d'évacuateur), voir la figure 8.7. La méthode nécessite une alimentation électrique, mais elle ne doit être mise en marche que pendant les périodes où la capacité de l'évacuateur de crues est requise.

- Emplacement de l'évacuateur de crues - Lors de la conception de nouveaux évacuateurs de crues, il faut étudier la topographie et la configuration spécifique des vents avant de déterminer l'emplacement de l'évacuateur afin de réduire les problèmes d'accumulation de neige.

Dans les régions steppiques de Russie, les barrières à neige constituées de rangées d'arbres plantés ou d'arbustes locaux sont courantes. Il s'agit d'une solution respectueuse de l'environnement pour éviter que les ouvrages ne soient recouverts et submergées par la neige.

Fig. 8.6
Toiture en béton sur le canal latéral de l'évacuateur de crues à Sønstevatn en Norvège, vue depuis l'amont (Lia, 1998)

Fig. 8.7
(A) Chute Allard, Hydro-Québec
a) et b): Été 2010: Installation de plaques chauffantes sur 3 barrages gonflables

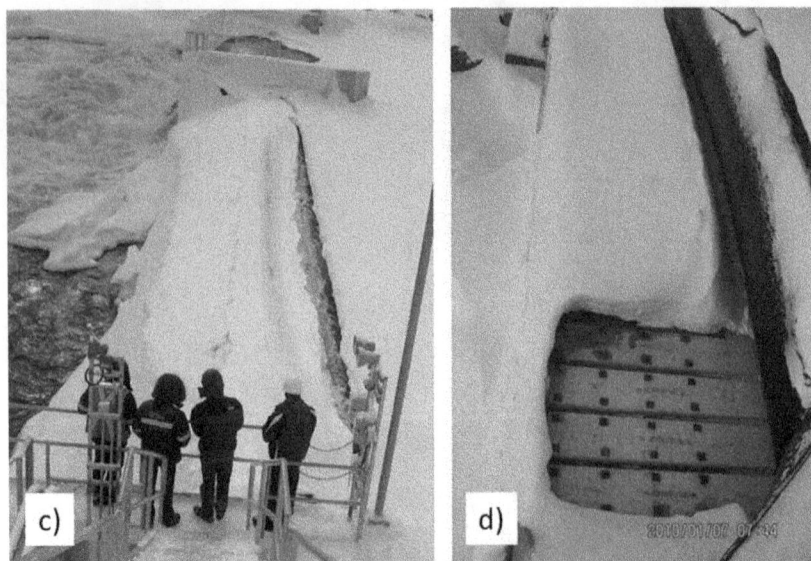

Fig. 8.7
(B) Chute Allard, Hydro-Québec
c) *Évacuateur en hiver - Neige accumulée en aval*
d) Plaques chauffantes testées durant l'hiver

8.2.4. *Dérive de la glace*

La couverture de glace sur les rivières ou les lacs se brise lorsque le débit d'eau augmente. Des morceaux de glace sont alors transportés par l'écoulement et peuvent causer un colmatage important des déversoirs. La structure de l'évacuateur de crues doit être conçue pour supporter la charge provenant des blocs de glace à la dérive.

8.3. I CHARGE INDUITE PAR LA GLACE SUR LA STRUCTURE DES DÉVERSOIRS

Afin d'assurer un fonctionnement fiable des évacuateurs de crues, des ouvrages d'adduction d'eau et des prises d'eau dans des conditions climatiques rigoureuses, l'impact de la glace sur la structure, particulièrement sur les vannes et les équipements mécaniques, devrait être réduit au minimum ou pris en compte. Les impacts et les charges induites par la glace sur les structures sont examinés par Gosstroy (1989).

Les évacuateurs de crue avec un écoulement sous vanne situé plus bas que le niveau du plan d'eau aval sont à éviter car il est pratiquement impossible d'empêcher un impact de glace sur la vanne depuis l'aval.

Pour limiter l'impact de la glace sur la partie amont des équipements mécaniques (vannes), on peut installer des barrages flottants et des parois déflectrices, comme le montrent les figures 8.2 et 8.3. L'installation de barrières flottantes spas sous le niveau de flottaison de la glace sert à assurer l'absence d'impact direct de la glace sur la vanne, c'est-à-dire que la barrière est installée sous la profondeur de pénétration du gel et sous la profondeur de la couche de glace. L'effet de la structure du barrage sur la capacité d'écoulement doit être pris en compte lors de la conception. La structure des barrages flottants reprend la pression de la glace et doivent donc être dimensionnés pour résister à cette charge. Les barrages flottants ou les parois déflectrices peuvent également servir de protection contre les débris flottants.

8.4. AUTRES DÉFIS CONCERNANT LES DIFFÉRENTES PARTIES DE L'ÉVACUATEUR DE CRUES

8.4.1. Ouvrages avec vannes

Il faut éviter la formation de glace dans les glissières ou derrière la vanne engendrée par le gel des fuites d'eau. Le poids de la vanne peut alors devenir beaucoup plus important et les vannes peuvent rester complètement bloquées par la glace ou mettre hors service le système. Les mesures suivantes peuvent être prises pour réduire ce risque :

- des systèmes de chauffage et d'isolation thermique peuvent être prévus pour éviter le gel des vannes et la formation de glace au niveau des glissières de la vanne. L'épaisseur de la couche d'isolation thermique doit être calculée au moyen des méthodes de physique thermique basée sur la conductivité thermique du matériau. Le chauffage des glissières et de la vanne avec des électrodes intégrées permet d'éliminer la glace de la structure. Pour un chauffage électrique fiable et sans coupure, le circuit électrique doit être doublé ;

- lors du choix du système de levage, il faut tenir compte du poids supplémentaire de l'isolation thermique et de l'éventuelle adhérence de glace sur la structure de la vanne ; et

- la mise en place d'un écran thermique permet de maintenir des températures positives à l'intérieur de la structure ce qui a un effet positif sur le fonctionnement des vannes, sur le système de levage et sur la structure dans son ensemble. Les écrans thermiques peuvent être utilisés dans les tunnels, les conduites d'évacuation, les coursiers à ciel ouvert ou derrière l'évacuateur de crues. Les fuites éventuelles par les joints d'étanchéité des vannes devraient être collectées et déviées vers le bassin aval par un système de drainage dont la sortie est située sous le niveau de pénétration du gel.

8.4.2. Tunnels et puits avec revêtement

La pénétration du gel à l'intérieur des revêtements et blindages de tunnel ne doit pas être acceptée. Dans les régions où les variations de température sont importantes (de négatives à positives), les revêtements en acier des tunnels et des puits subissent des déformations considérables. La présence d'un puits et d'un déversoir non vanné comme structure d'entrée rend plus difficile la protection contre le gel. Compte tenu de la différence entre la pression atmosphérique à l'entrée et à la sortie du tunnel et de la température naturelle du sol située sous la profondeur de pénétration du gel, la mise en place d'un écran thermique au niveau de la structure d'entrée permet de maintenir des températures positives dans le tunnel et le puits.

8.4.3. Coursier d'évacuateur

Dans un coursier d'évacuateur à ciel ouvert, l'épaisseur du radier et des murs doit être conçue en fonction des propriétés d'isolation thermique du béton, afin d'empêcher la pénétration du gel sous le radier et derrière les murs. Si l'épaisseur du béton n'est pas suffisante, des matériaux calorifuges peuvent être ajoutés dans la structure. La pénétration de gel dans le drainage sous le radier ou derrière les murs ne devrait normalement pas être admise. Le drainage doit de préférence être situé sous la profondeur de pénétration du gel dans le sol (voir la figure 8.3b).

8.4.4. Brouillard et givre

Dans les régions froides, les crues peuvent commencer au milieu de la saison froide en raison d'un réchauffement brutal à court terme ou lorsque les températures quotidiennes moyennes sont inférieures à 0° C. Si la crue atteint l'ouvrage au moment où le dégel est terminé et que les températures quotidiennes moyennes sont devenues inférieures à zéro, le passage de la crue peut générer du brouillard et du givre.

Si la température au milieu de la journée est supérieure à 0° C alors que la température nocturne est inférieure à 0° C, un ouvrage à forte aération (p. ex. dans la section de transition de l'évacuateur de crues ou dans un dissipateur d'énergie / saut de ski) peut causer du brouillard et du givre.

Le brouillard peut se propager sur de très longues distances et peut déposer du givre non seulement sur l'évacuateur de crues, mais aussi sur d'autres ouvrages à proximité. Pour les ouvrages situés dans les régions très froides, il faut éviter une conception qui entraîne la formation de brouillard d'eau dans l'évacuateur de crues. Un exemple frappant de formation de glace sur l'évacuateur de crues est la centrale hydroélectrique de Sayano-Chushenskoye, en Russie, où l'écoulement rapide de l'eau à travers les ouvrages en hiver a entraîné la défaillance et la mise hors service de la centrale hydroélectrique.

8.4.5. *Dissipateur d'énergie*

La conception des ouvrages de dissipation d'énergie dans des conditions climatiques très froides nécessite des approches particulières. La conception doit tenir compte du gel et du dégel possibles des structures et de l'effet de la glace sur celles-ci. Dans ce contexte, les dispositifs de dissipation d'énergie (seuils dentés, blocs dissipateurs, etc.) doivent être situés sous le niveau naturel de l'eau et sous la profondeur de pénétration du gel. Cette configuration va aider le dissipateur d'énergie à remplir sa fonction et à le protéger contre différentes dégradations (effritement, éclatement, fissuration).

8.5. CONCLUSION

Lors de la conception et la construction d'ouvrages d'évacuation de crues dans des conditions climatiques très froides, il faut toujours avoir à l'esprit la période caractérisée par des températures quotidiennes moyennes négatives (en degrés Celsius), à savoir la période hivernale, peut durer 6 à 9 mois. La conception et la construction dans les régions à pergélisol ne sont pas prises en compte dans le présent document, car il s'agit d'un sujet distinct nécessitant une approche particulière. Des températures négatives lors de la construction et l'exploitation des déversoirs imposent des exigences spécifiques lors de la conception et du choix des matériaux de construction - les matériaux doivent résister à de basses températures sans perdre leurs propriétés et, surtout, résister à des cycles alternés de gel et de dégel sans altération, c'est-à-dire être "résistants au gel".

Tout d'abord, cela s'applique au béton utilisé pour les structures, les enrochements, les éléments de soutènement ou de drainage. Une attention particulière doit être portée aux éléments de structures situés dans la zone de marnage - alternativement dans l'eau et à l'air libre. Ces zones sont les plus critiques.

Dans le cas de conditions climatiques froides, lors de la conception de l'évacuateurs de crues, il faut éviter des structures à parois minces qui gèlent entièrement et n'assurent pas une bonne isolation thermique des blocs de béton, du sol et des structures situés derrière elles.

Dans les environnements rudes, les structures en acier doivent conserver leur géométrie sur une large plage de température de +40˚C à -40˚C voir plus.

8.6. RÉFÉRENCES

Le chapitre 8 est principalement basé sur :

Shakirov, R. (2013) – Quelques particularités dans l'exploitation des déversoirs et des ouvrages d'adduction d'eau dans des conditions climatiques difficile, Contribution au Comité Technique de l'ICOLD sur l'Hydraulique des barrages.

Lia, L., (1998) – Blocage des tunnels par la neige et la glace, Université norvégienne de la Science et de la Technologie

Autres références :

Ashton, G. D. (1986) – Rivière, lac, glace et ingénierie. Water Resources Publications, Littleton Colorado, USA Gosstroy (1989) – « Code pratique russe dans le domaine de la conception et de la construction", SNiP 2.06.04-82* Charges et effets sur les ouvrages hydrauliques dus aux vagues, à la glace et aux embarcations, Moscou, URSS.

Schohl, G. A. and R. Ettema, (1986) – Croissance de la glace des Naled, Institut de Recherche en Hydraulique de l'Iowa, Rapport n° 297, Iowa City, IA, USA

Sokolov, B. L., (1973) – Différentes caractéristiques de la rupture structurelle et mécanique des naleds, leurs importances dans l'estimation du ruissellement des nales, Siberian Naleds 1973, Transaction of Naledi Sibiri, Moscou, URSS, Nauka, 1969, p. 206-226

Sommerfeld, R. A., (1989) – La perméabilité de Darcy au travers de la neige compact à grain fin, Colloque sur la neige de la Région Est de 1989, p. 121-128

Tuthill, A. M. (2002) – Éléments impactés par la glace dans les écluses et barrages, Rapport Technique ERDC/CRREL TR-02-4, Corps des Ingénieurs de l'Armée des États-Unis.

9.   MODÉLISATION HYBRIDE DES OUVRAGES HYDRAULIQUES

9.1.   INTRODUCTION

Depuis de nombreuses années, la modélisation hydraulique physique est un outil standard de conception pour l'ingénierie hydraulique des barrages et de leurs structures annexes. La capacité d'un ingénieur à rassembler et à analyser des données issues de modèles physiques continue de s'améliorer avec le développement d'une instrumentation électronique plus sophistiquée, d'ordinateurs plus puissants et de logiciels dédiés. Ces mêmes progrès en informatique ont permis le développement de techniques de modélisation mathématique plus avancées et plusieurs puissants algorithmes bidimensionnels et tridimensionnels sont aujourd'hui disponibles dans le commerce pour les ingénieurs hydrauliciens.

La modélisation physique utilise souvent la modélisation numérique dans le cadre du processus général de modélisation hydraulique. Par exemple, il est courant que le concepteur utilise un modèle numérique pour l'aider lors du processus de conception d'une structure hydraulique. La modélisation numérique peut représenter un gain de temps important et un gain en termes de précision lorsqu'elle se base sur des données de terrain ou issues d'une modélisation physique. Lorsqu'un modèle numérique est calé sur des conditions réelles d'écoulement ou sur les résultats d'une modélisation physique, le modèle numérique peut être utilisé pendant des années comme un outil de conception et d'exploitation précieux. Pourtant, la modélisation numérique est limitée lorsqu'il s'agit d'écoulements multiphasiques (mélanges air/eau), d'écoulements très turbulents avec des vortex et d'écoulements pour lesquels l'affouillement et/ou la sédimentation sont une problématique importante.

La tendance actuelle pour la modélisation des ouvrages hydrauliques est d'utiliser un modèle numérique et un modèle physique en parallèle pendant une étude (aussi connu sous le nom de modélisation « composite » ou « hybride »). La modélisation hybride est l'utilisation de la modélisation physique et numérique, utilisée en série ou en parallèle pour résoudre des problèmes hydrauliques complexes.

Bien que la modélisation physique soit la norme éprouvée pour la modélisation des ouvrages hydrauliques et qu'elle soit utilisée avec succès depuis des décennies, elle comporte des limites et des contraintes. De même, la modélisation numérique, qui est relativement nouvelle, comporte un certain nombre de limites et de contraintes ainsi que des avantages que l'on ne retrouve pas dans la modélisation physique. Cependant, lorsque les deux techniques de modélisation sont utilisées ensemble au cours d'une étude hydraulique, plusieurs des limites d'une technique peuvent être compensées par l'autre et vice versa. Par conséquent, lorsque les chercheurs comprennent les avantages et les limites de chaque technique de modélisation, ils peuvent élaborer un plan de recherche qui fait appel à des approches de modélisation composite où les deux techniques de modélisation sont utilisées ensemble pour accroître l'efficacité et l'efficience du processus de modélisation.

Par exemple, un modèle numérique peut d'abord être utilisé afin d'affiner la conception initiale de l'ouvrage. Un modèle physique est ensuite généralement construit à partir de ce premier dimensionnement. Les données sont recueillies à partir du modèle physique et sont utilisées pour caler le modèle numérique afin d'en assurer la précision. Le modèle physique est ensuite utilisé pour optimiser l'efficacité hydraulique et minimiser les coûts de construction de la conception initiale. Avec un modèle numérique calé, il (le modèle numérique) peut ensuite être utilisé pendant de nombreuses années pour vérifier les conditions d'écoulement ou les procédures opérationnelles même après le démantèlement du modèle physique. La chose importante à retenir avec cette approche est que les modélisateurs numériques ont besoin de données réelles provenant soit d'une crue survenue sur l'ouvrage lui-même, soit d'une étude par modélisation physique. L'étude du modèle physique est essentielle au succès du modèle numérique. Actuellement, la tendance dans la modélisation est d'utiliser les avantages des deux pour la conception.

9.2.   GÉNÉRALITÉS CONCERNANT LA MODÉLISATION PHYSIQUE

Les modèles physiques des ouvrages hydrauliques sont généralement conçus à l'aide d'une similitude de Froude, qui est basée sur le rapport de l'inertie sur les forces de gravité. Le paramètre de similitude utilisé pour faire fonctionner le modèle est le nombre de Froude, défini comme suit :

$$Fr = \frac{v}{\sqrt{gy}} \tag{1}$$

où y est une longueur caractéristique, g est la constante d'accélération gravitationnelle et v est une vitesse caractéristique. La similarité est obtenue en faisant fonctionner le modèle physique de sorte que le nombre de Froude dans le modèle soit le même que le nombre de Froude à échelle réelle. Ceci permet de s'assurer que les forces d'inertie et de pesanteur soient correctement mises à l'échelle. La longueur, le débit, la vitesse, les pressions, le temps et la vitesse de rotation du modèle par rapport à l'échelle réelle peuvent être calculés en utilisant les relations de Froude.

Pour qu'un modèle puisse simuler correctement les conditions hydrauliques réelles d'écoulement, il est nécessaire de maintenir une similitude géométrique, cinématique et dynamique. La similitude géométrique est obtenue en construisant soigneusement le modèle selon un rapport d'échelle sélectionné.

Le choix de l'échelle du modèle a un impact direct sur sa capacité à maintenir une similitude cinématique et dynamique. Les modèles physiques sont toujours construits aussi grands que possible afin de minimiser les effets d'échelle attribués à d'autres forces, telles que les forces visqueuses et de tension superficielle. Lorsqu'un grand modèle physique est construit (avec un petit rapport d'échelle donc), alors les phénomènes d'écoulement tels que les sur-hauteurs (dévers), les intumescences, l'action des vagues, les ressauts hydrauliques, les vortex, les décollements de nappe, les érosions et les affouillements locaux, l'entraînement d'air, les ondes stationnaires, les rugosités, etc. peuvent être modélisés précisément. En plus de pouvoir simuler physiquement des conditions d'écoulement complexes, le modèle physique fournit une simulation pratique de travail que la plupart des équipes d'ingénierie trouvent essentielle au processus de conception. Certains des avantages et des limites les plus importants de la modélisation physique sont énumérés et résumés ci-dessous.

9.2.1.   *Avantages de la modélisation physique*

- Tant que le rapport d'échelle est correctement choisi et que le modèle est construit aussi grand que possible pour limiter d'éventuels effets d'échelle, la modélisation physique est une méthode éprouvée et fiable qui produira des résultats précis.

- De petites modifications du modèle physique peuvent être apportées très facilement à l'aide de structures temporaires en bois, plastique, mortier, sacs de sable, etc. afin que les modifications hydrauliques qui en résultent puissent être immédiatement visibles. Il peut être long de simuler un mur ou un talus surélevé dans le maillage d'un modèle numérique, surtout si le modèle est complexe et que la résolution du maillage est fine, mais cela ne prend que quelques minutes pour en voir les effets dans le modèle physique.

- Plusieurs débits peuvent être testés sur un seul modèle en très peu de temps. Par exemple, un calcul de laminage de crue ou le passage d'un hydrogramme de crue peuvent être simulés en quelques minutes. Les conditions d'écoulement peuvent être marquées par un phénomène d'hystérésis lors de la variation des débits. La modélisation physique peut facilement déterminer la présence et l'ampleur de ces phénomènes.

- Les problèmes hydrauliques spécifiques sont immédiatement apparents dans un modèle physique opérationnel et, bien souvent, de multiples problèmes apparaissent simultanément. Les problèmes tels que les vortex, les zones de décollement, les potentiels d'affouillement, les mauvaises conditions de dissipation d'énergie, la surverse sur les bajoyers, les effets d'aération ou l'action des vagues sont immédiatement notés dans un modèle physique.

- Une fois construit, un modèle physique peut rapidement déterminer la capacité d'évacuation d'un ouvrage de régulation de débit en faisant augmenter le débit d'alimentation du modèle jusqu'à ce qu'un niveau donné soit atteint dans le bassin. Selon la complexité de la géométrie de l'ouvrage de contrôle et du canal d'approche, l'effort de modélisation numérique pour déterminer la capacité de débit peut être beaucoup plus long.

- Les effets du niveau aval ou de l'ennoiement d'un ouvrage hydraulique sont facilement et rapidement observés dans un modèle physique en faisant simplement varier la hauteur de l'eau dans le canal de restitution.

- Les effets tridimensionnels, les effets d'aération, les zones très turbulentes, les ondes stationnaires et les conditions de dévers sont facilement modélisables dans un modèle physique. Cependant, il faut se méfier des effets d'échelle concernant la problématique de l'aération.

- Les modèles physiques facilitent l'ingénierie "en équipe". Les ingénieurs géotechniques, hydrauliques et de génie civil peuvent obtenir un retour d'expérience immédiat et faire des suggestions utiles à la conception de l'ouvrage hydraulique lorsqu'ils voient les conditions d'écoulement des eaux sur le modèle physique.

- Les modèles physiques permettent de filmer ou de photographier l'écoulement des eaux afin que les conditions d'écoulement modélisées puissent être comparées aux conditions réelles, ce qui constitue un outil précieux pour le contrôle de la qualité des études réalisées.

9.2.2. Limites de la modélisation physique

- Le modèle physique peut être limité par le coût et la taille de la structure requise, ainsi que par le temps requis pour le construire et pour recueillir les données. Le coût et le temps sont habituellement proportionnels à la quantité d'information requise et au temps qu'il faut pour recueillir les données. Il est souvent difficile de modéliser physiquement toute l'étendue de la zone d'intérêt, du canal d'écoulement et les conditions d'approche, en raison de l'espace limité au sol du laboratoire. Les modifications majeures d'un modèle physique sont souvent longues et coûteuses en temps et en main-d'œuvre.

- Les mesures de pression, de débit et de vitesses effectuées dans le modèle physique introduisent certaines incertitudes en fonction de la précision de l'instrumentation utilisée pour les mesures. Il peut être difficile d'effectuer des mesures de pression détaillées dans un modèle physique en raison de l'installation de prises de pression, de capteurs de pression et de la présence d'air non dissous dans le flux.

- Si l'échelle du modèle physique n'est pas choisie de manière pertinente, la qualité des résultats ou le coût du modèle en sont affectés. Si l'échelle du modèle est trop grande (modèle de petite taille), les effets d'échelle peuvent influencer la précision des résultats. Si l'échelle du modèle est trop petite (modèle de grande taille), le coût du modèle augmentera. Il faut toutefois souligner qu'un modèle physique doit être construit aussi grand que possible afin de réduire les problèmes d'échelle.

- Il est difficile de faire varier la rugosité aux limites d'un modèle physique et une étude de sensibilité distincte est souvent nécessaire pour évaluer l'effet de cette rugosité.

- L'utilisation de la similitude de Froude pour l'étude du modèle physique peut être problématique dans la mesure où les profils de vitesse ne sont pas distribués au regard de l'extension de la couche limite, le modèle ne fonctionnant pas au bon nombre de Reynolds.

- Le calendrier et le budget du projet peuvent réduire le nombre de configurations ou de modifications qu'il est possible de tester.

- Le modèle physique doit être démantelé à un moment donné et peut alors ne plus être disponible pour d'éventuelles modifications de conception.

- Il existe des processus hydrodynamiques qui sont quasiment impossibles à simuler sur modèles physiques. Il s'agit par exemple de la formation de nuages eau-air-glace dans le canal de restitution d'un évacuateur de crues caractérisé par une aération importante de l'écoulement (par exemple un déversoir en surplomb avec une évacuation en jet).

9.3. GÉNÉRALITÉS CONCERNANT LA MODÉLISATION NUMÉRIQUE

Bien qu'il existe de multiples approches de modélisation numérique telles que les méthodes 1D et 2D, la méthode 3D-CFD (communément appelée *computational fluid dynamics*) est celle dont il est question dans le présent chapitre. Les programmes CFD les plus courants permettent de résoudre les équations de Navier-Stokes en moyennant le nombre de Reynolds (Reynolds-Averaged Navier-Stokes [RANS]) selon différentes méthodes. La plupart des programmes commerciaux de CFD permettent de calculer les écoulements à surface libre communément observés sur les ouvrages hydrauliques à l'aide de la méthode VOF (*Volume-of-Fluid*) ou une variante de cette dernière.

Les codes CFD résolvent les équations RANS en utilisant une méthode de volume fini appliquée à un domaine de calcul qui fait l'objet d'un maillage. Les maillages sont habituellement structurés (le maillage épouse la géométrie de l'ouvrage, méthode BFC [Body Fitted Coordinates]) ou non-structurés (technique de porosité pour laquelle le solide est importé dans le maillage existant et bloque les flux du fluide). Chaque méthode a ses avantages et ses inconvénients ; les BFC ont tendance à être plus efficaces sur le plan du calcul, en particulier pour résoudre les problèmes hydrauliques liés à la couche limite, mais la construction du maillage peut être complexe, alors que les méthodes de porosité sont plus faciles à réaliser mais moins efficaces sur le plan du calcul. De plus, la méthode de porosité se prête bien à l'importation de données d'étude (la géométrie de l'ouvrage) dans le modèle.

L'une des difficultés de la résolution numérique des problèmes d'écoulement dans un ouvrage hydraulique est la présence d'une surface libre variable dans le temps. Cela est particulièrement difficile lorsque la surface de l'eau évolue rapidement avec une courbure importante, par exemple lorsque l'écoulement passe d'un écoulement fluvial à un écoulement torrentiel au passage d'une crête de déversoir.

Une autre technique couramment utilisée dans la modélisation CFD consiste à approximer les surfaces courbes avec des surfaces planes. Ainsi, l'utilisation de maillages plus fins permet une meilleure approximation d'une surface courbe. Par conséquent, afin d'améliorer cette approximation, il est possible d'affiner le maillage au contact de géométries à forte courbure ou dans des zones où les conditions hydrauliques varient rapidement. L'inconvénient des maillages plus fins est l'augmentation du temps de calcul.

Pour résoudre numériquement les équations RANS à travers le domaine maillé, il existe une variété d'approches numériques qui peuvent être utilisées et chacune peut être définie à l'aide de plusieurs paramètres. Les données d'entrée de base comprennent la définition des conditions aux limites, le modèle de turbulence et ses paramètres, la rugosité du modèle et l'approche numérique pour résoudre les équations (implicite ou explicite, 1er, 2ème ou 3ème ordre, etc.). Certains paramètres, comme ceux liés aux conditions limites, sont essentiels pour simuler correctement les conditions réelles d'écoulement. D'autres paramètres apportent des changements moins significatifs. Il est essentiel de comprendre l'effet des diverses méthodes, approches et paramètres pour s'assurer que le modèle représente correctement les conditions réelles d'écoulement.

Des articles évalués par des pairs ont validé l'utilisation de la CFD pour modéliser les écoulements à surface libre pour de nombreux cas différents, en particulier pour les écoulements fluviaux. Cependant, la modélisation CFD n'a pas été validée pour tous les types d'écoulement et, comme pour toute méthode numérique, les résultats obtenus par modélisation ne peuvent être plus précis que les données et les hypothèses sur lesquelles reposent les techniques et méthodes de calcul : c'est-à-dire qu'il faut accepter que les valeurs calculées soient une approximation acceptable mais pas une vérité absolue. Un résumé des principaux avantages et des principales limites de la modélisation numérique est proposé ci-dessous.

9.3.1. Avantages de la modélisation numérique

- Un modèle numérique est un bon choix lorsque des changements importants sont apportés à la topographie du site ou à l'emplacement de l'ouvrage de contrôle.

- La modélisation numérique offre la possibilité de tester une configuration sans construire de structure physique, ce qui permet d'étudier plusieurs configurations en même temps à l'aide de plusieurs ordinateurs. La modélisation numérique permet l'optimisation automatisée selon différents critères d'éléments de conception.

- Les résultats de la modélisation numérique permettent de déterminer les vitesses, les pressions, les paramètres de turbulence et la surface libre à chaque pas de temps et à n'importe quel endroit dans le domaine de l'écoulement.

- La modélisation numérique aide le chercheur à avoir une vue d'ensemble ou, en d'autres termes, à déterminer où se situent les zones problématiques. Les configurations d'écoulement telles que les profils de vitesse et les vortex peuvent être facilement cartographiés à l'aide de tracés vectoriels, de lignes de courant.

- Le modèle numérique permet de s'affranchir de modèles physiques en coupe [exemple : section d'un évacuateur de crues construite à une plus petite échelle (modèle plus grand) afin de pouvoir estimer les pressions exercées à la surface de l'évacuateur de crues]. Le modèle numérique permet d'évaluer de manière satisfaisante les fluctuations de pression ainsi que l'ampleur des pressions négatives le long d'une structure.

- Les modèles RANS CFD ont du mal à simuler les fluctuations de pression associées aux tourbillons dynamiques qui sont petites et donc lissées (par exemple, les fluctuations de pression présentes dans un ressaut hydraulique). Les modèles RANS CFD sont bons pour évaluer les pressions de surface si les lignes de courant sont parallèles près de la surface et relativement stables dans le temps. Les modèles de simulation des grands écoulements tourbillonnants (Large Eddy Simulation [LES]) et des vortex isolés (Detached Eddy Simulation [DES]) éliminent en grande partie ce problème, mais sont coûteux sur le plan calculatoire et ne sont pas (encore) largement utilisés pour les écoulements à surface libre.

- Les changements importants apportés à une configuration donnée sont simples à réaliser sur ordinateur et n'entraînent que des coûts de main-d'œuvre et de licence de logiciel.

- Les coûts d'une modélisation numérique sont souvent inférieurs à ceux d'un modèle physique.

- Les modèles numériques peuvent être stockés et utilisés pour une utilisation ultérieure ou des changements futurs.

- Des animations de l'écoulement, y compris des animations de la surface libre, peuvent être construites en utilisant une variable d'intérêt telle que la vitesse ou la pression.

- Un modèle numérique est d'autant meilleur quand il s'appuie pour son calage sur des données de terrain ou des données issues d'une modélisation physique, mais ces dernières ne sont pas toujours disponibles ou obligatoires.

- Les modèles numériques sont limités dans le choix des paramètres d'entrée et de fonctionnement lorsqu'aucunes données de terrain ou issues d'une modélisation physique ne sont disponibles. De plus, les modélisateurs n'ayant pas reçu une formation adéquate ou n'ayant pas une grande expérience dans ces domaines spécifiques peuvent être à l'origine d'erreurs.

- Les modèles numériques sont limités lorsqu'il s'agit de rugosité géométrique et d'estimation du profil des vitesses au contact des parois (loi de paroi), particulièrement dans les écoulements torrentiels à forte turbulence. Selon la taille de la maille adoptée pour le modèle numérique, les vitesses réelles près de la paroi peuvent ne pas être exactes puisque les approximations de la loi de paroi dépendent de la taille de la maille numérique ainsi que d'autres facteurs.

- La modélisation numérique est souvent limitée par le temps de maillage ou de calcul pour les modèles à grands domaines spatiaux et/ou temporels. Pour réduire les temps de calcul, certains codes CFD incluent des solutions mixtes qui permettent d'obtenir des approximations 2D pour de grandes zones où les gradients verticaux ne sont pas critiques et qui peuvent être combinés à une analyse hydraulique 3D complète dans des zones localisées.

- Le nombre de calculs requis pour évaluer le tracé d'un hydrogramme prendra beaucoup de temps lors d'une étude numérique. De même, simuler le laminage d'un hydrogramme complet peut exiger des temps de simulation irréalistes.

- Les effets d'hystérésis entre les débits croissants et décroissants d'un hydrogramme peuvent se produire et être importants. Les modèles numériques sont limités dans leur capacité à analyser ce type de phénomènes.

- Il n'est pas possible de filmer ou de photographier l'eau en mouvement d'un modèle numérique, de sorte que les conditions d'écoulement modélisées puissent être comparées aux conditions réelles d'écoulement comme on peut le faire avec un modèle physique. Ce type d'analyse est extrêmement utile pour le contrôle de la pertinence de la modélisation. Il est toutefois possible de créer des courts clips vidéo facilitant la comparaison avec les conditions d'écoulement grandeur nature.

- Un modèle numérique est limité dans sa capacité d'analyser la stabilité hydraulique et l'effet d'un écoulement instable en aval d'un ouvrage de contrôle.

- Les conditions d'écoulement comme celles relatives à un ressaut hydraulique peuvent être très sensibles aux changements de rugosité des limites numériques et l'application de la théorie offre une grande latitude au modélisateur si les données de terrain ou du modèle physique ne sont pas disponibles.

- Un modèle numérique est limité par la taille du domaine de calcul à la fois spatialement et temporairement. Un grand domaine physique peut nécessiter une solution 2D plutôt qu'une solution 3D, ou une solution hybride comme indiqué ci-dessus, réduisant ainsi potentiellement la précision de la solution.

- Les modèles numériques sont limités en raison du temps qu'il faut pour construire le maillage numérique et exécuter la simulation lorsque des conditions d'écoulement complexes sont simulées.

- Les écoulements peu profonds peuvent nécessiter des calculs intensifs nécessitant un maillage très fin pour simuler avec précision un profil de vitesse. Un exemple serait un débit extrêmement faible sur un seuil profilé. Heureusement, la plupart des problèmes de conception reposent sur des débits plus importants.

- Les conditions d'écoulement complexes résultant d'un écoulement turbulent, d'un écoulement de recirculation, d'un décollement et d'un dévers dans une courbe peuvent ne pas être calculées avec précision dans un modèle numérique en raison des limites de la prise en compte de la turbulence. Bien que les modèles de turbulence fonctionnent bien dans certains cas documentés, ils peuvent ne pas convenir à toutes les conditions d'écoulement.

- Des techniques de maillage incorrectes ou mal appliquées peuvent introduire des erreurs numériques. Des tests de sensibilité et de convergence des résultats en fonction de la taille des mailles doivent être réalisés.

- Il faut un temps comparativement très important pour simuler et analyser un très petit changement dans un modèle numérique par rapport au temps requis pour un modèle physique.

9.4.   GÉNÉRALITÉS CONCERNANT LA MODÉLISATION HYBRIDE

La modélisation hybride ou mixte est l'emploi complémentaire des modélisations physique et numérique, utilisées en série ou en parallèle afin de résoudre des problèmes hydrauliques complexes. Malgré les limites de ces deux types de modélisation utilisées indépendamment, la modélisation hybride s'est avérée extrêmement efficace et efficiente. Une approche hybride bien menée améliorera la précision des résultats et des mesures hydrauliques en s'appuyant sur les avantages des modélisations physique et numérique utilisées conjointement. Il convient toutefois de noter que la modélisation hybride ne représentera pas nécessairement un gain de coût global par rapport à l'utilisation de l'une ou l'autre technique. La plus-value apportée par la modélisation hybride est généralement plus technique (qualité des résultats, confiance des choix de conception...) que financière. L'approche hybride permet d'évaluer un plus grand nombre d'options de conception et, par conséquent, d'optimiser le dimensionnement des ouvrages.

La modélisation hybride peut être réalisée suivant 3 étapes :

1. La modélisation numérique est réalisée préalablement à la construction du modèle physique et aux essais,

2. La modélisation numérique est réalisée simultanément ou en parallèle du modèle physique,

3. La modélisation numérique est réalisée après la fin de l'étude sur modèle physique.

9.4.1. *Modélisation numérique menée avant la construction du modèle physique et les essais associés*

- Avant la construction d'un modèle physique, les résultats de la modélisation numérique peuvent être exploités afin que la géométrie d'approche du modèle physique puisse être alignée correctement avec les lignes de courant de l'écoulement dans le but d'optimiser les dimensions du bassin d'alimentation. Cela permet de réduire les coûts de construction du modèle physique et permet de construire un modèle de plus grande taille (à plus petite échelle donc), réduisant ainsi les effets d'échelle de Reynolds, particulièrement dans la zone d'intérêt.

- Le modèle numérique peut aider à réduire le nombre de configurations testées sur modèle physique en effectuant des pré-simulations. Le modèle numérique permet de simuler les conditions hydrauliques de la conception initiale et d'éventuels dimensionnements alternatifs permettant de dégager un consensus préalable à la construction du modèle physique.

- Les pré-simulations du modèle numérique permettent aux ingénieurs hydrauliques de mieux appréhender la problématique hydraulique générale du cas traité en identifiant notamment les zones où des écoulements complexes peuvent apparaître (décollement de l'écoulement, vitesses excessives, écoulement non uniforme...).

9.4.2. *Modélisation numérique menée en parallèle de la modélisation physique*

- L'un des avantages les plus importants de la modélisation hybride est la "modélisation du modèle", ce qui signifie que la géométrie exacte du modèle physique initial ou de référence est modélisée numériquement à l'échelle 1:1 afin que les erreurs de modélisation numérique puissent être évaluées et corrigées si possible. Cet effort de contrôle de la qualité est pertinent pour réduire ou éliminer les incertitudes du modèle numérique.

- Il est également possible de rééchelonner le modèle numérique étalonné pour modéliser l'ouvrage taille réelle afin d'évaluer les effets d'échelle potentiels de la modélisation physique. Encore une fois, ces effets d'échelle devraient être réduits si le modèle physique est suffisamment grand.

- Une fois que le modèle numérique a été calé ou ajusté aux résultats du modèle physique, il peut être utilisé afin d'évaluer une multitude de paramètres (débits, vitesses, lignes d'eau, pressions superficielles et contraintes de cisaillement) qui exigent beaucoup de travail et sont donc coûteux à estimer à l'aide du modèle physique. Les paramètres de calage du modèle numérique sur le modèle physique sont généralement la rugosité de surface, le modèle de turbulence et la finesse/définition du maillage.

- Avec un modèle numérique calé, tout changement structurel ou géométrique proposé qui nécessiterait des modifications majeures du modèle physique peut être simulé à l'aide du modèle numérique et comparé à la situation de référence simulée numériquement. Les résultats de ces modélisations numériques sont ensuite soigneusement analysés afin de choisir la configuration géométrique provisoire ou finale devant être modélisée physiquement.

- La configuration provisoire ou finale du modèle physique est ensuite construite afin de pouvoir tester différents limnigrammes ou courbes de tarage. Un modèle physique peut déterminer rapidement et précisément la courbe de tarage d'un ouvrage hydraulique, tandis qu'un modèle numérique demandera beaucoup plus de temps pour générer les points cote/débit nécessaires à la construction de cette même courbe de tarage. Encore une fois cependant, lorsque le modèle physique est construit pour une configuration donnée, le modèle numérique permet d'extraire du domaine fluide un nombre très important d'informations (lignes de courant, trajectoires...) ce qui nécessiterait un effort considérable dans le modèle physique.

- Le modèle physique peut être photographié et les conditions d'écoulement ainsi que les problématiques associées au dimensionnement structurel peuvent être filmées. Ces enregistrements sont également extrêmement utiles à des fins de communication auprès du grand public et des organismes de financement.

- Lors de cette phase de la modélisation hybride, les zones pour lesquelles les simulations numériques posent question sont documentées. Il peut s'agir de zones d'écoulement très turbulent, de zones à l'intérieur d'un modèle plus grand dans lesquelles les écoulements caractérisés par des profondeurs extrêmement faibles et des vitesses élevées se produisent, de conditions d'écoulement proches du régime critique, d'ondes transitoires ou d'oscillations de vagues, de dévers dus à des écoulements torrentiels le long de courbures marquées, de zones potentiellement affouillables et d'écoulements fortement aérés.

9.4.3. Modélisation numérique menée une fois la modélisation physique terminée

Après toute étude menée sur modèle physique, ce dernier doit être démonté afin de libérer de l'espace dans le laboratoire pour une nouvelle étude. A contrario, le modèle numérique ne prend pas de place et doit être conservé. Les avantages d'avoir un modèle numérique calé et à jour sont les suivants :

- Le modèle numérique peut être utilisé indéfiniment après le démantèlement du modèle physique.

- Des informations supplémentaires peuvent être extraites si nécessaire dans les zones d'écoulement (pressions et vitesses enregistrées dans chacune des mailles) après le démantèlement du modèle physique.

- Des modifications mineures de la conception géométrique de l'ouvrage peuvent être testées dans le modèle numérique calibré avec l'assurance de la pertinence des résultats.

- Le modèle numérique calé peut être utilisé pour la formation et les décisions opérationnelles après le démantèlement du modèle physique.

9.5. RÉSUMÉ ET CONCLUSIONS

La modélisation hybride, à savoir l'utilisation conjointe d'un modèle physique et d'un modèle numérique, permet de poursuivre l'effort de modélisation hydraulique et a prouvé son efficacité pour améliorer la pertinence et la précision des modélisations. La modélisation hybride s'est avérée être un outil important pour le dimensionnement des ouvrages hydrauliques.

La modélisation hybride offre aux chercheurs et aux ingénieurs une occasion unique de comprendre les incertitudes et les limites des modèles physique et numérique, puisque leur fonctionnement en parallèle permet une comparaison et un étalonnage quasi immédiats. Lorsque l'on traite de conditions d'écoulement tridimensionnelles complexes, la capacité des modèles numériques à simuler le champ d'écoulement dans tout le domaine fluide étudié est souvent limitée en comparaison avec les modèles physiques. Cela comprend des conditions d'écoulement avec des dévers ou des ressauts hydrauliques non symétriques, des ondes stationnaires proches d'un régime critique, des zones de dissipation d'énergie.

Si la modélisation numérique s'est montrée, par le passé, très adaptée pour des problématiques bidimensionnelles et certaines applications tridimensionnelles, bon nombre de problèmes associés à l'hydraulique des ouvrages sont particulièrement complexes (très turbulents et tridimensionnels), et la modélisation numérique peut, dans ces conditions d'écoulement, être questionnée. Par conséquent, la modélisation hybride permet de vérifier et de valider les valeurs de débits, les lignes d'eau et les vitesses ponctuelles dans le domaine fluide et, ainsi, permet de déterminer les zones spécifiques du modèle numérique qui ne sont pas simulées de manière pertinente.

La modélisation hybride offre un modèle numérique calé destiné à extraire de très grandes quantités de données détaillées dans le domaine fluide et qui peut être utilisé efficacement bien après le démantèlement du modèle physique. La modélisation hybride permet de caler le modèle numérique sur les conditions d'écoulement de l'ouvrage réel à l'échelle 1:1 et/ou du modèle physique à l'échelle 1:1. Cela apporte une importante plus-value pour la qualité des études.

L'utilisation de la modélisation numérique comme partie intégrante du processus de dimensionnement s'est montrée à la fois rentable et rapide pour l'ingénieur hydraulique. L'utilisation préalable d'un modèle numérique pour définir les caractéristiques du bassin d'entrée du modèle physique et les conditions d'alimentation associées est un outil précieux dans le processus général de modélisation. Cette pré-modélisation fournit des informations importantes afin que le modèle physique puisse être construit aussi grand que possible, minimisant ainsi les éventuels effets d'échelle. Cette pré-modélisation fournit également des informations relatives à la façon dont l'alimentation en eau du modèle physique doit être conçue afin de s'assurer des bonnes conditions hydrauliques amont.

L'utilisation de la modélisation numérique comme partie intégrante du processus de dimensionnement s'est montrée à la fois rentable et rapide pour l'ingénieur hydraulique. L'utilisation préalable d'un modèle numérique pour définir les caractéristiques du bassin d'entrée du modèle physique et les conditions d'alimentation associées est un outil précieux dans le processus général de modélisation. Cette pré-modélisation fournit des informations importantes afin que le modèle physique puisse être construit aussi grand que possible, minimisant ainsi les éventuels effets d'échelle. Cette pré-modélisation fournit également des informations relatives à la façon dont l'alimentation en eau du modèle physique doit être conçue afin de s'assurer des bonnes conditions hydrauliques amont.

9.6.　RÉFÉRENCES

Willey, J., T. Ewing, R. Wark and E Lesleighter, (2012) – "Complementary use of physical and numerical modelling techniques in spillway design refinement", *ICOLD 25th Congress on Large Dams*, Kyoto, Japan.

Erpicum, S., B. J. Dewals, J-M. Vuillot, P. Archambeau and M. Pirotton, (2012) - "Coupling physical and numerical models: example of the Taoussa Project (Mali)", *4th IAHR International Symposium on Hydraulic Structures*, Porto, Portugal.

Rahmeyer, W., S. Barfuss and B. Savage, (2011) - "Composite Modeling of Hydraulic Structures", *Dam Safety 2011*, National Harbor, MD, USA.

Paxson, G., B. Crookston, B. Savage, B. Tullis, and F. Lux III,(2008) - The Hydraulic Design Toolbox- Theory and Modeling for the Lake Townsend Spillway Replacement Project, Assoc. of State Dam Safety Officials (ASDSO), Indian Wells, CA, USA.

Savage, B.M., M.C. Johnson and B. Towler, (2009) - "Hydrodynamic forces on a spillway: can we calculate them?", *Dam Safety 2009*, Hollywood, FL, USA.

10. ÉCONOMIE, RISQUE ET SÛRETÉ DANS LA CONCEPTION DES ÉVACUATEURS DE CRUE

10.1. INTRODUCTION

Les évacuateurs de crue (EVC) sont les éléments principaux qui assurent la sécurité des barrages quel que soit leur type. Pour cette raison, toute tentative de réduction directe des coûts sur ces éléments conduit, dans bien des cas, à une augmentation du risque et à une diminution de la sécurité globale du projet de barrage. Pour de nombreux projets, le coût direct des EVC représente une proportion conséquente d'un projet et sa réduction peut représenter une cible attrayante pour les concepteurs. Cependant, du fait de leur rôle fondamental dans la sécurité, les actions et mesures visant à réduire le coût du barrage dans la conception et la réalisation des EVC doivent être évaluées précautionneusement en gardant à l'esprit leur rôle. Malgré tout, l'économie et l'optimisation des coûts sur les EVC restent possibles et sont même souhaitables. Ce chapitre aborde ces aspects plus en détails.

On n'insistera jamais assez sur l'importance des EVC : beaucoup de ruptures de barrages ont été causées par des insuffisances dans la conception ou dans le dimensionnement des EVC. Une bonne capacité d'évacuation est d'une importance primordiale pour les barrages en terre ou en maçonnerie, qui sont plus facilement détruits en cas de débordement, alors que les barrages en béton seront capables de résister s'ils surversent modérément. Habituellement, une augmentation du coût direct des EVC n'est pas directement proportionnelle à l'augmentation de sa capacité d'évacuation. Souvent, dans les projets neufs de barrages en béton, le coût d'un EVC de bonne capacité ne va être que modérément supérieur à celui d'une capacité bien inférieure.

Pour de nombreux projets de barrages, les considérations économiques conduisent à une conception utilisant une capacité de réservoir au-dessus de la cote maximale d'exploitation RN. Afin d'établir la combinaison la plus économique entre une surcharge de réservoir et la capacité d'évacuation des EVC, des études de débit de crue et une considération des coûts des différentes combinaisons de barrage-EVC sont nécessaires. Cependant, la conduite de ces études doit garder en considération la taille minimale des EVC permettant de garantir la sureté de l'ouvrage. Dans de nombreux cas, les données des hydrogrammes de crue utilisée pour définir la crue de projet contraignent le pic de crue mais non le volume de crue. Souvent, de telles crues ont beau avoir le débit de pointe le plus grand, elles n'ont pas pour autant le plus grand volume. Lorsque des EVC de petites capacités sont dimensionnés pour ces pics de crue, des précautions doivent être prises afin de s'assurer que les EVC soit aptes à : (i) évacuer les surcharges du réservoir sans que le barrage ne surverse pour des tempêtes ordinaires ; (ii) empêcher une surcharge du réservoir due à une charge permanente qui provoquerait une insuffisance des EVC pour des ruissellements dont les pics ne dépasseraient pas le pic de crue.

Les réductions de coût dans la construction et l'exploitation des barrages et leurs éléments de contrôle des crues ont fait l'objet de nombreux comités et bulletin CIGB, mais le mot d'ordre reste le devoir de maintenir un niveau de sécurité convenable. Le concepteur a la charge d'évaluer les effets induits par ces questions et de maintenir les risques à un niveau acceptable.

Le bulletin CIGB n°152 « *Réduction des coûts dans les barrages* » traite des critères de conception des barrages et des structures attenantes, et aborde plus particulièrement les EVC, du point de vue économique ainsi que des risques liés aux réductions de coûts. Les sujets discutés incluent la détermination des crues de projet, les difficultés d'assigner une probabilité de rupture à un type d'EVC ainsi que la tendance des conceptions qui ne considèrent que la performance sans se soucier des coûts.

Ces dernières années, des approches innovantes ont été développées dans la conception des EVC pour répondre aux exigences de sureté des barrages à moindre coût. Celles-ci incluent : (i) l'utilisation d'une protection contre le débordement au niveau des remblais afin d'éviter la rupture lors des crues importantes ; (ii) l'élévation des barrages avec des parapets afin d'améliorer la limitation des pics de crue et d'augmenter la capacité de stockage ; (iii) l'application d'une procédure d'évaluation appropriée des risques humains et matériels en cas de rupture du barrage. Ce dernier point est abordé plus en détails dans le bulletin CIGB n°154 « Gestion de la sécurité des barrages en exploitation ».

De plus, des méthodologies et des recommandations ont été élaborées concernant la possibilité de diminuer les équipements de sécurité par des critères de conception des EVC lorsque les risques incrémentales (les dégâts) liés à une rupture du barrage deviennent négligeables devant ceux causés naturellement à l'aval en cas de crue. De telles approches, particulièrement adaptées sur les barrages existants pour l'ajustement de la capacité des EVC au niveau des exigences de sécurité, ont été mises en pratique dans quelques cas. Des analyses avantages-coûts ont été utilisées pour appuyer l'acceptabilité de ne pas augmenter la capacité des EVC lorsque le coût de leur réhabilitation était considéré disproportionné par rapport à l'amélioration ainsi gagnée.

Comme décrit dans les chapitres précédents de ce communiqué, de nouvelles alternatives efficaces d'EVC, développées et mises en pratique, améliorent les solutions traditionnelles et ce, à moindre coût. Ce chapitre expose les stratégies liées à la conception des EVC dans les projets de barrage en utilisant différentes possibilités, alternatives et méthodes pour tirer tous les bénéfices économiques. En règle générale, aucune mesure de réduction des coûts ne devrait être prise dans le but de diminuer la capacité d'évacuation des EVC en transférant une partie de cette tâche à d'autres structures ou dispositifs.

10.2. CRUES DE PROJET ET CRUES DE DANGER

Une revue des projets existants montre que, pour de nombreux projets modernes, le système d'EVC est conçu pour une valeur de débit de pointe, basé sur un critère considérant traditionnellement la crue de projet comme le débit entrant qui peut être évacué dans des conditions normales, avec une marge de sécurité donnée par une limite de plan d'eau acceptable. D'autre part, la crue de danger est définie comme étant le débit maximal qui ne cause pas la destruction du barrage et au-delà duquel la sécurité ne peut être garantie (communiqué n°82 du CIGB, 1992). La sécurité du projet repose, en définitive, sur la capacité des EVC à évacuer la crue de danger. Cette approche est également le critère de référence dans de nombreux pays où une recommandation officielle est en vigueur pour la conception des barrages.

I En général, la crue de danger possède une probabilité d'occurrence plus faible que la crue de projet. Dans de nombreux cas, elle est supposée égale à la crue probable maximale (PMF) pouvant arriver sur le barrage. Du fait de sa faible probabilité, le déversement du barrage lors de la crue de danger est évité, mais un certain niveau de dégradation de la structure est tolérable. Comme évoqué dans le bulletin n° 152 du CIGB (p. 133-135), il est possible de réduire le coût des EVC en augmentant la différence entre la crue de danger et la crue de projet, c'est-à-dire en réduisant la crue de projet sans diminuer la sécurité, qui est assurée par sa capacité à évacuer de la crue de danger. Cette capacité est associée à la possibilité d'augmenter l'élévation du plan d'eau du réservoir. Même si dans de nombreux projets, ceci correspond à empiéter sur le plan d'eau normal, l'élévation de la crête du barrage (en prévoyant les coûts correspondants) permettant un plus grand plan d'eau est une technique également utilisée. L'augmentation de l'élévation du réservoir permet une certaine atténuation du pic de crue, ce qui est détaillé dans la partie suivante.

10.3. LIMITATION DU PIC DE CRUE

La mise à disposition d'un volume supplémentaire, au-dessus de la cote maximale d'exploitation RN, pour stocker une partie de l'hydrogramme de crue et ainsi permettre une réduction du débit maximum évacué par les EVC, est une stratégie courante mise en place pour réduire directement le coût des EVC sans altérer la sécurité du barrage. Cependant, ceci n'est faisable uniquement si l'aire du réservoir est suffisamment grande pour fournir un grand volume avec une élévation limitée. L'élévation considérée lors de la conception, qui est atteinte lorsque le débit de crue entrant dans le réservoir est maximal, est généralement appelée PHE (Plus Hautes Eaux). Cette élévation est plus haute que la cote maximale d'exploitation.

D'un point de vue financier, le bénéfice de l'atténuation du pic de crue n'est réel que si le coefficient d'hydrogramme (défini comme le ratio entre le volume d'hydrogramme de crue divisé par le produit de l'aire de la surface du réservoir et la hauteur du barrage) est très petit devant un (Bouvard, 1988).

Dans tous les cas, une revanche au-delà du niveau PHE, qu'il coïncide ou non avec la cote à RN, devrait être imposé pour tous les projets de barrage. Cette disposition vise à protéger le barrage des vagues produites dans le réservoir et est normalement dimensionné à l'aide de critères standards bien définis qui tiennent compte du vent probable maximal dans la direction de la plus grande longueur du réservoir, avec l'eau stockée dans le réservoir au niveau maximal de crue. La probabilité d'une occurrence simultanée de la crue maximale avec le réservoir à son niveau maximal et le vent maximum produisant la hauteur de vague maximale, peut être considérée comme étant bien inférieure à la probabilité isolée de l'occurrence de chacun de ces évènements. Ceci a justifié le fait de considérer dans de nombreux pays (comme la Chine) un niveau de réservoir de crue de danger comme étant équivalent, ou très peu inférieur, à l'élévation de la crête du barrage. A noter à ce sujet que dans le cas d'un barrage avec une partie en béton et un remblai, il est d'usage de définir l'élévation de la crête du remblai à environ 1m (ou plus) au-dessus de l'élévation de la crête en béton afin de s'assurer que dans le cas d'un déversement, la partie en béton déverse en premier et que le risque de déversement du remblai soit au minimum réduit.

Cette considération signifie que la revanche au-dessus de la crue de danger sera réduite en lien avec la marge correspondant à la cote d'exploitation maximale RN. Comme précisé précédemment, l'économie réalisée sur le système d'EVC est le résultat d'une plus petite crue de projet à évacuer lorsque le niveau du réservoir est à RN. Cette réduction de la crue de projet est permise par une élévation du plan d'eau plus grande et par l'atténuation du pic de l'hydrogramme de crue.

10.4. DIVISION DES EVC EN DIFFERENTES STRUCTURES

Une des solutions pour réduire les coûts des EVC pour les grands barrages est de combiner différents systèmes d'évacuation. Il y a différents types d'EVC et, malgré le fait que certains types soient plus adaptés à certains projets de barrage que d'autres, il est possible de réduire les coûts d'investissement en considérant différentes combinaisons d'EVC ou variantes d'un même type. L'économie visée ne doit cependant pas être obtenue au détriment de la sécurité du projet; ceci signifie que les solutions de conception considérées doivent garder la même capacité d'évacuation en toute sécurité des quantités d'eau correspondant à la crue de projet.

Étant donné que les occurrences des crues et le débit devant être évacué varient, une approche stratégique d'un point de vue économique consiste à diviser la capacité totale d'évacuation du barrage en différents types d'EVC et à dimensionner chacun d'entre eux au regard de leur sécurité et de leur objectif. Dans de tel cas, des procédures claires doivent être définies. Pour des projets de barrage en béton, il est possible d'avoir une répartition des évacuations à travers des EVC de surface et de fond dans le corps du barrage et des EVC de rive avec coursier séparé, comme illustré dans le barrage de Karun III et du projet HPP (Figure 10.1) où une crue de 20 000 m³/s est évacuée. D'ailleurs, comme mentionné dans le Chapitre 2 de ce bulletin, dans le cas du barrage Karun III, du fait d'une excavation profonde en rive gauche, un bassin d'amortissement linéaire a été considéré avec un bassin d'amortissement de 36 m de haut construit à l'aval. Dans ce cas, l'EVC de rive avec coursier, fait office d'EVC de service, les orifices d'EVC auxiliaires et la crête du barrage d'EVC d'urgence comme illustré dans la Figure 10.1.

Fig. 10.1
Barrage de Karun III (Iran) : barrage voute en béton de 205m de haut avec des EVC de service, auxiliaires et d'urgence

10.4.1. EVC de service

L'EVC de service est celui qui est chargé d'évacuer les débits entrants, de façon continue ou fréquente de manière contrôlée ou non contrôlée depuis un réservoir, sans provoquer de dégâts au barrage, à la digue ou à toute autre structure annexe, jusqu'à et y compris son débit de projet. Lorsqu'un seul EVC est présent, son débit de dimensionnement doit être égal à la crue de projet. Les EVC de service peuvent globalement être classés en deux catégories : les EVC de surface et les EVC de (demi)fond en charge (ou en galerie).

Le choix d'utiliser un EVC de surface ou un EVC en charge est généralement dicté par les caractéristiques spécifiques du site. Dans un canyon étroit, il est souvent difficile d'intégrer des EVC de surface ou des EVC de fond et les galeries sont le seul choix logique. Un EVC de fond a l'inconvénient d'avoir une évacuation en fonction de la racine carrée de la hauteur d'eau ($H^{1/2}$) alors que celle d'un EVC de surface est en fonction de la puissance un et demi de la hauteur d'eau ($H^{3/2}$). Pour cette raison, les EVC de fond se montrent généralement efficaces seulement pour les barrages hauts comprenant de hautes hauteurs d'eau.

Les EVC de surface peuvent comporter des vannes ou non. Les caractéristiques du site, l'ampleur de la crue de projet, le risque estimé et le coût des différentes alternatives sont les critères principaux dans le choix d'utilisation d'un type d'EVC ou d'un autre. Les EVC sans vanne sont sans doute les plus sécuritaires. Ils sont également moins susceptibles d'être entravés par des embâcles. Puisqu'ils ne nécessitent pas d'opérateur, leur manœuvre est moins perturbée par de possibles erreurs opérationnelles. Cependant, ils sont également et généralement plus chers que les EVC avec vanne pour un débit maximal donné, étant donné qu'ils nécessitent de longues crêtes et donc des chutes et/ ou des diamètres de conduits plus importants. Un contre-exemple à cette règle générale sur les coûts relatifs est celui de l'évacuateur de crête en labyrinthe. Dans les cas où l'aire à la surface du réservoir est grande devant le volume d'eau entrant, un EVC en labyrinthe représente une structure sure et économique. Les EVC de service sans vanne possèdent l'avantage supplémentaire de n'impliquer aucun équipement de manœuvre et donc, de nécessiter de peu de travaux de maintenance réguliers.

Les EVC avec vanne sont généralement choisis lorsque le site est limité ou que l'ampleur de la crue de projet est très grande, et qu'il n'est physiquement ou économiquement pas envisageable de construire une crête d'une longueur nécessaire sans vanne. Le contrôle du débit en aval, ou la maximisation du stockage, nécessitent un contrôle plus flexible que celui permis par les EVC sans vanne. Pour les crues de projet importantes, le coût total direct d'investissement d'un EVC avec vanne sera généralement inférieur à celui d'un EVC sans vanne d'une capacité équivalente. Les pertuis des EVC de service devraient toujours être conçus et construits avec les équipements permettant d'installer des batardeaux en amont de la vanne de manière à ce que celle-ci puisse être entretenue au sec. Des exemples de problèmes pouvant survenir et impliquant des vannes dans des situations d'urgence sont détaillés dans le Chapitre 2 de ce bulletin.

10.4.2. EVC auxiliaires

Lorsque les conditions de site sont favorables, la possibilité de réduire les coûts en utilisant des EVC auxiliaires en association avec les EVC de service peut être considérée. Dans de tels cas, les EVC de service seront conçus pour gérer les crues fréquentes et les EVC auxiliaires pour prendre le relais lors des crues plus importantes.

Il doit cependant être noté que le concept d'EVC auxiliaire en complément des EVC de service, nécessaires pour évacuer la crue de projet en conditions normales tout en gardant une marge de sécurité en considérant un plan d'eau maximal toléré, ne fait pas l'unanimité. Dans de nombreux endroits, les EVC auxiliaires sont considérés comme des EVC d'urgence, construits pour gérer les crues plus importantes que la crue de projet.

En considérant les EVC auxiliaires comme faisant partie intégrante du système d'évacuation, en complément des EVC de service de manière à ce que les deux types soient nécessaires pour évacuer la crue de projet, le critère de sécurité concernant la manœuvre des vannes (dans le cas d'EVC auxiliaires avec vannes) tel que le cumul de leur alimentation en énergie, doit être le même pour toutes les vannes, et aucune réduction de coût ne doit être considérée pour cette partie de l'EVC auxiliaire. Cependant, en règle générale, du fait de son utilisation moins fréquente, un certain niveau de détérioration structurelle ou d'érosion à l'aval peut être accepté, à condition que ces dégâts n'affectent pas la sécurité de la structure permanente du projet, et soient localisés dans des endroits accessibles pour être réparés après le passage de la crue.

Une solution courante utilisée pour réduire les coûts dans le cadre de grands projets est de réduire, voire de faire l'impasse, sur le revêtement en béton du canal de sortie des EVC auxiliaires. Cependant, la décision de réduire les coûts en étant plus souple sur ces critères de revêtement de canaux des EVC auxiliaires devant évacuer de grandes crues, doit être mise soigneusement en perspective avec la possibilité de dégâts environnementaux tels que la création de forts gradients d'excavation ou de glissements de terrain. Des exemples de projets pour lesquels ces solutions ont été retenues sont détaillés dans le Chapitre 2 de ce bulletin.

Une considération importante dont il faut tenir compte dans la conception des EVC de service et auxiliaires en étant plus souple pour les EVC auxiliaires est dans la définition de la capacité maximale d'évacuation des crues des EVC de service, qui doivent être capables d'évacuer toutes les crues fréquentes. Les grands projets construits avec ce genre de solution ont considéré des EVC de service avec une capacité correspondant à des périodes de retour allant de 100 à 200 ans.

Dans de nombreux cas, les EVC auxiliaires sont localisés loin des EVC de service mais il y a également d'autres cas où ces deux structures sont côte à côte (Figure 10.2). Dans ce premier cas, les conditions favorables pour le choix d'EVC auxiliaires sont l'existence d'un col ou d'une dépression le long du bord du réservoir conduisant à une voie navigable, loin du barrage ou d'autres éléments de corps du barrage, permettant ainsi la construction d'une conduite de sortie plus économique et sans prendre le risque d'endommager le barrage.

Fig. 10.2
Xingo Project (Brésil) : EVC de service et auxiliaires. L'évacuation de la conduite de l'EVC auxiliaire est partiellement revêtue. Leur capacité d'évacuation est égale à 16 500 m³/s

10.4.3. EVC d'urgence

Il s'agit d'un EVC conçu pour fournir une protection supplémentaire au projet contre le déversement du barrage, et a pour objectif d'être utilisé en conditions extrêmes lors de crues plus importantes que la crue de projet ou dans le cas de disfonctionnement des EVC de service ou auxiliaires. Comme l'indique son nom, les EVC d'urgence sont également présents pour augmenter la sécurité, étant donné que les urgences ne sont pas prises en compte dans les hypothèses normales de conception. Ces situations peuvent avoir lieues dans le cas d'indisponibilité de certains EVC, du disfonctionnement des vannes des EVC, du blocage des EVC par des corps flottants, ou de l'impossibilité d'utiliser les EVC habituels en cas de dommage ou de défaillance de ces derniers. Une urgence peut survenir lorsque les débits de crue sont traités principalement par un surremplissage du réservoir et qu'une crue usuelle survient en complément avant que les EVC de service ou les vidanges de fond n'aient évacué la précédente crue. En fonctionnement normal, les EVC d'urgence ne sont jamais utilisés.

En ce qui concerne les projets pour lesquels la conception prévoit des EVC de service seuls ou associés à des EVC auxiliaires pour évacuer entièrement la crue de projet, les EVC d'urgence ne devraient pas servir à justifier des économies ou des réductions de la capacité d'évacuation du système principal. Il y a cependant des projets pour lesquels l'utilisation des EVC auxiliaires et d'urgence sont conjointes, comme illustré dans les exemples de la Figure 10.5. Dans ces cas, les EVC de service ont une capacité d'évacuation moindre, ce qui permet une réduction de l'investissement du projet.

Tous les projets ne possèdent pas d'EVC d'urgence mais, comme évoqué dans le bulletin CIGB n°142, tous les projets devraient envisager la venue de crues plus importantes que les crues de danger.

Il existe différents types d'urgence et le bulletin n°142 du CIGB en décrit les plus communs. Le concept de fonctionnement des EVC d'urgence consiste en une structure en terre ou en béton, détruite ou retirée, lorsque la charge hydraulique en fonction de l'élévation du plan d'eau augmente au-delà d'une certaine limite. Cette destruction ou ce retrait permet une augmentation du débit sortant, rétablissant un niveau d'eau acceptable et évitant ainsi le débordement du barrage.

Par conséquent, les crêtes de contrôles des EVC d'urgence, doivent être positionnées au moins au niveau des PHE du réservoir. L'exigence de revanche du barrage est basée en relation avec le niveau pour lequelle la digue d'urgence ou la structure en béton sont effacées. Cependant, dans de telles situations, et en fonction de la hauteur de remblai détruite, il peut être impossible de re-remplir le réservoir jusqu'à son niveau de fonctionnement tant que la digue fusible n'est pas reconstruite. Ceci peut constituer un réel inconvénient, même dans l'hypothèse où un tel incident est très peu probable.

10.5. OPTIMISATION GLOBALE DE L'AMÉNAGEMENT DU PROJET

Dans la proposition de disposition générale du projet, l'analyse des différents types d'EVC possibles et des différentes dispositions de ces structures, pour les mêmes critères de sécurité, peuvent avoir un impact significativement différent sur le coût global du projet. Différentes configurations d'EVC, avec ou sans vanne ou encore une combinaison des deux, ainsi que la division du système d'évacuation en EVC de service, auxiliaires et/ou d'urgence, peut amener à des économies non négligeables pour les projets de barrage. Les trois cas décrits dans les paragraphes qui suivent sont des exemples de solutions économiques obtenues en optimisant la conception générale du projet et la disposition des différentes structures, en réduisant la charge d'évacuation des EVC de service et en utilisant une combinaison d'EVC auxiliaires et d'urgence pour pouvoir faire face à des crues plus importantes.

Le premier exemple est celui des barrages en terre de Shamil et Nian qui mesurent 35m de haut, avec un réservoir commun d'une capacité de 69.10^6 m³. La longueur des crêtes des barrages de Shamil et Nian sont respectivement 320 m et 530 m. Cet ouvrage fournit principalement de l'eau domestique à la capitale de la province de Hormozgan, dans le sud de l'Iran. Lors de la conception du projet, la PMF était de 13 565 m³/s. Plusieurs alternatives pour les structures d'évacuation des crues ont été considérées. Les alternatives les plus économiques ont été l'utilisation d'un système d'évacuation composé d'EVC de service, auxiliaire et d'urgence. Ce système permet d'assurer la sécurité de l'ouvrage.

Les EVC de service possèdent des vannes et peuvent gérer des crues avec une période de retour allant jusqu'à 150 ans sans nécessiter l'aide des autres EVC. Si la crue excède cette charge, l'eau commencerait à se déverser par les EVC auxiliaires, avec une crête déversante de 300 m de long. Avec ces deux systèmes en service, le barrage peut gérer des crues avec une période de retour allant jusqu'à 10 000 ans. Si la crue est encore plus importante, une digue fusible peut rompre et gérer l'évacuation de la charge d'eau restante. Le système possède la capacité de gérer la PMF sans que le barrage ne surverse.

La probabilité d'apparition de la PMF est toujours très faible et lorsque sa valeur est plusieurs fois plus importante que la crue de projet, des EVC de service conçus pour gérer la crue de projet peuvent être combinés avec des EVC auxiliaires et d'urgence. Un exemple de ce cas de figure est le projet Mnjoli au Swaziland (Engels ans Sheerman-Chase, 1985). Ce projet possède un EVC sans vanne en béton conçu pour gérer des crues centennales avec un pic de 950 m³/s et une PMF estimée pour produire un pic de crue de 7.650 m³/s, plus de huit fois plus grande que la crue de projet. Une digue fusible a été construite pour évacuer l'éventuelle PMF, avec une unique structure composée de deux niveaux et d'un mur séparateur. Une partie doit déverser lorsque le réservoir atteint le niveau correspondant à une crue centennale et l'autre, de 60cm plus haut, doit rompre dans le cas de crues plus importantes. En janvier 1984, un cyclone tropical s'est abattu dessus et les épis de protection érodables des digues fusibles ont été effacés, ce qui a permis le déversement de la charge d'eau excédante par les lits et évitant ainsi le débordement du barrage.

Le troisième exemple est celui du projet hydro-électrique de Mrica 180-MW en Indonésie (Soerachmad, 1988). Ce projet est situé dans un pays tropical et montagneux avec une hauteur moyenne de précipitation annuelle de 3.839 mm et une augmentation très rapide du débit aux abords du site du projet. Le barrage est une structure en terre et en maçonnerie de 109 m de hauteur. Les études de crues ont mené aux périodes de retour suivantes :

Périodes de retour (années)	Débit de crue (m³/s)
100	4200
1000	6100
10000	7400
PMF	9300

Le système d'évacuation conçu comporte des EVC de service avec vanne, une conduite de vidange et une digue fusible faisant office d'EVC d'urgence. Ces deux premiers éléments peuvent évacuer conjointement une crue millénale d'un débit de 6100 m³/s. Lors de la conception, il a été prévu que les crues au-delà de la millénale et jusqu'à la PMF soient évacuées par les EVC de service, la conduite de vidange et la digue fusible une fois détruite, évitant ainsi que le barrage ne surverse. Dans le cas où toutes les vannes seraient endommagées et bloquées fermées, l'EVC d'urgence peut évacuer une crue centennale en toute sécurité. Cette solution a été basée sur des études économiques et d'évaluation des risques. Le projet a été achevé en 1990 et, pendant la construction en mars 1986, une crue avec un pic à 4486 m³/s a eu lieu, causant ainsi la révision des études hydrologiques et des systèmes d'évacuation, mais n'a pas modifié la conception même du projet.

10.6. LES APPROCHES DE CONCEPTION BASEÉ SUR LES RISQUES

Les risques de crues sur les barrages sont le résultat de la combinaison des aléas hydrologiques, de la vulnérabilité ou la sensibilité du barrage, et des conséquences en cas de rupture due à une crue (CIGB 2014, n° 154). Les approches de conception basée sur les risques permettent de considérer différents critères pour la conception selon le niveau de danger à l'aval (et parfois également à l'amont) en cas de rupture du barrage, ce qui ne dépend pas seulement du barrage en lui-même et des caractéristiques de son réservoir.

Des méthodes incrémentales d'évaluation des dégâts et des analyses avantages-coûts au sein des approches incrémentales, peuvent faire office de méthode d'optimisation afin de trouver les critères de conception adaptés au niveau de risque de crue, et ce particulièrement pour les barrages existants. Pour ces derniers, les analyses avantages-coûts peuvent être déclinées sous la forme d'estimation de la valeur actuelle nette d'un projet d'amélioration d'EVC, les bénéfices étant exprimés en évitant les dégâts et les conséquences.

La considération d'un critère d'acceptabilité ou de tolérabilité du risque lié à différentes catégories de conséquences et de dégâts (dégâts matériels, humains, économiques) se trouvent dans l'annexe B du bulletin n°154 du CIGB, même si celui-ci n'est pas spécifiquement dédié à la gestion de la sécurité des crues. Des comités nationaux ont travaillé (Australie, Canada) ou travaillent actuellement (France) sur ces questions et les approches possibles. La référence des publications de ces comités sont fournies à titre indicatif.

10.7. CONCLUSION

Ce chapitre a tenté de démontrer que la diminution des coûts dans la conception des EVC était possible sans altérer la sécurité du projet, qui reste primordiale. Il a également été montré que l'élaboration d'aménagements alternatifs des évacuateurs au sein de la configuration globale du projet pouvait conduire à des économies significatives.

10.8. RÉFÉRENCES

ICOLD (2011) - *Cost Savings in Dams*, Bulletin 152.

ICOLD (1992) - Selection of design flood- Current methods, Bulletin 82.

ICOLD (2012) - *Safe passage of extreme floods*, Bulletin 142.

Bouvard, M. (1988) - "Design flood and operational flood control", *General Report Question 63, Sixteenth Congress on Large Dams*, San Francisco, USA.

Engels, E. T. and A. Sheerman-Chase (1985) - "Design and operation of a fuse-plug spillway in Swaziland", *Water Power & Dam Construction*, pgs. 26-28, No. 6.

Soerachmad, S. (1988) - "Mrica Dam, Indonesia – Criteria for choice of spillways", R.48, Q.63, *Sixteenth Congress on Large Dams*, San Francisco, USA.

For Product Safety Concerns and Information please contact our EU
representative GPSR@taylorandfrancis.com
Taylor & Francis Verlag GmbH, Kaufingerstraße 24, 80331 München, Germany